DIESEL

AND HIGH COMPRESSION GAS ENGINES

3rd edition

Edgar J. Kates

Past Lecturer, Brooklyn Polytechnic Institute and Columbia University. Former Chairman, ASME Oil and Gas Power Division; Founder-Member, Oil and Gas Power Cost Committee. Former Director, Engineers Joint Council.

REVISED BY

William E. Luck

Director of Technical Education and Professor at University of Northern Iowa. Consultant-Examiner, North-Central Association of Colleges and Universities.

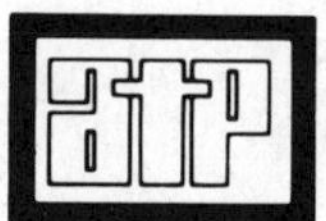

American Technical Publishers, Inc.
Alsip, Illinois 60658

Library of Congress Catalog Card Number: 73-84847
ISBN 0-8269-0203-0

First Edition

1st Printing 1954
2nd Printing 1955
3rd Printing 1959
4th Printing 1961
5th Printing 1962
6th Printing 1964

Second Edition (Revised)

7th Printing 1965
8th Printing 1966
9th Printing 1968
10th Printing 1970
11th Printing 1972
12th Printing 1973

Third Edition (Revised)

13th Printing 1974
14th Printing 1975
15th Printing 1977
16th Printing 1978
17th Printing 1981

PRINTED IN THE UNITED STATES OF AMERICA

Preface

Because of the widespread usage of diesel engines, diesel mechanics work in many diversified fields. Diesels are used as a power source for transportation equipment such as trucks, buses, locomotives, boats and ships. They are also used in construction machinery such as earth-movers, bulldozers, backhoes and cranes. On the farm, diesels are used in cotton pickers, binders, and cornpickers and for driving farm tractors. In addition, they may drive generators, compressors, pumps, and other steady-running equipment.

By the 70's, a full 120,000 persons were needed for diesel repair and maintenance alone. From one end of the country to the other, and all over the world, diesel-powered machines are doing the big jobs and the tough jobs.

In every major city, and in the smallest towns, the numbers of diesel-powered trucks and buses are increasing daily. With this extensive and varied use of diesel engines, the opportunities for diesel mechanics are expanding every day.

For twenty years or more, *Diesel and High Compression Gas Engines*, 1st and 2nd Editions, have provided a basic understanding of the construction and operation of diesel engines. As a book on fundamentals, this text tells what a diesel engine is, how it works, and the many different uses to which it has been put.

The new *3rd Edition* has been completely reorganized and updated. It clarifies diesel engines for the student, discussing their stationary parts — the frame, cylinders and heads—and then it explains in simple terms the major moving parts. The student then studies pistons, crankshafts, and connecting rods, and he learns the uses of rings, bearings, dampers, and flywheels—and the lubrication of all these parts.

After studying the basic terms in engineering, physics, the chemistry of combustion, and the principles of gases, the student will then learn the characteristics of oil and gaseous fuels.

Engine power is then revealed in terms of its historical and modern usage. Engine rating and measuring performance follow.

Further important studies cover supercharging and scavenging, terms for representative intake and exhaust methods, and the necessity for supplying large amounts of clean air to match precisely metered amounts of fuel. The complex injection fuel systems and the shape of chambers for burning of fuel follow. Also, the subject of governors is treated fully and simply, including illustrations of current models.

A new chapter on Testing, Instrumentation, and Design has been added to this *3rd Edition* to give the text wider usage. Along with extensive maintenance and troubleshooting, a full section on auxiliary systems is included.

The author has added more than a hundred and fifty illustrations, and he has updated the text to meet today's practical requirements. Because of the thoroughness, clarity, and up-to-date coverage, each student will benefit from the text's outstanding presentation.

The Publishers

Cummins Engine Company, Inc.

Contents

Page

Chapter 1. DISTINGUISHING FEATURES OF DIESEL ENGINES 1

Recognizing A Diesel Engine 1
Diesels Differ from Each Other 4
Automotive Service 9
Farm Power Equipment 10
Mobile Service 12
Railroad Service 15
Marine Service 17
Stationary Service 19

Chapter 2. WHAT A DIESEL ENGINE IS 25

Basic Parts 27
What Happens Inside the Engine 29
Four-Cycle Diesel Engine 30
Compression Ratio 34
Two-Cycle Diesel Engine 35
Diesel Engines Differ 39
Advantages of Diesel Engines 40
Disadvantages of Diesel Engines 42
How Diesels Are Used 43
Why It Is Called a Diesel Engine 44

Chapter 3. BASIC CONSTRUCTION OF A DIESEL ENGINE 46

Parts Needed in a Diesel 46
How the Assembled Parts Look 48
How the Individual Parts Look 52

Page

Chapter 4. CLASSIFICATION OF DIESEL ENGINES 65
Engines Classified by Design 65
Engines Classified by Use 70

Chapter 5. STATIONARY PARTS—FRAMES, CYLINDERS, AND HEADS 88
Engines Structure and Requirements 89
Cylinders and Liners 93
Cylinder Heads 96

Chapter 6. MAJOR MOVING PARTS 99
Pistons 99
Piston Rings 106
Connecting Rods 114
Wristpins 117
Crankshafts 120
Balancer Shafts and Vibration Damper 121
Bearings 122
Flywheels 123

Chapter 7. LUBRICATING THE DIESEL 125
Lubricating Principles 125
Basic Requirements of a Lubricant 128
Diesel Engine Lubrication Systems 133
Properties of Lubricating Oils 133
Selection of Lubricating Oils 136

Chapter 8. BASIC TERMS OF PHYSICS AND ENGINEERING 138
Basic Units of Measurement 138
Derived Units of Measurement 139
Work and Power 145
Energy 146

Chapter 9. HEAT AND COMBUSTION 149
What Heat Is 149
Heat Flow 150
Gases—Pressure and Volume 152
Gas Laws 154
Basic Terms of Chemistry 156
Chemistry of Engine-Fuel Combustion 157
Heat Quantities from Combustion 161

Chapter 10. OIL AND GASEOUS FUELS 164
What Oil Is 164

Page

How Oil Is Refined 165
Properties of Diesel Fuels 167
Ignition 171
Dual-Fuel Engines Burn Gas and Oil 173
Gaseous Fuels 173

Chapter 11. ENGINE POWER AND FUEL CONSUMPTION 175

Indicated Power 175
Brake Horsepower 177
Torque 179
Brake Mean Effective Pressure 181
Efficiency and Fuel Consumption 182
How Volumetric Efficiency Affects Engine Power 186
Effect of Compression Ratio on Thermal Efficiency 187
Where the Lost Heat Goes 189

Chapter 12. ENGINE RATING AND PERFORMANCE 190

Engine Rating 191
Combustion and Cooling Limit Power Rating 191
Lubrication and Inertia Limit Speed Rating 192
Standard Ratings 196
Other Power Ratings 199
Fuel Consumption 200

Chapter 13. INTAKE AND EXHAUST SYSTEMS—SCAVENGING AND SUPERCHARGING 205

Intake and Exhaust Systems 205
Valve-Actuating Gear 210
Valve Timing 214
Air-Intake System 215
Inlet Manifolds 219
Scavenging 219
Supercharging 226

Chapter 14. INJECTING FUEL 240

Fuel Injection System 240
Multiple Plunger Injection System 241
Distribution Type Fuel Injection Pump 253
Unit Injectors 267
Injectors and Nozzles 276

Chapter 15. BURNING THE FUEL 281

Diesel Combustion 281
Solid Fuel Injection System 284
Special Design Combustion Chambers 286

Page

Chapter 16. GOVERNING 293
Governors 294
Speed Governors 296
Definition of Terms 300
Mechanical Governors 303
Hydraulic Governor Principles 307
Hydraulic Governor with Permanent Speed Droop 309
Isochronous Hydraulic Governor 314
Governor Modifications 319
How Governors Are Used 331

Chapter 17. DIESEL ENGINE TESTING, INSTRUMENTATION AND DESIGN 336
Engine Testing—Torque, Horsepower and Loading Devices 336
Pressure Measurement 347
Other Engine Tests and Calculations 355
Diesel Engine Design 361

Chapter 18. HIGH COMPRESSION GAS-BURNING ENGINES 368
High Compression Increases Burning Efficiency 368
Gas-Diesel Engines 370
Ignition and Combustion in Gas-Burning Engines 372
Dual-Fuel Engines 377
High-Compression Spark-Ignited Gas Engines 387
Uses of High-Compression Gas-Burning Engines 396

Chapter 19. AUXILIARY SYSTEMS 400
Lubricating System—Large Engines 400
Cooling System 406
Fuel-Supply System 417
Air-Intake System 422
Exhaust System 426
Starting System 428
Electric Ignition Systems for Gas Engines 431
Alarm and Shutdown Systems 438
Automatic Starting and Load-Control Systems 440

Chapter 20. OPERATION AND MAINTENANCE 443
Operation 443
Operating Procedures 445
Performance Records 448
Operator As Trouble Shooter 449
Fundamental Problems 458

Glossary 474

Index 481

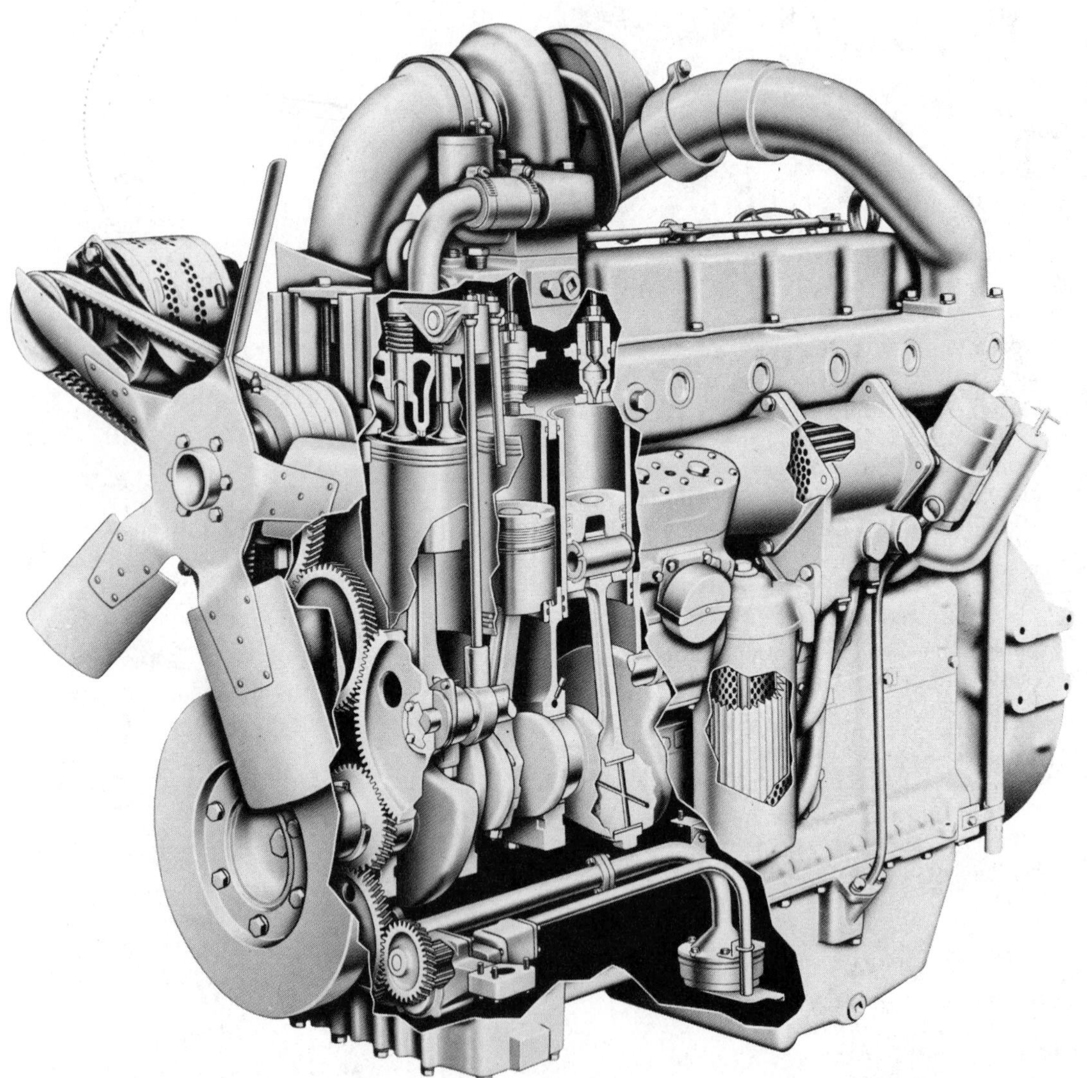

Caterpillar Tractor Co.

Distinguishing Features of Diesel Engines

Chapter

1

You are going to learn how to recognize a diesel engine when you see one. All diesels have certain features which distinguish them from other kinds of engines; you will find out what these are.

Having learned how to recognize a diesel engine, you will next see how various kinds of diesel engines differ from each other. You will also learn the reasons why they differ.

Finally, you will discover that diesel-engine applications fall into several classes. You will learn what these classes are, and in what ways the engines must be constructed differently to suit the needs of the different applications.

Recognizing A Diesel Engine

Whenever you see an engine you should be able to tell quickly whether or not it is a diesel. Many diesels, at first glance, look like gasoline engines, but it's easy to distinguish them if you keep in mind three essential differences in the way they work. These are: (1) the diesel ignites its fuel solely by the heat of compression, and therefore has *no external ignition system;* (2) the diesel draws in air alone when it fills its cylinders, and therefore uses *no carburetor;* (3) the die-

TABLE 1-1. DIFFERENCES BETWEEN A DIESEL AND A GASOLINE ENGINE

Features	Diesel Engine	Gasoline Engine
Ignition system...	Has no visible ignition system	Has spark plugs Has spark coil or magneto
Carburetion......	Cylinders are filled with air alone Has no carburetor	Has carburetor to mix gasoline and air before they enter cylinders
Fuel injection....	Has fuel-injection pump to spray oil into cylinders under high pressure Has spray nozzle in each cylinder	Has no fuel-injection pump or spray nozzles in cylinders

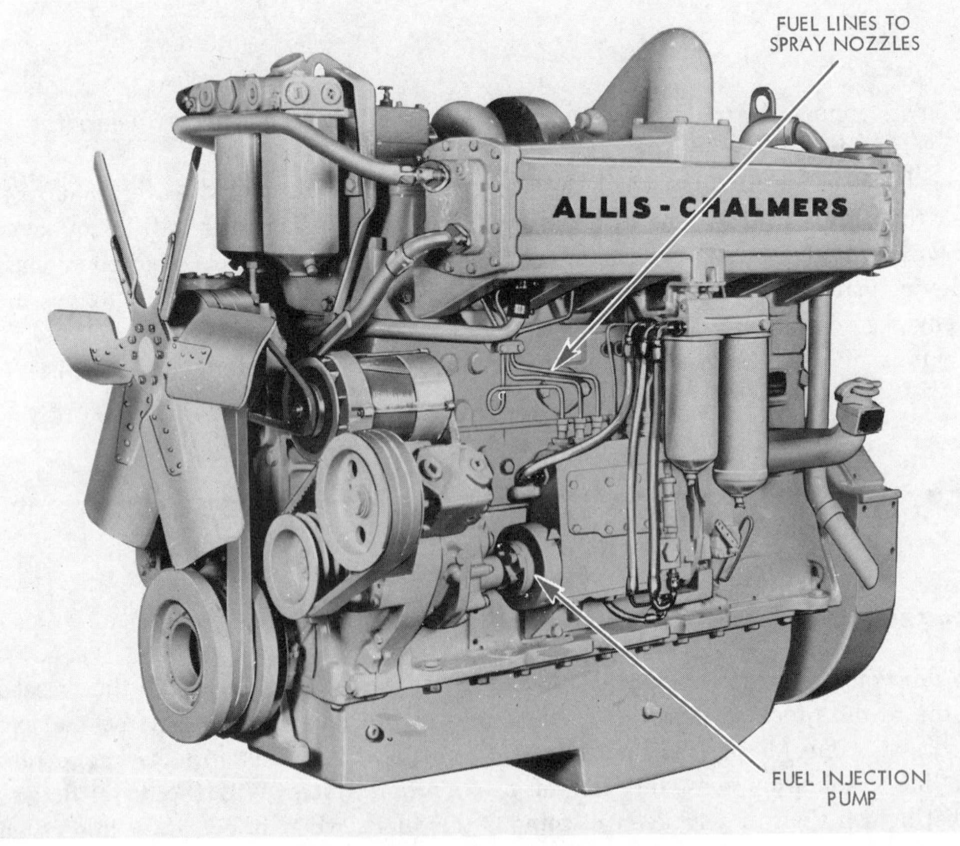

Fig. 1-1. High-speed diesel engine. (Allis-Chalmers, Engine Div.)

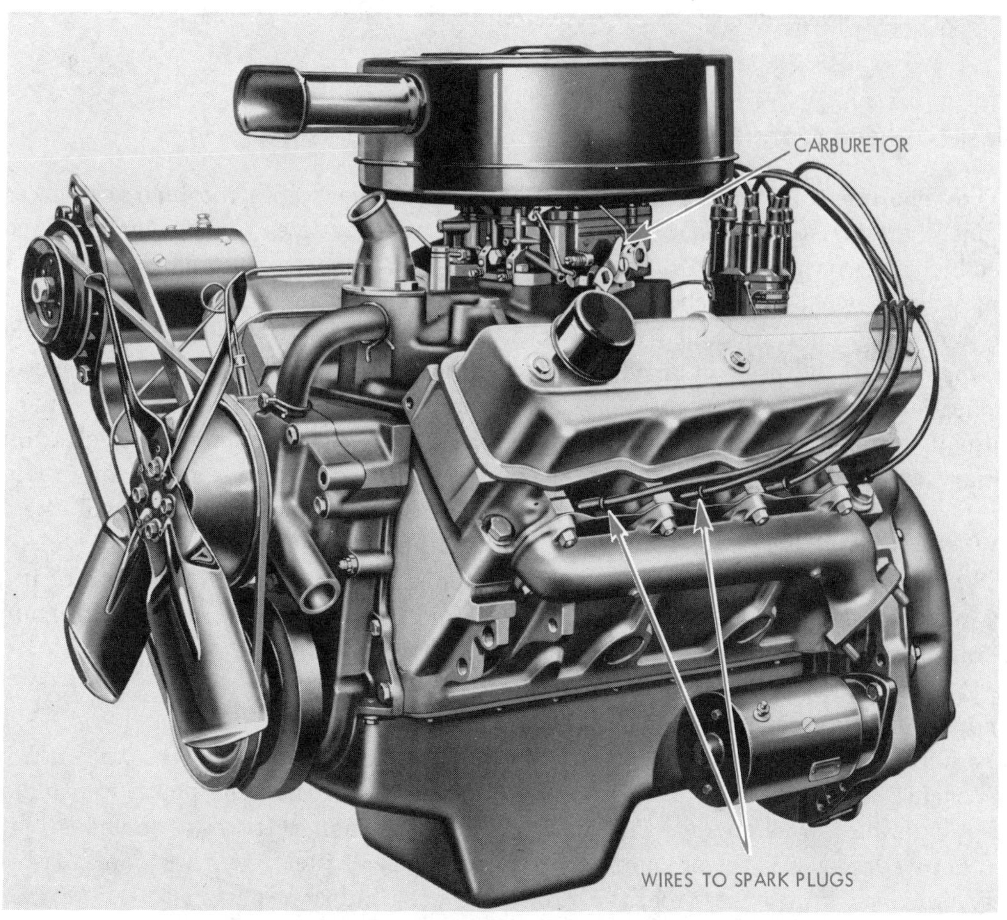

Fig. 1-2. High-speed gasoline engine. (Plymouth Div., Chrysler Corp.)

sel injects its fuel into the cylinder in the form of a spray at high pressure, and therefore uses a *fuel-injection pump.*

Table 1-1 shows the essential differences between a diesel and a gasoline engine.

Now look at Figs. 1-1 and 1-2, which show two high-speed engines, one diesel, the other gasoline. Note the fuel-injection pump on the diesel; note the carburetor and spark plugs on the gasoline engine. There's another important difference between diesel and gasoline engines—the diesel is built heavier because it must withstand higher pressures, but you can hardly see this from the outside.

Some vertical steam engines of late design look remarkably like diesels, which is not surprising because the designers of these steam engines deliberately copied some of the mechanical features of diesel construction. But the large steam supply pipe gives them away, as does the absence of a fuel-injection pump.

Diesels Differ From Each Other

Different uses call for different diesel engine characteristics as regards weight, speed, cost, governing, kind of fuel, etc. You will see what these characteristics are and which ones are called for when an engine is to be used in a particular application, such as automotive, mobile, railroad, or marine service, as well as stationary service.

High-Speed Engines Have Small Cylinders

First, you should understand the important connection between speed and weight. Each time a diesel piston makes a power stroke, a certain amount of energy is produced. But how much *power* the engine produces depends upon how *often* it develops this energy. Putting it another way, power is the amount of energy produced in a given time, the *rate* at which energy is produced.

For example, consider two identical engines, one of which runs twice as fast as the other. The faster engine will make twice as many power strokes, and will therefore develop twice as much energy in a minute as the slower one; consequently, it will have twice the power.

So if we want to get a lot of power out of a particular engine, we run it as fast as it can go safely. But there is a limit to how fast we can run a given engine. As the speed goes up, the pistons slide faster against the cylinder walls and the *rubbing speed* increases. Also, the forces created by motion, called *inertial forces*, go up.

If the speed goes too high, the excessive rubbing speed and inertial forces cause damage to the engine parts. But if pistons have smaller diameters and shorter strokes, the inertial forces and rubbing speeds are less. For this reason designers of high-speed engines make them with a large number of small cylinders, thus obtaining engines of greater power and less weight. You will therefore find that if a lightweight diesel engine is needed for a certain kind of use, it will be a high-speed engine with many small cylinders.

Diesels Require More Air

The more air and fuel we can burn in the cylinders, the more power we can get from the engine. It's easy enough to put in plenty of fuel, but if we want the fuel to burn fully, we must also put in more air. Putting in more air isn't easy.

The air around us is always under pressure, which we call *atmospheric pressure* or *barometric pressure*. This pressure is produced by the weight of the air which surrounds the earth (air really has weight—12 cubic feet of air weighs about a pound). At the outer limit of the atmosphere, the air is under no pressure. As we approach the earth's surface, however, the weight of all the air above presses down with more and more pressure until at sea level the atmospheric pressure is about 15 pounds per square inch (15 psi). You don't feel this pressure on your body, because the pressure is inside you as well as outside, and so

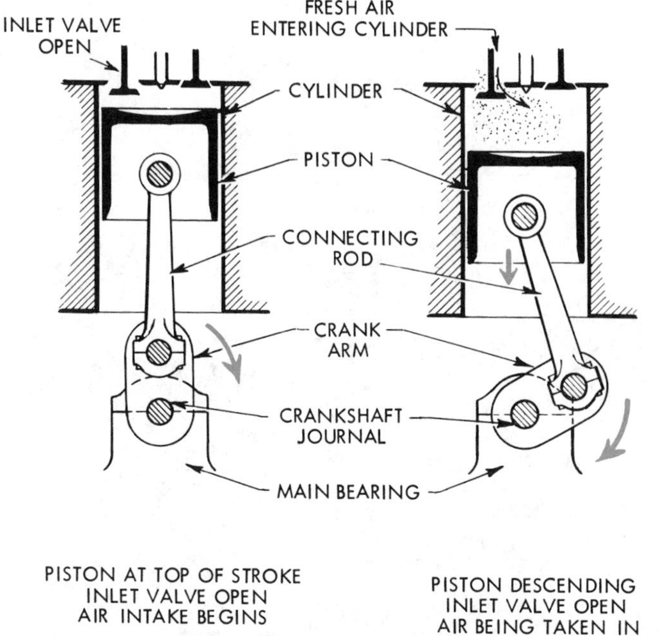

Fig. 1-3. Schematic—intake stroke of a four-stroke cycle engine.

is balanced. But it is this *atmospheric pressure* which forces air into the cylinders of an ordinary four-cycle engine which is normally aspirated (normal breathing).

Let's look at the ordinary four-cycle engine, Fig. 1-3. Just how does the air get into the cylinders? Notice that the piston descends on the inlet stroke and the inlet valve is open, so that air seems pulled into the cylinder. That's a simple way to say it, but really it isn't pulled in at all. What actually happens is that as the descending piston makes a *partial vacuum* it lowers the pressure *inside* the cylinder below that of the outside atmosphere. The outside pressure (atmospheric pressure) then *pushes* air into the cylinder, filling it to or near outside pressure. If the outside pressure is *higher* than atmospheric, it will fill the cylinder to a higher pressure, thus putting more air into the cylinder. This is just what a supercharger does. By putting in more air, a supercharger increases the engine's power.

A supercharger is simply a kind of air pump that takes air out of the surrounding atmosphere, compresses it to a higher pressure, and then feeds it to the engine's inlet valves. It is called a *supercharger* because it charges the engine with air at a *super pressure*.

Superchargers are of two general types. The Roots-type supercharger shown in Fig. 1-4 is a positive-displacement compressor which is driven by chain belt or gear from the engine. This type is often used with two-stroke engines. Notice in Fig. 1-4 that the supercharger, or blower,

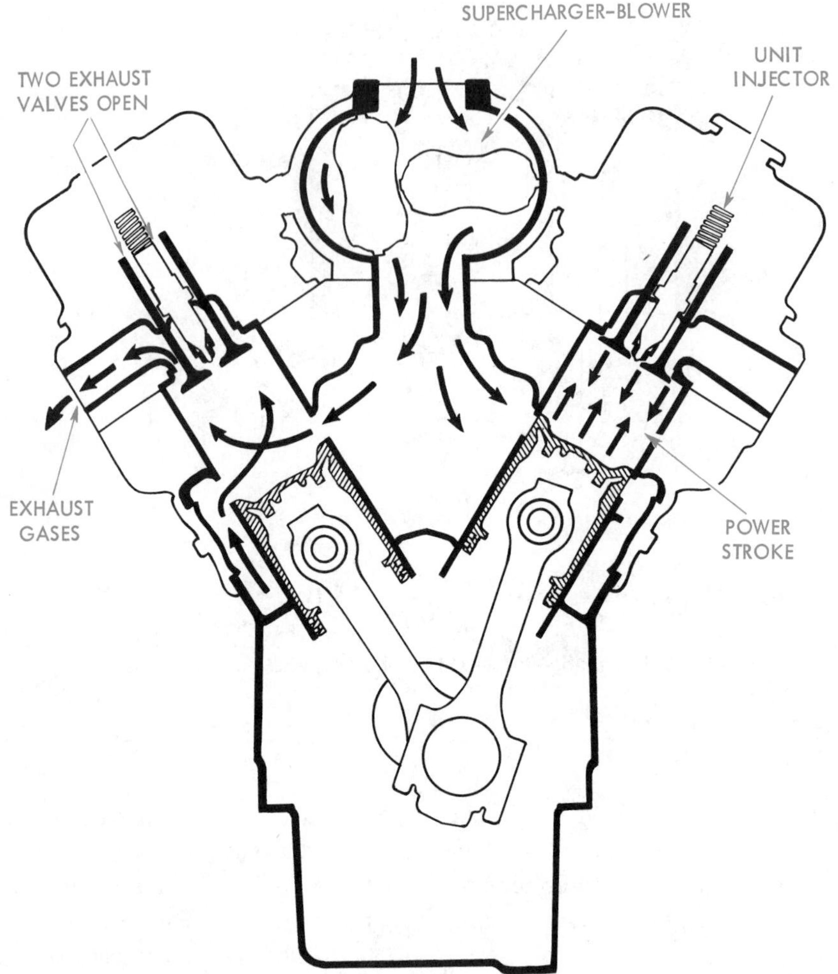

Fig. 1-4. Supercharger (blower) on a two-stroke V-type engine. (General Motors Corp.)

sits above and between the cylinders on the two-stroke V-type engine in this case.

In operation, air enters at the top (inlet side) due to the vacuum created by the rotation of the rotors. The air passes between the lobes of the rotors and the blower housing to the outlet or pressure side at the bottom of the blower. In this illustration the left piston is at the bottom of its stroke and has uncovered the holes in the cylinder liner so the fresh air can blow the exhaust gases out the two exhaust valves which are in their open positions.

The right piston is going down on the last part of the power stroke, the holes in the liner are about to become uncovered and exhaust valves about to open, allowing the air from the blower to come in above the piston.

Fig. 1-5 illustrates a centrifugal-type supercharger. This type of supercharger is usually driven by high-pressure exhaust gases from the engine and is generally called a *turbocharger*. In operation, the turbocharger rotates at very high speed, usually 40,000 or more revolutions per minute (rpm).

Referring to Fig. 1-5, you will see that a turbocharger is basically two wheels connected by a shaft. Both wheels carry vanes which make them act like fans. The hot exhaust gases from the engine cylinders blow on one wheel and make it turn at high speed, driving the other wheel. The fan action of the second wheel draws in outside air and increases its pressure before packing it into the engine cylinders. The increased weight of air combines with more fuel to produce more power.

You also need to know that superchargers are used for *two* important purposes. The first is to get more power out of a given engine by feeding it with more air (and more fuel also) than it would take directly from the atmosphere in normal operation. Thus we get an engine of

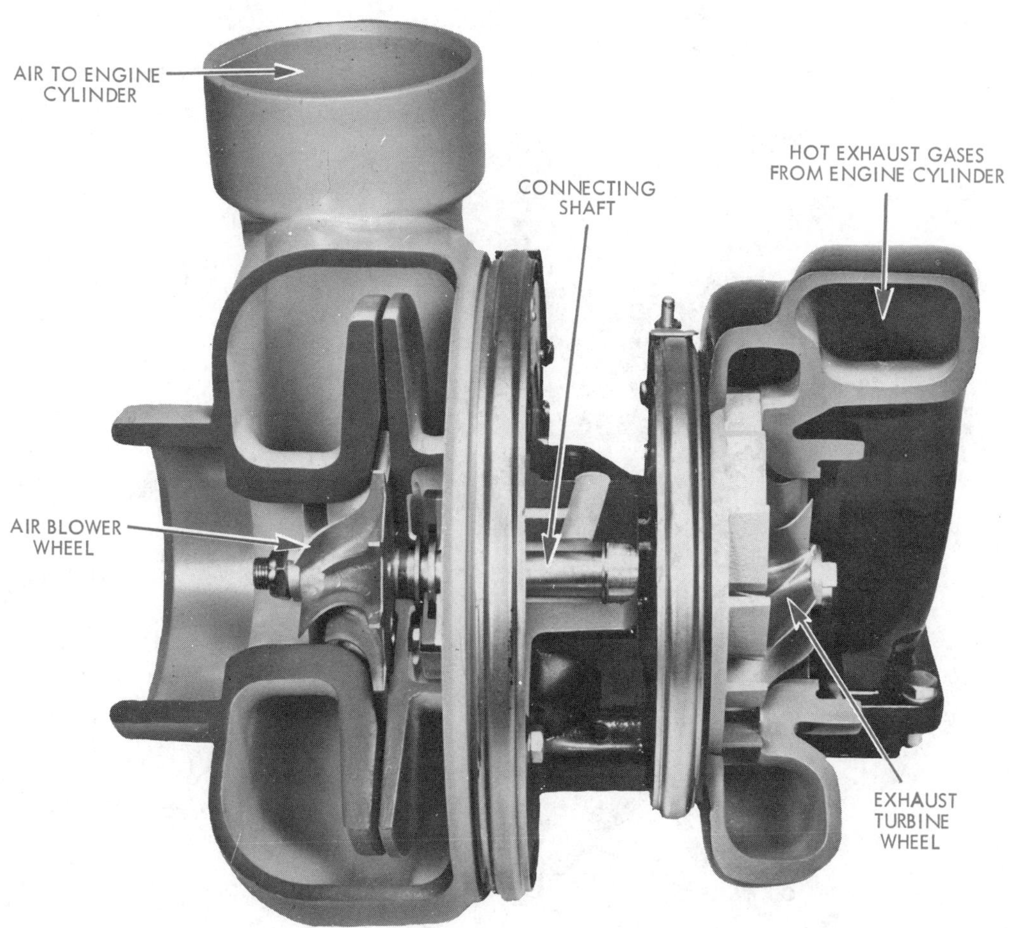

Fig. 1-5. Turbocharger. (Caterpillar Tractor Co.)

greater power for the same dimensions and almost the same weight. The second purpose is to restore some of the power which any engine loses when it is taken above sea level to a high altitude.

The pressure of the atmosphere is less the higher you go, so there is less pressure to force air into the cylinder, and the engine becomes weaker for lack of air. (An engine's power at 8000 feet above sea level is only 75 percent of its power at sea level.) Here's where the supercharger becomes essential. It can increase the low atmospheric pressure of high altitudes to sea-level pressure and thus restore most of the lost power. (It can't restore *all* the lost power because some of the regained power must be used to drive the supercharger.)

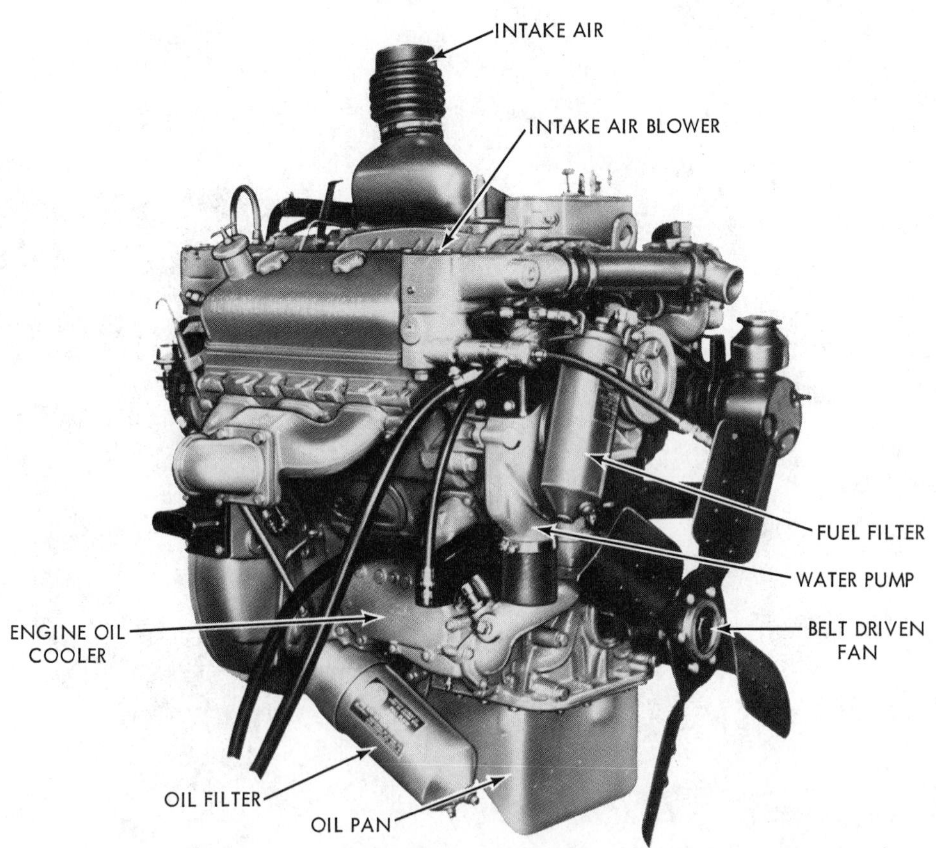

Fig. 1-6. A diesel of the V-6-71 series. (Detroit Diesel Allison Div., General Motors Corp.)

Automotive Service

Although the word *automotive* means *self-propelling,* something which causes itself to travel, the term *automotive diesels* is generally used to refer only to diesels which propel highway vehicles such as automobiles, trucks, and buses.

Trucks and buses are much heavier than passenger cars, and therefore, diesel power adds little more proportional to their weight and cost. Also, trucks and buses are in motion many more hours per year and are fitted with much larger engines. As a result, diesels reduce the fuel cost by so many dollars that the extra first cost over gasoline engines is quickly repaid. Today we find diesel engines commonly used on larger trucks and buses and coming into use on taxicabs. See Figs. 1-6 and 1-7.

What kind of diesel engines are used? First, they must be lightweight, which means they are high-speed, small cylinder engines. Second, they must run well at many speeds; they must be *flexible.* Also, they should produce little smoke or offensive odors under all operating conditions. Finally, they must be designed to start readily in cold weather.

Fig. 1-7. Two-stroke cycle diesel, V-6-71 series. (Detroit Diesel Allison Div., General Motors Corp.)

Because of the big market for automotive diesels, great efforts have been made to develop satisfactory engines, and many manufacturers are engaged in building them to fit the various power user requirements.

Farm Power Equipment

The modern farmer has recently become one of the chief users of diesel engines. Diesels are used in farm tractors for plowing, seeding, and harvesting as well as for other farm work. The trend toward larger tractors, more power, and greater traction is in evidence in Fig. 1-8 as this four-wheel-drive farm tractor pulls a plow which turns over eight 16-inch strips of soil at the same time.

Fig. 1-8. Four-wheel drive farm tractor, diesel drive. (Deere & Company)

Fig. 1-9. Diesel powered two-wheel drive tractor, pulling a planter. (Deere & Company)

At planting time, huge acreages must be covered in the minimum of time with a greatly reduced labor force. Fig. 1-9 illustrates a diesel-powered farm tractor pulling a planter.

With increased power available, driver comfort has become increasingly important. Fig. 1-10 illustrates a harvesting operation in which the driver sits in a dust-free and air-conditioned cab.

Fig. 1-10. Modern diesel-powered harvester. (Deere & Company)

Fig. 1-11. Self-propelled diesel-powered cotton picker. (Deere & Company)

The harvester shown in Fig. 1-10 is power-driven at two wheels. The diesel engine is six-cylinder, 404 cubic inch (cu in) displacement and has 128 horsepower (hp) at 2500 rpm. Another specialized machine for the large farmer is the cotton picker shown in Fig. 1-11. It is powered by a six-cylinder diesel engine, 329 cu in displacement and has 105 hp at 2500 rpm.

Mobile Service

By *mobile diesel* we mean a diesel which is mounted on a piece of equipment which can be moved from place to place; the machine is *mobile*. The diesel's power may or may not be used to move the machine—its main purpose is to supply power to the machine to do the work for which the machine is designed. Such machines are widely used for all kinds of construction work.

We find mobile diesels used for power hoists, drag-lines, tractors, bull-dozers, power shovels, graders, scrapers, air-compressors, drills, and portable electric generators.

The diesel engines used in mobile work are similar to those used for the automotive applications previously discussed, but since more power is usually needed, the mobile diesels are often larger, heavier, and somewhat slower in speed. In applications where the same amount of power is needed, the mobile diesels closely resemble automotive diesels. See Figs. 1-12 through 1-15.

Fig. 1-12 shows an Caterpillar Tractor Co. mobile generating unit mounted on its own moveable trailer, ready for emergency service anywhere at any time.

Four diesel powered electric generator sets of 250 kw provide a 1000 kw power package for operating a portable asphalt road building plant are mounted on a flat-bed, as in Fig. 1-13. The Allis-Chalmers Model 2500 diesels are turbocharged and intercooled.

Hydraulic operated power shovels such as the one in Fig. 1-14 are used in the building construction, road construction, mining, and many other industries. This

Fig. 1-12. Mobile generating unit (housing removed). (Caterpillar Tractor Co.)

Fig. 1-13. Four diesel-powered electric generator sets. (Allis-Chalmers, Engine Div.)

unit is powered by a six-cylinder diesel engine, 404 cu in displacement, and has 146 hp at 2200 rpm.

The track-type back-hoe in Fig. 1-15 makes heavy duty ditch-digging an easy job. This machine is powered by a six-cylinder 404 cu in displacement diesel, which develops 175 hp at 2100 rpm.

As is obvious, the diesel engine may be a power source for driving the particular unit, or it may be a separate engine, mounted on a trailer, and having but the single function of generating electric power. In some cases, the diesel may be used for locomotion, and for driving a separate mechanism.

Fig. 1-14. Hydraulic-operated power shovel. (Deere & Company)

Fig. 1-15. Diesel-powered heavy-duty back-hoe. (Deere & Company)

Railroad Service

Practically all railroad locomotives now being built in the United States are powered with diesel engines. These locomotives are of several types; switchers, road-switchers, and road locomotives of both passenger and freight types. Switchers generally use diesels of 600 to 1200 hp, while road locomotives use larger engines of 1800 to 3350 hp each.

The diesels in most cases drive electric generators which supply current to electric motors at the wheels; the locomotives are therefore called *diesel-electric*. The main advantages of the electric drive are the avoidance of transmission gears, easy control, and great pulling power when starting the train.

Diesels used in railroad locomotives must not be too heavy; therefore, they have many cylinders and run at medium high speeds. One might say that they are a cross-breed between low-speed stationary engines and high-speed automotive engines.

Fig. 1-16 shows two 1125 hp, twelve-cylinder engines being installed on the

Fig. 1-16. Two 1125 hp diesel engines and generators being installed on underframe of passenger locomotive. (General Motors Corp.)

Fig. 1-17. A 4800 hp engine being lowered into locomotive. (Colt Industries, Fairbanks Morse Motor & Generator Operation)

underframe of a passenger locomotive. Fig. 1-17 shows a 4800 hp engine being lowered into a locomotive.

The railroads also use diesel railcars for runs where traffic is light. Each railcar carries its own power plant; therefore a single car can be run by itself. The diesels used are of the automotive type. Sometimes they are connected to wheels by electric drive, as used in diesel-electric locomotives; in other designs, hydraulic transmissions are used similar to those used in modern automobiles.

Small diesel switching locomotives are used by industrial concerns for moving freight cars in their own yards. The engines used are similar to the mobile diesels previously discussed.

Marine Service

Diesel engines are used to propel all kinds of marine craft, from motor boats to ocean-going motor ships. In between are tugs, barges, etc. Examples of high-powered diesel ships are U.S. Coast Guard icebreakers with 13,000 total hp,

Fig. 1-18. Turbocharged marine diesel engine in dredge. (White Diesel Engine Div.)

and U.S. Navy submarine tenders powered with 15,500 hp diesels.

Marine diesels vary greatly in size and type. But they all have one common characteristic which distinguishes them from diesels used on land—they must be able to reverse the propellers which they drive. On small vessels like inboard powered motor boats, reversal is generally accomplished by ordinary reverse gears such as are common in gasoline motor boats. The diesels used here are of the automotive type, but must have special provision for salt-water cooling when used on coastal waters which are salt.

Reversing of medium-size and large-size marine diesels is often accomplished by fitting the engines with an extra set of cams for operating the valves. Thus, the engine can be run in either direction; such engines are called *direct-reversing engines.*

Other large marine engines have electric drive to their propellers; this arrangement not only provides for reversing but also permits the use of an engine which runs at higher speed than the propeller. Reduction gears are also sometimes used

Fig. 1-19. One of two ten-cylinder opposed-piston engines in river towboat. (Colt Industries, Fairbanks Morse Motor & Generator Operations)

Fig. 1-20. Shop view of 10,800 hp two-cycle, twelve-cylinder marine diesel engine. (Nordberg Manufacturing Co.)

with direct-reversing engines for the same purpose, to permit the use of a faster engine of less size and weight.

The wide range in size and type of marine diesels is clearly shown in Figs. 1-18 to 1-20.

Stationary Service

Diesels used in stationary service are usually set on foundations resting on the ground. We don't worry much about the weight of the engine. Also we're not too much concerned about the space they occupy, although there are exceptions, such as where a diesel power plant is to be installed in close quarters in the basement of a city office building of very modest size, or where foundation supports must be vibration proof.

Continuous-Run Engines

One common use of stationary diesels is to drive electric generators to produce a continuous supply of electric power. Typical examples are electric utility plants and factory power plants. Here we're interested particularly in high reliability, low fuel consumption, low cost of repairs, long life and good governing.

The reason is that such engines must

run many hours per day, for many years, and they must be thoroughly dependable because a breakdown might interrupt important electric service. Since the engines run so much of the time, they consume a lot of fuel in a year. Therefore, they should be as efficient as possible. They should consume the minimum amount of fuel per unit of power produced because at the end of a year the total saving in fuel will amount to many dollars.

The same is true as to cost of repairs. If such engines were to wear out rapidly, they would need frequent overhauls. We want engines that will run for a long time without major repairs or upkeep.

All of these factors help to keep down the cost of producing electric power, which is the reason why most of these engines are installed in the first place—otherwise the owners would buy the electricity from a power company.

The type of diesel engine which best meets the above requirements is a sturdy engine of low or medium speed, whose parts are ample in size for the work they have to do, and whose fuel-injection system is designed to consume the least amount of fuel. Such engines cost more to buy than lighter-built engines, but their cost is justified by their lower operating cost.

Such engines are always fitted with good governors. A *governor* is a device

Fig. 1-21. Municipal power plant, Columbus, Wisconsin. (Worthington Corp.)

which keeps the engine running at a constant speed even when the load on the engine changes. It does this by continually changing the amount of oil supplied to the power cylinders. The more accurate the governor, the closer it controls the fuel to match the changes in load.

Load changes are frequent in most plants where engines are used to produce electric power; we don't want the voltage or frequency of the electric current to change every time the load changes. So we fit the engine with an accurate governor which will keep the engine's speed the same at all times. Some typical stationary diesels used for electric generation are shown in Fig. 1-21.

Fig. 1-22. Emergency diesel-electric set in Cleveland, Ohio, hospital. An eight-cylinder, 900 rpm diesel engine drives a 750 kw AC generator. (Chicago Pneumatic Tool Co.)

Short-Run Engines

We have been discussing the usual applications of diesels for producing electric power where the engines run continuously for long periods. But not all electric units are required to run continuously — an important exception is the *standby* or *emergency unit.* An example of this type of engine is illustrated in Fig. 1-22.

Such units are often used in places like telephone exchanges, hospitals, and airports, where the regular source of electric power is from the public supply lines. The diesel unit is installed only to supply power when an interruption occurs in the regular supply. Consequently, emergency units are called upon to run only at in-

Fig. 1-23. Diesel engine belted to cotton gin. (Waukesha Motor Co.)

frequent intervals and they rarely run many hours in a year.

Emergency units, therefore, need not have the low fuel consumption or long life required of engines in continuous operation. Consequently, they are often lighter-built, inexpensive engines of the automotive type.

Other kinds of electric service have different requirements as to space available, hours of operation, nature of load, etc. Consequently, almost every type of diesel has been used at some time to drive an electric generator.

Mechanical Drive

Many stationary diesels are used, not to deliver power in the form of electricity, but to drive machines *mechanically*, either by connection directly from the engine shaft or through belts or chains, Fig. 1-23. Some are intended to run continuously, others only occasionally. Consequently all kinds of engines are used, depending upon how much they are to run. For the reasons previously given, we select slow-speed, sturdy, efficient engines for steady service, and high-

Fig. 1-24. Gas engine driving built-in gas compressor in gas pipeline pumping station. (Ingersoll-Rand Co.)

speed, inexpensive engines for intermittent service.

The pumping stations of oil pipelines contain diesel engines directly connected to the big oil pumps which force oil through pipes hundreds or thousands of miles from the oil wells to the refineries. Heavy-duty, highly reliable engines are used because they must be able to run, fully loaded, continuously for months.

High-Compression Gas Engines

Natural gas is the least expensive fuel in many parts of the United States. So is sewage gas, which is a by-product of the sewage treatment plants used by many municipalities. These cheap gases can be efficiently burned in the modern high-compression gas engine which was developed from the diesel engine. You will learn about these in Chapter 18.

The many gas pipelines which traverse the country are powered by thousands of these gas-burning engines. The engine and the gas compressor which forces the gas through the pipeline, are often combined into a common unit, as shown in Fig. 1-24.

Checking On Your Knowledge

The following questions give you the opportunity to check up on yourself. If you have read the chapter carefully, you should be able to answer the questions. If you have any difficulty, read the chapter over once more so that you have the information well in mind before you go on with your reading.

DO YOU KNOW

1. Why is a diesel usually built heavier than a gasoline engine?
2. Why do high-speed diesel engines usually have small cylinders and more cylinders than slow-speed diesel engines?
3. Why are diesel engine manufacturers concerned with piston speed?
4. What is enertia and how does it affect engines?
5. What does atmospheric pressure mean?
6. When a diesel engine is a normally aspirated engine, what does this mean?
7. What is the purpose of a supercharger?
8. What is the difference between a supercharger and a turbocharger?
9. Why is a turbocharger so important at higher elevations?
10. What are some of the specific features of different diesel engines which make them better engines for continuous-run engines, short-run engines, automotive type, railroad type, marine type, or farm tractor type engines?

What a Diesel Engine Is

Chapter

2

The purpose of this chapter is to further introduce you to the subject of diesel and high-compression gas engines. You will learn what a diesel engine is mechanically; then you will want to know how it works and how it differs from a gasoline engine. After learning these facts, you will find out the many ways in which diesel engines are put to use. Knowing these things will make it easier to progress in your study of a type of engine somewhat like the diesels—the high-compression engine that burns gaseous fuels.

While reading this chapter you will pick up the meanings of some words and phrases that are used often, such as *internal combustion, compression, two-cycle*, and *four-cycle*. You'll also find out how the diesel engine came to be named *diesel*.

Let's pass up the technical definitions of a diesel engine. Simply, a diesel engine is a machine which produces power by burning oil in a body of air which has been squeezed to a high pressure by a moving piston. Since it is a machine that produces power, it is called an *engine;* and since the burning or *combustion* takes place within the engine itself, it is called an *internal* combustion engine. A steam engine, by contrast, uses steam, burning fuel *outside* the engine; this is called *external* combustion. In both, heat is converted to power.

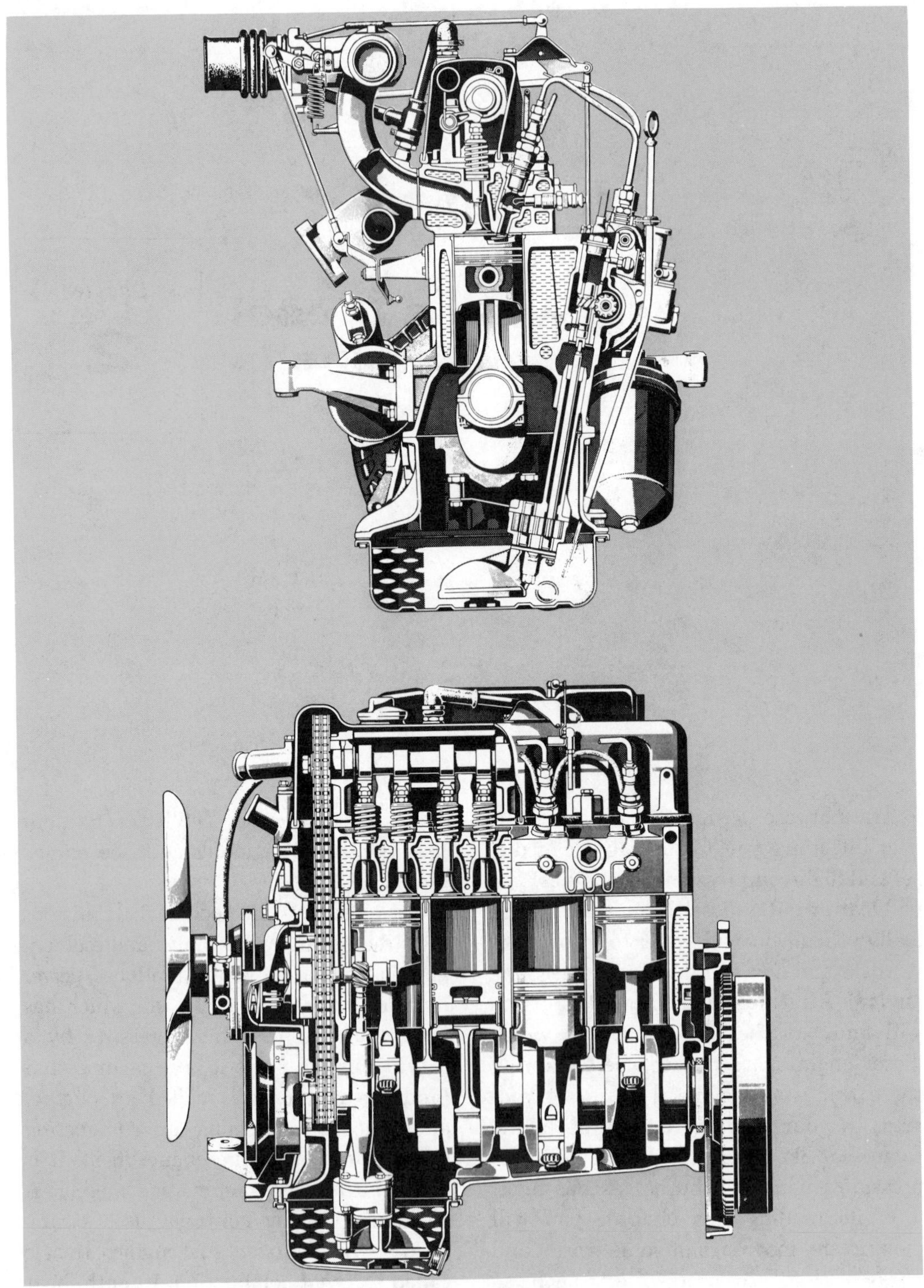

Mercedes-Benz 240-D diesel engine used in North America.

Basic Parts

To compress the air, put in the oil, and produce power, every diesel engine must have certain basic parts. See Fig. 2-1. It must have a round sleeve or *cylinder*, in the *bore* of which a close-fitting plug or *piston* can slide in and out to make *strokes*. The piston must be connected to a mechanism which controls its sliding. For this purpose ordinary engines use a crank mechanism. This consists of a round bar or *shaft* which can turn or *revolve* in circular guides called *bearings*

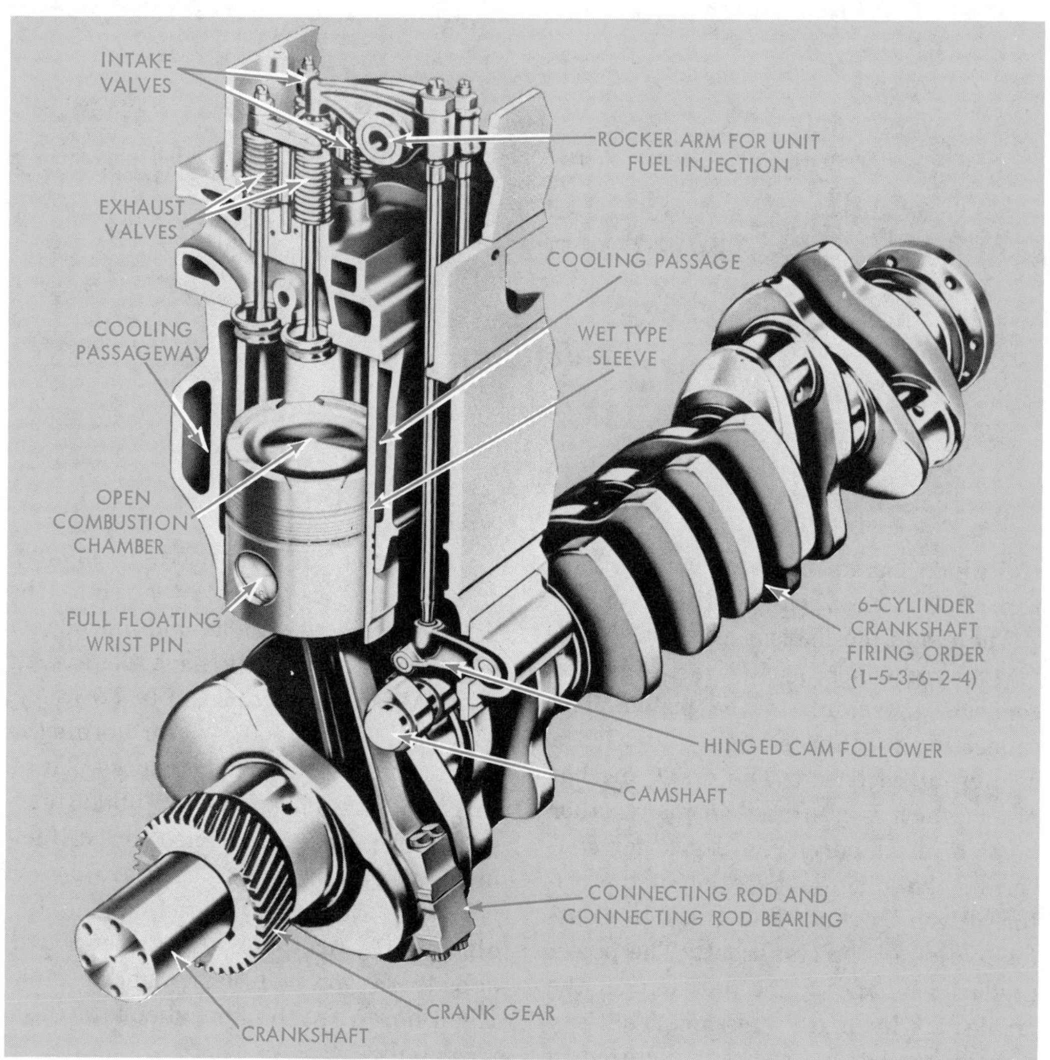

Fig. 2-1. Some basic parts of a diesel engine.

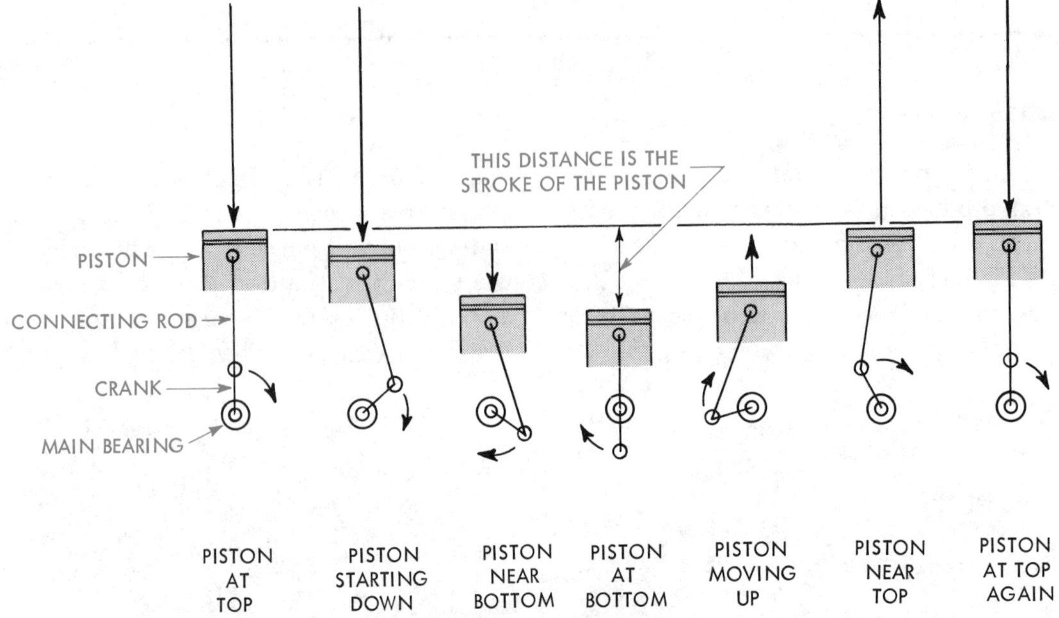

Fig. 2-2. How a crank mechanism works.

and which has an offset or *crank* which turns in a circle when the shaft revolves.

The crank mechanism also has a *connecting rod* which, as the name implies, connects the crank to the piston. The connecting rod is a straight rod with a bearing at each end. The crank mechanism (which is also used in many other kinds of machinery) is a device for converting the in-and-out motion of the piston, driven by expanding gases, to rotating motion of the crankshaft. The power produced by the engine is taken off the crankshaft by a transmission. The way a crank mechanism works is pictured in Fig. 2-2.

Other basic parts are necessary. *Valves* or *ports* are needed to let air into the cylinder, and also to let out burned or spent gases after they have done their work. Also a *spray nozzle*, or *fuel injector*, is needed to deliver the oil for burning in the form of a finely divided spray. To do this the oil must be forced in under pressure. This is accomplished by a fuel-injection pump.

Even the simplest diesel engine has other parts than those just mentioned, and these you will learn about later. Right now we're thinking about the basic parts only.

What Happens Inside the Engine

Let's see what happens inside an engine, step by step. First, air must be gotten into the cylinder because no fuel will burn without air.

Burning or combustion is the process of uniting a fuel or *combustible* with the oxygen in the air. The process is a chemical one, which means that fuel and oxygen when united under the proper conditions change into new substances.

Then the air must be squeezed or *compressed* to a high pressure. There are two reasons for compressing the air. One reason is that if the combustible mixture has been compressed to a high pressure before it starts to burn, it will produce more power. (This will be explained fully in Chapter 11.) The other reason is that when air (or any gas) is compressed, its temperature goes up — the higher the pressure, the higher the temperature. In a diesel engine, air is compressed so much that it becomes hot, so hot that it will *ignite* oil that is sprayed into it.

This is a fundamental difference between a diesel engine and a gasoline engine. In a gasoline engine, a spark is used to ignite the combustible mixture, while in a diesel engine the fuel ignites by itself from contact with air—air that is very hot because it has been highly compressed.

Next, the fuel must be squirted into the cylinder in the form of a fine spray. The oil is squirted in *after* the air has been compressed and thus heated to a high temperature. It must be in the form of a fine spray so that a cloud of oil droplets will spread through all of the air. This produces a thorough or *homogeneous* mixture of oil and air, which is needed for quick and complete combustion. (It all must happen so fast in an engine that there's not even tenths of a second to waste.)

After that, combustion takes place immediately after the oil is sprayed into the cylinder; this generates a large amount of heat. The burning gaseous mixture gets hotter and tries to *expand.* It pushes on the piston, which in turn transmits the force through the connecting rod to the crank on the crankshaft. This makes the crankshaft revolve and thus deliver *power* to whatever machine the engine is driving.

Finally, when the piston has finished its power stroke, and the gases in the cylinder have lost their pressure, the spent gases must be pushed out, or *exhausted.*

Meaning Of A Cycle

When the cylinder is clear of spent gases, it is ready to receive a fresh charge of air and start the *cycle* all over again. A cycle, thus, is a full series of separate steps or *events.*

Now what is meant by the common terms of *two-cycle* engines and *four-cycle* engines? These terms are really abbreviations for *two-stroke* cycle and *four-stroke cycle,* which make more sense.

But, almost everyone now uses the shorter terms.

In a two-cycle engine, it takes *two* strokes of the piston (that is, one upstroke and one downstroke) to go through one complete cycle of events. In a four-cycle engine, a complete cycle requires *four* strokes of the piston (one up, one down, one up and one down). These terms apply to all kinds of internal-combustion engines, not merely to diesels.

Four-Cycle Diesel Engine

We have seen in the previous section each of the actions that must take place inside a diesel engine as it goes through one cycle of events. Now we want to learn how these actions come about. We want to study how a typical diesel engine works.

Fig. 2-3 is an elementary diagram which shows the basic moving parts of a simple four-cycle diesel engine with only one cylinder. (Most diesel engines have more than one cylinder, but all the cylinders work alike). The illustration is divided into several views, each view showing the same engine, but with its moving parts in a different position. By studying the different positions in order, we learn how the engine works.

The Intake Stroke

Starting at Fig. 2-3A, we find the engine with its piston at the top of the cylinder (*top center*) and ready to draw in a charge of air. The *inlet valve is open;* the other valves are closed. The crankshaft is turning. As the shaft turns to the right and downward, it pulls down on the connecting rod, and the rod in turn pulls down on the piston. Thus the piston descends in the cylinder as shown in Fig. 2-3B. It draws in a charge of fresh air through the inlet valve.

When the piston has descended to the bottom of its stroke (bottom center), as shown in Fig. 2-3C, the inlet valve closes. The cylinder is now full of fresh air. Fig. 2-3A, B, and C show the start, middle, and end of the intake stroke.

The Compression Stroke

Fig. 2-3C also shows the position where the piston is about to compress the air charge. As the crank turns to the left and upward, the connecting rod likewise pushes the piston upward, as shown at Fig. 2-3D. Since all the valves are closed, the air cannot escape; the air is therefore forced into a smaller space. This increases the pressure of the air and also its temperature, so that when the piston has reached the top of its stroke (Fig. 2-3E) the air occupies only about 1/16 of its original space and its temperature has risen to 1000°F or more. We call the action shown in Fig. 2-3C, D and E the *compression stroke.*

The Power Stroke

The air is now so hot that it will automatically ignite oil if oil is sprayed into it. That is what now happens when the fuel is squirted through the fuel-injection nozzle, as indicated in Fig. 2-3E. The oil burns quickly because it is well mixed with hot air. The burning process produces more heat, which makes the burning mixture still hotter. Because this hot body of gases is confined in the small space between the top of the piston and the top of the cylinder its pressure goes up too. This pressure is exerted on the top of the piston and pushes it down on the *power stroke.*

By the time the piston reaches the position shown at Fig. 2-3F, the oil has all been sprayed in and has finished burning. The hot gases now occupy a larger space (because of the descending piston), and we therefore say that they are *expanding.* This is sometimes called the *expansion stroke* instead of the power stroke.

The piston continues to descend with the hot gases pushing on it, but the expanding gases lose pressure and become cooler. Meanwhile, the pressure of the gases on the piston is transmitted through the connecting rod to the crank. The force on the crank makes the crankshaft turn.

Shortly before the piston reaches bottom center on the power stroke, the exhaust valve is pushed open mechanically and the spent gases, which have now lost most of their temperature and pressure because of the expansion, start to blow out through the exhaust valve. (The spent gases are generally blown away outdoors.) See Fig. 2-3G. This is the end of the power stroke, which started with the piston at top center, as shown in Fig. 2-3D.

The Exhaust Stroke

When the exhaust valve opens at the end of the power stroke, the spent gases in the cylinder escape only until their pressure falls to a little above the pressure of the outside air. The cylinder still remains full of dead gas which must be cleared out to make way for a fresh charge of air. The engine does this on the next stroke, when the piston rises with the *exhaust valve open,* as shown at Fig. 2-3H, and pushes out the spent gases through the exhaust valve. When the piston has risen to top center, the spent gases have all been expelled from the cylinder, and the *exhaust stroke* has been completed.

This ends one cycle. The inlet valve opens again and the engine is back in the position shown in Fig. 2-3A, ready to start on another cycle.

The Flywheel

Now we mention another basic part of a diesel engine, the *flywheel.* This is a heavy wheel fastened firmly to the crankshaft. Its purpose is to keep the engine running smoothly from the time of one power stroke to the next power stroke. You have noticed that it is only during the power strokes that the piston delivers power to the crankshaft. At other times the crankshaft must return power to the piston, which is then taking in fresh air, compressing the air and pushing out the exhaust gases.

The flywheel returns some of the engine power to the piston by first picking up a little extra speed on the power stroke and then losing it on the other

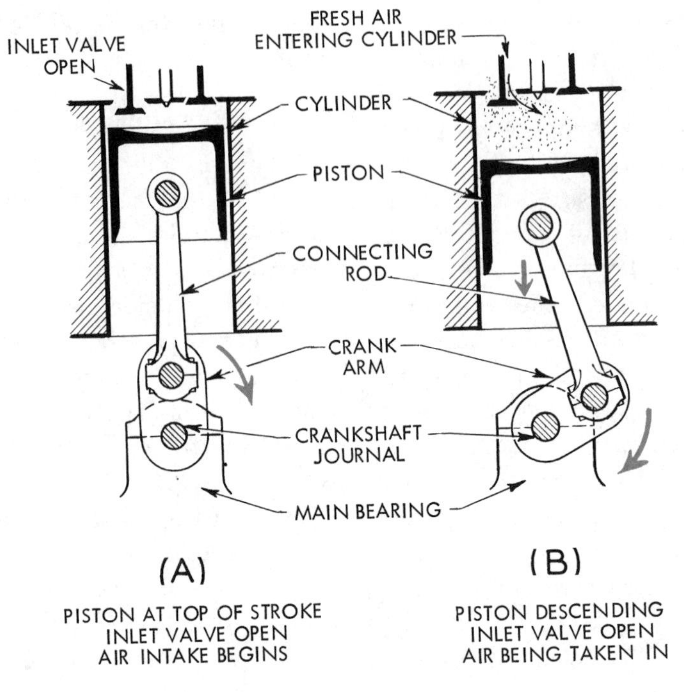

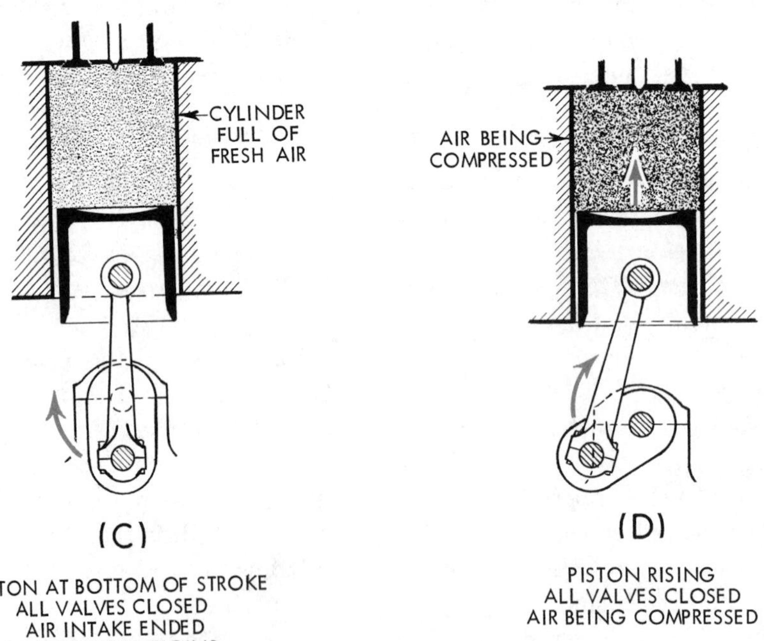

Fig. 2-3. How a four-cycle diesel engine works. (A-B-C show intake stroke; D compression stroke;

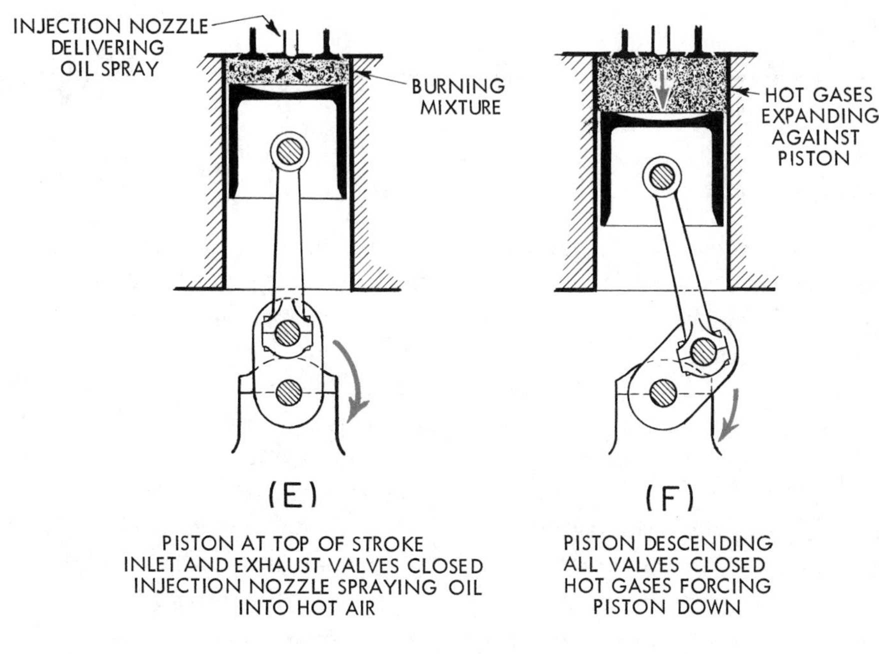

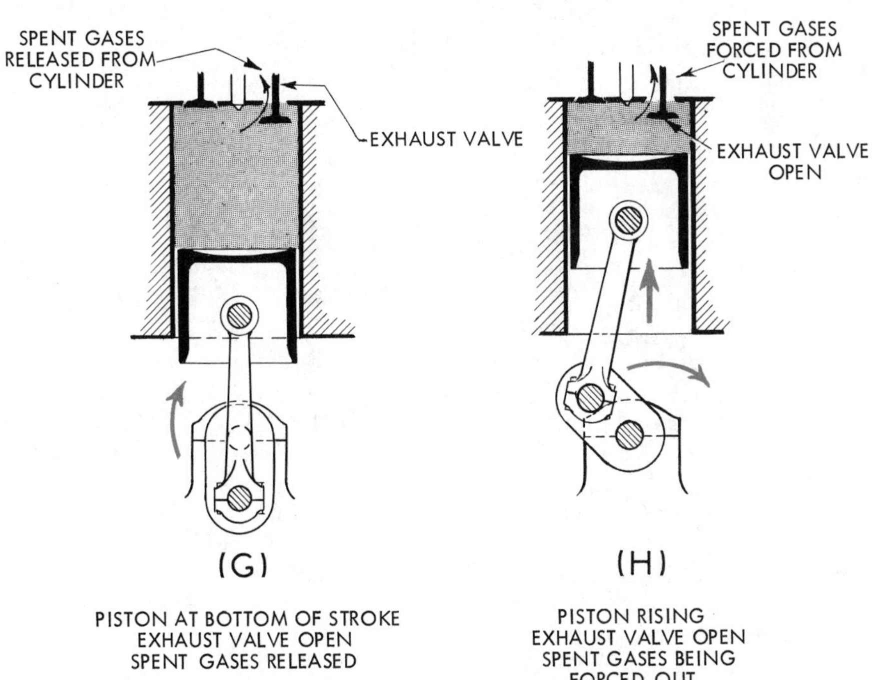

Fig. 2-3. cont'd—E-F power stroke; G-H exhaust stroke).

strokes. Like any moving object, when it turns faster it gains *momentum*, which is a form of energy, and when it slows down it gives up this energy.

Compression Ratio

As you have already noticed, the process of squeezing or *compressing* the air in the cylinder before the fuel is burned is fundamental not only to diesel engines, but also to gasoline engines. We shall soon be thinking about the *amount* of this compression, which we generally call the *compression ratio.*

Referring to Fig. 2-4A, the piston is in its lowest position, about to start upward on the compression stroke. The volume of the cylinder is now at its greatest. Fig. 2-4B shows the piston at its highest position. Now the contents of the cylinder have been squeezed into a small space above the top of the piston, and the volume of the cylinder is at its least.

The *compression ratio* is the ratio of the whole cylinder volume to the least cylinder volume. We could measure these volumes by filling the cylinder full of oil and then measuring the amount of oil. For example, if the whole cylinder volume were 160 cubic inches (piston at the bottom), and the least cylinder volume were 10 cubic inches (piston at the top), the compression ratio would be 160 to 10, or 16. In other words, when the piston is at the top, its contents have been squeezed to 1/16 of its former volume.

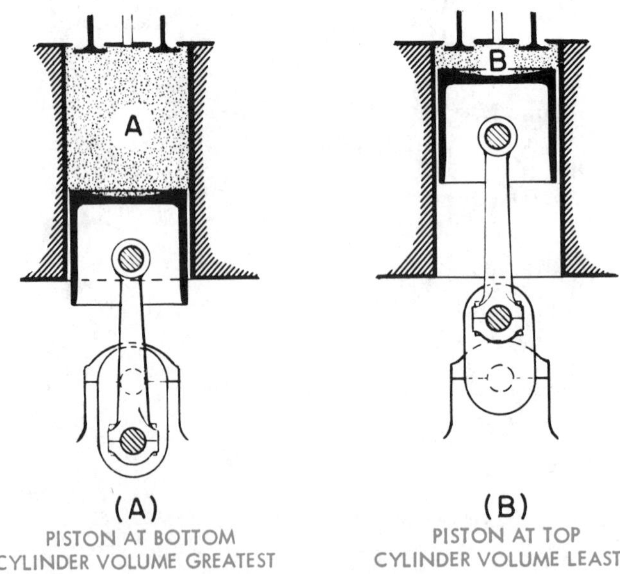

Fig. 2-4. Meaning of compression ratio.

Two-Cycle Diesel Engine

Early in this chapter you learned how a four-cycle diesel works. Remember that it needs *four* full strokes of the piston to go through one cycle: intake, compression, power, and exhaust. Since there are two strokes for each complete rotation or *revolution* of the crankshaft, it takes two revolutions to complete a cycle. Thus there is only one power stroke for every two revolutions.

The principle of a two-cycle engine is to complete the cycle in only *two* strokes of the piston, that is, in only *one* revolution of the crankshaft. Thus a two-cycle engine running at the same speed (revolutions per minute) as a four-cycle engine will have twice as many power strokes. So here's another way of getting more power out of the same size cylinders, or getting an engine weighing less for the same power production.

Function of Air Pump

The two strokes we must have in any diesel engine are the *compression* and *power* strokes. The other two strokes are almost wasted strokes which we would like to eliminate. They are the *intake* and *exhaust* strokes. During these two strokes, the piston of the four-cycle engine is really acting as an air pump.

The way we eliminate the pumping strokes in the two-cycle engine is by using a *separate* air pump to take in atmospheric air, compress it, and feed it to the engine cylinder. This separate air pump on a two-cycle engine is sometimes a rotary *blower*, and might be mistaken for a supercharger on a four-cycle engine (which you will learn about in a later chapter). However, there's an important difference: a supercharger is just an accessory added to give the engine more power, whereas, an air pump *must* be used on a two-cycle engine or it won't get any air at all. Air pumps for two-cycle engines are made in many forms, but we won't go into detail now; you will learn about them in Chapter 13.

Valve-In-Head Exhaust

Now, let's trace what happens in a typical two-cycle diesel engine. Fig. 2-5 shows the arrangement of its parts. This engine uses a rotary blower which takes in outside air and delivers it at a low pressure to an air chamber. The blower consists of two gears which mesh with each other and which rotate inside of a casing. As the gears revolve, the spaces between the gear teeth and the casing carry the air picked up at the blower entrance port around to the discharge ports where it is delivered at the required pressure.

Although there are two valves in the cylinder head, as in four-cycle engine, *both* are exhaust valves, instead of one inlet and one exhaust.

Then how do we get the air into the cylinder? We cut a row of holes, or *ports*, in the cylinder wall. These *inlet ports* are covered by the piston and therefore closed off most of the time, but they are uncovered when the piston nears the bottom of its stroke. The ports are con-

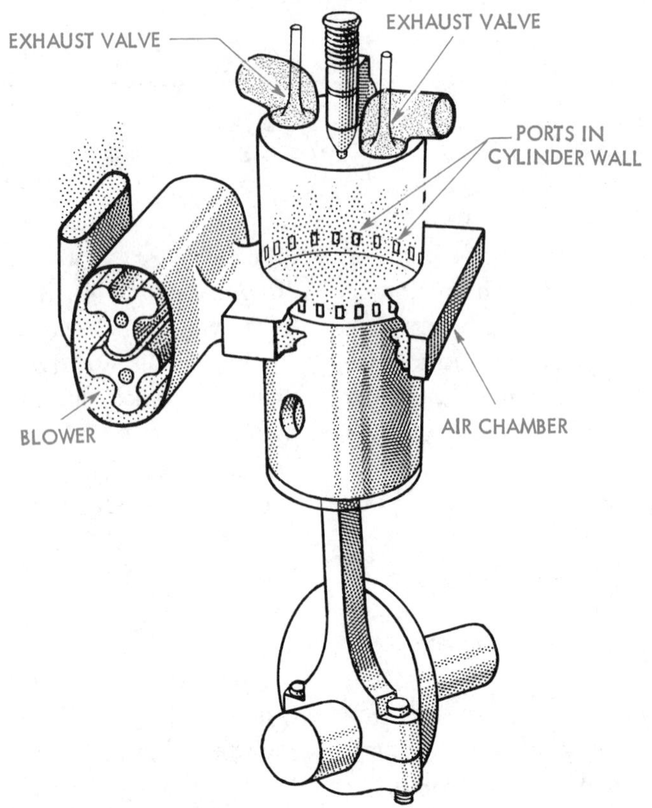

Fig. 2-5. Elements of two-cycle engine with rotary blower. (General Motors Corp.)

nected to the blower outlet by the air chamber.

Let's begin with the piston at the bottom of its stroke, just starting up. See Fig. 2-6. The intake ports are open; so are the exhaust valves. Air is being blown in through the inlet ports and pushes exhaust gases, left over from the previous cycle, out through the exhaust valves.

When the piston has risen about a quarter of the way up, the exhaust valves will close and the inlet ports will be covered. The exhaust gases will have all been blown out and the cylinder will be full of fresh air. The rest of the upward stroke, as in Fig. 2-7, is an ordinary compression stroke which squeezes the air into a small space at the top.

Just before the piston reaches its top position (top dead center) the fuel nozzle shoots a spray of fuel into the chamber which is full of hot compressed air. Ignition and expansion take place just as in the four-cycle engine, and the piston starts down on its power stroke, Fig. 2-8. About three-quarters of the way down, the exhaust valves open and the burned gases (which are still under some pressure) start to escape. As the piston descends farther, it uncovers the inlet ports, and fresh air is again blown into the cylinder, Fig. 2-9. As before, this

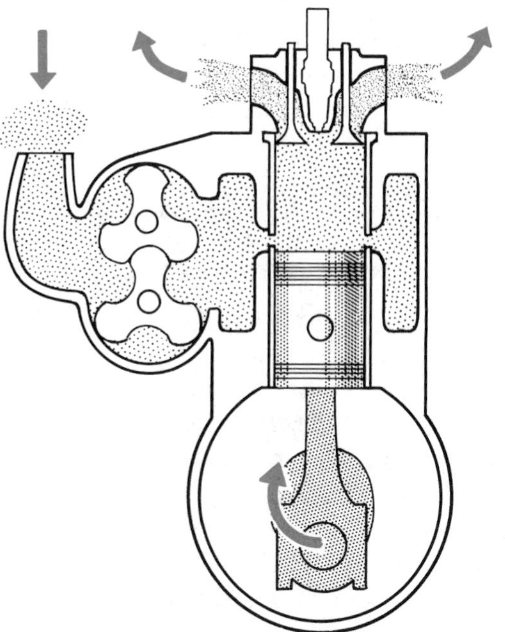

Fig. 2-6. Intake and exhaust of two-cycle engine. (General Motors Corp.)

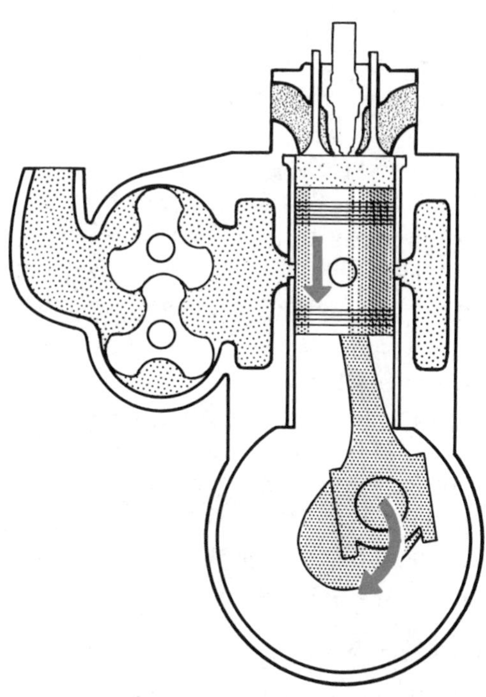

Fig. 2-8. Power, two-cycle engine. (General Motors Corp.)

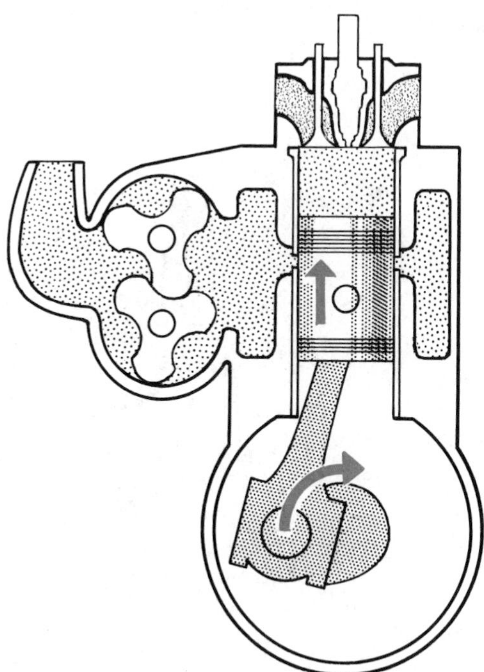

Fig. 2-7. Compression, two-cycle engine. (General Motors Corp.)

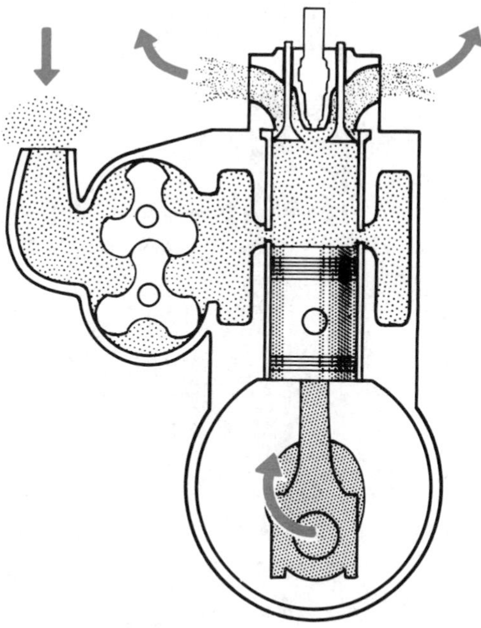

Fig. 2-9. Exhaust and intake, two-cycle engine. (General Motors Corp.)

helps to push the exhaust gases out, or *scavenges* them. It also fills the cylinder with fresh air. The piston reaches its bottom position and the cycle is complete, all in one revolution.

We can see that the compression and power strokes are not much different from those in four-cycle engines. But the exhaust and intake actions take place more or less together and in a much shorter length of time. We cannot use the piston to push the exhaust gases out, so we have to blow them out with air under pressure. We cannot draw the air in by the movement of the piston, so we have to force it in. The blower takes care of both of these requirements.

Port Exhaust

Don't get the idea from the foregoing that exhaust valves are used on *all* two-cycle diesels, nor that rotary blowers are always used to deliver the air. Many two-cycle engines have two rows of ports. In addition to a row of inlet ports which admit air, they have a row of exhaust ports to let out the exhaust gases. See Fig. 2-10.

Exhaust ports, like the inlet ports, are located low down in the cylinder and are uncovered by the piston near the bottom of its stroke. However, the tops of the exhaust ports are higher than the tops of the inlet ports; consequently the descending piston uncovers the exhaust ports first and lets some of the exhaust gases escape before admitting fresh air through the inlet ports. This release of the exhaust gases reduces the pressure in the cylinder, and therefore lets the fresh air enter more freely.

Crankcase Scavenging

Instead of rotary blowers, many two-cycle diesels employ what is called *crankcase-scavenging*. That is, they use the

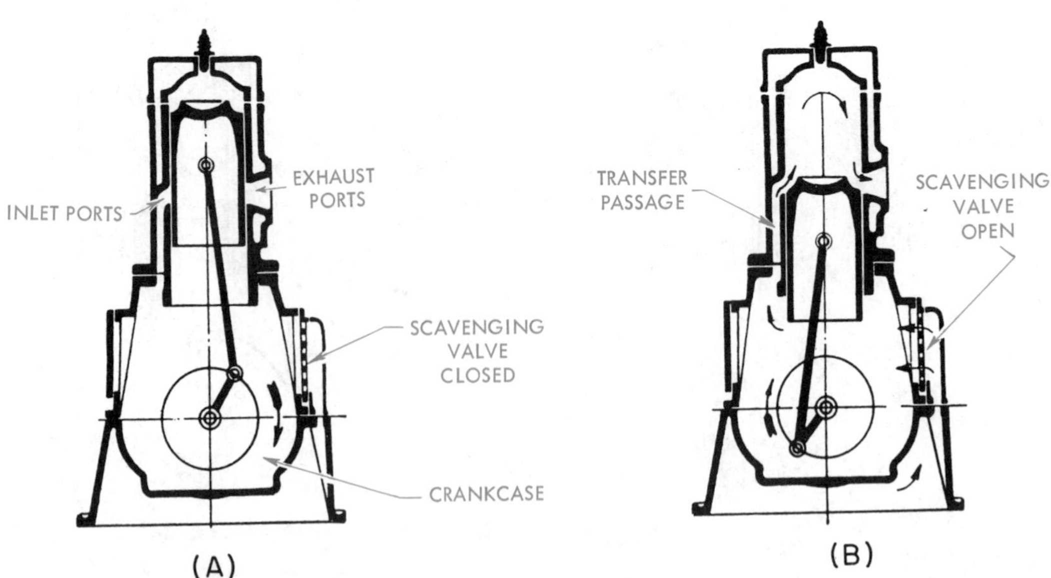

Fig. 2-10. Crankshaft scavenging engine with inlet and exhaust ports. (Colt Industries, Fairbanks Morse Motor & Generator Operation)

lower side of the piston as an air-pump to pump air in and out of the crankcase.

Fig. 2-10 shows an engine with exhaust and inlet ports, and which uses crankcase scavenging. In Fig. 2-10A the fuel has been injected and burned; the gas pressure is pushing the piston downward. As it approaches the bottom of its stroke, the piston uncovers the exhaust ports, permitting the burned gases to leave the cylinder.

When the piston descends a little farther, it uncovers the inlet ports, and air from the crankcase under a slight pressure then flows through the cylinder, clearing out the remainder of the burned gases and filling the cylinder with fresh air.

Fig. 2-10B shows the piston moving upward and about to cover the inlet and exhaust ports. After the ports are covered, the rising piston compresses the entrapped air. Just before it reaches the top dead center, fuel is injected and ignition begins as the piston reaches the top of its stroke. Thus in two strokes the cycle has been completed.

Diesel Engines Differ

If you know something about ordinary gasoline engines, such as those used in automobiles, you will have noticed that diesel engines, in many respects, work in the same way. Before we see how they differ, let's see how they are alike.

Both types of engines run on liquid fuels. In certain special cases they use exactly the same fuel. Gasoline engines have been made to run on kerosene, and so have diesel engines. Gasoline, kerosene, and diesel oil are all produced from natural petroleum (crude oil), and are distinguished from each other mainly by their *volatility* or the ease with which they can be changed from liquid to vapor. Gasoline is quite volatile, it evaporates or *vaporizes* at a low temperature. Kerosene needs more heat to make it vaporize, while diesel oil requires still more heat.

Both types of engines are *internal-combustion* engines, that is, they burn the fuel inside their cylinders. Most gasoline engines and many diesel engines work on the four-stroke cycle: the piston makes a suction stroke (down), a compression stroke (up), a power stroke (down), and an exhaust stroke (up).

What are the main *differences* between diesel engines and gasoline engines? A diesel engine has no ignition system—it has no spark-plug to be fed with high-tension electricity from a distributor, spark-coil, timer, battery, or magneto. None of these are needed on a diesel engine because the fuel is ignited simply by contact with very hot air which has been highly compressed in the cylinder.

A diesel engine draws into its cylinder *air alone*, and it compresses this air on its compression stroke *before* any fuel enters the cylinder. On the other hand, a gasoline engine mixes air with fuel in a carburetor *outside* the cylinder before it enters the engine through the inlet valve during the suction stroke.

Diesel engines also use greater compression than gasoline engines. In a gasoline engine, the amount of compression or *compression ratio* that can be used is limited because *fuel,* as well as air, is being compressed. If the combustible fuel-air mixture is compressed too much, it gets so hot that it will ignite by itself. In other words, the mixture will *preignite* before the piston has completed its compression stroke, and will try to stop the piston. Even a little more compression than the right amount will cause *detonation* or *knocking* because some of the mixture burns all at once before the flame from the spark gets to it.

As we shall learn later, a mixture of a certain amount of fuel and air will produce more power the more it has been compressed. Therefore, the *efficiency* of a gasoline engine, the amount of power produced from a given amount of fuel, is limited by the permissible compression ratio, which is about 9 to 1 (a little more if highly leaded premium gasoline is used.)

The compression in a diesel engine is not limited by the possibility of preignition because a diesel engine compresses air only. Therefore, diesel engines use compression ratios of about 16 to 1, and so achieve greater efficiency in the use of fuel.

Diesel engines use less volatile, heavier liquid fuels than gasoline engines. Gasoline engines must use a more volatile fuel because only a fuel which will evaporate at low temperature will form a uniform mixture with the rapid current of air flowing through the carburetor.

Diesel engines use fuel pumps and injection nozzles to inject the oil into the cylinder in the form of a fine spray. Gasoline engines, on the other hand, mix the fuel and air in a carburetor.

Diesel engines are heavier than gasoline engines of the same size because they work against greater pressures, and consequently their parts must be stronger. The greater strength is obtained (for the same materials) by making the parts thicker and therefore heavier.

Advantages of Diesel Engines

Why are diesel engines used so much? Not merely because they can produce power — there are many other ways of producing power. Besides diesel engines, there are gasoline engines, gas turbines, steam engines, steam turbines, and water wheels (known as hydraulic turbines), to name some though not all such examples. These machines are known as *prime movers.*

You can often buy power also in the form of electricity which has come over wires; this is power which first has been produced by a prime mover and then converted into electricity. So it is a fair question—why, with all these different sources of power, are diesel engines often preferred to the others? Sometimes the question is easily answered; sometimes it demands professional engineering study.

It's not the purpose of this book to explain the work of professional engineers, so let's just list the main advantages that diesel engines have over other forms of power, for certain uses.

Small Consumption Of Fuel. The diesel engine is one of the most highly efficient heat-engines. It gets more power out of the fuel it burns than any other prime mover, except the largest and most efficient steam turbines. Its fuel consumption is much less than that of a gasoline engine. It is an engine of high *economy*.

Relatively Cheap Fuel. The diesel engine uses fuels costing less than gasoline.

Economy at Light Loads. The diesel engine is not only efficient when it is fully loaded, but also when it is only partly loaded (which is the way most engines run most of the time). When running at half load, the diesel engine consumes *only* about 10 percent more fuel *per unit* of power produced than it does at full load. The efficiency of other engines drops off greatly when the load is reduced.

Greater Safety. Diesel fuel is nonexplosive and is far less flammable than gasoline. In fact, it requires special effort to make it start to burn.

This feature alone makes the diesel attractive in the large motor boat field. Also, diesel exhaust gases are less poisonous than those from gasoline engines because they contain less carbon monoxide.

Economy In Small Sizes. In great contrast to a steam power plant, a small diesel engine is nearly as economical to run as a large one. This makes it possible to build an efficient power plant just large enough to meet present needs, and to enlarge it with additional units as the load grows. At all stages of growth, plant efficiency is high.

Sustained Economy in Service. Again in contrast to a steam power plant, diesel efficiency falls off little during thousands of hours of use between overhauls. Steam-power plant efficiency depends on continuous control of furnace conditions and gradually falls off as boiler tubes become fouled. These factors are not present with diesel engines—as long as the exhaust gases are clear, the fuel economy must be good.

Independence of Water Supply. An efficient steam plant requires great quantities of condensing water, and must generally be located near a large body of water. A diesel engine requires little water. As it operates well above atmospheric temperatures it can be successfully used in the most arid regions.

Lightness and Compactness. The diesel engine is a complete power plant in itself; it can be made light in weight and it takes up a small amount of space. It is therefore well suited to portable and mobile installations.

Quick Starting. A cold diesel engine can be started instantly and made to carry its full load in a few minutes. It is therefore ideal for supplying emergency power.

Easily Reversible. By adding an extra set of cams, a diesel engine can be made to run at full power in either direction. This is important in marine applications.

Economy in Labor. No fireroom force is needed, compared with that used in steam installations.

Freedom From Nuisance. There are no ashes to be disposed of, no noisy and dusty coal-handling and pulverizing equipment to maintain, and no smoke. Noise can be easily eliminated.

Disadvantages Of Diesel Engines

We have just learned the main advantages that diesel engines have over other forms of power producers. But diesel engines also have certain disadvantages, and you might as well learn right now what they are, so that you will understand why such engines do not fit all applications.

Cost. Diesel engines, because of the higher pressures at which they work, require sturdier construction, better materials, and closer fit of parts than do gasoline engines. Therefore they cost more to build.

Weight. Because of the sturdier construction, diesel engines weigh more than gasoline engines of the same power.

Attendance. A diesel engine requires more attention than an electric motor running on purchased current. It also requires more attention per unit of power produced than a large steam turbine or water wheel.

Fuel Cost. The type of oil which most diesel engines use for fuel is often more costly than coal when compared on a heat-value basis. A dollar's worth of coal contains more heat units than a dollar's worth of diesel oil. Ordinarily, this difference in fuel price is more than offset by the small amount of oil consumed by the diesel compared with the amount of coal burned in a steam-power plant of the same power. But if coal is unusually cheap compared to oil, a steam-power plant may produce power at a lower cost for fuel than a diesel could.

A steam-power plant may also be able to cut its fuel cost below that of a diesel if its exhaust steam (which contains most of the heat) can be fully utilized for some heating process (as in a laundry).

Of course, no fuel at all is needed to drive a water wheel in a water-power plant. (The main items of cost in a water-power plant are the charges on the investment.)

Space. For large power outputs, diesel engines occupy much more space than steam turbines.

You can now see why, for one or more of the above reasons, other forms of power are preferred for such uses as:

Airplanes: Gasoline and jet engines are lighter.

Ordinary automobiles: Gasoline engines are cheaper to build.

Electric-power plants in large cities: Large steam turbines occupy less space, require less attendance, and can use cheap coal efficiently.

How Diesels Are Used

Now let's look at the applications in which diesel engines are superior and thus are widely used. The reason they are preferred in these applications is that they offer some advantage, or as a whole have superior economy.

Trucks, Buses, Taxis, Tractors, Power Shovels, Construction Machinery, and Logging Equipment

Diesel engines have come into general use for all these applications. The principal reason is the saving in cost of fuel—the diesel engine uses fewer gallons of a less costly fuel than the gasoline engine does. True, the diesel engine costs more to begin with, but if it is used in a class of service which keeps it busy most of the time, the saving in cost of fuel soon pays back the extra investment.

Another advantage of the diesel is its greater pulling or lugging power when it slows down under a heavy load. In other words, the diesel loses less power at reduced speed than the gasoline engine.

Locomotives and Railcars

Almost all railroad locomotives now being built are powered with diesel engines, and steam locomotives have been replaced with diesels. Locomotives for pulling trains and for switching service are mostly *diesel-electric*—that is, the diesel engine drives an electric generator which supplies electric power to electric motors connected to the wheels. You might say it is an electric locomotive which carries its own power plant along with it.

Where passenger traffic is light, and only one or two cars are needed, diesel-electric railcars are used. They work like the locomotives, but each car carries its own diesel-electric power plant as well as the electric motors to drive the wheels.

The reason why the railroads have almost universally adopted the diesel engine is to save money. Compared to steam locomotives, diesels save money by using much less fuel and by being available for service for much more of the time.

Compared to straight electric locomotives, diesels save the heavy investment required for overhead wires or third rails. Passengers prefer diesel-electric locomotives to steam because trains start more smoothly, travel more comfortably at higher speeds, and give off little smoke.

Locomotives for Mines and Tunnels

Diesel locomotives are now preferred to electric locomotives for mine haulage and for tunnel construction because they are less expensive in both first cost and operating cost. Their exhaust gases contain little poisonous monoxide gas, which is an important objection to gasoline engines in mines.

Marine Uses

Diesels are now widely used in marine service of many kinds, such as sea-going vessels (both passenger and freight), motor boats, ferry boats, tugs, naval vessels, and icebreakers. The main reason for these uses of diesels is, again, lower cost of fuel compared to steam.

Submarines formerly were powered with gasoline engines; now (unless nuclear-powered) they invariably use diesels, not only because of the greater range of travel due to consuming less fuel, but also because of the reduced fire hazard. The latter is also an impelling reason for the use of diesels on ships and motor boats.

Stationary Power Plants

Diesel engines are employed in a great many kinds of stationary power plants. The reasons are many; the chief ones are saving in cost of fuel compared to small steam or gasoline power plants, and lower total cost than that of purchased electric power.

Additional advantages enter into certain special applications such as isolated service stations, railway water stations, vacation resorts, lumber camps, mine power plants, oil-well drilling, and emergency power plants. Here the following advantages of the diesel are important: independence of water supply, lightness and compactness, freedom from fire hazard, and ability to start quickly.

Why It Is Called A Diesel Engine

You have learned earlier in this chapter that a diesel engine is one form of internal-combustion engine, the latter being an engine from which work is obtained by the combustion of fuel within the engine cylinders. A diesel engine is that type of internal-combustion engine which injects fuel oil in a finely divided state into a cylinder within which air has been compressed to a high pressure and temperature. The temperature of the air is high enough to ignite the particles of injected fuel; no other means are used for ignition.

Because of this method of ignition, diesel engines are sometimes called *compression-ignition* engines. This sets them apart from the other internal combustion engines, called *spark-ignition* engines. These latter engines use gasoline or gas as fuel, and the mixture of fuel and air is ignited by an electric spark.

Why do we call this compression-ignition engine by the name *diesel?* Simply because a man whose name was Rudolf Diesel originated in Germany and obtained patents in 1892 on a high-compression, self-ignition engine originally intended to burn powdered coal. The first commercial engines constructed according to these patents, and using oil fuel, were built both in Germany and in the United States in 1898, and were called *Diesel engines*.

Meanwhile, an Englishman, Herbert Ackroyd Stuart, had been granted a patent earlier in 1888, on an engine which ignited oil fuel by compressing the mixture of oil and air in contact with a hot wall in the combustion chamber. Commercial production of these engines under the name *Hornsby-Ackroyd*, began in England and the United States in 1893. As time went on, the compression pressure of the Hornsby-Ackroyd engine was increased step by step until a hot surface was no longer needed for igniting the oil. Modern engines, now called *diesel* (with a small *d*), are practically all derived from these two parallel developments.

Checking On Your Knowledge

The following questions give you the opportunity to check up on yourself. If you have read the chapter carefully, you should be able to answer the questions. If you have any difficulty, read the chapter over once more so that you have the information well in mind before you go on with your reading.

DO YOU KNOW

1. What are the main differences between a diesel and a gasoline engine?
2. Why is a diesel engine called a compression-ignition engine?
3. What is the meaning of an engine cycle?
4. Why does the temperature rise on the compression stroke and to what extent in degrees?
5. Explain what the differences are between two-stroke and four-stroke diesel engines.
6. What is the purpose of the flywheel?
7. What is compression ratio?
8. List some of the advantages of diesel engines.
9. What are some of the disadvantages of diesel engines?
10. How are diesel engines used?
11. What does lugging power mean in regard to diesel engines?

Chapter 3 Basic Construction of a Diesel Engine

In the two preceding chapters you've learned what a diesel engine is, how it works and how it is used. The purpose of this chapter is to acquaint you with the various parts that make up actual engines. You'll learn the names of the essential parts and the purpose that each one serves. Next you'll see some pictures of the inside as well as the outside of actual engines with the various parts labeled, so that you'll learn where the parts go and how they fit together in a real engine.

Then we'll take the parts out of these engines and see how each part looks by itself. Thus, you'll learn to recognize the various parts of a real engine, and to understand why they are constructed the way they are.

In this chapter we'll deal with fundamental construction only. This will give you the basic information you need before you examine, in succeeding chapters, the differences in details that are found in various types and makes of engines. This basic information will also help you when you learn how to keep the engine parts in good working order.

Parts Needed in a Diesel

In Chapter 1 you learned that the power of a diesel engine originates in the cylinder. Oil fuel is sprayed into the cylinder by means of the pressure produced by the fuel-injection pump. There it meets air which was compressed by the piston while the piston was being pushed upward in the cylinder. The oil burns, producing heat and more pressure. The hot confined gases press the piston downward with greater force than was used to push it upward. The piston in turn

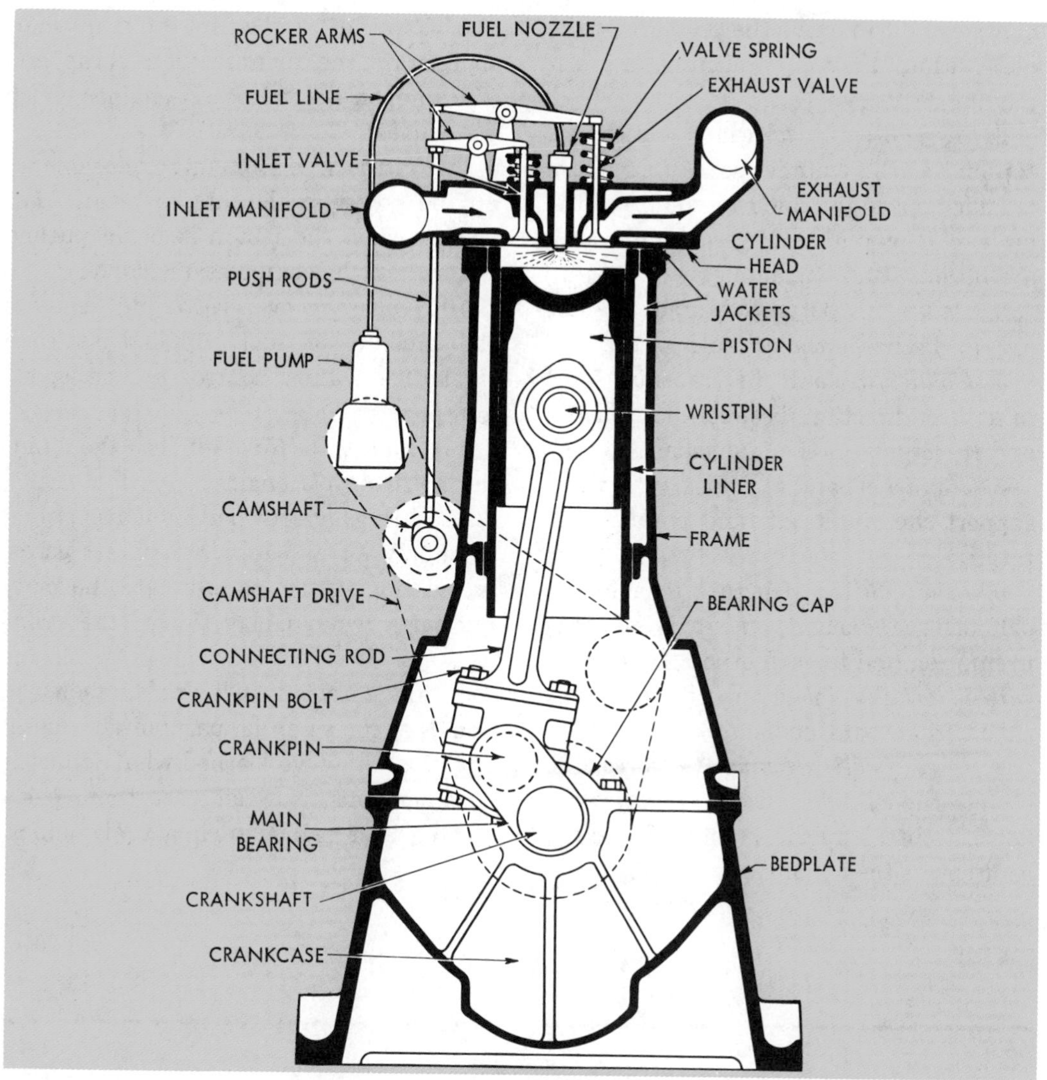

Fig. 3-1. Parts of a four-cycle diesel engine.

pushes on the connecting rod which connects the piston to the crank on the crankshaft. The force on the crank makes the crankshaft turn in its bearings, and the rotating crankshaft supplies power to whatever machinery the engine drives. This is the way in which the diesel engine changes the heat energy of oil into mechanical power.

Now let's note what essential parts the engine must have to carry out these operations. Fig. 3-1 will help you identify the parts; it is a simplified picture of the inside of a typical, vertical, four-cycle diesel engine. The important parts are labeled.

1. A *piston* sliding in a *cylinder*. The piston has two jobs: to compress the air

charge, and to receive the pressure of the gases while they are burning and expanding.

2. A *cylinder head* which closes the top end of the cylinder so as to make a confined space in which to compress the air and to confine the gases while they are burning and expanding.

3. *Valves* or *ports* to admit the air and to discharge the exhaust gases.

4. *Connecting rod* to transmit force in either direction between the piston and the crank on the crankshaft.

5. *Crankshaft* and *main bearings* which support the crankshaft and permit it to rotate.

6. A supporting structure to hold the cylinders, crankshaft, and main bearings in firm relation to each other. This structure is usually made up of two parts, called *frame* and *bedplate.*

7. *Fuel-injection pump* to force the oil into the cylinder; also a *fuel-injection nozzle* to break up the oil into a fine spray as it enters the cylinder.

8. *Camshaft,* driven by the crankshaft to operate the fuel-injection pump and also to open the valves (in engines which use valves).

9. *Flywheel,* to store up surplus energy on the power stroke and to return that energy when the piston is being pushed upward on the compression stroke.

10. *Governor* or *throttle,* to regulate the amount of fuel supplied at each stroke, and thus control engine speed and power.

11. *Blower,* to force air into the cylinder of *two-cycle* engines.

12. Miscellaneous parts, such as piping to supply air and remove exhaust gases, lubricating system to lubricate the moving parts, water jacket to cool the cylinder, etc.

In the foregoing list you've seen the names of the essential parts of any diesel engine and you've learned what job each part performs. Now let's see these parts put together as they are in actual engines.

How the Assembled Parts Look

To see how the various parts fit together in an actual engine, let's examine the construction of some well-known four-cycle diesel engines such as are widely used to drive tractors, trucks and many kinds of mobile equipment.

Exterior Appearance

Fig. 3-2 shows the exterior of an engine. A picture of the outside, like this one, shows only a few of the main parts, the others being hidden. For instance, you

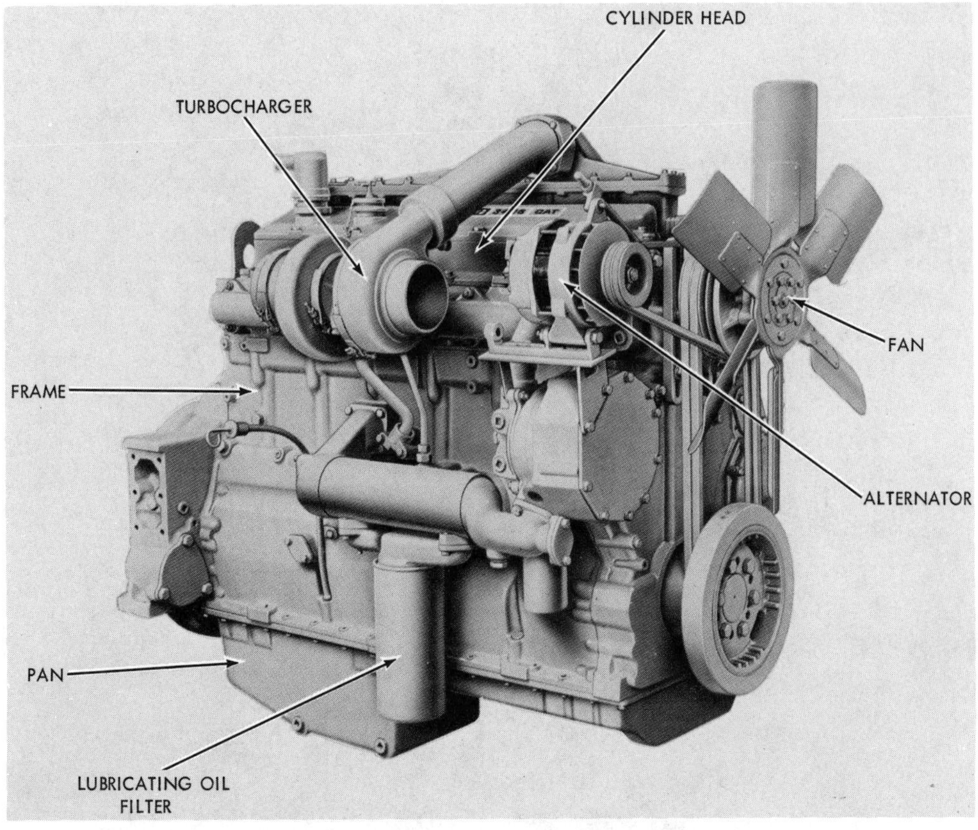

Fig. 3-2. Exterior of a modern six-cylinder turbocharged truck engine. (Caterpillar Tractor Co.)

can see the *frame*, the *cylinder head* which covers the tops of all four cylinders, and the *pan*, which in other designs supports the engine and is called the *base* or *bed-plate*.

Most of the other parts shown in this picture are accessories, the most prominent of which are the turbocharger and the lubricating oil filter. The fuel-injection pump is mounted on the other side of the engine, and is not visible in this view. Other illustrations included in this book show the pump.

Interior Appearance

Because exterior views of engines show so few of the main parts, cross-section views are frequently used. A cross-section view of an engine is a picture showing how it would look if it were cut through in one or more places. Fig. 3-3 shows a similar engine, but one cut through at several places in a vertical direction in line with the crankshaft; it shows the opposite side of the engine from Fig. 3-2.

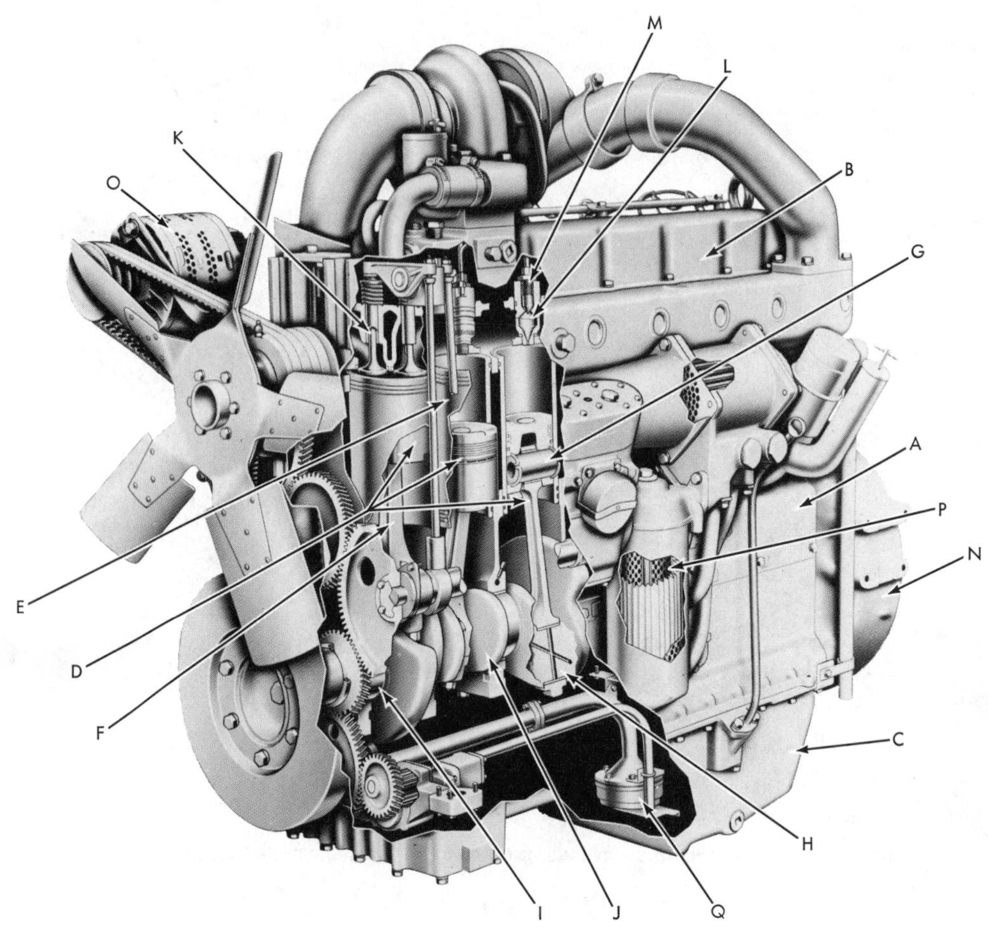

Fig. 3-3. Cross-section of diesel engine in-line with crankshaft. (Caterpillar Tractor Co.)

Fig. 3-4 is another cross-section of the same engine, but this time it is cut through vertically across the crankshaft. Cross-sections are helpful in showing the inside appearance of all sorts of things. Since you will find many cross-sections in this book, you should accustom yourself to understanding them as well as you understand ordinary exterior pictures.

Fig. 3-3 A-R shows a lot more of the engine. Again you see (A) the *frame*, (B) the *cylinder head*, and (C) the *pan*, all of which appeared in Fig. 3-1. Three of the six *pistons* appear in Fig. 3-3 D, while Fig. 3-3 E is the *cylinder sleeve* or *liner*, so-called because it is removable.

Fig. 3-3 F is the *connecting rod*, which carries at its upper end the (G) *wristpin bearing*. At its lower end is the *crankpin bearing*, Fig. 3-3 H. Fig. 3-3 I is the *crankshaft*, to which the forces on the piston are transmitted through the con-

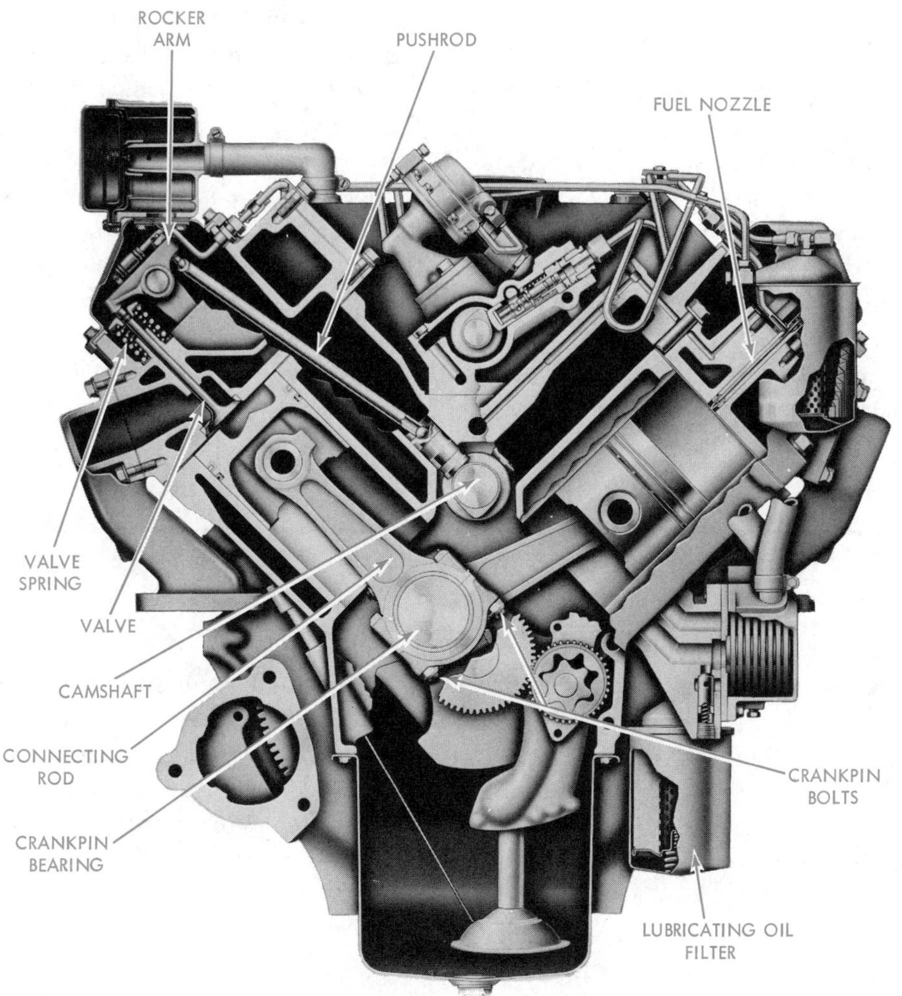

Fig. 3-4. Cross-section of a V-type diesel engine across the crankshaft. (Caterpillar Tractor Co.)

necting rod. The crankshaft rotates in a series of *main bearings*, as illustrated in Fig. 3-3 J. The enclosed space formed by the frame and the pan, within which the crankshaft runs, is called the *crankcase*.

The *valves*, Fig. 3-3 K, are located in the cylinder head. One is the *air inlet valve*, the other the exhaust valve. Fig. 3-3 L is the *pre-combustion* chamber into which the fuel oil is injected through fuel nozzle, Fig. 3-3 M, and is partially burned (in this type of engine) before the main combustion takes place in the space between the bottom of the cylinder head and the top of the piston. The *flywheel* which is a heavy wheel fastened on the end of the crankshaft, is housed in Fig. 3-3 N.

Several important accessories also appear in this illustration. Fig. 3-3 O, P, and Q show the *alternator* which supplies the electrical requirements, the *fuel filter*, and the *oil pump* which circulates the lubricating oil.

The *camshaft*, which opens the valves, does not appear in Fig. 3-3, so we'll look at another engine cross-section, Fig. 3-4. Here the camshaft is shown, and a *push rod* which, when lifted by the cam on the camshaft, opens the *valve* by means of the *rocker arm*. The valve is closed by *valve springs*, also shown in Fig. 3-4.

In Fig. 3-4 you see a different kind of *fuel nozzle*. Also in this view you can see the *crankpin bolts* which clamp together the two halves of the *crankpin bearing* at the the lower end of the *connecting rod*. Another accessory, the *lubricating-oil filter*, appears in Fig. 3-4.

How the Individual Parts Look

In the preceding section you saw how the parts of an engine appear when they are fitted together and ready to go to work. Now let's take an engine apart and look at each of the parts by itself. Since the parts shown do not belong to one particular make and size of engine, they naturally look different in details than if all were from the same engine. Later on you will learn about the varied designs used in engines of other makes, types and sizes.

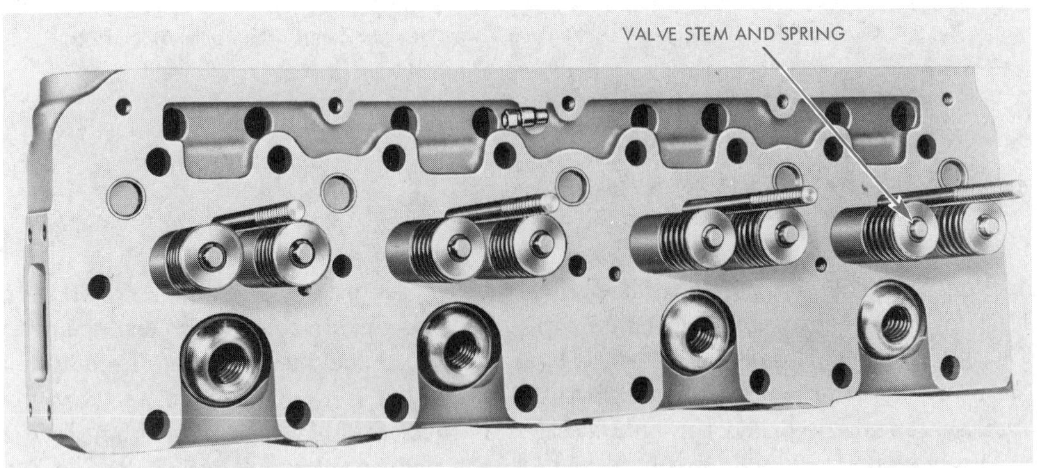

Fig. 3-5. Cylinder head, top view, showing valve stems and springs. (Caterpillar Tractor Co.)

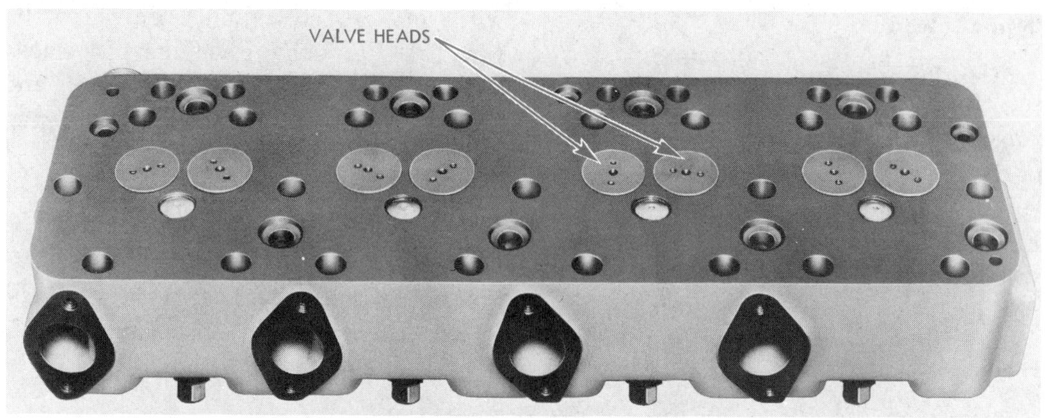

Fig. 3-6. Cylinder head, bottom view, showing valve heads. (Caterpillar Tractor Co.)

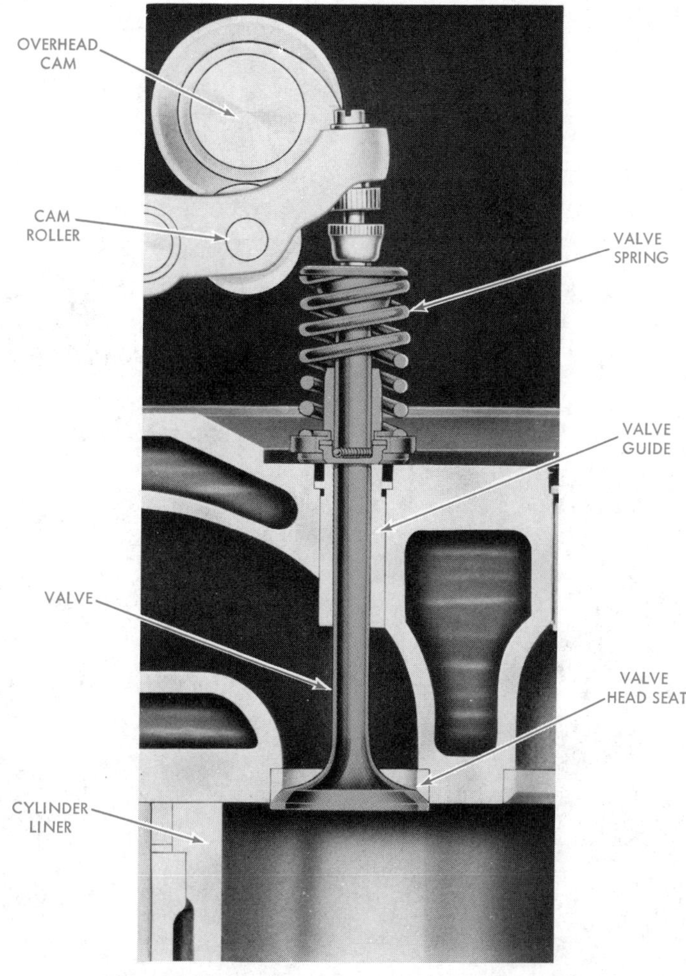

Fig. 3-7. Overhead cam with valve, valve guide, and valve spring. (Caterpillar Tractor Co.)

Cylinder Head and Valves

Fig. 3-5 shows the cylinder head as seen from the top; also the stems of the valves and the valve springs. The bottom of the cylinder head, with the heads of the valves on their seats, appears in Fig. 3-6. Fig. 3-7 shows an overhead cam with valve, valve guide and valve spring in operating position.

Frame and Liners

In Fig. 3-8 you see the engine frame, with one of the cylinder liners partly withdrawn and the other liners in place. The studs at the top of the frame are used to fasten the cylinder head to the frame.

The cylinder liner by itself is illustrated in Fig. 3-9. The flange at the top is used to hold the liner in the frame; the grooves at the bottom carry rubber rings which seal the water space around the outside of the liner.

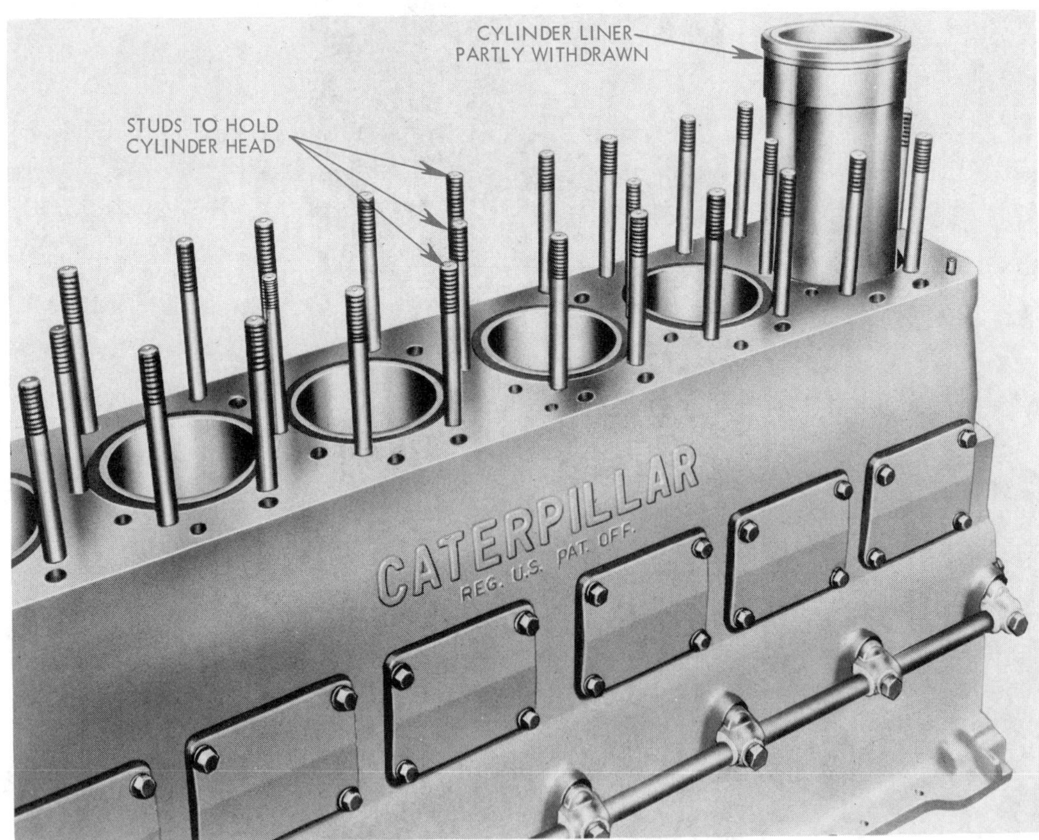

Fig. 3-8. Engine frame and liners. (Caterpillar Tractor Co.)

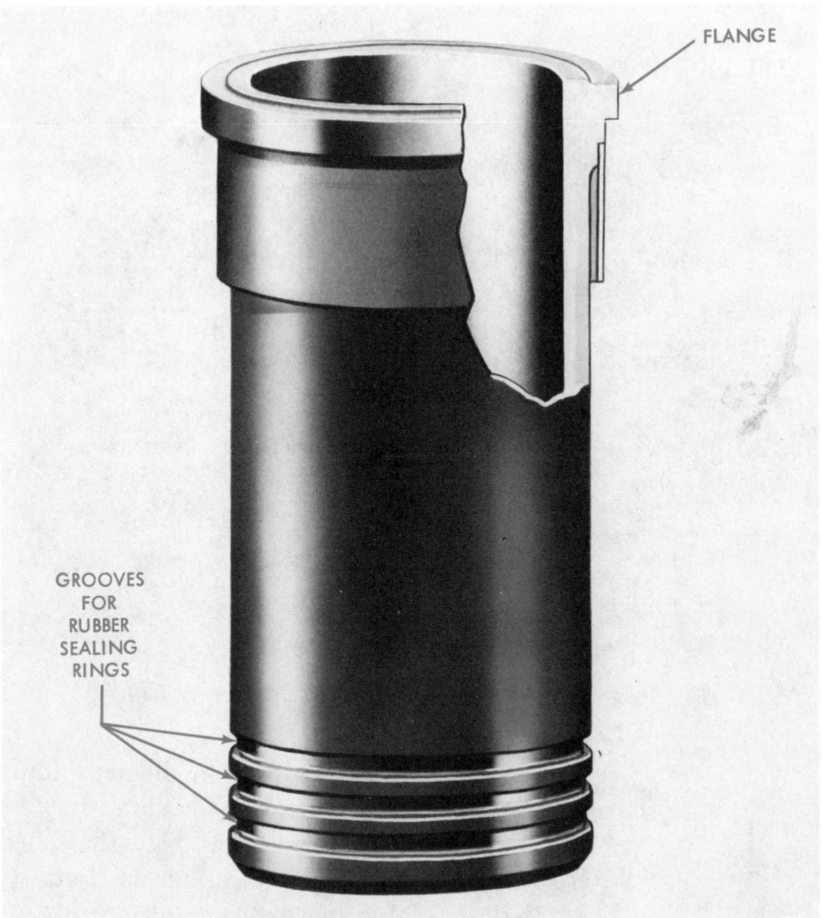

Fig. 3-9. Cylinder liner. (Caterpillar Tractor Co.)

In Fig. 3-10 the frame is seen bottom up, so that the main bearings are visible. One of the main bearings is shown with its cap removed.

Fig. 3-11 shows a pair of main bearing shells whose inside surface supports the rotating crankshaft and is therefore made of anti-friction metal. These shells seat in the main bearing housings, and are removable, so they can be replaced when they wear out.

Crankshaft

Fig. 3-12 shows how the crankshaft is made up of a series of bearing surfaces called *journals*. There are two kinds of journals: *main-bearing journals* and *crank journals*. The latter are often called *crankpins*. The main bearing journals are all in the same line; the crank journals are offset from the main-bearing journals and are supported by the crankarms,

Fig. 3-10. Frame, shown bottom up. (Caterpillar Tractor Co.)

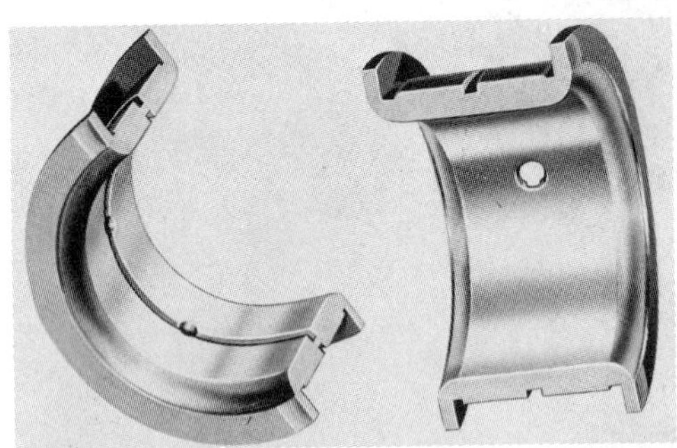

Fig. 3-11. Main bearing shells. (Caterpillar Tractor Co.)

which project in different directions in order to space the power impulses of the various cylinders. Note the oil hole, which carries lubricating oil from the main bearing to the crank bearing.

Camshaft

As you see in Fig. 3-13, the camshaft is a long slender shaft with a number of carefully formed projections called *cams*.

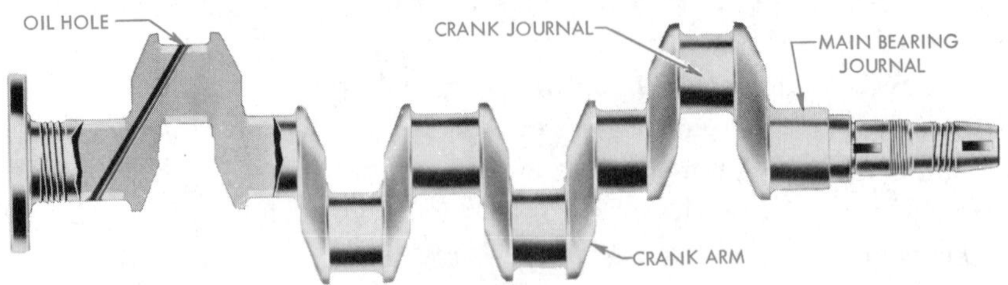

Fig. 3-12. Crankshaft. (Caterpillar Tractor Co.)

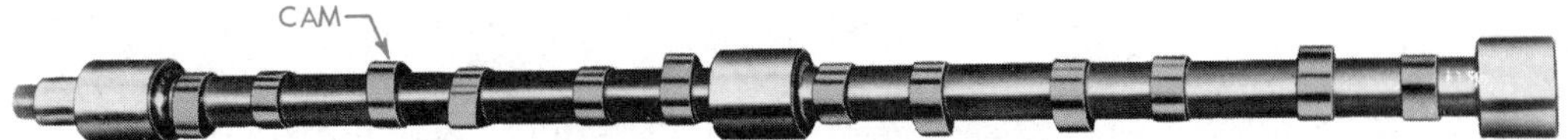

Fig. 3-13. Camshaft. (Caterpillar Tractor Co.)

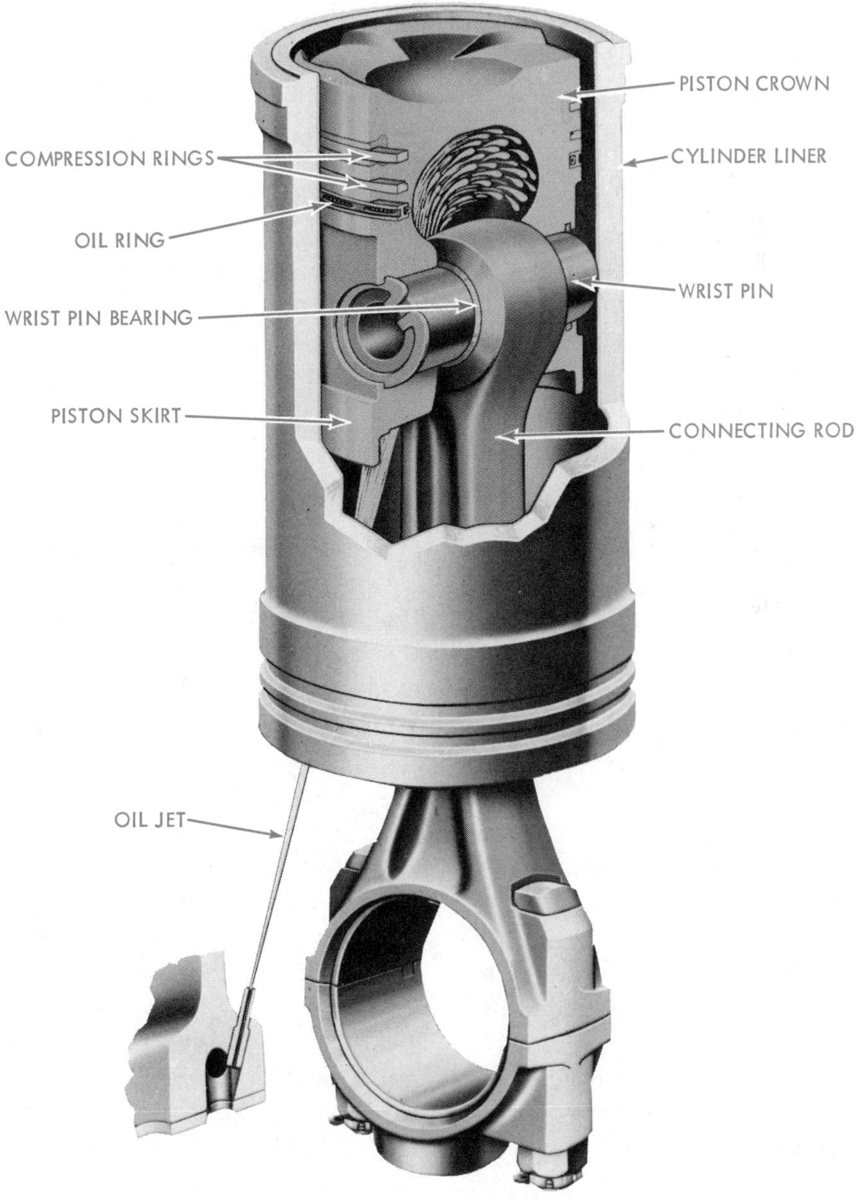

Fig. 3-14. Piston and connecting rod in cylinder liner. (Caterpillar Tractor Co.)

The camshaft is connected to the crankshaft by gears or chain, and rotates at half the crankshaft speed (in four-cycle engines). As it rotates, the cams come under the valve push rods and lift the valves.

Piston Assembly

Fig. 3-14 shows, partly in exterior view and partly in cross-section, the piston and connecting rod assembled in the cylinder liner. Note the piston rings which are inserted in the upper part of the piston. These rings are flexible and springy so that they bear firmly on the cylinder wall surrounding the piston.

The two top rings are *sealing* or *compression rings* which hold back the gas pressure on the top of the piston. The bottom ring is an *oil ring*, the purpose of which is to wipe excess lubricating oil off the cylinder walls so that the excess oil is carried down again into the crankcase. Otherwise, the excess oil would be exposed to the burning gases and would itself burn, causing not only a loss of oil but also deposits of carbon.

This picture also shows how the upper bearing of the connecting rod (the wristpin bearing) rides on the wristpin which is fitted into the piston. The vertical jet of lubricating oil cools the piston crown, which is exposed to the burning gases. The oil also lubricates the wristpin bearing.

Note how thick the upper walls of the piston are. This thickness is needed not only to withstand the gas pressure, but also to carry away the heat transferred to the piston crown by the burning gases.

Fig. 3-15 is the connecting rod by itself. The wristpin bearing (in this engine) is a single circular shell, but the crankpin bearing consists of two half-shells held together by the cap and the crankpin bolts. This split construction is necessary to permit assembly and disassembly of the crankpin bearing on the crankpin. The inner surface of the shells is of antifriction metal.

Fuel-Injection Pump

In Fig. 3-16 you see a cross-section through the fuel-injection pump for one of the cylinders. The significant features of a diesel fuel-injection pump are: the *cam* which lifts the *roller,* which in turn lifts the *pump plunger;* and the regulating mechanism which rotates the pump plunger and thus controls the amount of fuel injected. The pump *discharge valve* is located above the pump plunger.

Governor

The purpose of the governor, Fig. 3-17, is to keep the engine at a constant speed even though the load changes. The two governor flyweights are pivoted on a shaft that is geared to the crankshaft. Thus, the flyweights whirl around at a speed that is proportional to the engine speed. When the speed exceeds the normal speed, the increased whirling speed of the flyweights tends to make them move outward from their shaft. This movement of the flyweights is carried through several links to the regulating mechanism of the fuel-injection pump, where it reduces the amount of fuel injected and thus brings the engine down to normal speed again.

The governor shown is a simple mechanical type. Larger engines requiring close speed control are often fitted with more sensitive and more powerful governors of the hydraulic type, which will be described in Chapter 16.

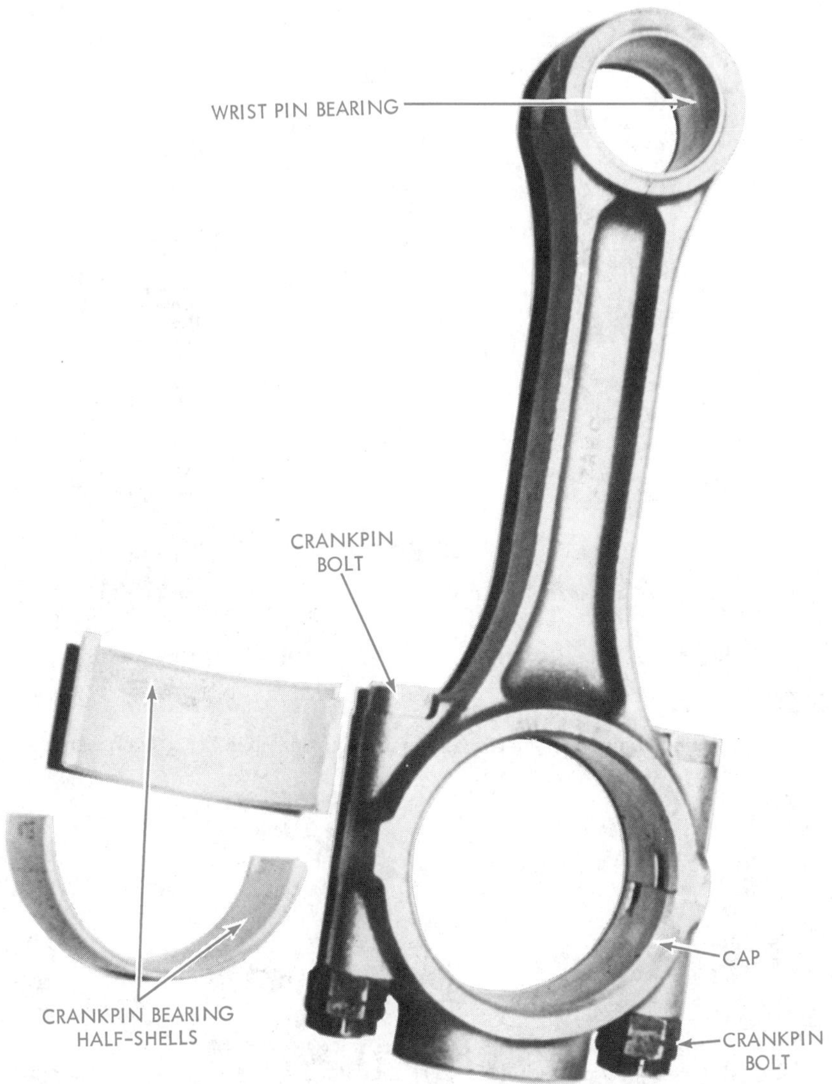

Fig. 3-15. Connecting rod and crankpin bearing shells. (Caterpillar Tractor Co.)

Fuel-Injection Valve and Pre-combustion Chamber

The fuel delivered by the injection pump is piped to the fuel-injection valve, Fig. 3-18, located in the cylinder head. The injection starts when the pump has built up enough pressure to open the fuel-injection valve.

A measured quantity of fuel is sprayed through the nozzle below the valve, and enters the pre-combustion chamber in

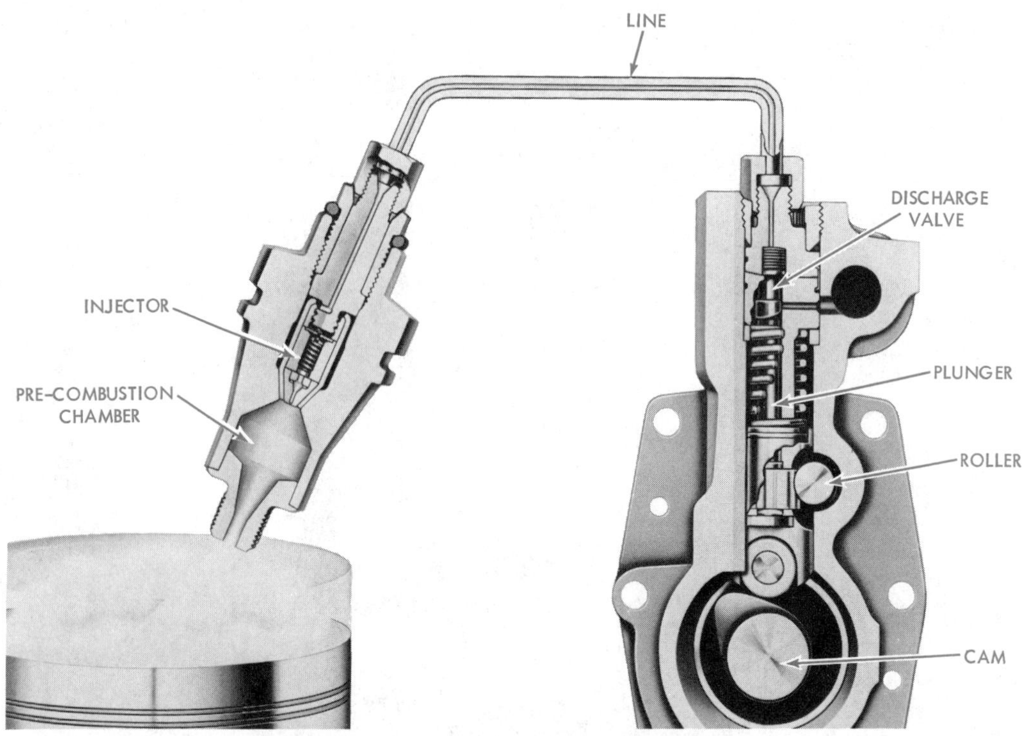

Fig. 3-16. Fuel-injection pump, line, injector and precombustion chamber. (Caterpillar Tractor Co.)

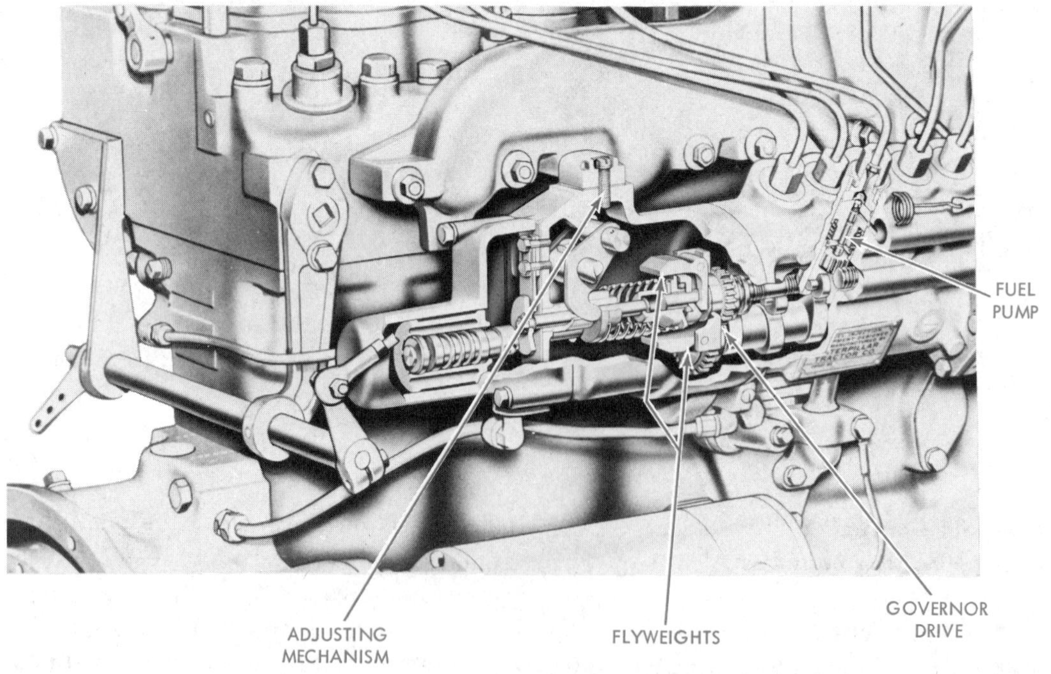

Fig. 3-17. Governor system connected to fuel pump group. (Caterpillar Tractor Co.)

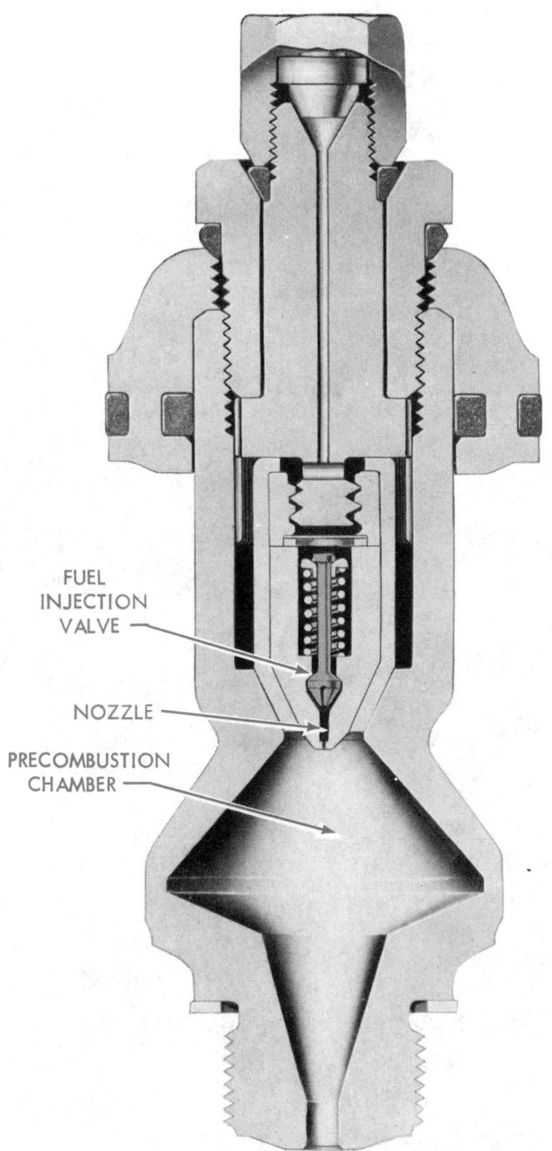

Fig. 3-18. Fuel-injection valve and pre-combustion chamber. (Caterpillar Tractor Co.)

atomized form. The limited amount of air in the pre-combustion chamber, at the high temperature created by the heat of compression, ignites the fuel and allows partial burning. Then, with pressure and temperature raised still further, the gasified mixture blows with great force into the main combustion space above the

Fig. 3-19. Timing gears. (Caterpillar Tractor Co.)

piston, where the surplus of air mixes thoroughly with it and completes the burning of the fuel. The pre-combustion system used in this engine is one of several ways in which complete combustion is obtained, as you will learn when we discuss combustion chambers in a later chapter.

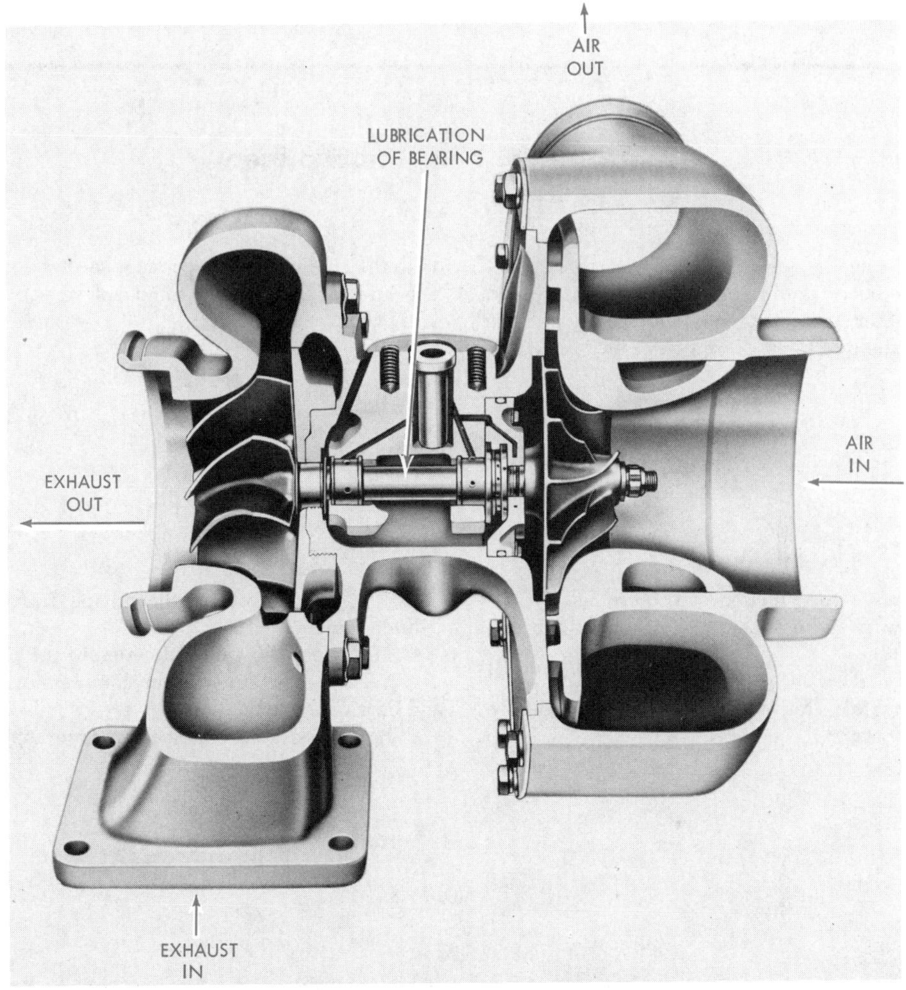

Fig. 3-20. Cutaway of a turbocharger. (Caterpillar Tractor Co.)

Timing Gears

The timing gears, Fig. 3-19, are located at one end of the engine. They drive the camshaft, the fuel-injection pump, and the governor (and also certain accessories) by means of a train of gears which starts with a gear mounted on the crankshaft. Note that the camshaft gear has twice as many teeth as the crankshaft gear; thus the camshaft makes one complete revolution while the crankshaft makes two revolutions. This is necessary, because in a four-stroke cycle engine a complete cycle takes two crankshaft revolutions. Fig. 3-20 shows a typical turbocharger which is cut away to show the design.

Checking On Your Knowledge

The following questions give you the opportunity to check up on yourself. If you have read the chapter carefully, you should be able to answer the questions. If you have any difficulty, read the chapter over once more so that you have the information well in mind before you go on with your reading.

DO YOU KNOW

1. What is the function of the crankshaft?
2. How is the crankshaft connected with the camshaft?
3. What does valve timing mean?
4. What is a pre-combustion chamber and how does it operate?
5. What is meant by the engine frame and cylinder liners?
6. How does the camshaft operate the valves?
7. What is the function of compression rings and oil rings?
8. How does a mechanical governor operate?

Classifications of Diesel Engines

Chapter

4

The purpose of this chapter is to acquaint you with the various classifications of diesel engines in respect to design and specific applications. Each part of an engine has a specific job to do. The nature of its task, and the qualities needed to perform it satisfactorily, determine to a large extent the shape, size, and material of the part.

In addition, engines may be grouped or classified in respect to speed (rpm), and size (hp per cylinder) both having a direct relationship to their application or use.

Examples of actual engines which illustrate some of the various classifications and significant features will be examined.

Engines Classified by Design

Although certain classifications of diesel engines were introduced in Chapter 1, our purpose at this time is to investigate in much more detail so that a certain amount of variation on this subject may be understood before further reading in this text.

Diesel engines may be classified as to design in the following way: (1) operating cycle, (2) piston action, (3) piston connection, (4) cylinder arrangement, and (5) although not quite in the same category as the other four items within this group, *materials of construction* does relate to design features.

Operating Cycle

Diesel engines can be divided into two groups, based on the number of piston strokes per cycle, either four or two. As you learned in Chapter 2, an engine which needs four strokes to complete one cycle is a four-stroke cycle engine or, for short, a *four-cycle engine.* If it needs only

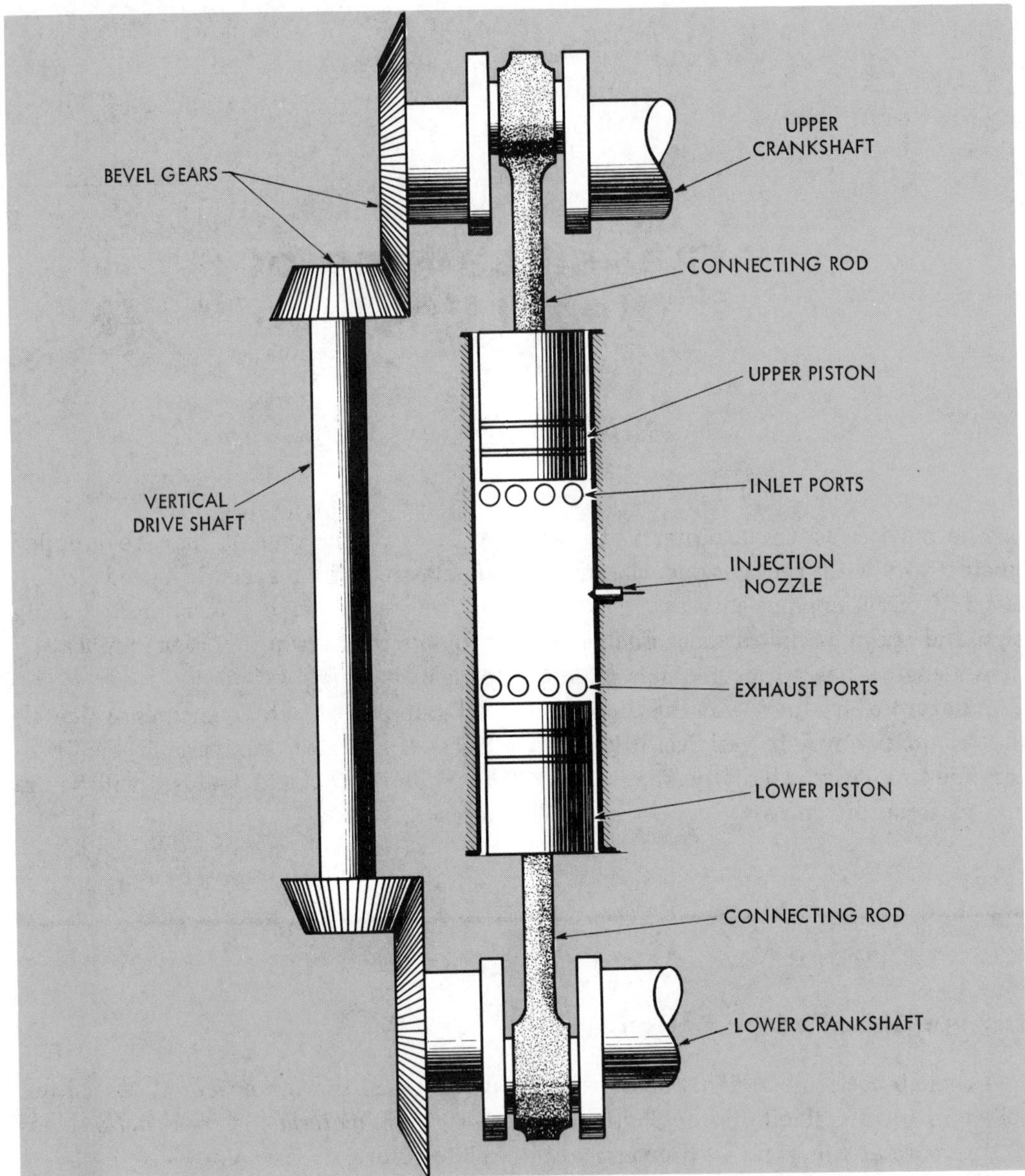

Fig. 4-1. Opposed-piston action.

two strokes to complete a cycle, we call it a two-stroke cycle engine or, for short, a *two-cycle engine.* Thus, a two-cycle engine fires twice as often as a four-cycle engine.

Two-cycle engines are further classified according to the method used to supply the engine with fresh air. You learned (Chapter 2) that air at a pressure slightly above atmospheric must be forced into the working cylinder to blow out the spent gases and to fill the cylinder with fresh air. This scavenging air is obtained, in some two-cycle engines, by using the crankcase and underside of the piston as an air compressor. (See Fig. 2-10.) This is called *crankcase scavenging*. Scavenging air in other engines is supplied by blowers or pumps; this is called *blower scavenging* or *pump scavenging*, respectively.

Piston Action

An engine's piston action may be classified as (1) single-acting, (2) double-acting, or (3) opposed-piston.

Single-acting engines use only one end of the cylinder and one face of the piston to develop power. This working space is at the end away from the crankshaft (that is, at the upper end of a vertical engine).

Double-acting engines use both ends of the cylinder and both faces of the piston to develop power on the upstroke as well as on the downstroke. The construction is complicated, therefore double-acting engines are built only in large and comparatively low-speed units, generally to power motor ships.

An *opposed-piston engine* has cylinders in each of which two pistons travel in opposite directions as in Fig. 4-1. The combustion space is in the middle of the cylinder between the pistons. There are two crankshafts; the upper pistons drive one, the lower pistons the other. Note that each piston is single-acting; it develops power with only one face of the piston.

Piston Connection

The piston may be connected to the upper end of the connecting rod either directly (trunk-piston type), or indirectly (crosshead type).

In *trunk-piston* engines, a horizontal pin within the piston is encircled by the upper end of the connecting rod. This

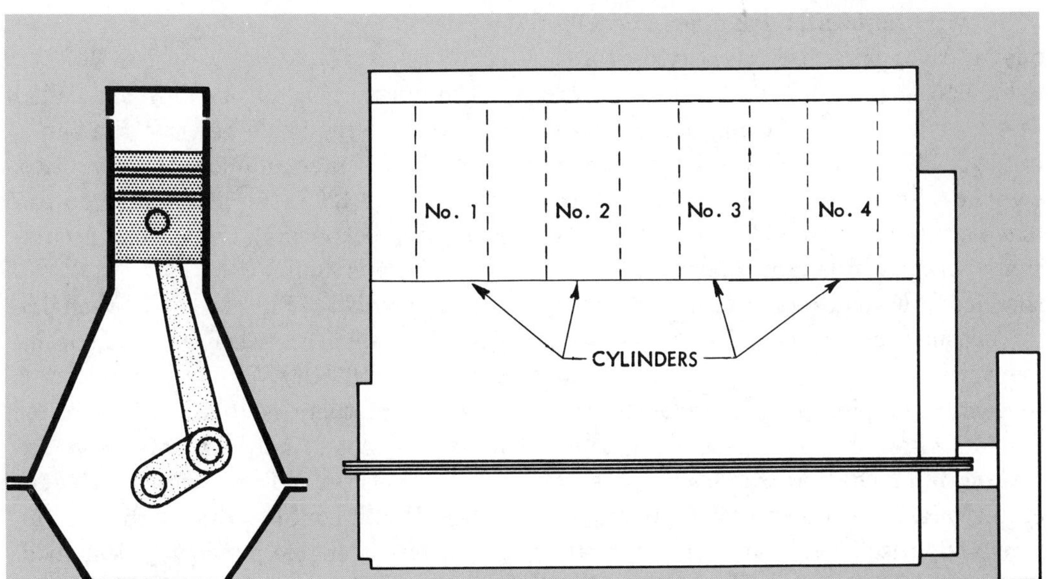

Fig. 4-2. In-line cylinder arrangement.

is by far the most common construction.

In *crosshead-type* engines, the piston fastens to a vertical piston rod whose lower end is attached to a sliding member called a *crosshead,* which slides up and down in guides. The crosshead carries a crosshead pin which is encircled by the upper end of the connecting rod. This more complicated construction is required in double-acting engines. It is also used in some large, slow-speed, single-acting engines.

Cylinder Arrangement

A diesel or gas-burning engine has its cylinders arranged in one of several ways. Four basic cylinder arrangements are: (1) cylinders-in-line, (2) V-arrangement, (3) flat, and (4) radial.

A *cylinders-in-line* arrangement is shown in Fig. 4-2. This is the simplest and most common arrangement, with all cylinders arranged vertically in line. This construction is used for engines having up to twelve cylinders. Engines are also built with horizontal cylinders, usually one or two, in a few cases with three cylinders.

A *V-arrangement* of cylinders is shown in Fig. 4-3. If an engine has more than eight cylinders, it becomes difficult to make a sufficiently rigid frame and crankshaft with an in-line arrangement. Also, the engine becomes quite long and takes up considerable space. The V-arrangement, with two connecting rods attached to each crankpin, permits reducing the engine length by almost one-half, thus making it much more rigid, with a stiff crankshaft. It also costs less to manufacture and install. This is a common arrangement for engines with eight, twelve, and sixteen cylinders.

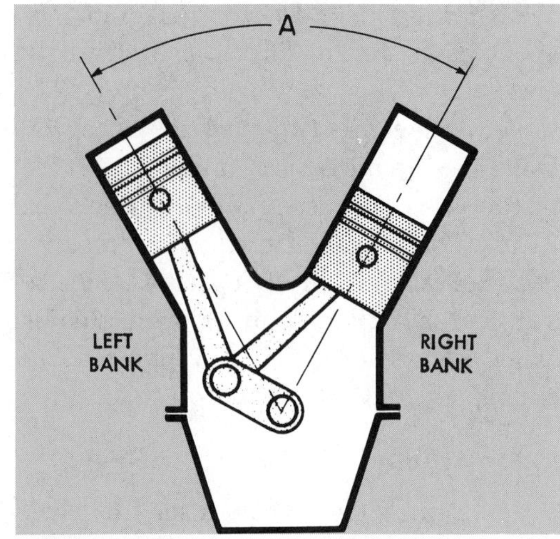

Fig. 4-3. V-arrangement of cylinders.

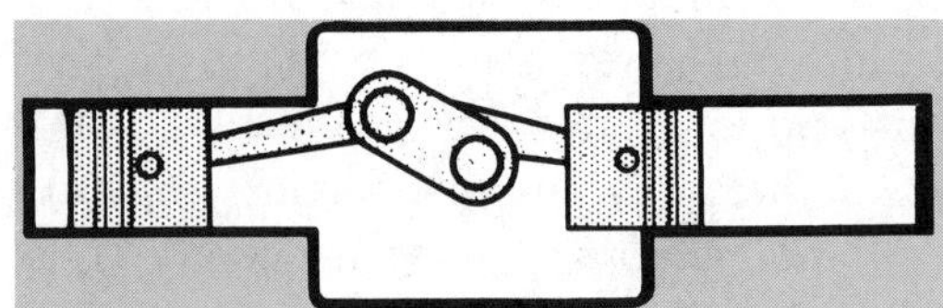

Fig. 4-4. Flat arrangement of cylinders.

Cylinders lying in one line are called a bank, and the angle between the banks may vary, in manufacturing practice, from 30° to 120°, the most common angles being between 40° and 75°. (A complete circle is 360°.)

A *flat engine,* Fig. 4-4, is a V-engine with the angle between the banks increased to 180 deg. This arrangement is used where there is little headroom, as in trucks, buses, and rail cars. Flat engines are also called *opposed-cylinder* engines. Don't confuse this with the *opposed-piston* engine previously described.

In a *radial* engine, Fig. 4-5, all the cylinders are set in a circle and all point

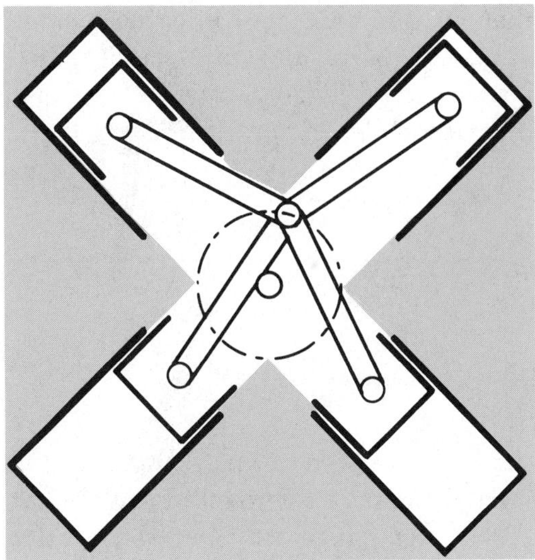

Fig. 4-5. Radial arrangement of cylinders.

toward the center of the circle. The connecting rods of all the pistons work on a single crankpin, which rotates around the center of the circle. The name *radial* is given because each cylinder is set on a radius of the circle. Such a radial engine occupies little floor space. By attaching the connecting rods to a master disk surrounding the crankpin, as many as twelve cylinders have been made to work on a single crankpin.

Materials of Construction

The main materials are cast-iron, steel, aluminum, brass, and copper. None of these materials is a simple substance; they are all alloys. They contain, besides the iron, aluminum, copper, etc. by which we name them, other materials such as nickel, manganese, silicon, chromium, molybdenum, and zinc (in brass) to give them strength, stiffness, wear resistance, freedom from corrosion, minimum thermal expansion (small changes in size for changes in temperature), and other desired properties.

The properties of these materials vary also with the process by which they are produced. Most of them start with a molten metal which is cast—poured into a mold—and then allowed to harden into solid form. *Cast-iron* parts are thus made in molds which have the shape of the desired part. The working surfaces of the cast-iron part are then machined smooth to the desired dimensions by a machine tool which shaves off the metal.

Steel parts are sometimes used directly in the form of cast steel. More often, however, the steel goes through a forging process. Here the steel, which has been cast in the form of a large bar or ingot, is first softened by heating it white hot and is then squeezed by means of a hammer, press, or rollers into the form of bars, sheets, and special shapes.

Crankshafts, for instance, are usually forgings made by hammering a block of white hot steel under a power hammer or by squeezing it into a die which has the shape desired. After the forging operation, the crankshaft is machined to size. Many crankshafts are steel castings.

Heat treatment is also used to improve the properties of certain materials. For instance, the steel used for such important applications as crankpin bolts is always heat-treated. This consists of heating the material to red heat and then *quenching* it, which means cooling it in a liquid which will cause it to cool at the desired rate so that the material will acquire great strength and toughness. *Annealing* is a form of heat treatment whereby the material is made less brittle through changing the grain size of the metal crystals.

It is not necessary for you to understand just how these processes work, for that is the specialty of highly skilled metallurgists, but you should appreciate that all this skill goes into the making of engine parts and that an unskilled person cannot judge the quality of a piece of material by simply inspecting it.

Engines Classified by Use

Often one hears that a particular diesel engine belongs to a certain type, or class. Actually, all diesel engines are more alike than different, and it is not easy to put them into types or classes.

Engines are frequently grouped according to the use to which they may be put—marine, automotive, railroad, stationary, etc. But this is unsatisfactory, because the same engine, with only minor changes, might be used for several different applications.

A similar difficulty arises when engines are classified according to the kind of duty required of them—*heavy duty, intermittent duty*, etc. For example, an engine intended primarily for intermittent service (automotive, mobile machinery, etc.) can be applied successfully to continuous service (electric power generation, marine propulsion, etc.) by *de-rating* it. We de-rate an engine when we run it at less than its rated speed, or load it to less brake mean effective pressure (bmep) than its *rated* bmep. (See Chapter 12 on rating.)

Even such a seemingly simple classification as *speed* (rpm) is more complex than it appears. You might suppose that a high-speed engine must be designed to withstand more severe inertial forces than a low-speed engine. Actually, the inertial forces of a low-speed engine that has a long stroke may be even greater than those of a high-speed engine that has a short stroke. However, engine speed is so common a classification that it cannot be ignored. So, where engine speed is mentioned in this book, it is intended to mean the following:

Low speed .... below 350 rpm
Medium speed .... 350 to 1200 rpm
High speed .... above 1200 rpm

Engine size is another hazy kind of classification, because it involves so many different factors—cylinder size, mean effective pressure, number of cylinders, and speed. However, in order to give some idea of size in the following engine descriptions, we will take as a measure, *horsepower per cylinder* (hp/cyl) at rated speed, and use the following rough definitions:

Small size .... below 25 hp/cyl
Medium size .... 25 to 200 hp/cyl
Large size .... above 200 hp/cyl

In most cases, an engine model represents a specific design in a specific cylinder size. For economy in manufacture, most builders limit their line to a relatively small number of models, which can be applied to a wide range of capacity requirements by changes in the number of cylinders and in the speed.

Now let's look at the cross-sections of a few typical diesel engines and note some of their significant features.

Small Four-Cycle Engines

The numerous makes of small four-cycle diesel engines all bear a striking external resemblance to ordinary automotive gasoline engines. However, their

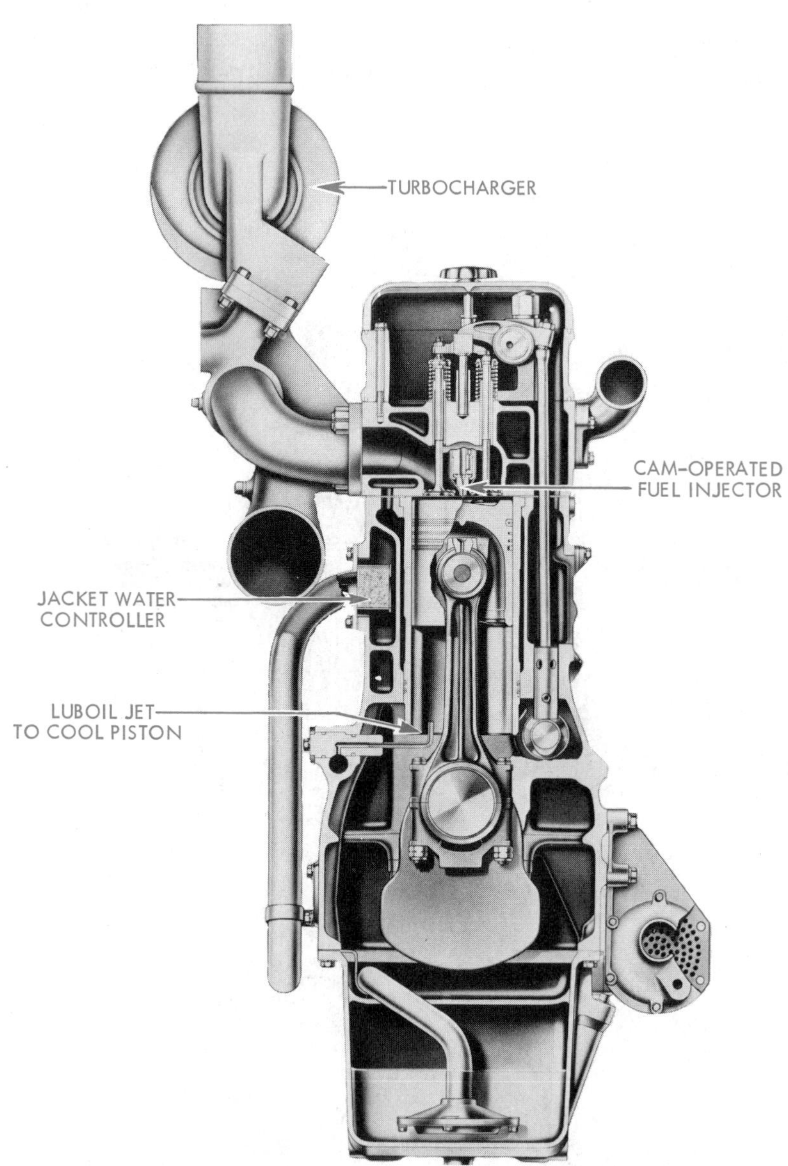

Fig. 4-6. International Harvester engine. (International Harvester Corp.)

construction is much stronger and heavier. They use various fuel-injection and combustion systems. Features of typical injection and combustion systems follow.

The Caterpillar engine was shown in Figs. 3-2, 3-3, and 3-4. All important parts were identified, including the pre-combustion chamber. Note the removable wet cylinder liners, with a sliding joint at the lower end, sealed by rubber rings. The fuel-injection system is the jerk-pump type, with individual pump plungers for each cylinder.

Fig. 4-6 is the cross-section of an International Harvester engine which uses the direct-injection fuel system; the fuel is sprayed directly into the space above the piston. The fuel passes through two pumps—the first meters and delivers it to the injector in the cylinder head which contains a cam-operated plunger. At the right moment the plunger descends

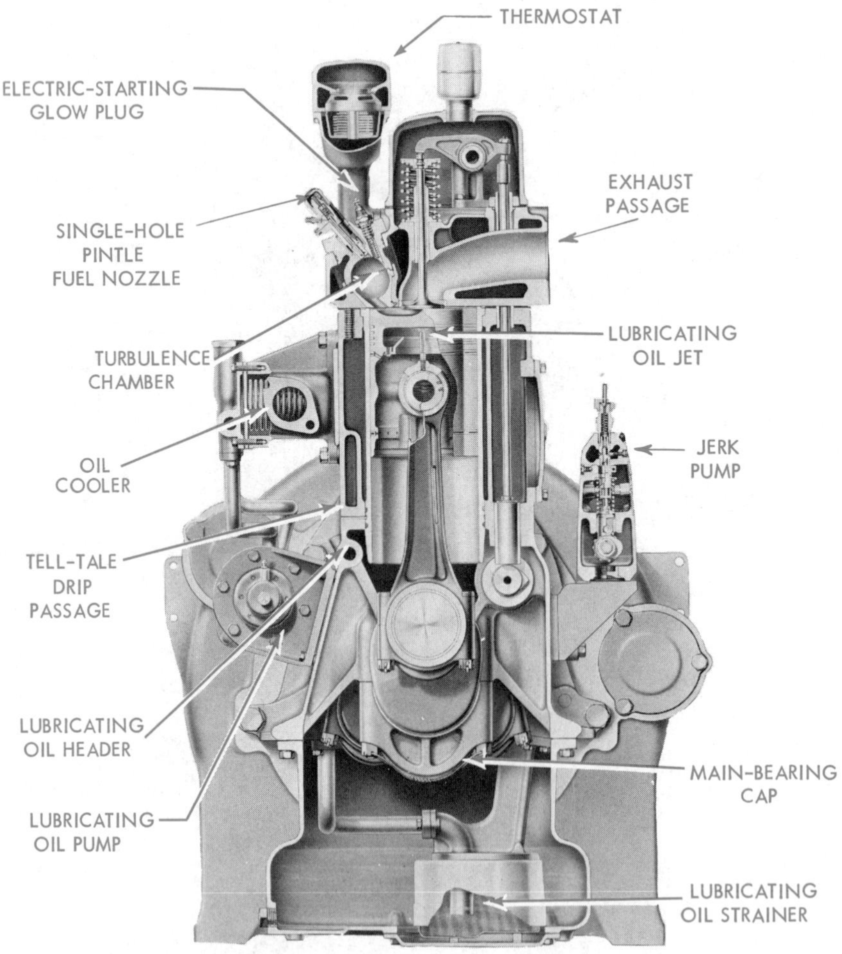

Fig. 4-7. Waukesha engine. (Waukesha Motor Co.)

quickly, building up a high pressure which forces the fuel through the nozzle orifice and produces a fine spray. Note the turbocharger. The piston is cooled by a jet of lubricating oil.

The Waukesha engine, Fig. 4-7, employs a turbulence chamber of spherical shape, located in the cylinder head. Pistons are aluminum alloy and are cooled by a jet of lubricating oil delivered through the connecting rod and wristpin bearing. Note the tell-tale drip passage to show if any water leaks past the liner seal rings. As in most small engines, the

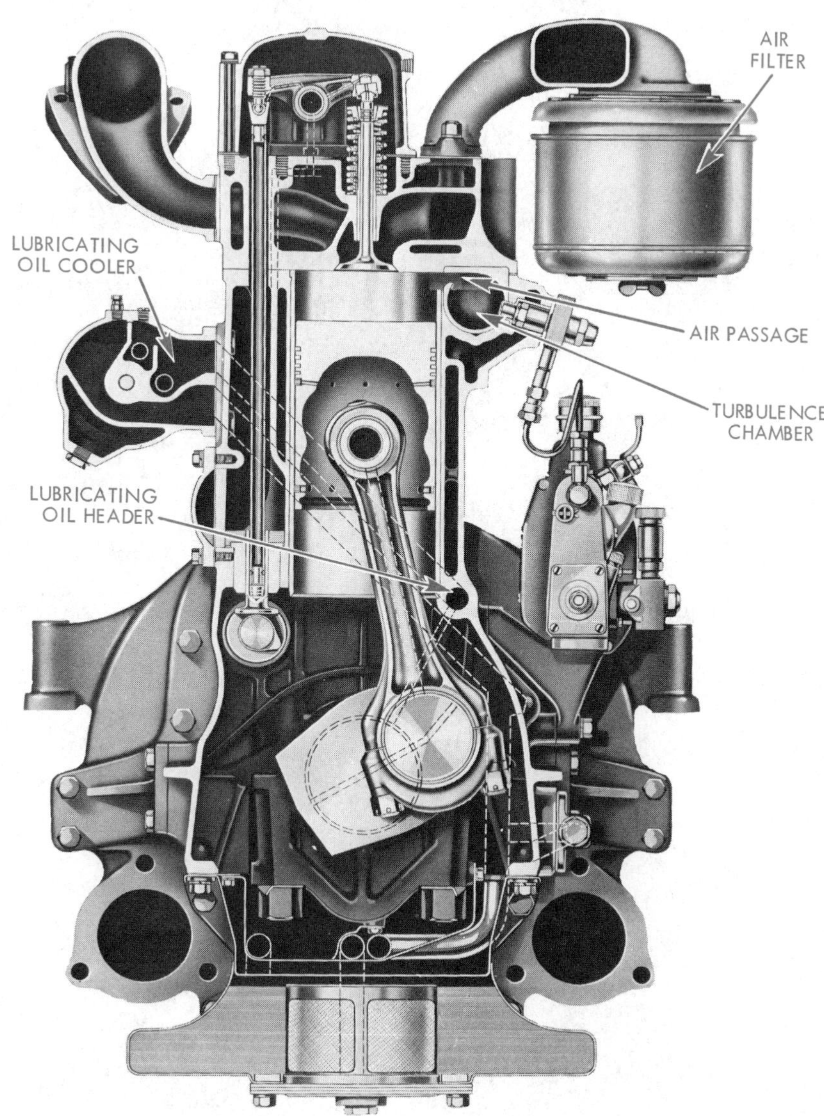

Fig. 4-8. Transverse cross-section of Hercules engine. (Hercules Motor Corp.)

crankshaft is underslung; that is, it rests on the caps of the main bearings and is removed downwards.

The Hercules engine, Fig. 4-8, also uses a spherical turbulence chamber with controlled air flow. The connecting passage between the turbulence chamber and the cylinder is so arranged that as the piston nears the top of its stroke, it begins to cover the passage. This speeds up the air flow in the passage and creates greater turbulence in the chamber at the time the fuel is being injected and burned. Note the removable *dry* cylinder liner, and also

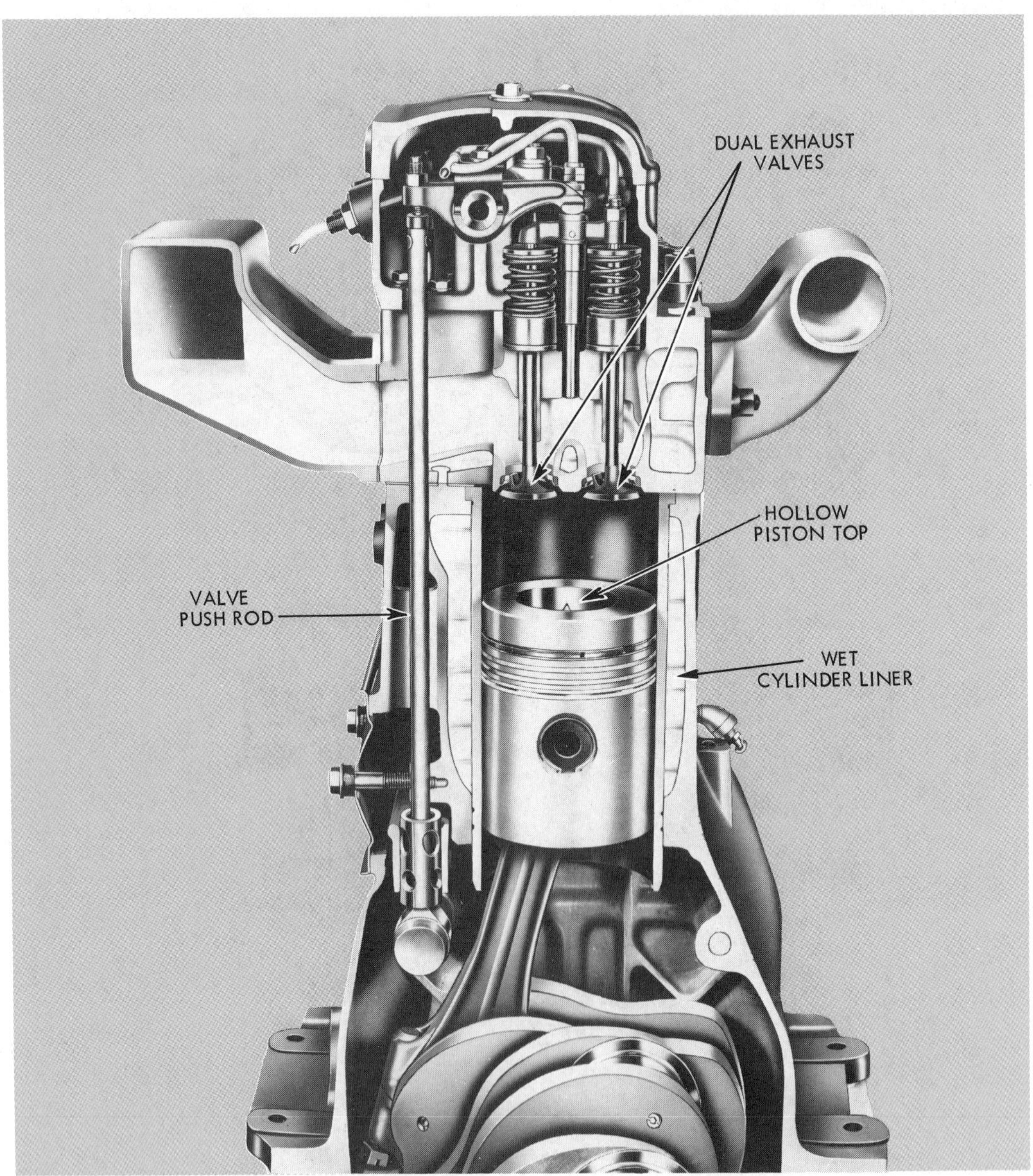

Fig. 4-9. Allis-Chalmers engine. (Allis-Chalmers, Engine Div.)

the thick crown of the aluminum piston to help carry off heat to the ring section.

The Allis-Chalmers engine, Fig. 4-9, employs a direct-injection, open-chamber combustion system. Each cylinder head contains two inlet and two exhaust valves, permitting good breathing at high speeds. It uses air swirl to cause complete combustion and thus increase the engine efficiency. This is done by directional ports in the cylinder head which give the incoming air a swirling motion around the cylinder. On the compression stroke the hollow piston top intensifies the rotary movement and squishes the air toward the center. At this moment the fuel is injected in a spray which mixes thoroughly with the violently turbulent air.

The John Deere three and four-cylinder models in Fig. 4-10, use a direct-injection fuel system. The three-cylinder engine which is cutaway has a displacement of 151.9 cu in, and an observed

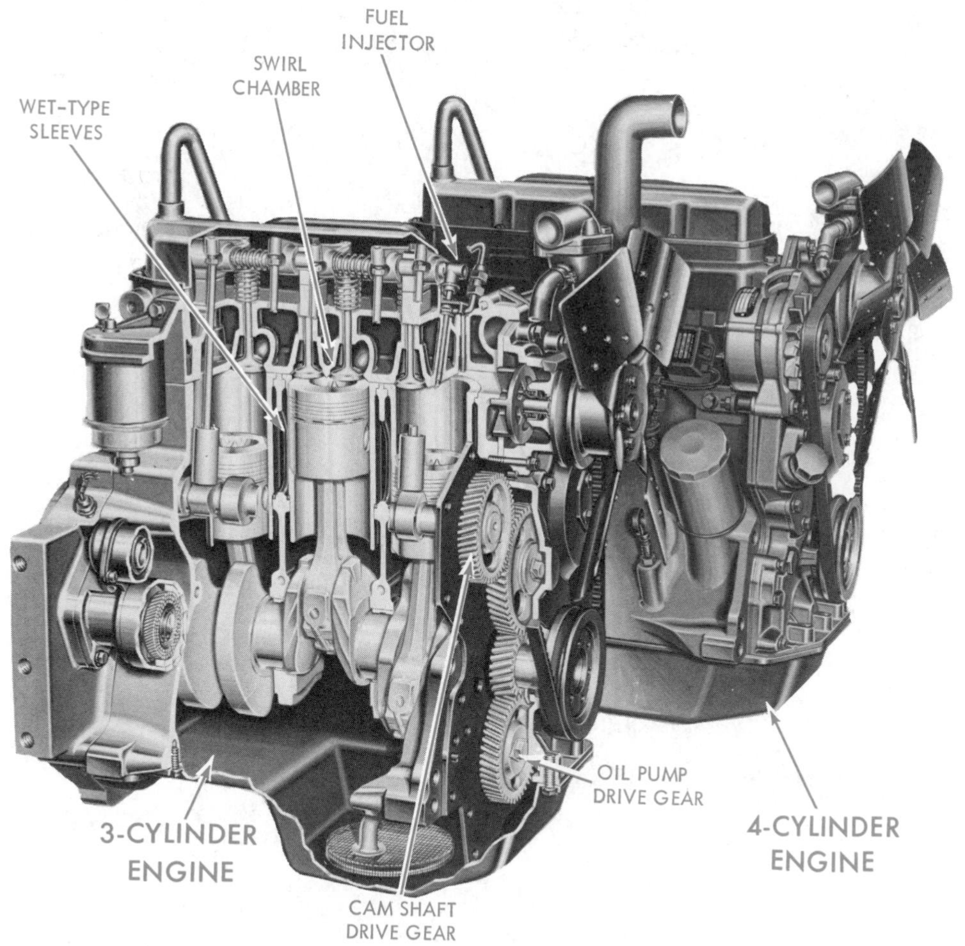

Fig. 4-10. Three and four-cylinder diesel engines. (Deere & Company)

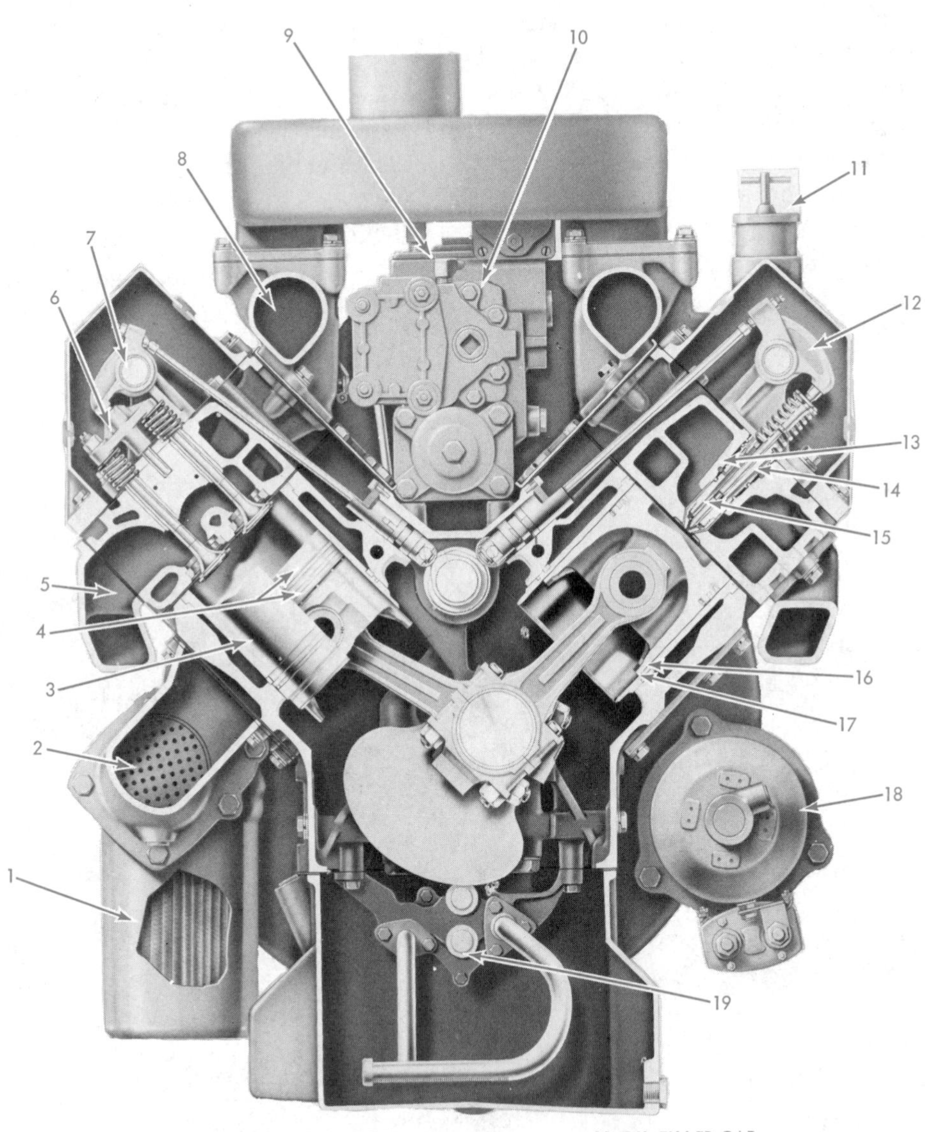

1 FULL-FLOW LUBRICATING OIL FILTER
2 OIL COOLER
3 CYLINDER LINER
4 PISTON RINGS
5 EXHAUST MANIFOLD
6 VALVE CROSSHEAD
7 VALVE ROCKER LEVER
8 AIR INTAKE MANIFOLD
9 FUEL SUCTION INLET TO PUMP
10 FUEL PUMP
11 OIL FILLER CAP
12 INJECTOR ROCKER LEVER
13 FUEL IN
14 FUEL OUT
15 INJECTOR
16 CREVICE SEAL
17 PACKING RING
18 CRANKING MOTOR
19 LUBRICATING OIL PUMP

Fig. 4-11. Cummins V-8 truck engine. (Cummins Engine Company)

horsepower of 31 at 2100 engine rpm. Notice the rugged construction of this small diesel engine. The center *wet-type* cylinder is clearly visible, showing the upper and lower seals which prevent coolant water from getting into the oil (crankcase) or into the combustion chamber. Note the pencil-type (Roosa-

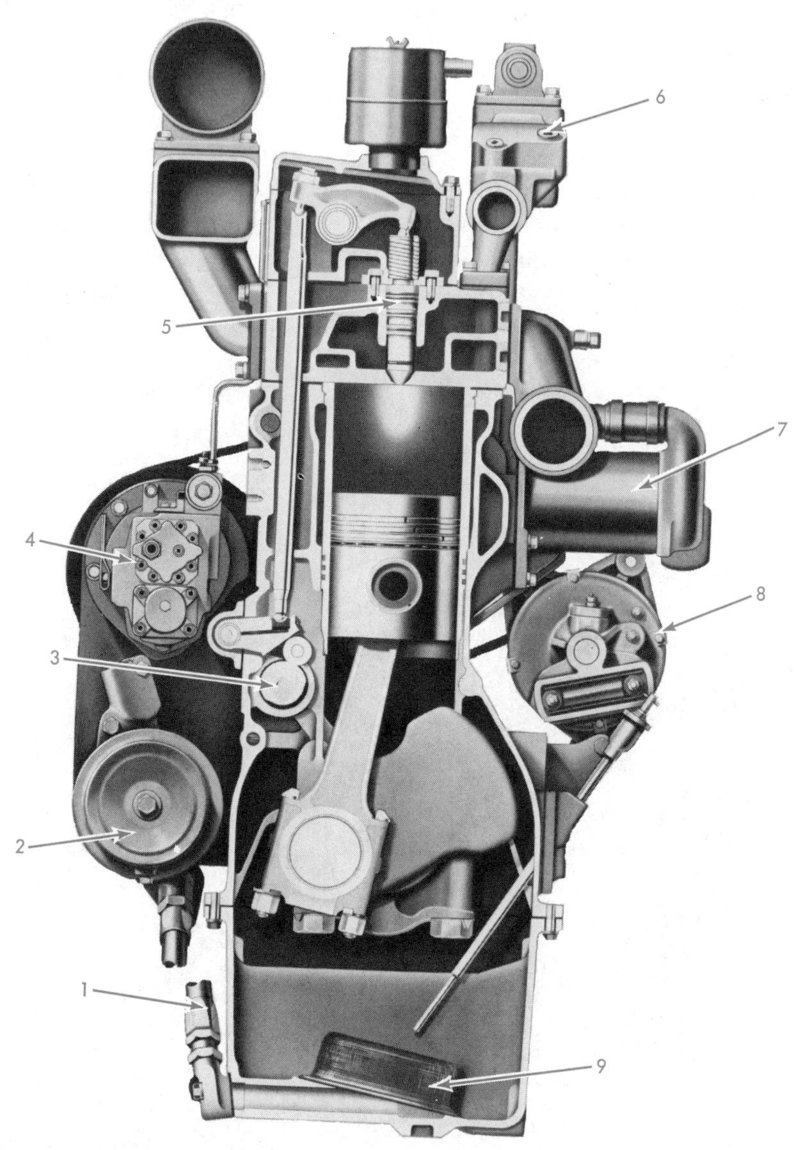

1 ATTACHES TO OIL PUMP
2 OIL FILTER
3 CAMSHAFT
4 FUEL PUMP
5 INJECTOR
6 WATER MANIFOLD THERMOSTAT HOUSING
7 OIL COOLER
8 ALTERNATOR
9 OIL PICKUP SCREEN

Fig. 4-12. Cummins NHC 250 six-cylinder truck engine. (Cummins Engine Company)

master) fuel injector in the front cylinder just to the right of the valve.

The camshaft gear, idler, crankshaft and oil pump gears are visible at the front of the engine. The Cummins V-8 truck engine in Fig. 4-11 is 504 cu in displacement. In addition to being used for trucks up to 72,000 pound gross vehicle weight, it is also used to power construction machinery and fire pumps.

The Cummins six-cylinder engine in Fig. 4-12 is rated from 250 to 265 hp, depending upon compression ratio and other accessories.

Small Two-Cycle Engines

The General Motors Series 71 diesel engine (Detroit Diesel Engine Division), shown in Fig. 4-13, was the first small two-cycle diesel put into large-scale production (1938). Its combustion system employs a unit injector spraying directly into an open combustion chamber. An engine-driven rotary blower with three helical (corkscrew) lobes supplies air that scavenges the cylinder on the *uniflow* principle. The scavenging air enters the cylinder through ports at the

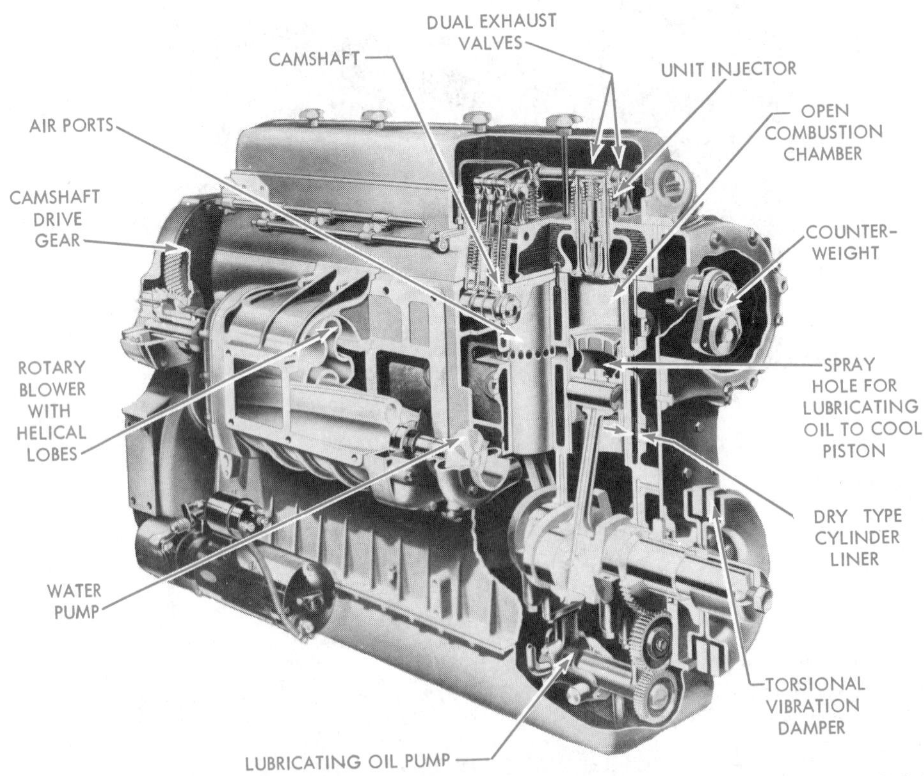

Fig. 4-13. General Motor Series 71 Detroit Diesel engine. (Detroit Diesel Allison Div., General Motors Corp.)

lower end, directed at an angle to create a swirling motion for greater turbulence.

Exhaust gases leave through cam-operated dual exhaust valves in the cylinder head. The piston crown is cooled by lubricating oil sprayed from the top of the connecting rod. The cylinder liner is the dry type. The larger engine of this line, known as the Series 110, is similar in design, but employs a gear-driven centrifugal blower for scavenging.

Fig. 4-14 shows the P&H (Harnishfeger) engine. Note how the wet cylinder liner is screwed into the cylinder head and encased in a water jacket to form a single assembly, which is removable as a unit. The combustion system uses a jerk-pump with a self-actuated injection valve which sprays into an open combustion chamber. Like the previously described engine, it employs uniflow scavenging; the air ports are at the lower end, and a single exhaust valve is in the cylinder head. The piston is aluminum.

Medium-Size Four-Cycle Engines

Fig. 4-15, showing the Worthington DR engine, typifies the medium-speed, me-

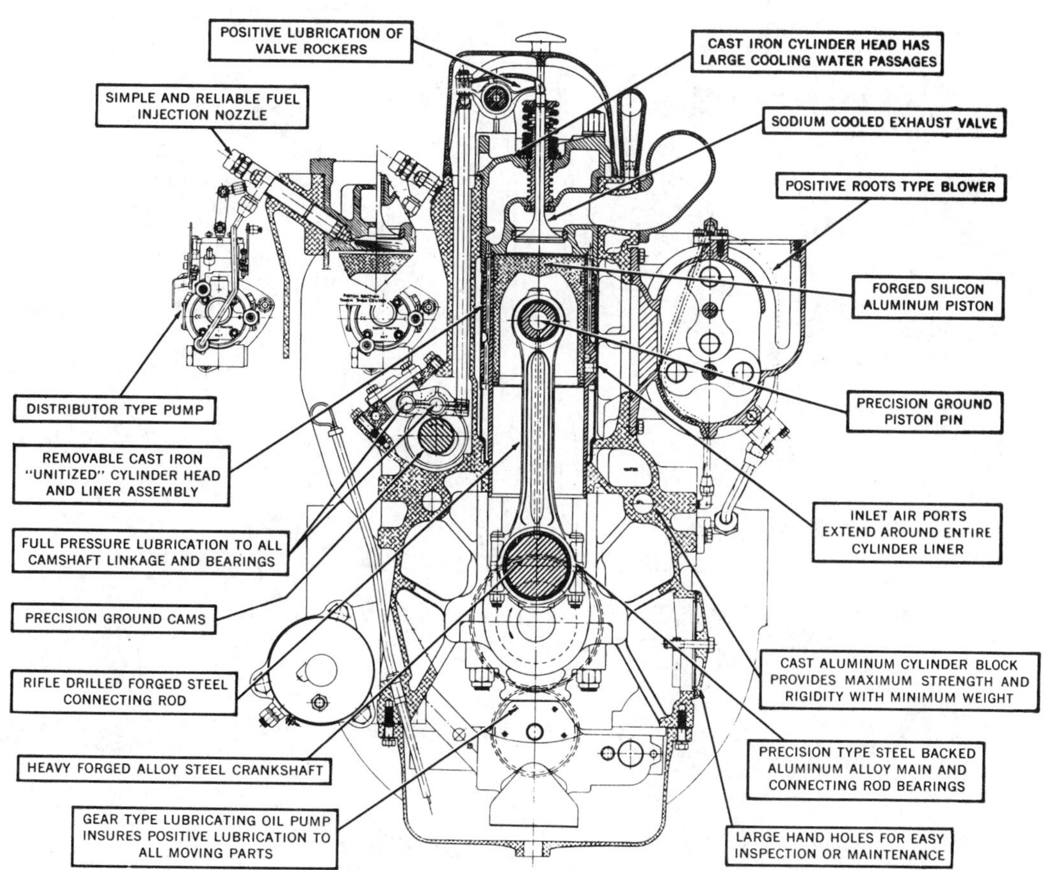

Fig. 4-14. P. & H. (Harnischfeger) engine. (Harnischfeger Corp.)

dium-size engine. The crankshaft seats in the bedplate and is removed upwards. It employs a two-piece frame, which is designed to enclose all working parts. Each cylinder has its own jerk-pump, which injects fuel into an open combustion

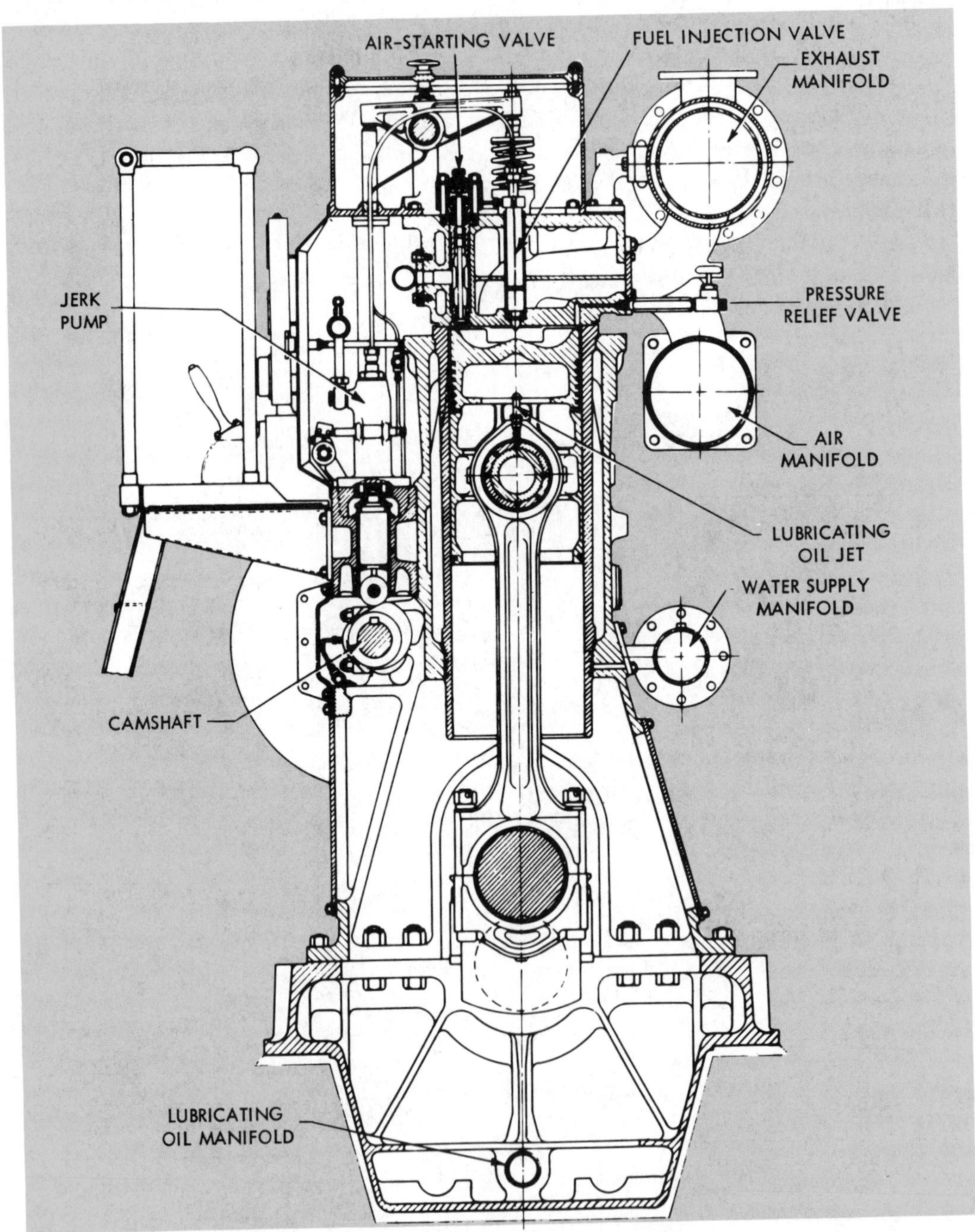

Fig. 4-15. Worthington DR engine. (Worthington Corp.)

chamber. A jet of lubricating oil from the top of the connecting rod cools the piston crown. The cylinder head carries an air-inlet valve, exhaust valve, fuel-injection valve, air-starting valve, and pressure-relief valve.

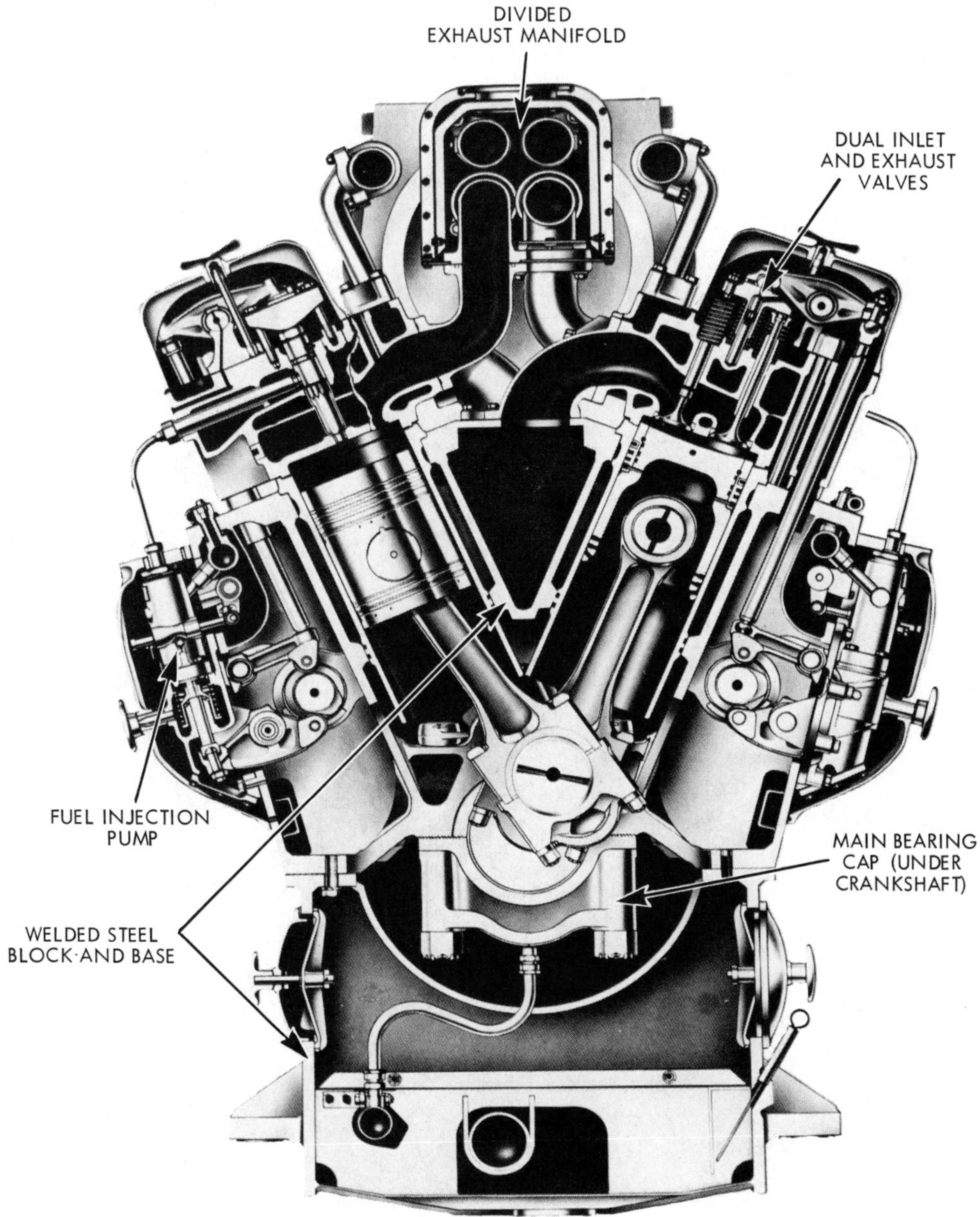

Fig. 4-16. Alco locomotive engine. (Alco Products, Inc.)

The Alco Model 251 engine, Fig. 4-16, is of the V-type, with either twelve or sixteen cylinders, and is specially designed for locomotive use to give maximum power for its weight and space. The engine block and base are made of steel plate and forgings, welded together. Turbocharging at high pressure raises its rated bmep to 185 psi. The pulse system is used, with the exhaust manifold divided into four branches.

Fig. 4-17 shows the General Motors Electro-Motive V-type engine. The frame is welded steel. Uniflow scavenging is employed, with ports in the cylinder walls for inlet, and valves in the cylinder head for exhaust. An engine-driven blower (not shown) supplies the air. The main bear-

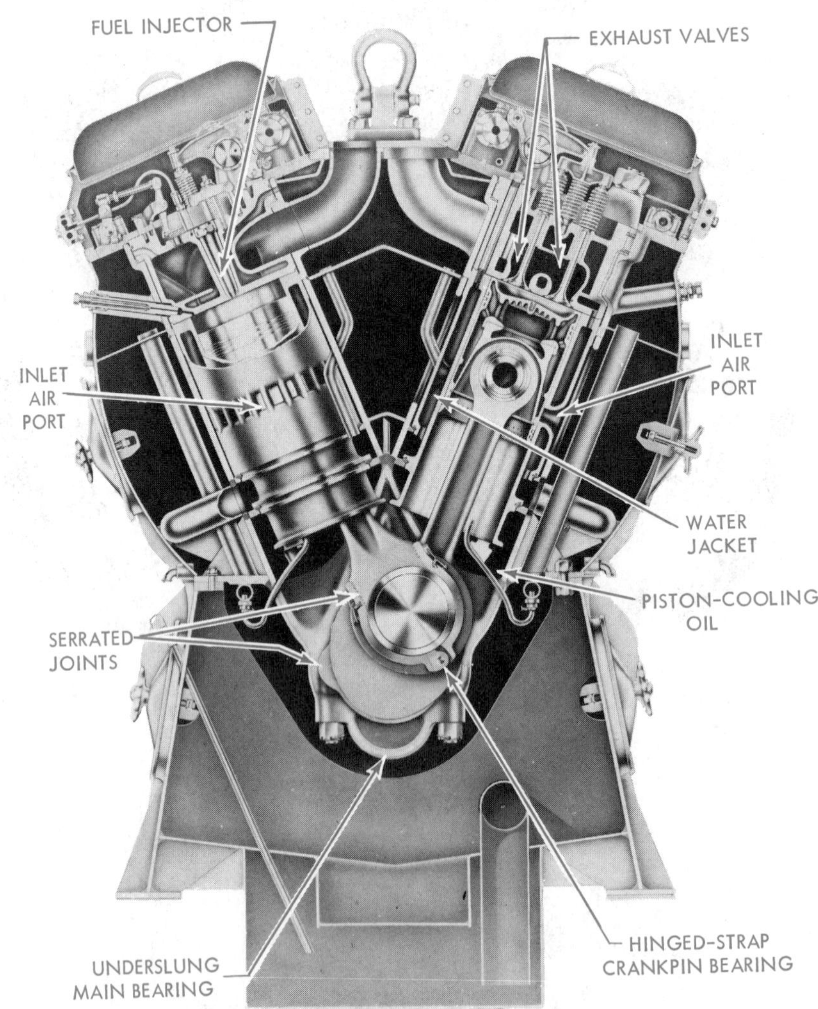

Fig. 4-17. General Motors Electro-Motive engine. (Electro-Motive Div., General Motors Corp.)

ings are underslung. The hinged-strap crankpin bearing design permits pulling the connecting rod through the cylinder despite the large-diameter crankpin. Pistons are cooled by jets of lubricating oil.

The opposed-piston engine, Fig. 4-18, was designed by Fairbanks, Morse and Company for applications where minimum weight and space are important, as in locomotives and submarines, but it is

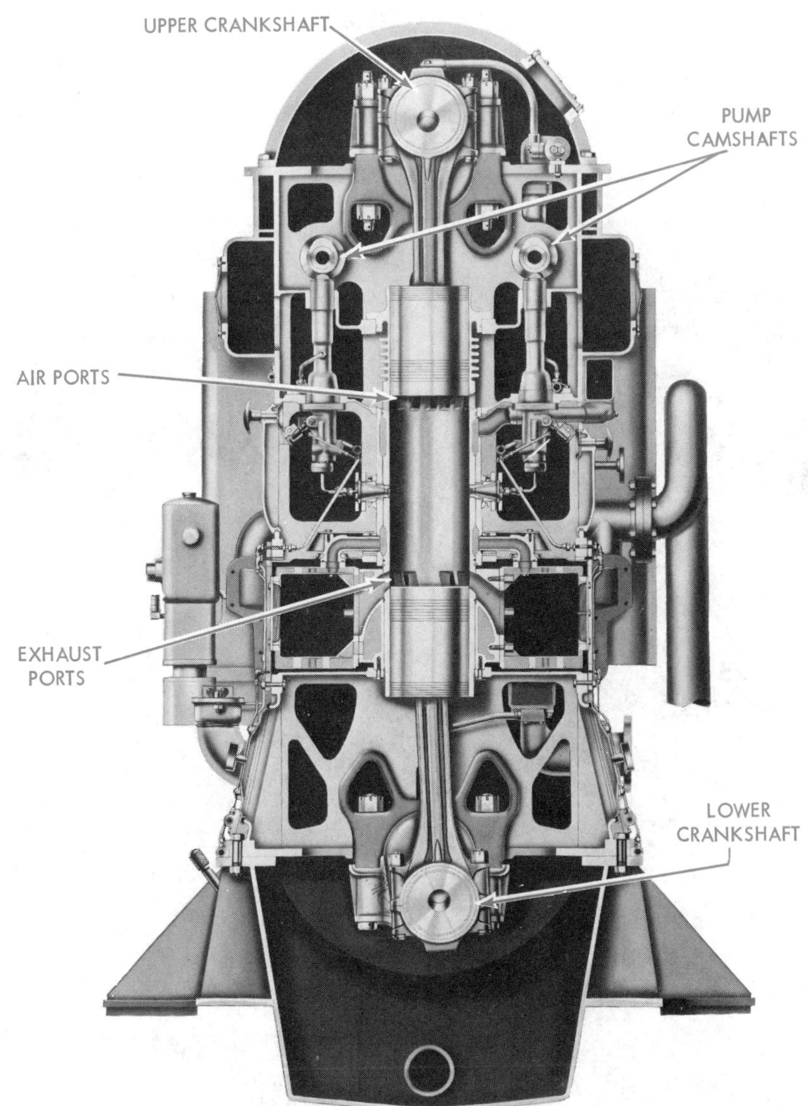

Fig. 4-18. Fairbanks-Morse opposed-piston engine. (Colt Industries, Fairbanks-Morse Motor & Generator Operations)

also widely used for power plants, emergency units, and all types of marine applications.

Two pistons work in the same cylinder—the space between the pistons forming the combustion chamber. There is no cylinder head. Two crankshafts are also used; the upper one is connected to the lower one by means of bevel gears and a vertical shaft. Power is taken off the lower crankshaft. The frame is welded steel. The scavenging is uniflow, with the intake ports at the top and the exhaust ports at the bottom (as fully explained in

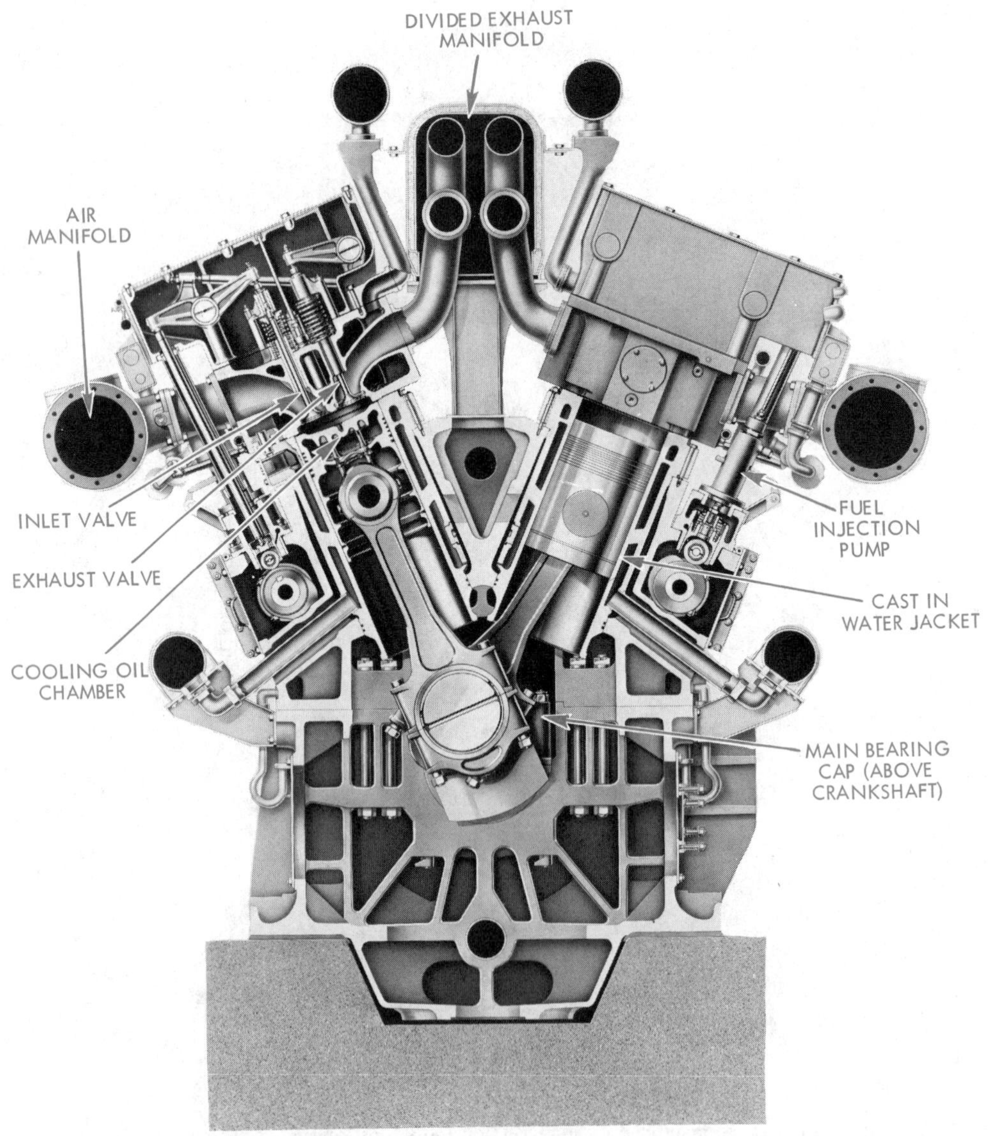

Fig. 4-19. Worthington SW four-cycle engine. (Worthington Corp.)

Chapter 13). The engine-driven rotary blower supplies scavenging air.

Large-Size Engines

Most large engines built in the United States have been Nordberg two-cycle, slow-speed engines with cylinders of large bore and stroke. However, advances in design have opened this field to four-cycle enginees up to power capacities of about 360 hp per cylinder. These four-cycle engines are highly turbocharged and run at

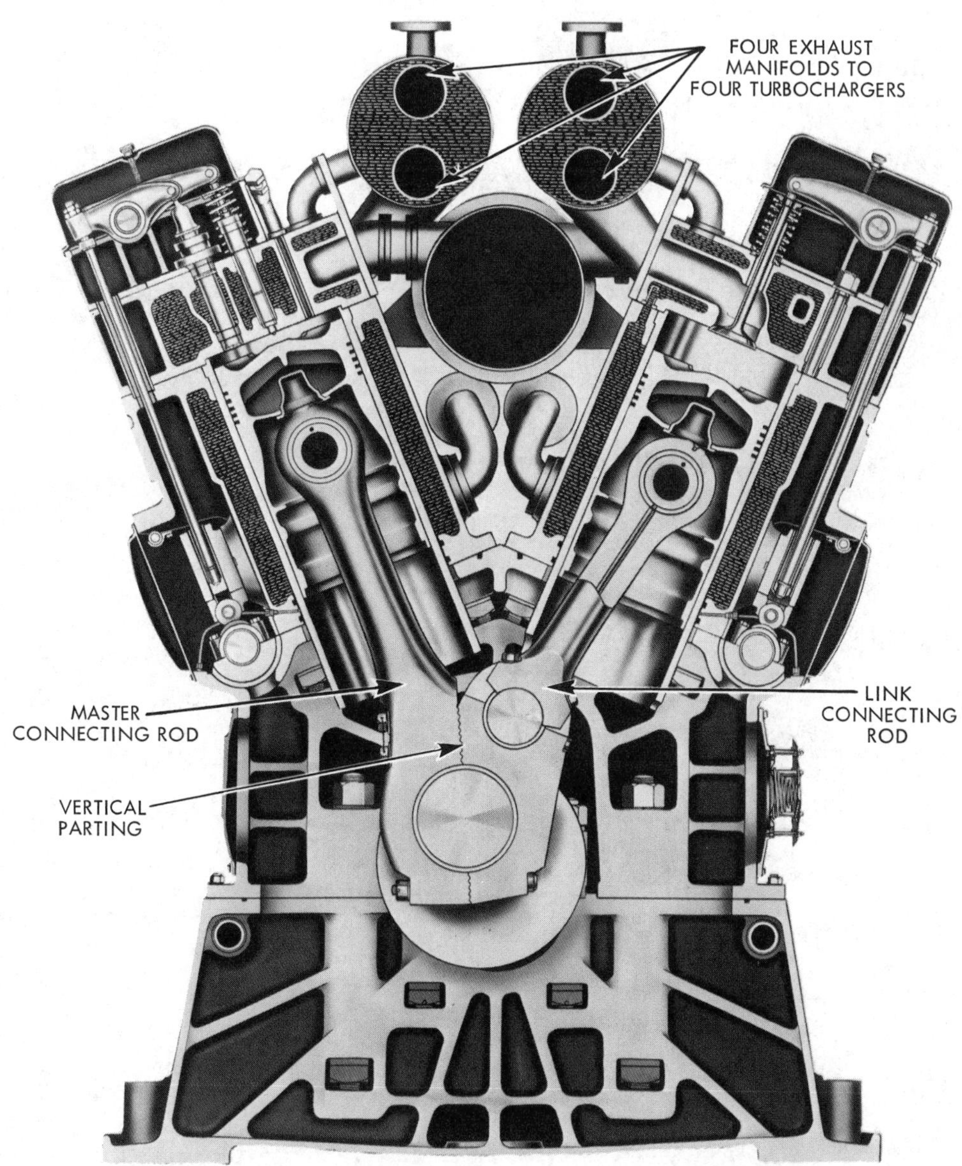

Fig. 4-20. Enterprise RV four-cycle engine. (Enterprise Engine & Machinery Co.)

greater speeds than their two-cycle counterparts. For maximum output in minimum space they are generally of the V-type, with twelve or sixteen cylinders, producing up to 5700 hp.

Fig. 4-19 shows a Worthington SW engine; it is of the V-type, four-cycle and highly supercharged by the pulse system. The cylinders have cast-in water jackets, thus eliminating sealing rings. An interior partition below the piston crown forms a chamber which is flooded with oil delivered through the drilled connecting rod, for cooling the pistons and rings. The

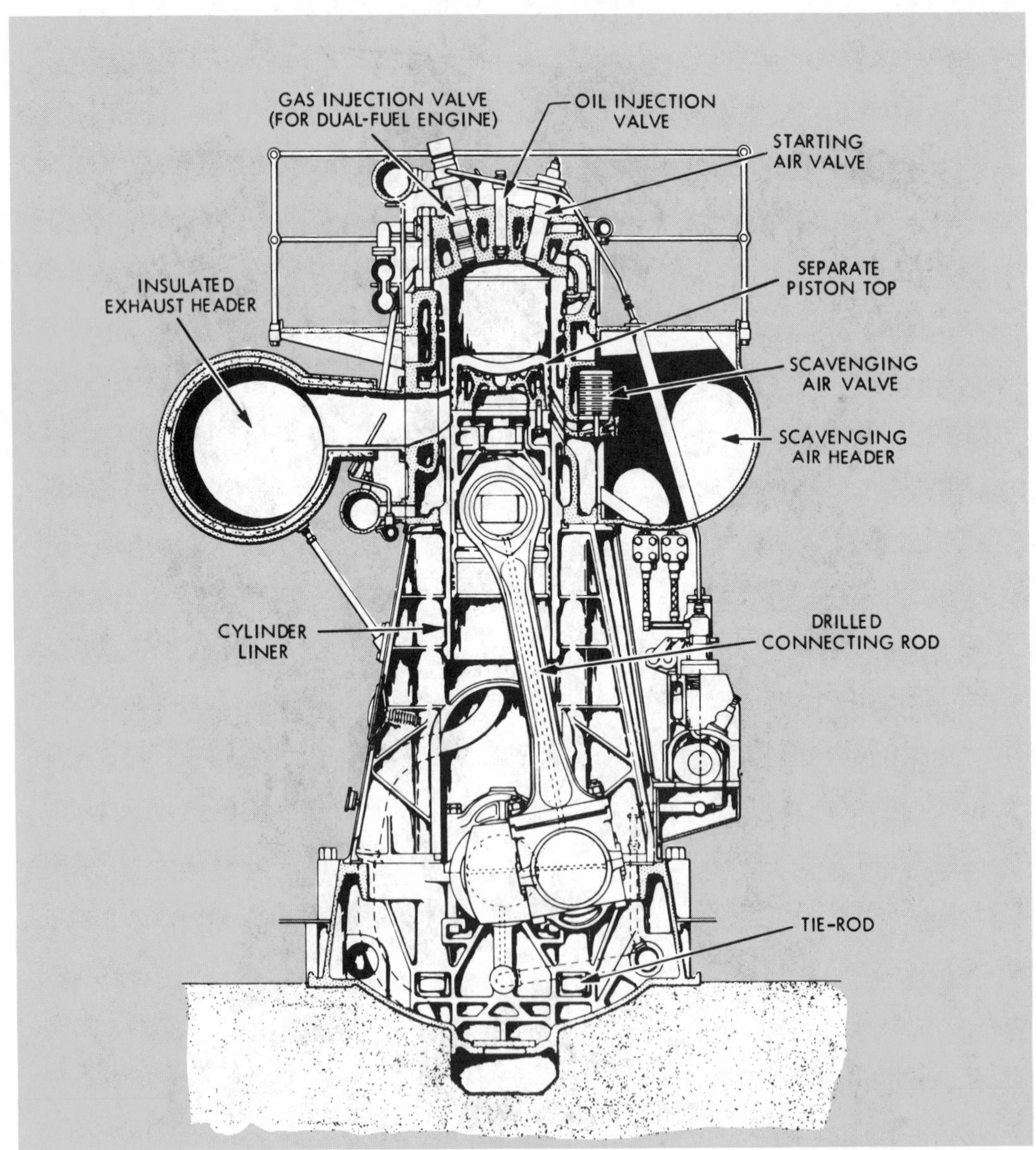

Fig. 4-21. Nordberg large two-cycle engine. (Nordberg Mfg. Co.)

connecting rods of opposite cylinders fit side by side. Each cylinder has two inlet and two exhaust valves.

Another large four-cycle engine is the Enterprise RV, shown in Fig. 4-20. Note the unique arrangement of the master-and-link connecting rod which permits opposite cylinders of the V to lie in the same plane, thus shortening the engine. The vertical parting of the connecting rod permits the assembly of piston and rod to be lifted through the cylinder liner. The sixteen-cylinder model uses four turbochargers, one at each end of each bank of the V.

Fig. 4-21 shows the two-cycle Nordberg engine, with cylinders of 21½″ bore by 31″ stroke, which develops more than 700 bhp/cylinder. Cylinders of 29″ bore by 40″ stroke develop over 1050 *bhp* at 200 rpm, and can be built up to twelve cylinders to produce 12,800 *bhp*.

The engine shown employs a trunk piston, but crosshead construction has also been used. As in all large slow-speed engines, tie-rods (instead of the frame) carry the pressure force on the cylinder head. The piston top, which carries the rings, is a separate piece from the piston skirt, and is cooled by a stream of oil fed through the connecting rod. Port scavenging is used, with automatic air valves in the air header to prevent exhaust gases flowing back into the scavenging air.

A scavenging pump is used on the smaller engines; the larger ones employ a motor-driven blower. The wristpin is *bolted* to the piston (as shown in Fig. 6-29). This permits removing the piston without taking out the connecting rod.

Checking On Your Knowledge

The following questions give you the opportunity to check up on yourself. If you have read the chapter carefully, you should be able to answer the questions. If you have any difficulty, read the chapter over once more so that you have the information well in mind before you go on with your reading.

DO YOU KNOW

1. How are diesel engines classified as to design?
2. What is a crosshead type piston?
3. Identify the important properties of the materials used in the following engine parts: (a) pistons, (b) crankshafts, (c) engine block, (d) valves and (e) cylinder liners.
4. Explain how diesel engines are classified as to use.
5. What is a *wet* type cylinder liner and what is a *dry* type cylinder liner?
6. What does scavenging mean in regard to diesel engine operation?
7. Explain the *uniflow* principle of scavenging.
8. What is an *open* combustion chamber?

Chapter

5 Stationary Parts–Frames' Cylinders and Heads

Chapter 3 introduced you to the basic construction of a diesel engine by telling you what parts are needed and how they look in one particular engine. Before we study the various designs used in different engines, let's review briefly what the parts do.

The cylinder head, liner, and piston form the *combustion space*, in which the air charge is compressed and into which the *fuel injection system* (pumps, fuel lines, nozzles) discharge a measured amount of fuel oil, properly atomized for good combustion. In the dual-fuel engine, which burns gas and oil, the fuel injection system handles the oil only, while the gas enters the cylinder with the air charge.

The expansive force produced by the burning fuel drives the piston down, and through the connecting rod delivers energy to the crankshaft. The *valve mechanism* of a four-cycle engine (camshaft and drive, push rods, rocker arms) actuates and times the movement of the valves, controlling the flow of fresh air into the cylinder and the discharge of the spent gases.

In a two-cycle engine the fresh air is brought in and the spent gases are blown out by the *scavenging system* (air pump or blower, cylinder ports, and sometimes valves).

You can see that the engine parts break down into four groups:

1. Structural parts (which do not move); bedplate, base or pan, frame, cylinder liners, cylinder heads.
2. Major moving parts: pistons, connecting rods, crankshaft, and their respective bearings.
3. Arrangements for getting air in and exhaust gases out: valves, valve mechanism, manifolds, scavenging, and supercharging systems.
4. Fuel injection system: pumps, nozzles, control devices.

You will understand what forces act on these parts, why they are constructed the way they are, and why certain materials are used to give strength and stiffness where these properties are needed.

Engine Structure and Requirements

Speaking in a general way, the engine structure includes all the fixed parts that hold the engine together. Its basic job is to support and keep in line the moving parts while resisting the forces set up by the engine's operation. In addition, the engine structure supports auxiliaries, provides jackets and passages for cooling water, a sump for lubricating oil, and a protecting enclosure for these parts.

Two principal kinds of forces act on the structure. The *firing pressure* tries to push the cylinder heads and the crankshaft bearings apart, producing a stretching force in the frame. The motion of the pistons, connecting rods, and crankshaft sets up *inertial forces* (Chapter 12), whose direction and amount vary from instant to instant. These inertial forces tend to twist and bend the frame and to move the engine on its foundation. The frame must be strong and rigid in order to resist these forces, but at the same time it must use only a reasonable amount of material and must be capable of low-cost manufacture.

In addition to pressure and inertial forces acting on the stationary structural parts of the engine, thermal stresses occur on the cylinder block, cylinders, heads, manifolds and other units due to conduction of heat from the hot to cooler areas.

Also the problem of noise from the diesel engine creates a need to design the basic structure of the diesel engine much heavier than the gasoline engine so that noise may be insulated by the greater mass.

Automotive-Type Frame

All diesel tractor, truck, and the smaller stationary engines use the automotive design in cylinder block structure. The cylinder block is a one-piece casting with the upper half of the main bearing housings being part of the main casting.

The main bearing caps bolt to the upper half, like the four-cylinder John Deere engine, Fig. 5-1. Notice in Fig. 5-1 the heavy webbing and the five main bearings used to reinforce the structure against firing pressures and inertial forces. In contrast to many of the larger engine designs shown later in this chapter, the lower section or oil pan on this engine can be made of light material because it is not subjected to stress.

The automotive-type V-6 diesel engine in Fig. 5-2 is similar in design to the four-cylinder engine. It is a series 53 Detroit diesel two-stroke engine. The casting as well as the machining of this engine structure is much more complicated than the four-cylinder example. The view on the left shows the engine from the rear and the right view shows the engine from the front. The block is bored to receive replaceable wet-type cylinder liners.

On this model, a water jacket surrounds the upper half of each cylinder liner. The water jacket and air box are

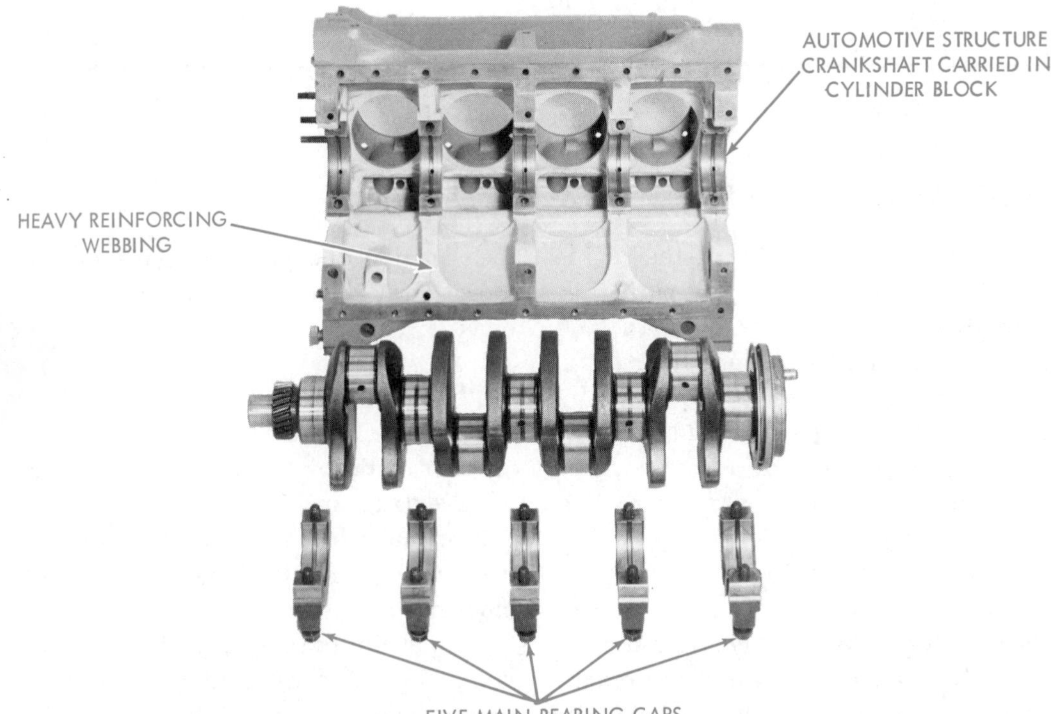

Fig. 5-1. Cast iron four-cylinder engine block. (Deere & Company)

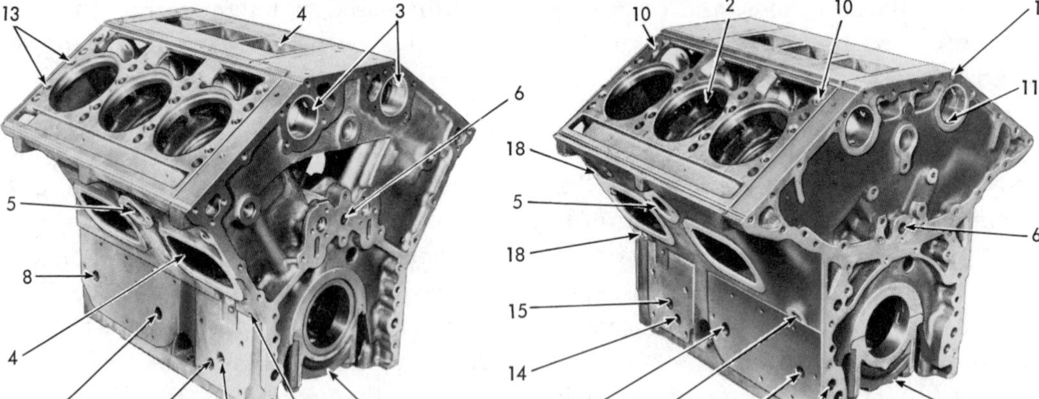

1 CYLINDER BLOCK
2 BORE TO CYLINDER LINER
3 CAMSHAFT BORE
4 AIR BOX
5 WATER PASSAGE FROM OIL COOLER TO BLOCK
6 OIL GALLERY-MAIN
7 OIL PASSAGE TO OIL COOLER
8 OIL PASSAGE FROM OIL COOLER
9 CAP-MAIN BEARING (FRONT)
10 OIL PASSAGE TO CYLINDER HEAD
11 OIL GALLERY TO CAMSHAFT
12 CAP MAIN BEARING (REAR)
13 WATER PASSAGE TO CYLINDER HEAD
14 OIL PASSAGE TO FILTER FROM PUMP
15 OIL PASSAGE FROM FILTER TO OIL COOLER
16 OIL PASSAGE TO MAIN GALLERY VIA FRONT COVER
17 OIL PASSAGE FROM OIL PUMP TO FILTER
18 OIL PRESSURE TAKE-OFF OPENING

Fig. 5-2. Cylinder block (6-V cast iron). (General Motors Corp.)

sealed off by a seal ring compressed between the liner and a groove in the block. As in all engines of this type, the main bearing bores are line-bored with the bearing caps in place to insure longitudinal alignment. Drilled passages in the block carry the lubricating oil to all moving parts of the engine to eliminate external piping.

Cast Frames for Stationary Engines

Typical frame designs for vertical engines appear in Figs. 5-3 to 5-6. For stationary engines, which rest on a substantial foundation, the two-piece construction shown in Fig. 5-3 is most widely used. The lower section, or *bedplate*, forms a base, supports the main bearings, encloses the lower part of the *crankcase* (the space in which the crankshaft turns), and forms a sump for lubricating oil.

The upper section, or *centerframe*, includes the upper part of the crankcase and the cylinder block in which the cylinders are supported. In order to get strength and rigidity (stiffness), both of these castings are made in boxlike form, with cross girders, webs, and ribs. Assembled, the two castings form a single

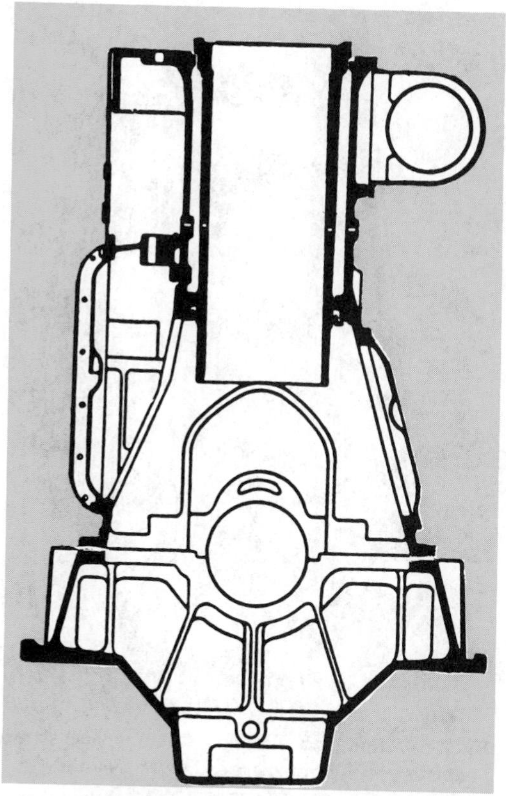

Fig. 5-3. Two-piece frame. Frame is split at bearing line; upper section carries cylinder liners and forms upper part of crankcase.

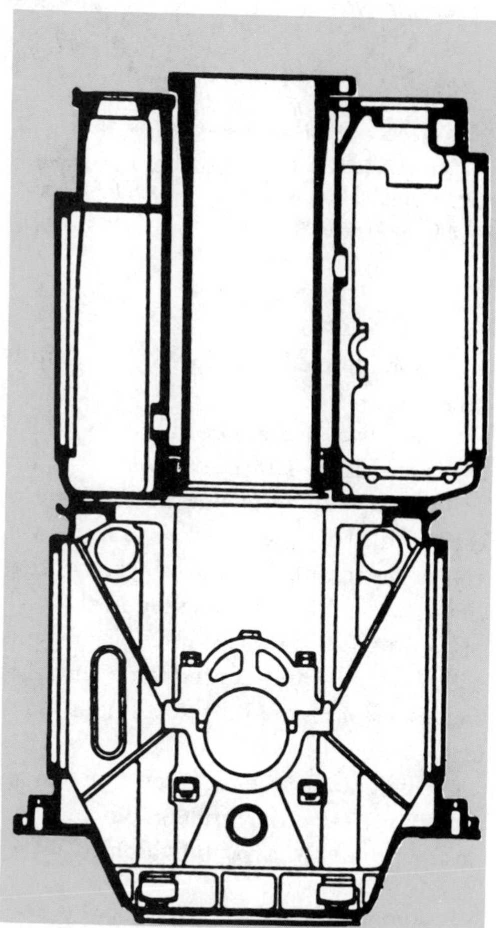

Fig. 5-4. Two-piece frame. Frame is split above bearings; lower section forms entire crankcase.

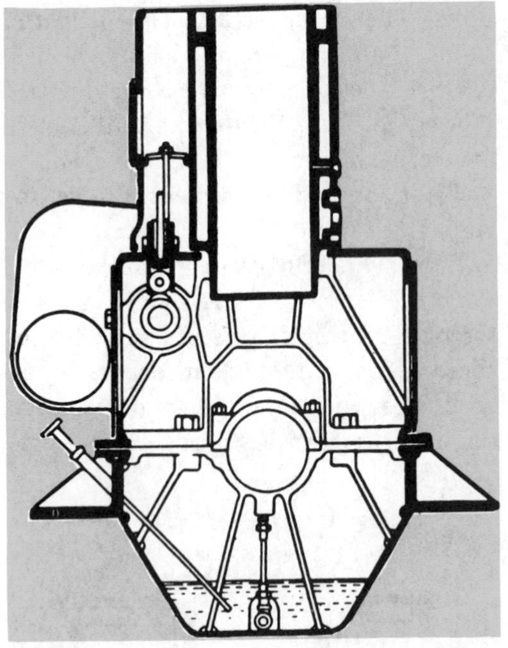

Fig. 5-5. Three-piece frame. Bedplate and center section are made of welded steel; cylinder blocks are individual castings.

heavily reinforced structure of great stiffness.

The design of Fig. 5-4 differs in principle from Fig. 5-3 in that, instead of the two castings being parted at the level of the crankshaft, they are parted well above the crankshaft bearings, so that the entire crankcase is located in the bedplate. Fig. 5-5 shows a three-piece frame with bedplate and centerframe made of welded steel plate and the cylinder block of cast iron.

Cylinders are generally cast in blocks containing several cylinders, but in a few designs they are cast individually.

A-Frame Construction

In large engines (for stationary and marine service), the A-frame construction, Fig. 5-6, is often used. In this design a cast bedplate is used to support frames which have the general shape of the letter A and which in turn support the cylinder block. The cylinders may be cast individually or in groups of two or three. In either case dowels and bolts are used to hold the cylinders together and thus make a stiff structure in the horizontal direction.

In addition, *through-bolts* run from bedplate to cylinder heads to tie the structure together in the vertical direction. When the engine is being erected, these through-bolts are tightened to keep the frame squeezed together at all times,

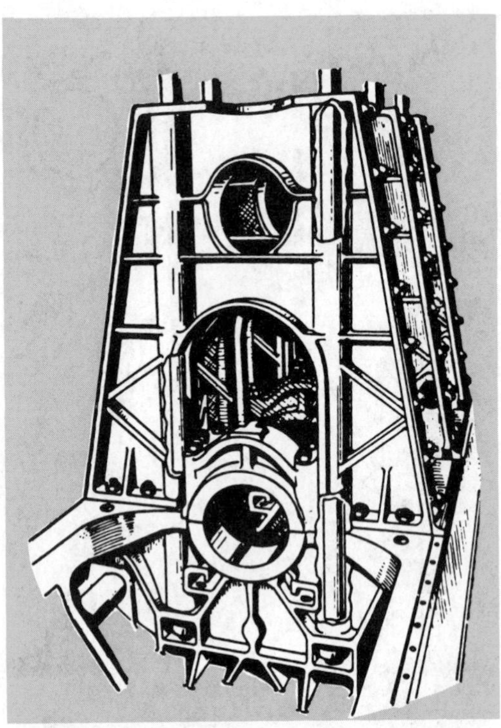

Fig. 5-6. A-frame construction. Cutaway view shows bedplate with A-frames spaced on it. Cylinder castings rest on the A-frames and are bolted together horizontally. Tie rods or through-bolts run from bedplate to cylinder head to take pulling forces.

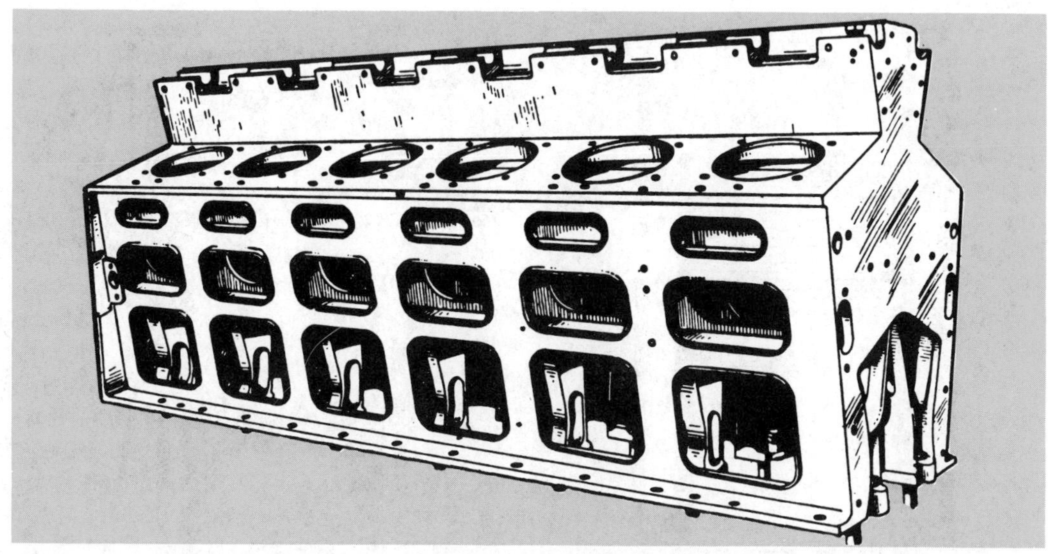

Fig. 5-7. Welded steel frame. Lightweight, rigid frame of twelve-cylinder V-engine is made by welding steel plate—forged or rolled sections.

while the bolts themselves take all the pulling forces produced by expanding gas pressures. This results in a good division of tasks; the cast sections provide stiffness, for which they are naturally fitted, and the bolts take the pulling forces or tension for which they are best suited and for which they can be designed accurately. The spaces between the A-frames are closed by bolted plates to complete the enclosure of the crankcase.

Through-bolts, because of the advantages explained above, are also coming into frequent use even in smaller engines where A-frame construction is not used. You can see them in Figs. 5-1, 5-2, and 5-4.

Welded Structures

Welded frames are frequently used when frames are made of steel instead of cast iron. The centerframe for a V-engine, Fig. 5-7, is built up by welding from forged and rolled-steel sections and steel plate; it shows what can be done by this method of fabrication. Designers still debate the relative merits of cast versus welded frames; both are used successfully.

Cylinders and Liners

The inside surface (bore) of the cylinder is subjected to the rubbing action of the piston rings and to high combustion temperatures, particularly at the upper end. The cylinder bore must also take the side thrust of the piston caused by the

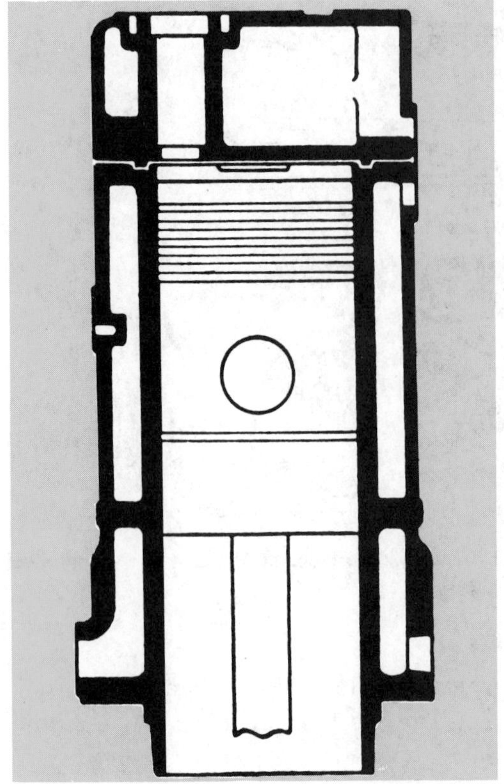

Fig. 5-8. Cylinder and water jacket cast in one piece. Integral construction—no liner used.

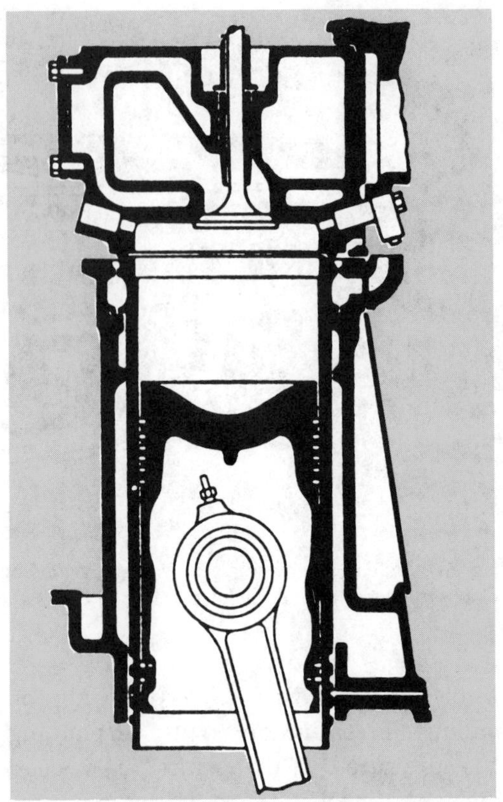

Fig. 5-9. Wet liner. Liner is inserted into cylinder casting to form water jacket.

connecting rod acting at an angle. Resistance to wear and adequate cooling are therefore prime requirements.

Figs. 5-8 to 5-10 show various cylinder and liner arrangements.

Although the water space for cooling the cylinder is sometimes cast in one piece with the cylinder proper, as in Fig. 5-8, the common scheme is to keep them separate, as in Fig. 5-9. The sleeve or *liner* is inserted in the cylinder-block casting, and the space between the liner and the block forms the *water jacket*. The liner usually carries a narrow flange at the top, which rests on a shoulder in the cylinder casting; the cylinder head is

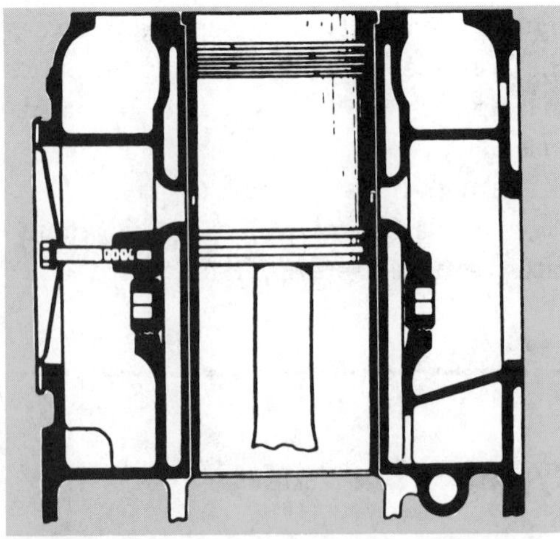

Fig. 5-10. Dry liner. Liner makes metal-to-metal contact with cylinder casting containing water jacket.

bolted down on this flange and thus holds the liner in place. The joint is made gas-tight either by a copper gasket or by an accurate metal-to-metal fit.

Liner Sealing

The liner must be free to expand and contract with temperature changes. For this reason the lower end of the liner usually makes a sliding fit with the cylinder casting; rubber rings or packing are used to make a watertight seal. Rubber rings are made of synthetic rubber (Neoprene) that is not affected by the oil which may reach the rings from the crankcase.

In two-cycle engines, with one or more rows of ports in the cylinder walls, the joints above and below the ports must be sealed to prevent water leakage into the ports. Since temperatures here are too high for rubber, a light press-fit is made, or copper rings are used.

Wet and Dry Liners

The liner shown in Fig. 5-9 is called a *wet liner* because its outside surface is wetted by the cooling water. In contrast, the *dry liner* of Fig. 5-10 does not touch the water; it makes metal-to-metal contact with the cylinder casting, which contains the water jackets.

Dry liners find application on small engines because they permit cylinder bores to be renewed while avoiding the

Fig. 5-11. Wet-type cylinder liner. (Allis-Chalmers, Engine Div.)

complication of the water seal needed with a wet liner. They can be made fairly thin because the cylinder casting itself resists all forces. Close contact between a dry liner and the cylinder casting is absolutely necessary; if it is missing, considerable resistance to the heat flow results.

Liner Materials

With removable liners, the engine designer is free to choose the material he considers best; ordinary cast iron, alloy cast iron, carbon steel, and alloy steel have all been used. Most liners are made of alloy cast iron, containing small amounts of nickel and manganese. Sometimes chromium and molybdenum are added. Liners are usually produced by centrifugal casting (where the mold is spun rapidly while the metal is poured into it). This yields a denser, more uniform product than ordinary casting in a stationary sand mold.

Fig. 5-11 illustrates one of the conventional wet-type liners being used. Three synthetic rubber O-rings fit in the three grooves to allow for expansion of the liner without the leakage of water or oil.

Liner bores are produced by precision machining, grinding, and honing. Various surface treatments are in use; many liners are plated with chromium to reduce the wear of liner and piston rings, the process being controlled to give a finely porous surface which can retain the lubricating oil.

Cylinder Heads

To withstand the firing pressures and the heat of combustion, cylinder heads must be strong and carefully cooled. Water-cooling prevents excessive temperatures which might crack the head and which would interfere with the operation of the fuel injection nozzle and all other valves.

Fig. 5-12 illustrates a fuel injection nozzle holder assembly installed in the cylinder head. The critical parts of the injector are adequately cooled by the large coolant passages in the head. Note the wet-type liner has coolant passages almost to the top of the cylinder.

Because valve openings must pass through the water space, the internal structure of a cylinder head is somewhat complicated; consequently cylinder heads are usually sand cast from alloy iron. Large engines use individual cylinder heads, small ones use groups of heads cast together. The space available for the cylinder head studs which hold the head to the cylinder block is limited, consequently these studs are generally made of alloy and are machined with close-pitch threads (which are stronger than standard threads).

In four-cycle engines the need to provide for inlet and exhaust valves and their port passages, plus an injection noz-

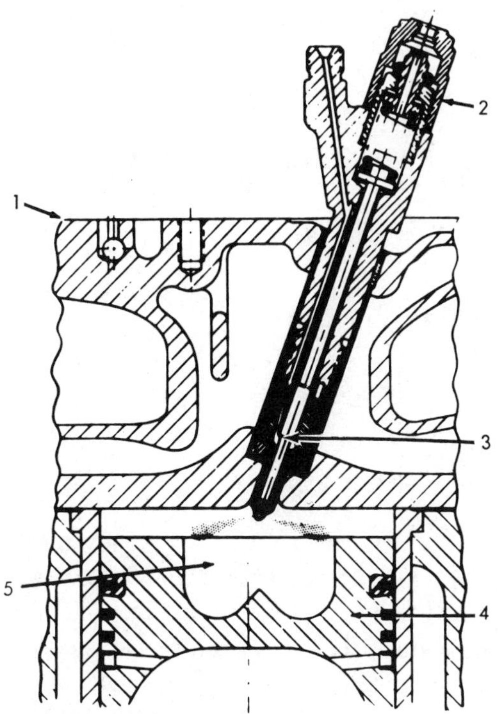

Fig. 5-12. Cutaway view of head and cylinder coolant passages. (Allis-Chalmers, Engine Div.)

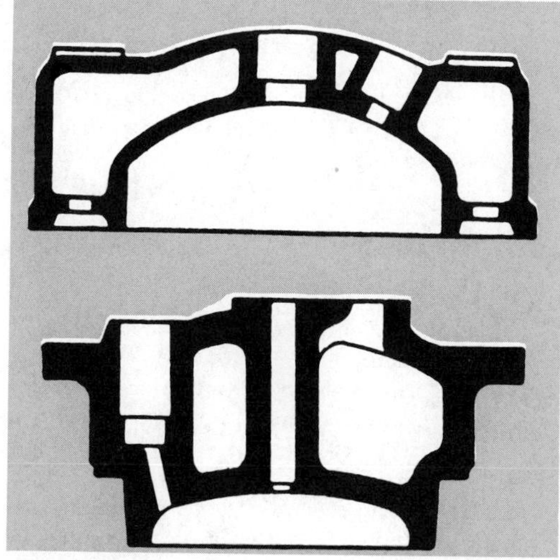

Fig. 5-13. Cylinder heads of two-cycle engines.

zle and sometimes air starting valve, relief valve, and auxiliary combustion chamber, makes the cylinder head an intricate casting. (Note Fig. 5-9.) In contrast, cylinder heads for two-cycle engines are usually simple and nearly symmetrical, as illustrated by Fig. 5-13.

Due to warping, many cylinder heads must be refaced. Although there are variations in the amount of stock that can be removed from the bottom (fire-deck) of the head, the .020″ maximum for this Series 53 Detroit diesel appears to be about average. Note the overall thickness for this particular head in Fig. 5-14 must not be under 4.376″.

It should be noted at this time that all engine manufacturers recommend a certain procedure, sequence, and torque wrench (foot-pound) setting on tightening head bolts. Usually they begin with the center bolts and in a circular-spiral sequence finish the tightening procedure at the extreme outside bolts.

Water Jackets

Nearly always the cooling water enters the engine at the lower end of the cylinder jackets where the cylinder wall temperature is moderate. Thus the coolest water meets the coolest metal, reducing temperature strains. The water then rises through the engine and passes from the upper end of the cylinder jacket to the cylinder head, either through gasketed passages between the cylinder head studs or through an external jumper pipe from block to head.

One advantage of having an external connector is that a water leak is quickly discovered and easily repaired. Water leaving the cylinder heads passes to the exhaust manifold or header if the exhaust

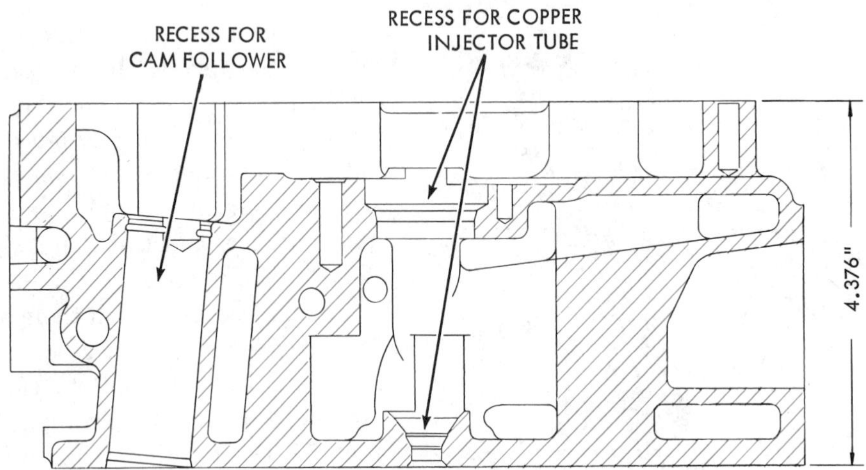

Fig. 5-14. Cylinder head cutaway. (General Motors Corp.)

manifold is water-jacketed, otherwise it leaves the engine through a discharge header.

The area of the cylinder where the cooling is critical is at the top of the cylinder where the combustion takes place. Various means are adopted to maintain positive and rapid circulation there and also in the head. Note, in Fig. 5-9, how the water jacket is carried to the very top of the liner; a bridge in the head jacket space forces incoming water to flow across the cylinder head and around the valve seats.

Checking On Your Knowledge

The following questions give you the opportunity to check up on yourself. If you have read the chapter carefully, you should be able to answer the questions. If you have any difficulty, read the chapter over once more so that you have the information well in mind before you go on with your reading.

DO YOU KNOW

1. What are the major requirements of a diesel engine structure—bedplate, pan, frame, cylinder liners and cylinder head?
2. What are the principle differences in framework between a large low-speed, a smaller medium-speed and a small high-speed diesel engine?
3. How is expansion of cylinder liners controlled in design of diesel engines?
4. How are cylinder liners sealed at the top and bottom?
5. List the advantages and disadvantages of *wet* and *dry* liners.
6. Check the cylinder head tightening procedure in two diesel engine manuals. Develop a table including procedure, sequence and foot-pound setting.
7. Describe the coolant water circulation path through a complete cycle on a diesel engine.
8. Where are the critical areas of an engine in respect to cooling?

Major Moving Parts

Chapter **6**

The purpose of this chapter is to help you understand the fundamentals about the major moving parts of engines: pistons, piston rings, connecting rods, wristpins, crankshafts, bearings, and flywheels. These are the moving parts that receive the gaseous energy produced in the combustion space and deliver it to the output in the form of useful power.

You will see how widely the designs vary and you'll learn the reasons why.

Piston rings will receive some extra attention because the operating man must deal with them continually. The satisfactory performance of the engine depends upon them.

Pistons

The piston, with its rings, seals the cylinder and transmits the gas pressure to the connecting rod. It absorbs heat from the gas, and this heat must be carried away if the metal temperature is to be held within safe limits. Also, the constant reversal of the piston in travel sets up inertial forces, which increase both with the weight of the piston and with its speed. For this reason, designers try to keep piston weight low, particularly in high-speed engines.

In uncooled pistons, the heat absorbed by the piston top travels through the pis-

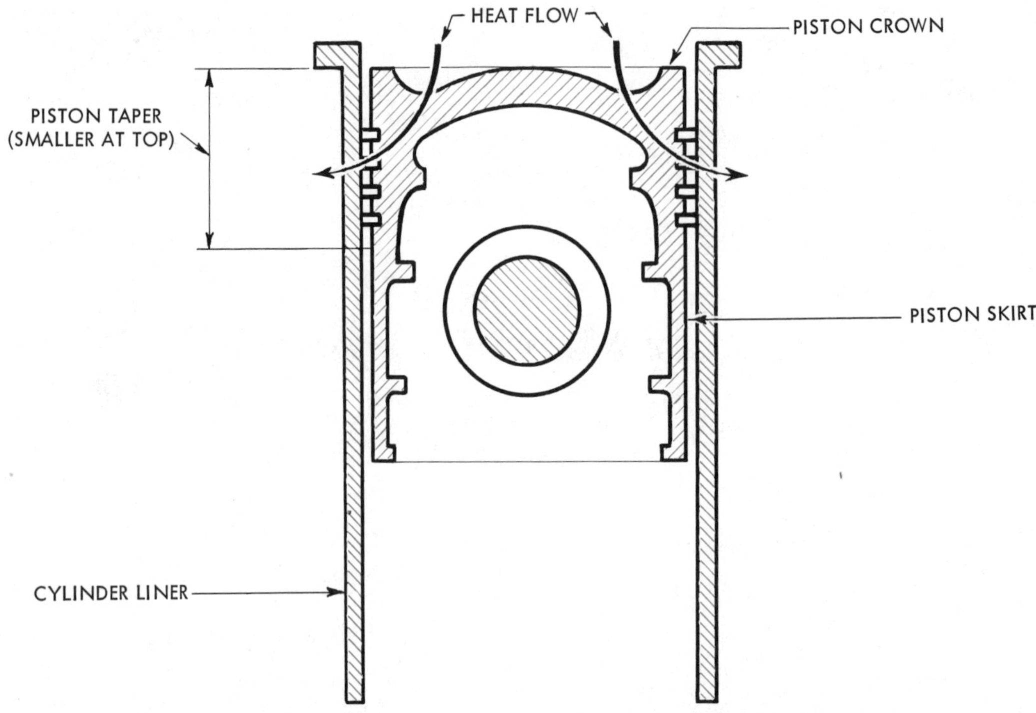

Fig. 6-1. Path of heat flow from piston crown to cooling water.

ton side-walls, then through the rings into the cylinder walls and finally to the jacket water surrounding the cylinder, Fig. 6-1. In highly loaded engines, additional means of heat removal are necessary; this is accomplished by a liquid, generally oil. These are termed *cooled pistons.*

The top section of a piston is called the *crown* and the lower section the *skirt.* The shape of the top of the crown depends largely on the design of the combustion chamber. The crown section is usually thick, in order to withstand the gas pressure without distorting and to provide a good path for the heat flow from the upper surface of the crown to the rings. When running, the piston diameter increases because of the increase in temperature. To take care of the greater expansion in the high-temperature zone, the crown is machined on a slight taper, the diameter being greatest where the crown meets the skirt and becoming less toward the top.

The crown carries the upper set of rings, which are the compression, or power rings. In most uncooled pistons, a division plate above the wristpin prevents lubricating oil from splashing against the hot under-surface of the crown, as in Fig. 6-2. The amount of splashed oil would be insufficient to cool the piston, thus the oil would overheat and carbon deposits would be formed.

The skirt of a trunk piston (Chapter 3 explains trunk pistons) takes the side-thrust of the connecting rod and prevents

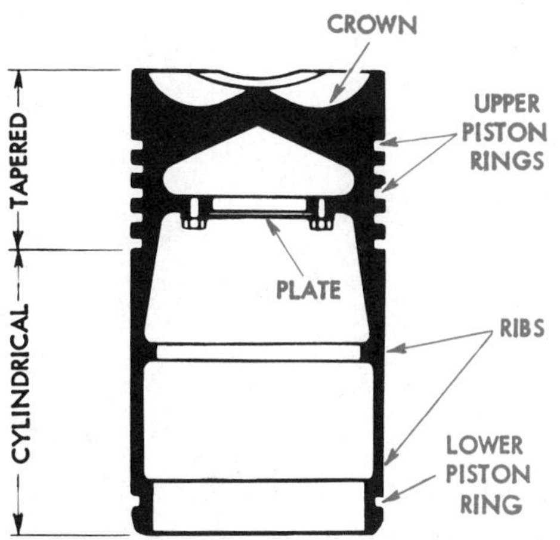

Fig. 6-2. Uncooled cast iron piston. Note steel plate fastened below crown to keep oil mist from striking hot crown. Unless prevented, latter action would crack the oil and cause carbon deposits.

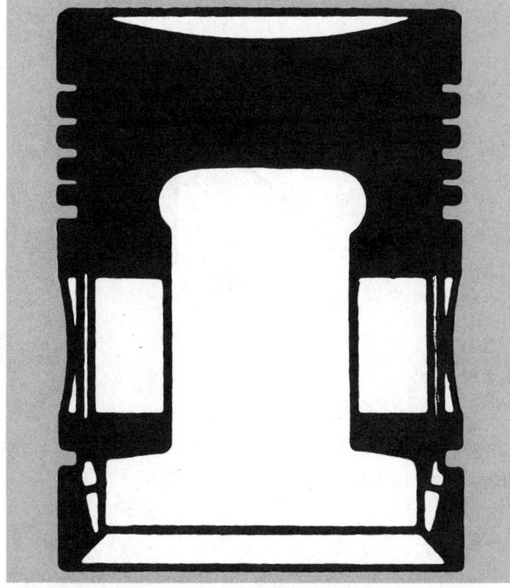

Fig. 6-3. Uncooled aluminum alloy piston. Note thick crown for good heat transfer; also note heavy bosses for stiffness.

the piston from rocking from side to side. It also carries the lower set of rings, which are the oil-control rings. Below the upper ring band, where heavy wall thickness is needed to carry the heat flow, the piston skirt becomes much thinner.

The running clearance between the skirt and the cylinder wall must be small; therefore, it is important that the thin skirt should stay truly round. For this reason, interior ribs are used to stiffen the skirts of cast pistons, as shown in the typical designs of Figs. 6-2 through 6-5; and 6-7 through 6-11.

Piston Materials

The ideal piston material would be light and strong, conduct heat well, expand only slightly when heated, resist wear, and be low in cost. Formerly, all pistons were made of cast iron, but today cast alloys of iron are widely used. These offer greater heat resistance and better wearing qualities. Such materials are not perfect as regards weight and heat-transfer ability, but nevertheless are quite satisfactory in slow-speed engines, particularly where the piston is effectively cooled by a liquid.

For equal strength, an aluminum-alloy piston weighs about half as much as an iron one; therefore such pistons are often used in high-speed engines in order to reduce the inertial forces. Aluminum conducts heat about three times as well as cast iron; consequently aluminum-alloy pistons can be made in large sizes without requiring liquid cooling. See Fig. 6-3.

Aluminum alloys, unfortunately, expand considerably when heated. Consequently, if used in an ordinary iron-cylinder, it is necessary to give the cold piston more clearance in the cylinder than would be used for a cast-iron piston. This makes for a noisy engine when starting or running at light load.

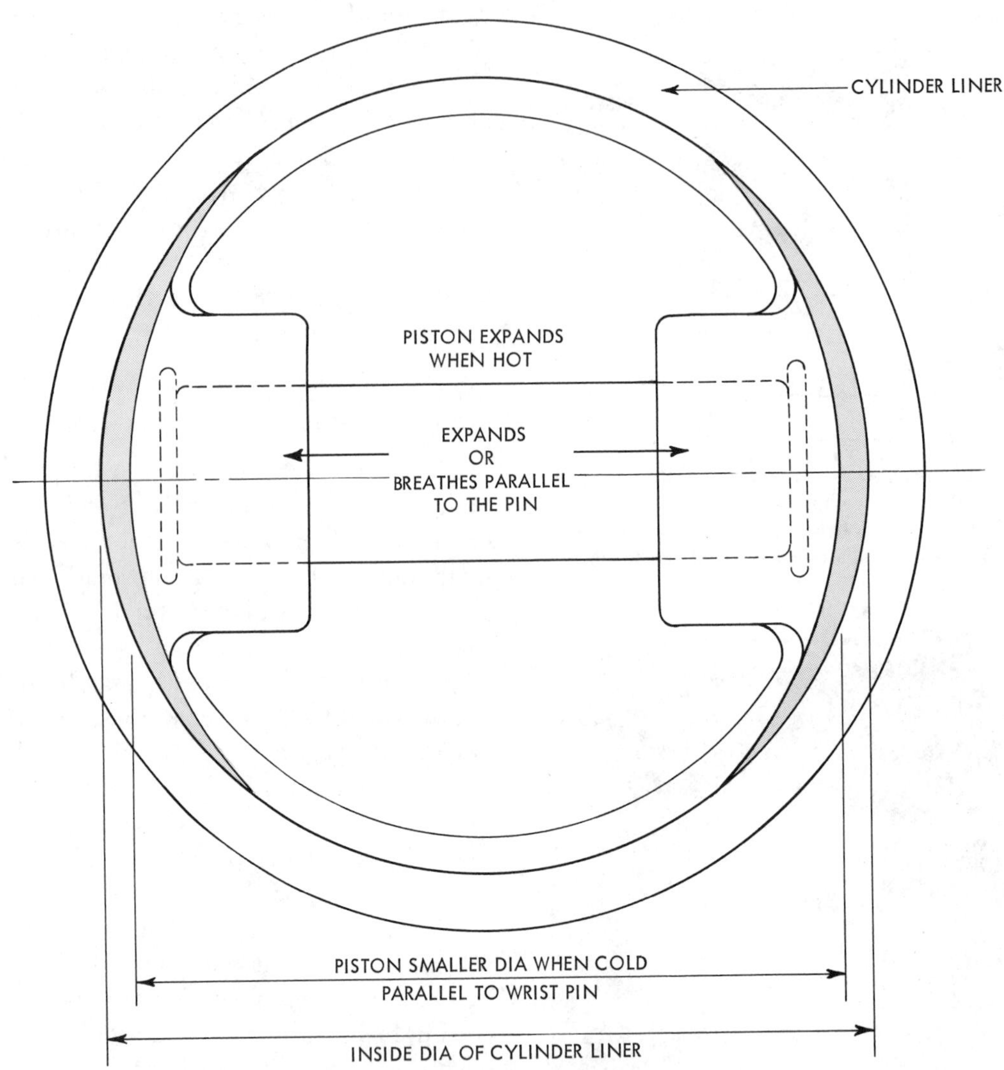

Fig. 6-4. Cam-ground piston—expansion characteristics.

An alternate solution is to install cam-ground pistons. A cam-ground piston has the diameter of the piston skirt (which is parallel with the wristpin) smaller in diameter by several thousandths of an inch than the diameter of the cylinder. The piston skirt, perpendicular to the wristpin, is machined to an easy sliding fit in the cylinder.

As the piston warms up under operation, the expansion will take place at the smallest diameter of the skirt and will become round. This is sometimes referred to as the piston breathing. See Fig. 6-4. Another solution of this difficulty is to use cylinder liners made of a nickel iron which expands at nearly the same rate as the aluminum alloy piston.

Fig. 6-5. A comparison of small engine pistons, gasoline and diesel. (Deere & Company)

The two pistons in Fig. 6-5 are forged aluminum alloy, weight controlled, and with cam-ground skirts. Both pistons are fitted with compression rings in the two top grooves and oil control rings in the lower groove. Where the gasoline piston on the left has only a slight depression cut-away on top, the diesel piston on the right has a deep cut-out swirl cup to swirl compressed air-fuel mixtures for complete combustion.

Composite Pistons

Forged steel crowns and cast-iron skirts have been combined in pistons for large engines. This is in order to obtain the strength and heat resistance of steel in the upper section, and the good wearing properties of cast iron in the lower section, where the piston bears against the cylinder.

Many cast-iron pistons are electroplated with metals like cadmium or tin. The purpose is to permit closer clearances (giving a quieter engine), assure proper run-in (that is, the seating of a new piston), eliminate seizing and provide longer life. In some designs, wear is reduced by using inserts of antifriction metal in the piston walls.

Trunk Pistons vs Crosshead

Most engines use *trunk pistons;* that is, the connecting rod acts directly on the piston. The resulting side-thrust causes the piston to press against the cylinder wall, first on one side, then on the other. At the top of the stroke, when the gas pressure is greatest, the angle made by the connecting rod is so small that the side-thrust is negligible; for this reason most of the wear from the rod thrust occurs near the middle of the stroke. Making the piston skirt relatively long spreads the side-thrust over a greater area, thus reducing the unit pressure.

In most trunk pistons, the unit pressure is so small that neither the piston nor the cylinder liner wears much. Side-wear of the grooves for the upper piston rings ordinarily determines a piston's useful life.

The advantages offered by use of a *crosshead*, Fig. 6-6, to take the side-thrust are: (1) easier lubrication and reduced cylinder and liner wear, (2) uniformly distributed clearance around the piston, and (3) simpler piston construction because the wristpin and its bearing are eliminated. These are offset by greater complication, added height and weight of engine, and need for careful adjustment.

Today, crosshead construction is generally used only when it can't be avoided,

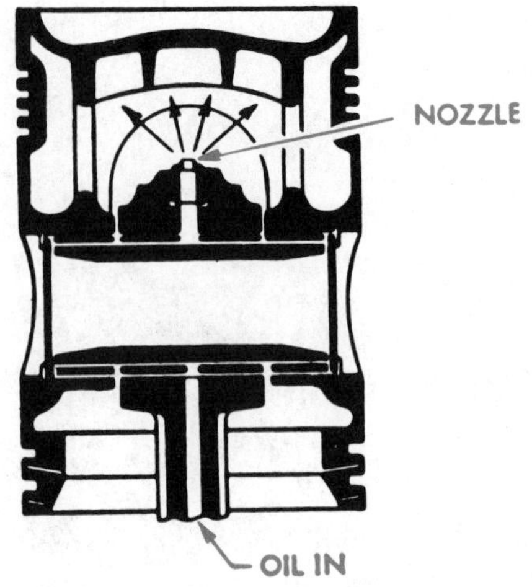

Fig. 6-7. Piston cooled by oil spray from top of connecting rod.

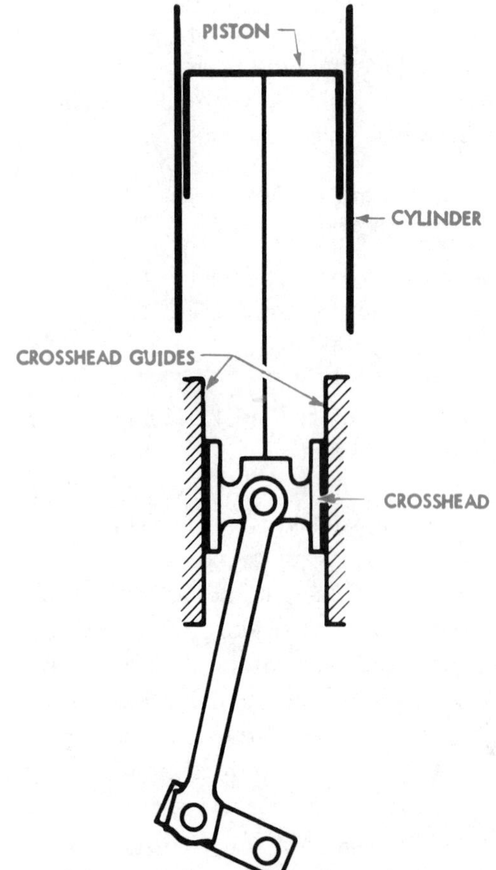

Fig. 6-6. Crosshead principle.

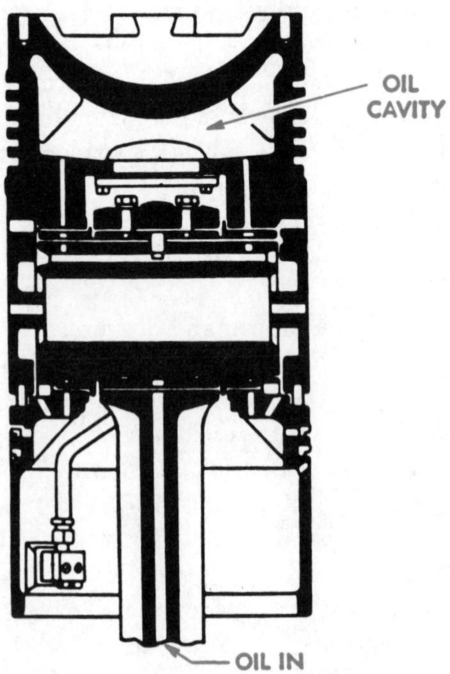

Fig. 6-8. Piston cooled by oil circulation in cavity under crown. Oil is supplied through passage in connecting rod.

such as on double-acting engines, and on certain two-cycle engines where the crankshaft end of the cylinder is enclosed for use as a scavenging pump. It is also used on some very large single-acting engines.

Piston Cooling

Large engines and engines operated at high ratings usually require piston cooling. Although water has been used for this purpose, water leakage into the crankcase gave trouble; consequently, oil is now used almost universally.

In the simplest system for oil-cooling, shown in Fig. 6-7, the lubricating oil forced through the hollow connecting rod to the upper end of the rod serves a double purpose. Part of the oil lubricates the wristpin bearing in the usual manner; the remainder passes through a nozzle in the end of the rod and sprays against the underside of the piston crown. After absorbing heat from the metal, the oil drips down the piston and falls back into the crankcase. Note the ribs in the piston crown which not only strengthen it but also help to transfer the heat to the oil.

In many cooled pistons, the oil circulates through an enclosed space under the crown. In the design of Fig. 6-8, the oil delivered by a pump first rises through

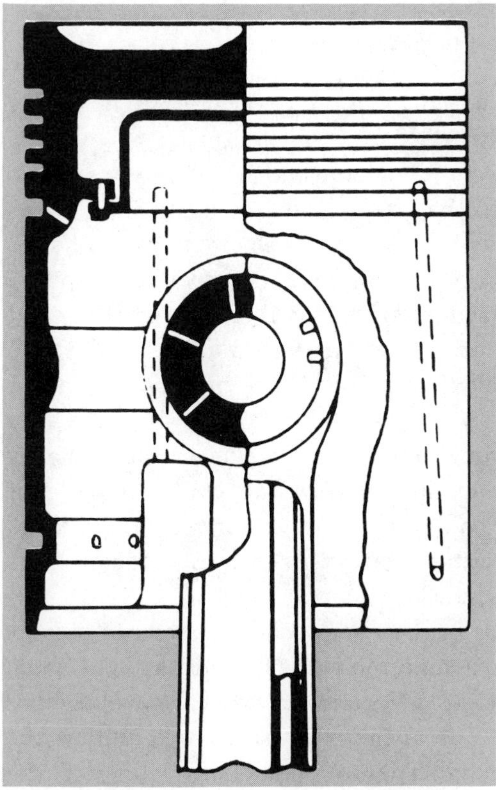

Fig. 6-9. Piston cooled by intermittent circulation. The oil discharge is timed by the piston travel.

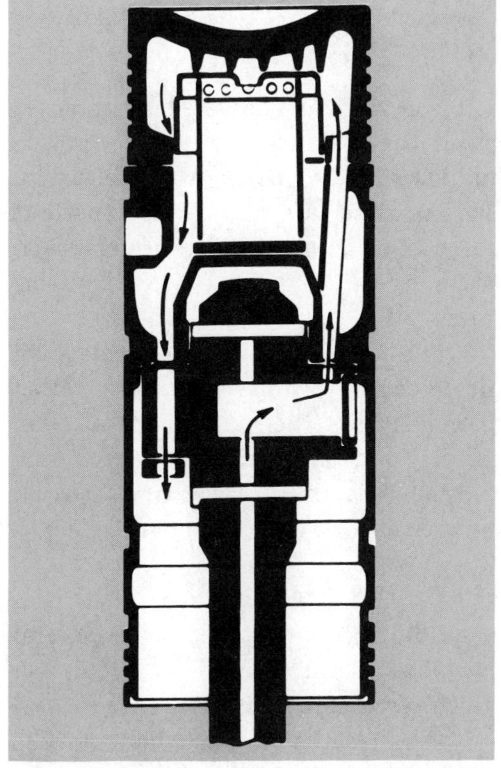

Fig. 6-10. Piston cooled by oil circulating at high speed. Discharged oil is carried off by telescoping pipes.

the connecting rod and lubricates the wristpin bearing; it then flows through drilled passages to the cavity under the piston crown. The warmed oil returns to the sump by gravity.

In the piston of Fig. 6-9 the circulation is intermittent. Oil enters the enclosed space under the crown from the hollow wristpin and is retained there until the piston has descended to the end of its stroke. At this point, a release passage in the piston skirt mates with a slot in the cylinder liner, and the oil then discharges into the crankcase.

Fig. 6-10 shows the construction for a large engine. Oil from the hollow wristpin is forced under pressure at high velocity around the ribs under the piston crown. After doing its work, the oil is carried off by a telescopic pipe below the wristpin and is delivered directly to an oil cooler. Note how the amount of oil carried in the piston is kept to a minimum to reduce the weight (and the inertial forces) of the piston.

Piston Rings

Piston rings are designed to fit in recesses that encircle the upper (crown) and lower (skirt) portions of the piston. Such rings serve a number of important purposes in an engine, and are classified according to the major service they perform.

Fig. 6-11 illustrates a typical small diesel engine piston. The cross-section view identifies the various parts. Note the Keystone type top compression ring and the variations in lower ring selection between six and eight-cylinder engines.

Compression Rings

At the top of the piston are several *compression rings* (usually four to six) which serve two purposes:

1. They seal the space between the piston and the liner, thus preventing the escape, down the liner, of the air charge during the compression stroke or the combustion gases during the power stroke.

2. They transmit heat from the piston to the water-cooled cylinder liner.

Compression rings are made of alloyed gray cast iron. Some types have special facings, such as a thin bearing surface of anti-friction metal or a chemical treatment, to facilitate the run-in or seating of new rings. Such facings cause the tiny rough spots on the iron surfaces of ring and liner to wear off gradually, so that good iron-to-iron contact is achieved without excess friction which might cause scuffing or scoring.

Another way of helping the run-in is to machine the ring face with a slight angle (1 to 2 degrees), so that good contact is made at first on a small area and, as the rough spots are gradually worn off, the entire ring face comes to a good seating.

The top compression rings are some-

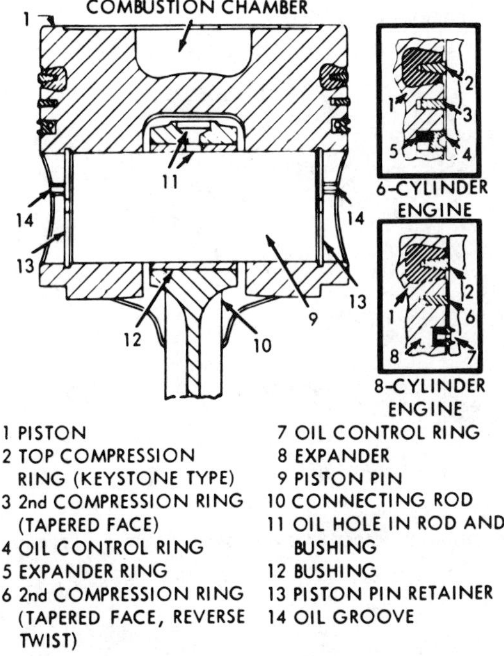

Fig. 6-11. Small diesel engine piston with rings. (General Motors Corp.)

times chromium-plated. During run-in, the hard ring surface smoothes the cylinder walls. Afterwards, the chromium facing reduces the wear not only of the rings but also that of the cylinder.

Plain compression rings are single-piece rings having simple joints as shown in Fig. 6-12, which illustrates the butt or straight-cut joint, the angle joint, and the lap or step joint. Butt or angle joints are the least expensive, the butt joint usually being preferred for narrow rings because of its greater strength. The lap joint is supposed to seal the gap better than a butt or angle joint, but little is gained by its more complicated shape.

The diameter of the ring, when free, is slightly larger than the cylinder bore. Consequently, when the ring is squeezed into the cylinder, it presses against the

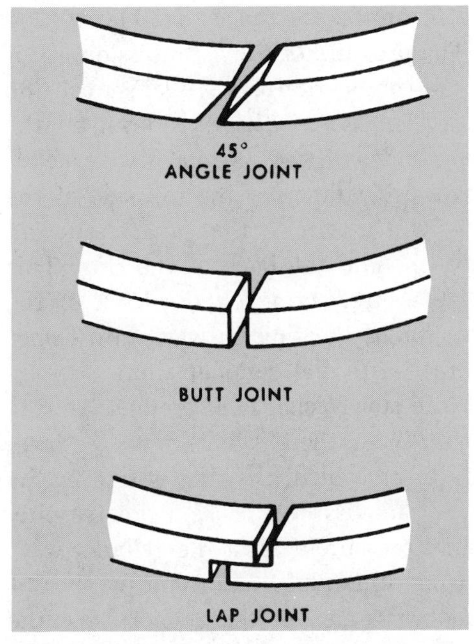

Fig. 6-12. Compression rings with simple joints. (Sealed Power Corp.)

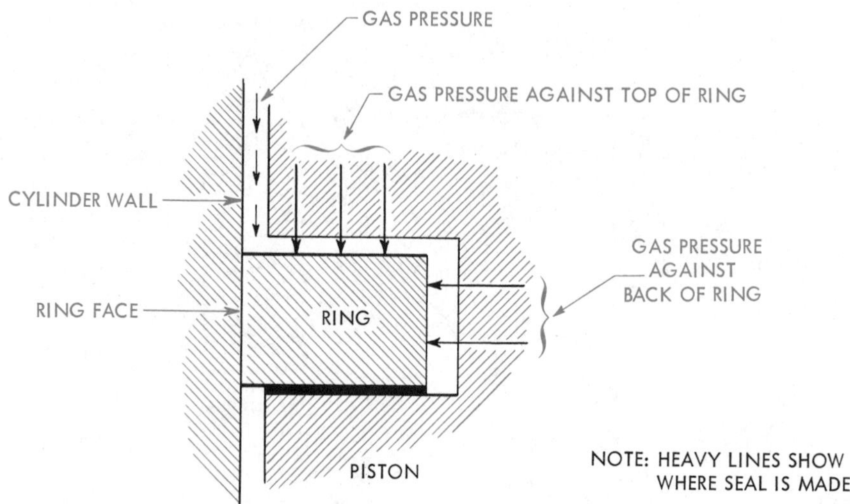

Fig. 6-13. How gas pressure forces a compression ring tighter against the cylinder wall.

cylinder wall and tends to seal it. This initial sealing action is greatly improved by the pressure of the engine gases, as shown in Fig. 6-13 (which exaggerates the clearances for the sake of clarity).

The pressure of the compression air or of the combustion gases against the top of the ring forces the ring down on the lower side of the piston groove. This leaves a clearance at the top side of the ring which permits the gas pressure to travel behind the back of the ring. This gas pressure, acting on the back of the ring, forces the ring outward into firmer contact with the cylinder wall.

Note how efficiently a compression ring works. When there is little or no gas pressure to be sealed, the ring is free in the groove and its own tension creates only a light pressure against the cylinder wall, causing minimum friction and wear. But when the gas pressure increases the ring presses correspondingly tighter both against the cylinder wall and against the piston groove, thus improving the seal and reducing leakage.

The cutaway view in Fig. 6-14 shows a diesel piston and liner assembly featuring two Keystone rings carried in a small cast-iron insert. The Keystone design helps prevent build-up of deposits; assures good ring contact for better oil control and greater reliability. A special process, *plasma coating*, deposits a hard, long-wearing surface on the rings.

Turbo-charged models use the two Keystone rings where the naturally aspirated diesels use a single Keystone ring in a cast-iron insert. (A naturally aspirated engine is one which uses atmospheric pressure only as a means of forcing air into the cylinder.)

Note the cylinder liners have a smooth machined surface at the lower end and the three O-rings are embedded in grooves cut into the engine block. (In Fig. 6-11 the O-ring grooves were cut into the liners.)

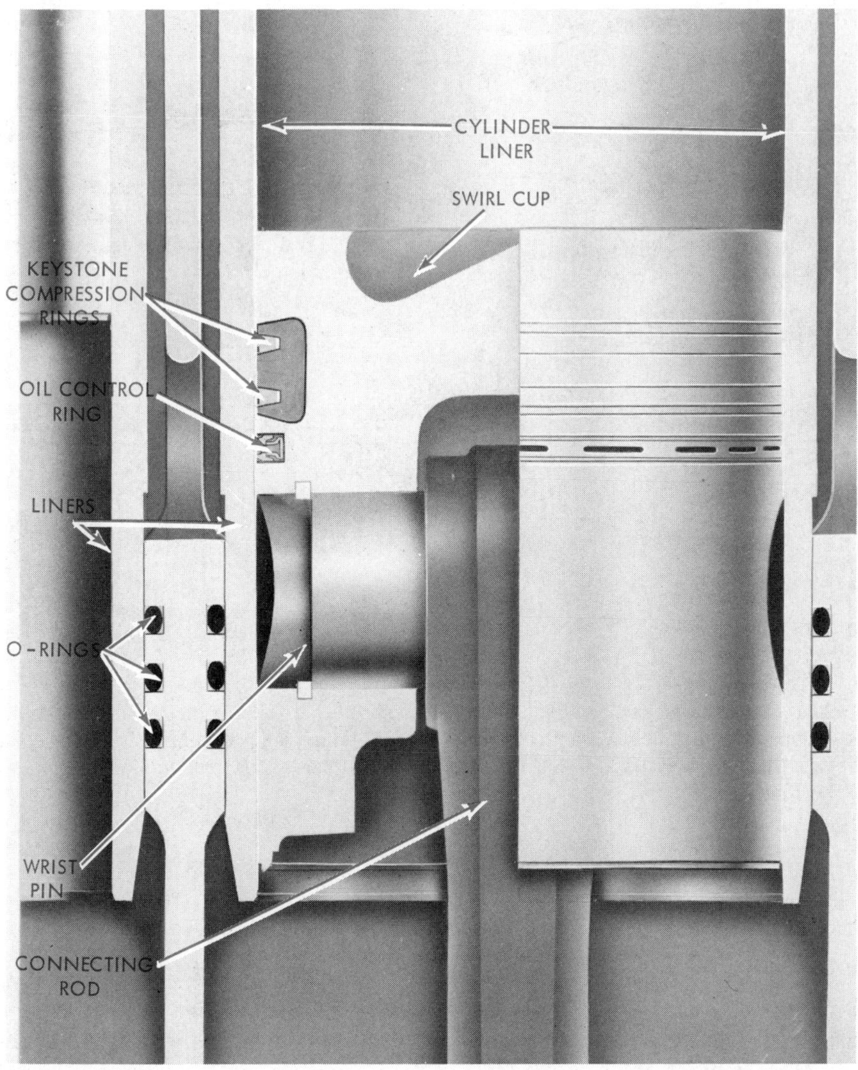

Fig. 6-14. Cutaway view of turbocharged diesel piston and liner assembly. (Deere & Company)

Seal Rings

The gap at the joint of a plain compression ring becomes larger when either the ring face wears down or the cylinder bore wears larger. To reduce the compression loss caused by leakage through the gap in such cases, special *seal rings* may be used. The gaps in such rings remain sealed even when the ring passes over the enlarged diameter of a worn liner.

The principle of a seal ring is shown in Figs. 6-15 and 6-16. Fig. 6-15 shows a one-piece, seal-cut ring in free position. When squeezed into the cylinder liner, the mating halves of the joint B overlap considerably but still leave unavoidable clearance spaces at A and C. Nevertheless, the shapes of the two joint sections

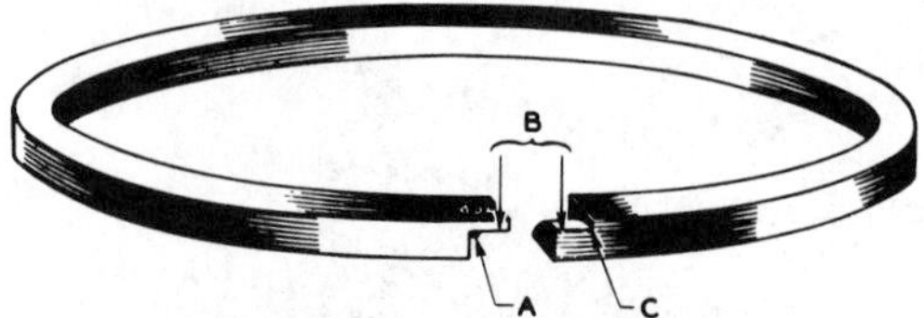

Fig. 6-15. Seal-cut compression ring. It is also called step-seal or joint-seal. (Wilkening Manufacturing Co.)

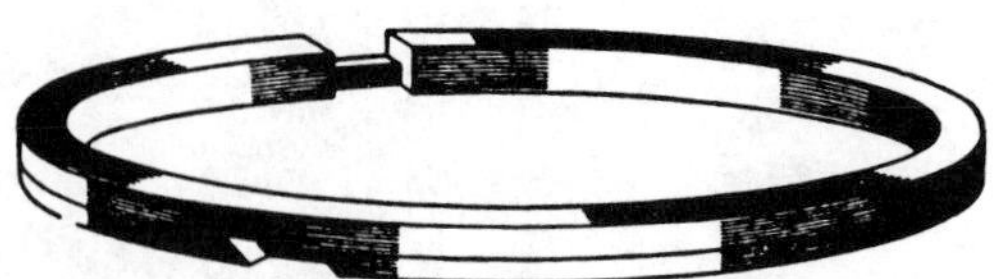

Fig. 6-17. Two-piece sealing ring. (Wilkening Manufacturing Co.)

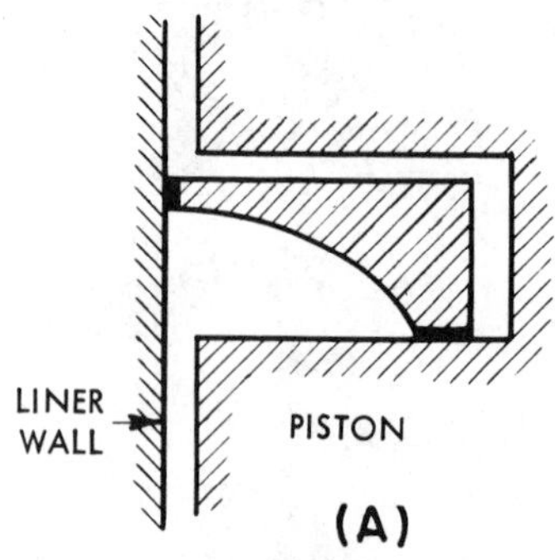

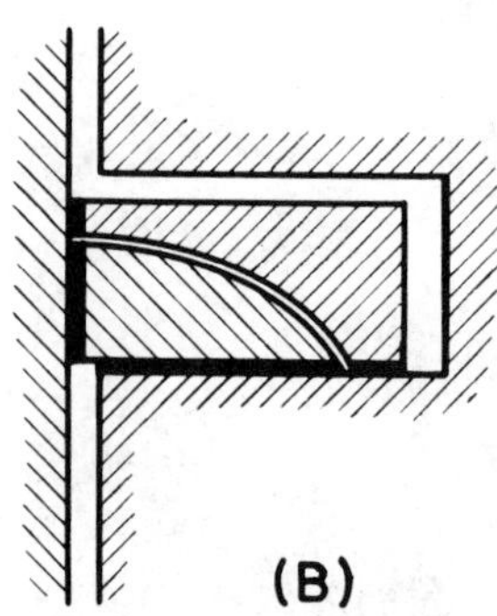

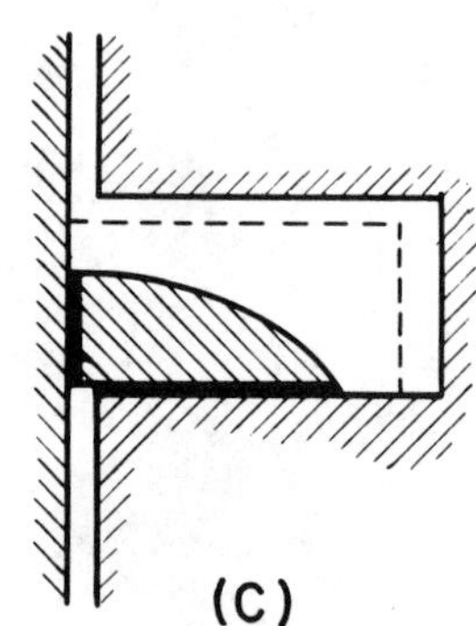

Fig. 6-16. How a seal-cut ring seals the joint. Heavy lines show where seal is made. The recessed section is shown in A alone; this section seals both the groove and the face. In B is shown that, where the sections overlap, the joint acts like the full ring. In C is shown the tongue section alone, which seals both the groove and the face.

are such that each section by itself makes a seal at both the groove and the face, preventing leakage either around the groove or down the cylinder bore, as illustrated in Fig. 6-16, A through C. Seal rings make a seal in only one direction; consequently the top side is marked to insure the ring's being installed top-side up.

Sealing rings are also made in two pieces, as shown in Fig. 6-17. This consists of a master ring of L-shaped cross-section and an outer ring which completely seals the joint opening in the master ring. Two-piece seal rings find their application on larger engines where considerable cylinder wear has taken place.

Oil-Control Rings

The rings classified as *oil-control rings* (one or two, occasionally three) are located on the piston skirt below the compression rings. They are designed to (1) scrape off on the downstroke most of the lubricating oil splashed into the cylinder by the crankshaft and connecting rod, and (2) ride over the remaining oil film on the way up. Their purpose is to prevent surplus oil from being carried up into the combustion chamber where it would burn incompletely and form carbon. But it must allow sufficient oil to be carried to the upper part of the liner during the upstroke to lubricate the piston surface and the compression rings.

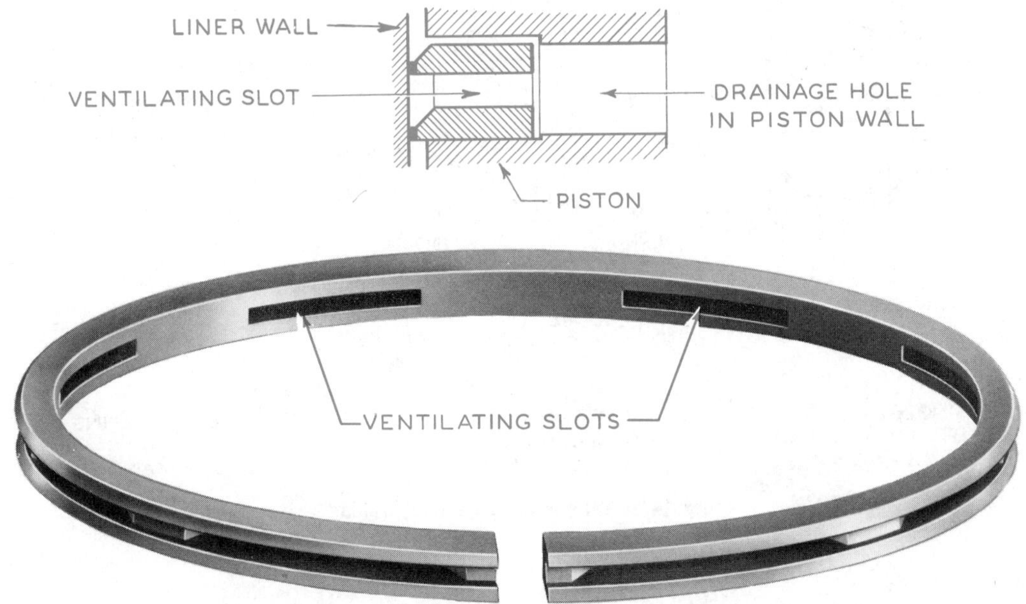

Fig. 6-18. Double-land beveled and ventilated oil ring. (Sealed Power Corp.)

Oil-control rings, usually made of cast iron, have a narrow face so as to obtain a higher wall pressure per unit of surface, and are often beveled to give a downward-scraping edge. The ring has one or two narrow scraping edges.

The oil scraped by the ring must be drained off immediately, otherwise it will build up a pressure which will force the ring back into its groove and stop the scraping action. A common design, known as a *double land beveled and ventilated oil ring,* is shown in Fig. 6-18. This is called a ventilated ring because of the ventilating slots between the two lands or faces, which drain the scraped oil to the bottom of the piston groove. Numerous holes drilled from the bottom of the piston groove to the inside of the piston permit the drained oil to escape into the piston, where it falls back into the crankcase.

Cross-sections of other designs of oil-control rings are shown in Fig. 6-19. It is important in all cases that the drainage from the piston grooves be complete. The illustration shows several drainage methods. One method is to machine an oil-collecting groove on the piston just below the ring and to drill a series of holes through the piston wall from this oil-collecting groove.

Another method may be used if the oil-control ring is at the bottom of the piston. Here the piston skirt is machined to a slightly smaller diameter all the way from the oil ring to the bottom of the piston, so that the scraped oil is discharged directly into the crankcase.

A third method provides double drain-

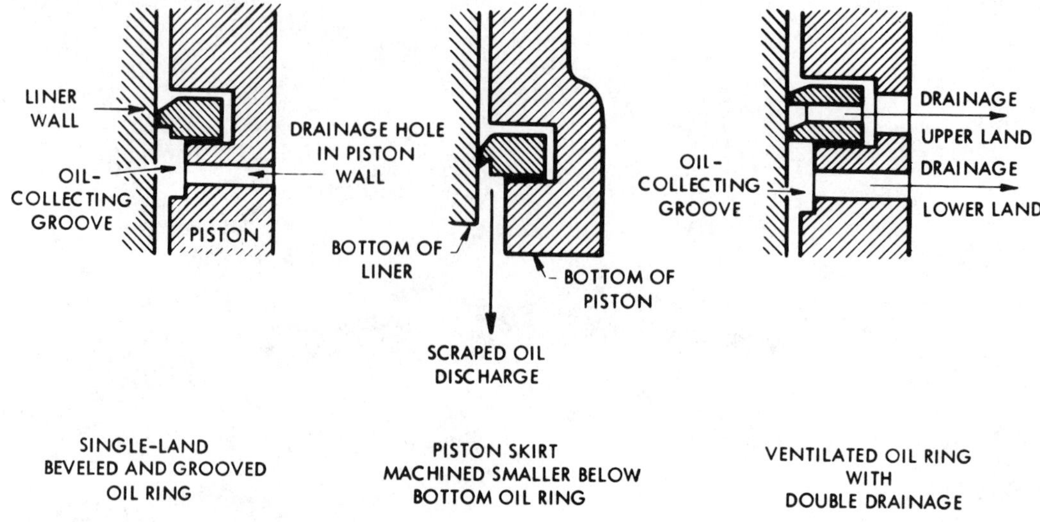

Fig. 6-19. Oil control rings and drainage.

age for ventilated rings. The holes from the bottom of the piston groove drain the upper scraping edge; the oil-collecting groove located on the piston below the oil ring drains the lower scraping edge through a second set of holes.

Since oil-control rings work in a region of little or no gas pressure, the pressure against the cylinder wall depends upon the tension of the ring itself. In order to increase the wall pressure and improve the scraping action, spring-steel expanders are sometimes used behind the rings to push them out.

Fig. 6-20. Conformable oil control ring. (Koppers Company, Inc.)

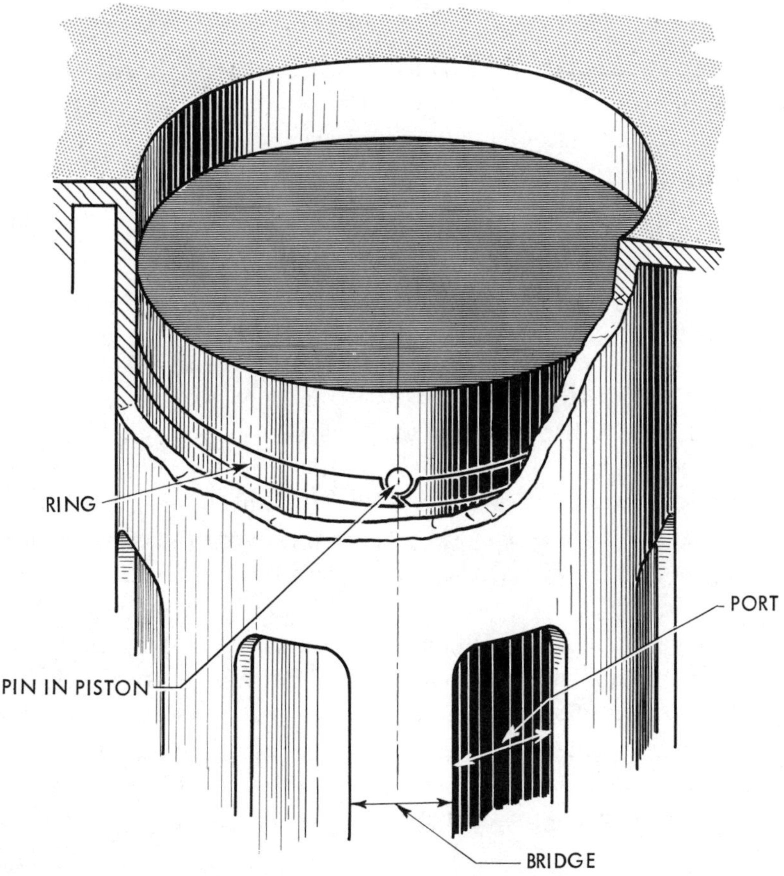

Fig. 6-21. Pinned ring for two-cycle engine ring joint slides over unbroken surface between ports.

Conformable Rings

When the operating conditions are unusually severe or when the pistons are out of round, *conformable* oil rings with expander springs may be used. Fig. 6-20 shows a conformable oil-control ring. It differs from an ordinary oil ring in that it has a thinner cross-section and an internal groove which holds a continuous coil spring expander. When in use, the squeezing of the spring exerts a high outward pressure against the ring, forcing it to conform to the cylinder wall even if that surface is out of shape.

Rings for Two-Cycle Engines

Piston rings for two-cycle engines are likely to break if the ring joints slide over the ports, because the free ends open up and catch in the slots. To prevent this, the rings are pinned and the ring is notched at the top to match with a pin inserted in the piston. Fig. 6-21 shows one of many designs. The ring ends can be kept in line by pinning the ring end to match up with a bridge or unbroken cylinder surface between the edges of the ports.

Connecting Rods

The connecting rod is a bar or strut with a bearing at each end. It's purpose is to transmit the piston thrust to the crankshaft. It must be strong, yet must not be too heavy because of adding to inertial forces, especially in high-speed engines. The rod is usually forged from alloy steel, frequently with an I or channel section to give it greatest stiffness for its weight.

Fig. 6-22 shows a diesel connecting rod and piston sawed in half. The drilled passage allows lubricating oil to reach the pin. On turbocharged models, sprays are provided to cool the underside of the piston. Note the high-strength rod bolts are relieved through the cap for easy assembly.

In large engines the crankpin end may consist of a separate bearing box (in two parts) bolted to a foot on the rod, as shown in Fig. 6-23. Shims between the foot and the box permit adjusting the cylinder compression by allowing the piston to move nearer to or farther from the cylinder head at top dead center. This is called a *marine* big end.

The simpler construction of Fig. 6-24, in which the upper half of the crankpin box is part of the rod itself, is preferred on smaller engines.

The need for large crankpin bearings in

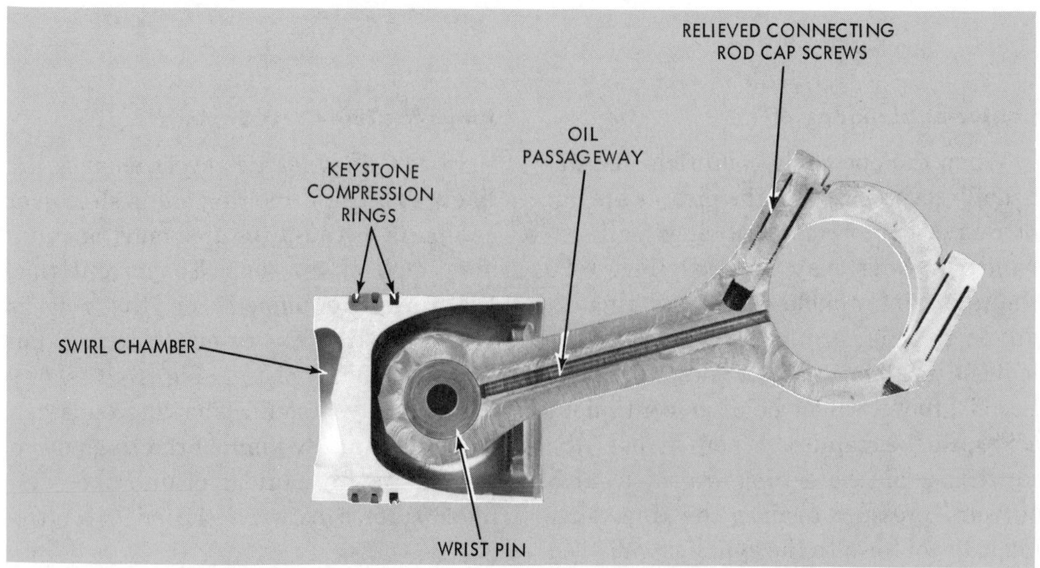

Fig. 6-22. Connecting rod and piston sawed in half. (Deere & Company)

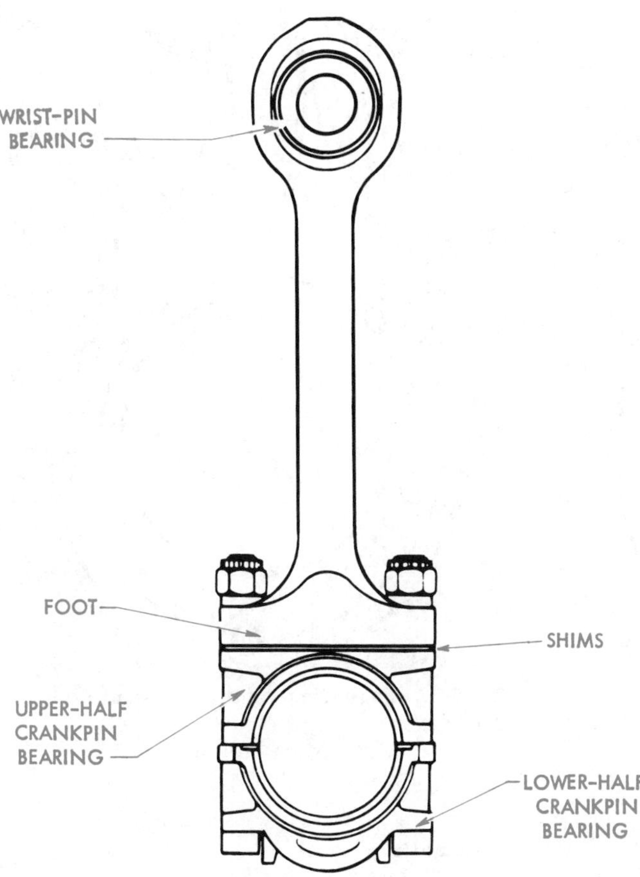

Fig. 6-23. Connecting rod with *marine* big end.

highly loaded engines makes it difficult to keep the crankpin end small enough to withdraw the rod through the cylinder (which is desirable to facilitate maintenance work on the piston and the rod bearings). To make the crankpin end more compact while still retaining the large bearing, the following constructions are used: (1) four rod bolts of smaller diameter instead of two larger bolts, (2) studs (which can be placed closer together) instead of bolts, Fig. 6-25; (3) splitting the crankpin box on an angle, Fig. 6-26; and (4) hinged-strap design, Fig. 6-27.

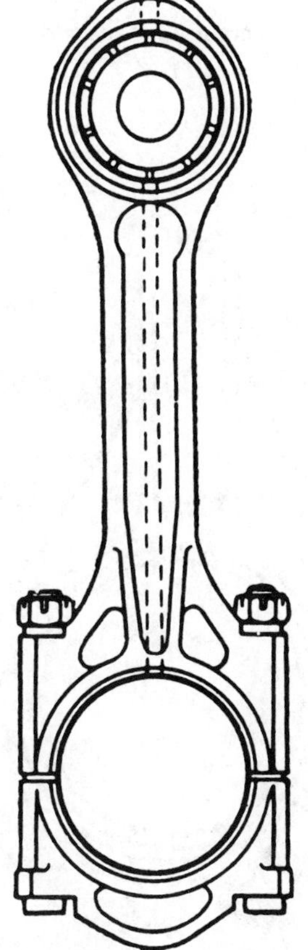

Fig. 6-24. Common design, upper half of crankpin box is part of connecting rod itself.

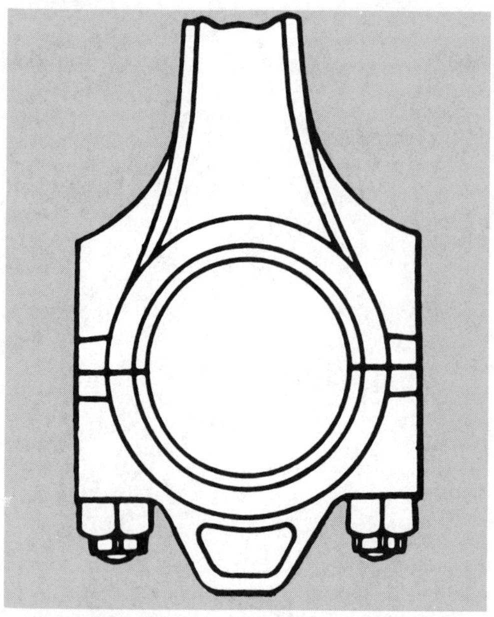

Fig. 6-25. Studs are used in place of bolts for compactness.

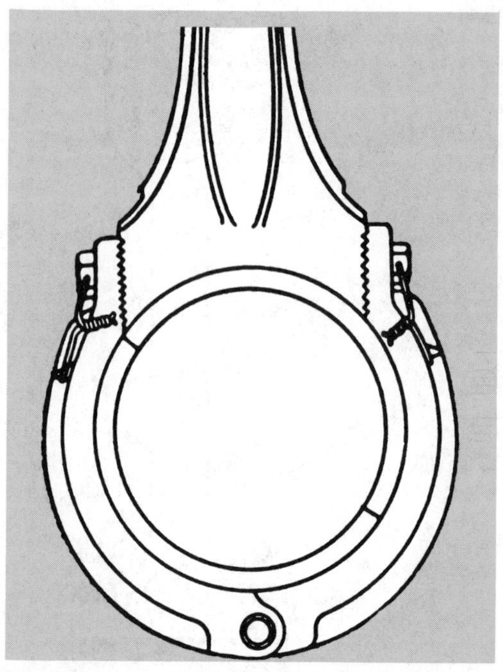

Fig. 6-27. Hinged-strap construction is also compact.

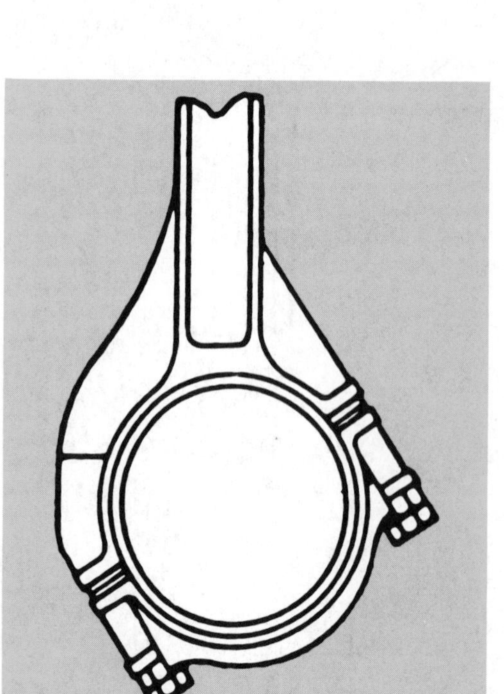

Fig. 6-26. Crankpin box split at angle is small enough for withdrawal through cylinder.

Rod Bolts

Connecting rod bolts, because of the need for great strength, are generally made of heat-treated alloy steel. Fine threads with close pitch are used to give maximum strength and to permit secure tightening. The threads are carefully cut and the entire bolt is smoothly finished in order to avoid any surface defects which might start *fatigue cracks*. Many materials tend to fail by gradual cracking when they are subjected to large intermittent forces for a long time; this is called *fatigue failure*. The fatigue crack always starts at some irregularity in the surface, such as the sharp bottom of a thread, a sharp shoulder, or even a heavy scratch.

The inertial forces which act sidewise

on the connecting rod tend to displace the crankpin bearing. In order to keep the parts in line, the crankpin bolts are snugly fitted into their holes. Also, to assist the bolts in resisting these side forces, a step or tongue-and-groove is often used at the joint between the bearing halves, or, at the joint between the foot of the rod and the upper half-bearing in the *marine* design.

Wristpins

The wristpin is the link between the connecting rod and the piston. Three arrangements, seen in Fig. 6-28 A, B and C. are possible: (1) The wristpin is secured in the piston and the bearing is held in the rod end. (2) The wristpin is fastened to the rod and the bearing is part of the piston. (3) The wristpin is free and bears against bearings in both the piston and the rod. The latter, *full-floating design*, is now most common.

In the full-floating design the piston bosses usually carry bushings for the pin, although in some engines using aluminum pistons, the pin bears directly on the aluminum. Snap rings or end-caps hold the pin from sliding against the cylinder walls. End-caps also keep the wristpin-bearing lubricating oil from leaking into the cylinder.

The fastened-in-the-rod construction may be used in such a way as to secure maximum bearing surface on the *top* of the pin, where it is most needed. Instead of the rod completely encircling the wristpin, the top of the rod is made with a pad or saddle which bolts to the pin, as shown in Fig. 6-28D. Thus, the entire top surface of the pin can be used to bear against a long bushing in the piston.

Fastening the pin in the piston (Fig. 6-28A) was a common design but gave difficulty because of the press fits needed to hold the pin snugly in the piston

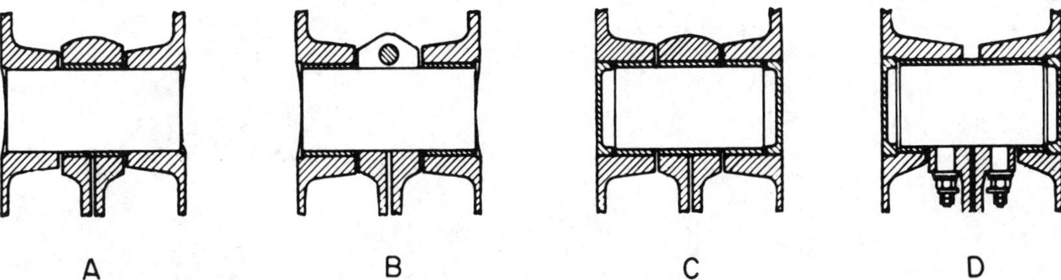

Fig. 6-28. Wristpin arrangements. At A, the pin is tight in piston; bearing is in rod. At B, pin is tight in rod; bearing is in piston. C shows full-floating design; pin bearings are in both piston and rod. D shows pin bolted to pad on top of rod; top of pin has long bearing in piston.

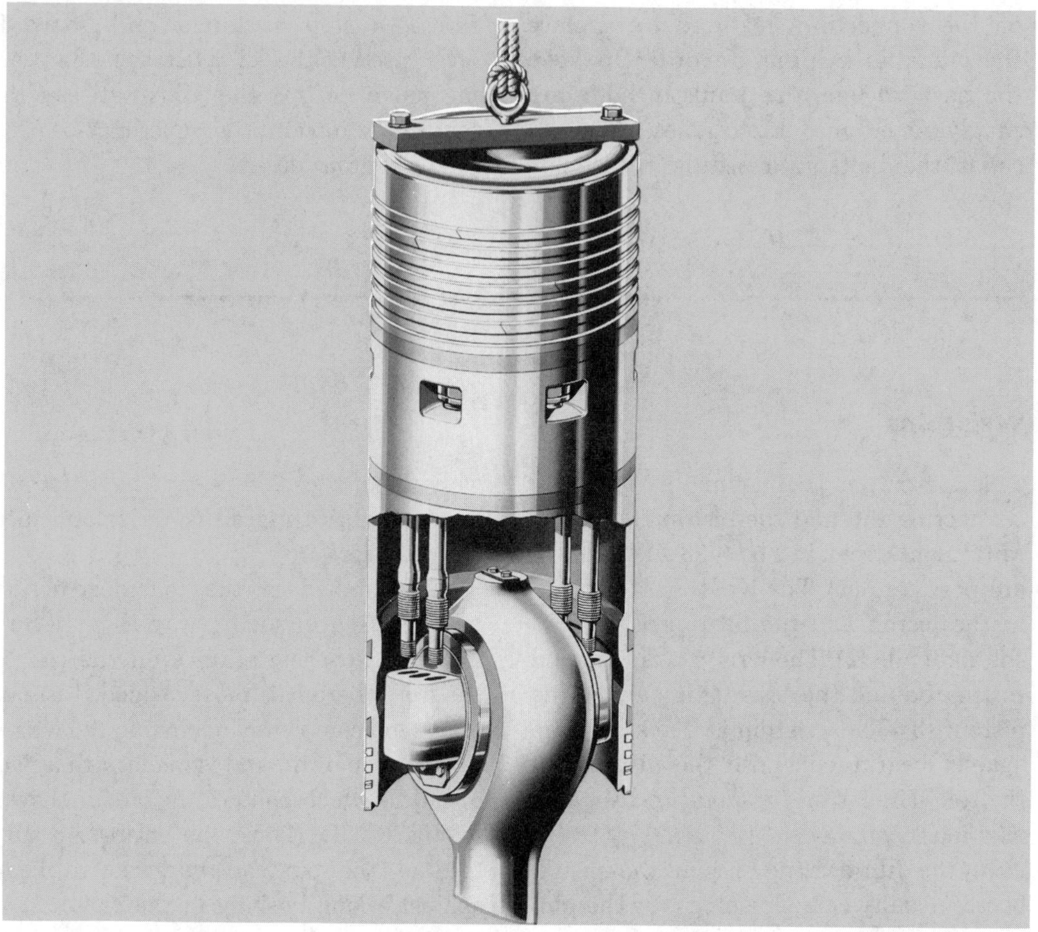

Fig. 6-29. Pin bolted to piston. The wristpin bearing is in the rod. (Nordberg Mfg. Co.)

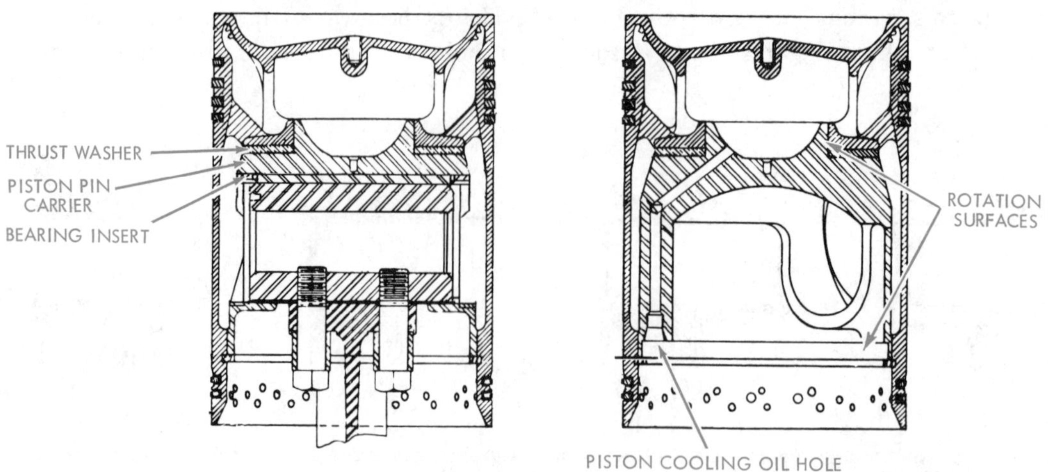

Fig. 6-30. Floating piston. (Electro-Motive Div., General Motors Corp.)

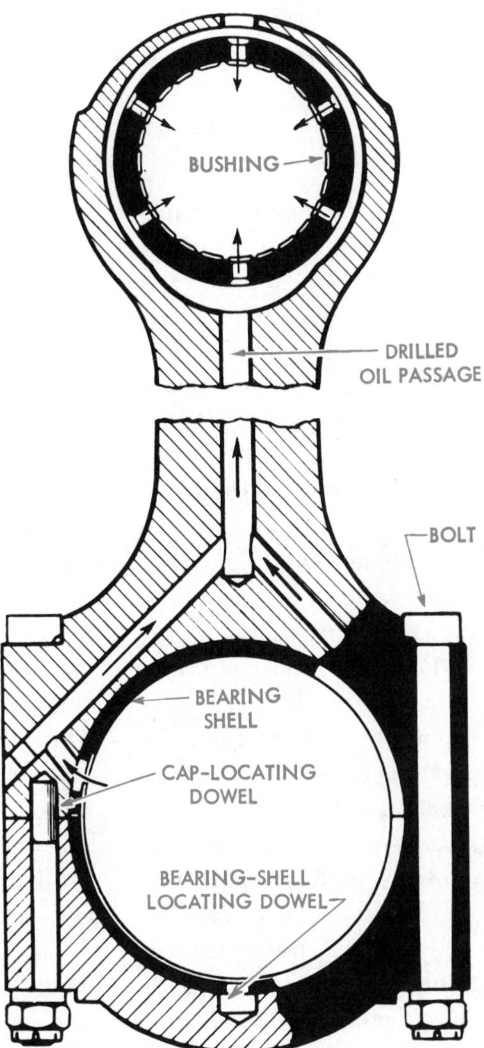

Fig. 6-31. Section of typical rod. Note the lubricating oil path as shown.

bosses. This almost always resulted in distorting the piston, and also made disassembly difficult. However, these drawbacks are avoided in the large piston of Fig. 6-29, where the wristpin is *bolted* to the piston. The General Motors Electro-Motive floating piston design, Fig. 6-30, supports the wristpin in a circular carrier around which the piston itself can rotate. This construction relieves the piston of heat distortions and equalizes the wear.

The wristpin bearing in the connecting rod is usually a solid bushing. It receives lubrication from the crankpin bearing through long holes drilled through the rod, as exemplified by Fig. 6-31. The wristpin itself is usually steel and is made hollow for lightness.

Crankshafts

The principal force an engine crankshaft must resist is the bending action of the connecting rod thrust when the piston is at top center. Then the maximum gas pressure acts straight down on the crankpin and tends to bend the shaft between the adjacent supporting bearings. The crankshaft must also withstand the torsional forces produced by the turning effort of the connecting rod.

Most modern engines have one-piece crankshafts forged from carbon steel or from alloy steel containing nickel, chromium, and molybdenum. In large, in-line engines of many cylinders, the crankshaft may be two forgings, flanged at the ends and bolted together. In extremely large engines, crankshafts too large to be made as single forgings are built up of individual webs *(throws)* and journals. The web forgings are bored slightly smaller than the journal fits and the parts are then shrunk firmly together. A few modern engines use cast crankshafts, made of special cast-iron containing chromium, nickel, and molybdenum.

It is customary to heat-treat forged shafts to relieve the internal strains caused by the forging process and to improve the strength of the material. Induction hardening, an electrical process which rapidly heats metal parts on the outside, is often applied to give the journals a hard, long-wearing surface but not affecting the tough inner core. In addition, the journal surfaces may be superfinished to make them extremely smooth.

The crankshaft in Fig. 6-32 is a six-cylinder in-line with seven-main bearings. This allows a main bearing on each side of every piston and connecting rod assembly for maximum support of the pressure created by the burning of fuel and the inertial forces of fast-moving piston and rod assembly.

Care in storing a crankshaft of this length is necessary or it will warp and not turn easily when torqued into its main

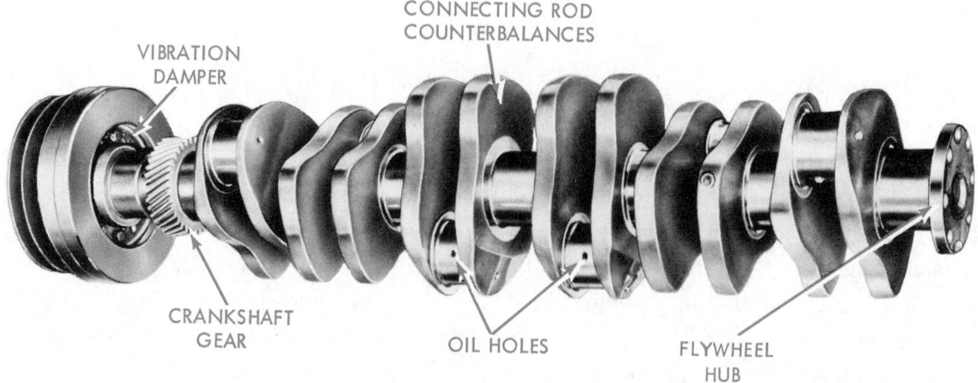

Fig. 6-32. A six-cylinder, seven-main-bearing crankshaft. (Allis-Chalmers, Engine Div.)

bearing saddles in the engine block. Crankshafts in storage should either be placed on end or hung from one end; never laid horizontally.

Counterweighting

Crankshafts are sometimes counterweighted by adding weights to the crank webs opposite to the crankpins. This relieves the loads on the main bearings by offsetting the inertial forces. In the case of one, two, and three-cylinder engines, it improves the balance of the engine itself. In crankcase-scavenging engines, counterweights serve a further purpose by occupying idle space in the crankcase and thus improving the pumping action of the bottom side of the descending piston.

Balancer Shafts and Vibration Damper

Fig. 6-33 shows one of a number of kinds of balancer shafts. Since a four-cylinder engine is inherently an unbalanced unit in operation, the two balancer shafts are mounted in the lower half of the block. The two shafts are driven at twice engine speed in opposite directions to smooth out the running of this four-cylinder engine.

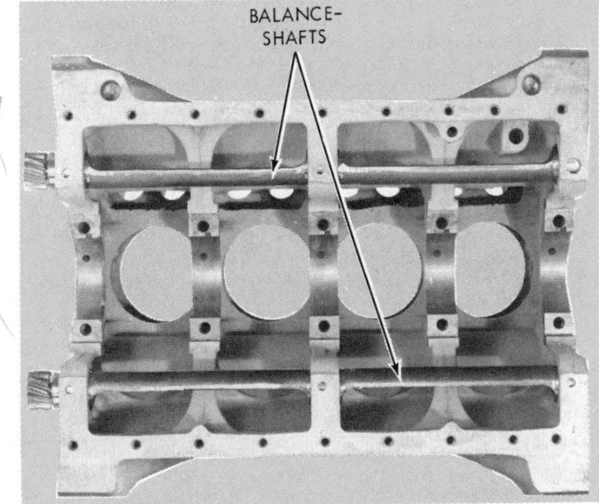

Fig. 6-33. Balancer shafts in a four-cylinder engine. (Deere & Company)

Engines which operate in the higher speed ranges are usually equipped with a *vibration damper* connected to the front end of the crankshaft. The damper is kind of a flywheel and operates to reduce the torsional stresses on the crankshaft caused by the power strokes and the loads on the engine.

There are basically two types of damper assemblies used, one an elastic type and the other a fluid or viscous type. The elastic type is usually incorporated with the fan pulley (Fig. 6-32) and consists of a rubber ring bonded to a heavy metal ring on one side and a stamped metal disc on the opposite side. The rubber allows some flexing between the heavy metal ring and the crankshaft.

The fluid type, Fig. 6-34, consists of a heavy metal disc suspended in fluid inside a sealed drum. Any movement of the inertial mass is resisted by the friction of the fluid, which tends to dampen excessive torsional vibrations in the crankshaft.

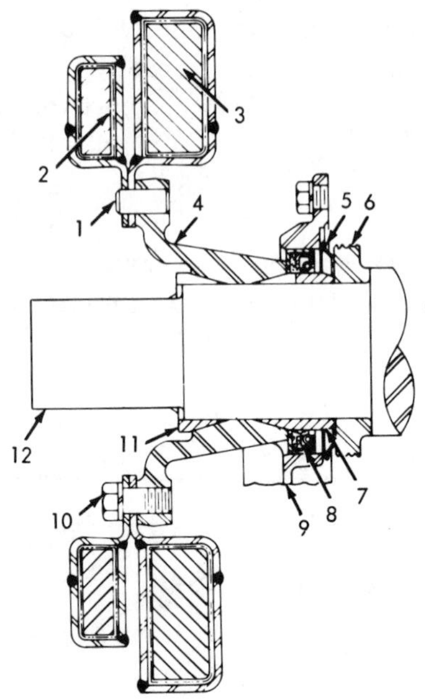

1 DOWEL
2 VIBRATION DAMPER-LIGHT
3 VIBRATION DAMPER-HEAVY
4 VIBRATION DAMPER-HUB
5 OIL SLINGER
6 OIL PUMP DRIVE GEAR
7 TAPERED CONE-INNER
8 OIL SEAL
9 CRANKSHAFT FRONT COVER
10 CAP SCREW
11 TAPERED CONE-OUTER
12 CRANKSHAFT

Fig. 6-34. Sectional view of a fluid-type vibration damper (General Motors Corp.)

Bearings

The object of bearings is to support rotating shafts and other moving parts, and to transmit loads from one engine part to another. To accomplish this purpose in a practical manner, bearings must (1) reduce the friction between the moving surfaces by separating them with a film of lubricant, and (2) carry away the heat produced by unavoidable friction.

Diesel and gas engines use sleeve bearings almost exclusively. In a sleeve bearing the shaft, or journal, is surrounded by a continuous surface of antifriction metal. This material may be a Babbit metal composed mostly of tin or lead (occasionally silver) or may be an aluminum alloy.

Main and crankpin bearings are split into halves, into which bearing shells are fitted. Grooves are generally used to distribute the lubricating oil but must be carefully located to avoid reducing the oil film thickness in high-pressure areas.

Precision bearings, now widely used, are shells which have been so accurately machined that they are inter-changeable without hand fitting. One construction is a thin back or shell, generally steel, and an extremely thin antifriction lining. Another design is *tri-metal,* with a steel back, a copper-lead intermediate layer, and a lead-tin overlay only one thousandth of an inch thick.

Flywheels

The *flywheel* is a wheel or disc of substantial mass which is firmly attached to the crankshaft. Because of its rotation the flywheel acquires kinetic energy (which is covered in more detail in Chapter 8). When the flywheel speeds up, it stores additional kinetic energy, and when it slows down it gives back that energy. The amount of energy which a flywheel will store for a given change in speed depends on its *inertia,* which in turn depends on its mass and its effective diameter.

The energy which the engine pistons deliver to the crankshaft fluctuates, being greatest when a piston has started on its power stroke, much less on the exhaust and suction strokes, and negative (losing energy) during the compression stroke. These fluctuations in energy *to* and *from* the crankshaft cause corresponding fluctuations in its speed. The effect of the flywheel is to reduce the speed fluctuations by storing energy when the crankshaft accelerates and giving it back when the shaft starts to slow down. The heavier the flywheel or the larger its diameter the smaller will be the speed changes.

For an engine of a given horsepower, the energy variations during a complete cycle are greatest if the engine has only one cylinder. Single-cylinder engines therefore require large flywheels to keep the momentary speed variations within reasonable limits.

In multicylinder engines the energy changes become less as the number of cylinders increases. The reason is that, not only are the cylinders smaller, but also that their impulses are more frequent. In the case of engines with many cylinders, one piston delivers power at the same time as another is on compression, consequently a small flywheel is sufficient. The cranks, crankpins, and large ends of the connecting rods have considerable rotating weight and exert the same inertial effect as a flywheel. So does the rotor of a connected electric generator. Therefore in some large multicylinder engines flywheels are not necessary and are not used.

Checking On Your Knowledge

The following questions give you the opportunity to check up on yourself. If you have read the chapter carefully, you should be able to answer the questions. If you have any difficulty, read the chapter over once more so that you have the information well in mind before you go on with your reading.

DO YOU KNOW

1. How are pistons cooled?
2. Trace the heat flow from the piston to the cooling water.
3. Where is the major thrust side of a piston?
4. What is meant by the statement that a piston breathes parallel to the wrist pin?
5. Explain the principles and advantages of Keystone rings.
6. What is the job of a compression ring and explain how it does it.
7. How do the oil-control rings perform ?
8. What are the differences in ring structure between two- and four-cycle engines?
9. Identify four different wristpin arrangements and identify advantages of each.
10. Explain the origin of the forces which a crankshaft must withstand.
11. What is the purpose of counterweights on crankshafts?
12. Explain how an elastic and a fluid type vibration damper operates.

APPROXIMATE DIMENSIONS:

LENGTH	56 IN (1422 mm)	56 IN (1422 mm)
WIDTH	32 IN (813 mm)	32 IN (813 mm)
HEIGHT	52 IN (1321 mm)	52 IN (1321 mm)
NET WEIGHT (MASS) (DRY)	2195 LB (996 kg)	21 LB (996 kg)

Detroit Diesel model 6-71T. Note metric dimensions. (Detroit Diesel Allison Div., General Motors Corp.)

Lubricating the Diesel

Chapter

7

The purpose of this chapter is to present the basic fundamentals of lubrication. For instance, what is friction in the diesel engine? What is the theory of lubrication, and what are the basic requirements of lubricant for diesel engines? We should understand what the operating problems of diesel engines are that affect lubrication.

We will then examine some of the diesel lubricating systems and some of their parts.

Lubricating oil for diesel engines will be investigated. Next we will look at what happens to the oils under operation. Finally, some understanding of how to maintain diesel engine lubrication systems will be presented.

Lubrication Principles

The science of lubrication involves a full knowledge of all fundamentals related to lubrication. It begins with *what is friction* and ends with *how is friction reduced* in a diesel engine?

Friction is defined as that force which acts at the surface of contact between two bodies to cause resistance to their moving upon each other. If the two bodies are solids the force is called solid friction. This may be of the sliding or rolling type. If at least one of the bodies is a fluid, such as a boat in the water, the force is called fluid friction.

Examples of sliding friction are a piston sliding in a cylinder, or a shaft revolving in a bearing. In each case, no lubricant must be present between the sliding surfaces.

If round shafts are replaced for one sliding surface, such as placing two round pipes under a heavy piece of flat plate, the plate could be moved more easily along the floor and we have rolling friction.

Cohesion and adhesion are two natural attractive forces which are directly related to the basic causes of friction between particles of matter in motion. *Cohesion* is the force that causes molecules of the *same substance* to stick together. A hard dense piece of steel where the molecules are close together has very high cohesive forces. Mercury is another example of very strong cohesion.

The second force, *adhesion,* causes molecules of different kinds to stick to each other. As previously stated, mercury has high cohesive forces, it sticks together but mercury will not stick to anything else so it has very low adhesive forces. Water has strong adhesive forces; it will cling to (or wet) most substances, but it is very weak in *cohesion.*

Lubricating oil is strong in both cohesion and adhesion; it adheres to metal bearing surfaces, and also maintains a relatively thick film even under pressure and heat conditions.

Heat and Power Consumption

Two fundamental facts which are important in regard to friction is that it always produces heat and consumes power. The amount varies greatly and, under the same pressures, is largely de-

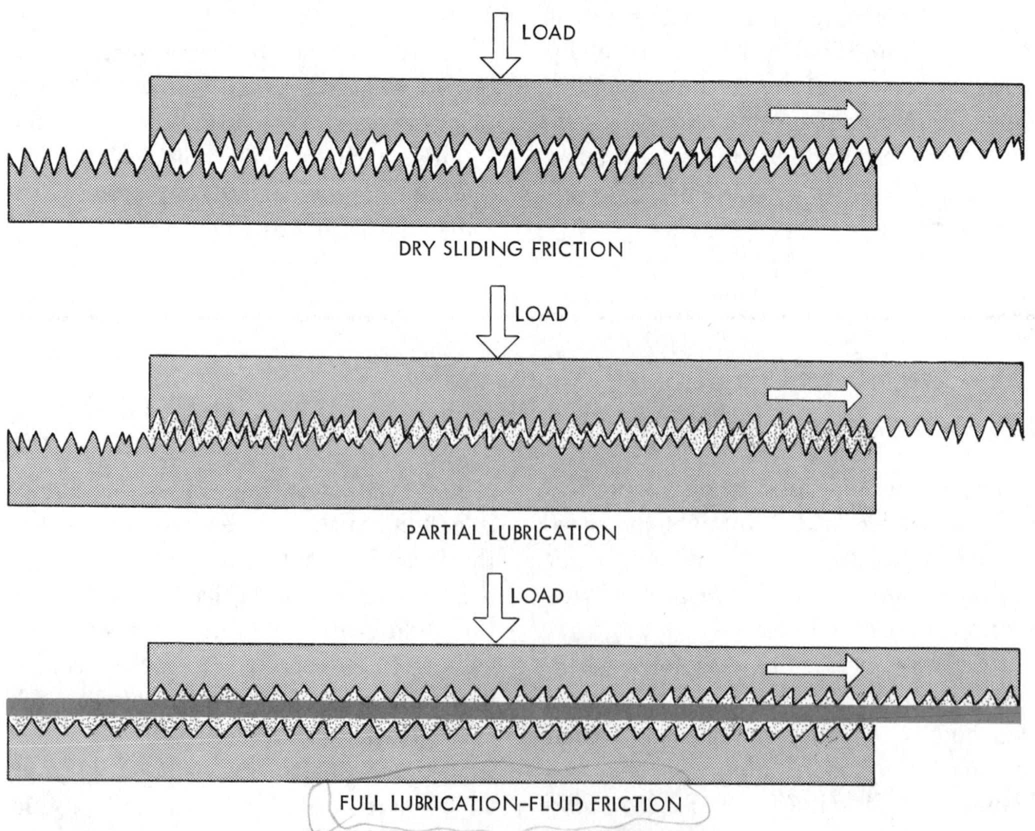

Fig. 7-1. Rough edges cause sliding friction.

pendent upon the type of friction involved. Sliding produces the most heat; rolling friction, considerably less; and fluid friction, the least.

Surface Condition and Load

In the case of sliding friction two factors govern the amount of friction produced; one of these is the degree of roughness of the opposing surfaces and the others is the amount of pressure (load) which holds these surfaces together. Fig. 7-1 illustrates the principles just covered.

A microscopic view of the surfaces of two flat pieces of metal would reveal tooth-like formations on the surface of each plate. These projections would cling together and be broken off if the plate were moved in opposite directions, and friction would be high as shown in Fig. 7-1; added load would cause even greater friction. If a lubricating fluid were introduced between the two surfaces and the projections kept apart by the fluid, they are said to be lubricated. The plates are lubricated. Thus, friction is reduced by substituting fluid friction for dry friction. Just how good this fluid friction is, will depend upon the cohesive and adhesive qualities of the fluid.

Oil-Film and Wedge Principle

When a crankshaft is at rest in a bearing the weight of the crankshaft forces out the oil film at the lower part of the bearings, Fig. 7-2 top. As the crankshaft turns, the oil entering the bearing from the top will be pulled along with the crank until the oil film lifts the crank off from the bottom and forms a film wedge. Position of the film wedge will vary with rpm of the journal, Fig. 7-2 bottom.

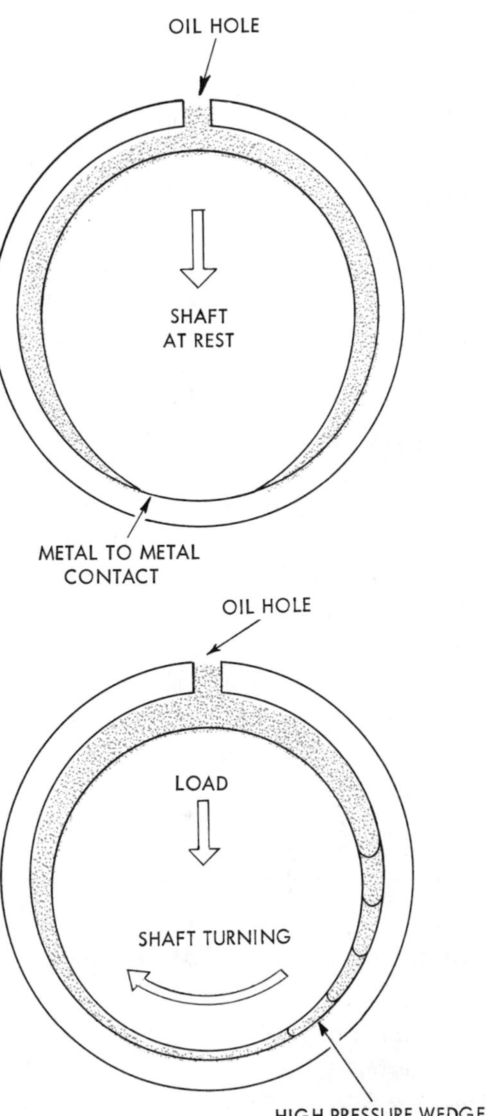

Fig. 7-2. Oil film wedge principle.

It is estimated the pressure builds to several hundred psi in the oil film wedge and without this principle, heavy crankshaft loads would cause considerable boundary-line lubrication.

Governing Factors in Lubrication

The amount of cohesive force between oil molecules determines the viscosity of the oil. In effect, the molecules of the more *viscous* oil hold together more tightly than the less viscous oil. The force of adhesion of the oil molecules to the moving surfaces of a journal in a bearing is not as great as the force of cohesion exerted by the oil molecules on each other. Therefore, adequate lubrication cannot be obtained by using a high viscosity oil. An oil with too low a viscosity, on the other hand, would not have enough cohesion between the molecules and would not build up a sufficient oil film to keep the metal surfaces apart.

Three factors govern the degree of viscosity needed to lubricate a normal bearing under normal operating conditions: (1) the rubbing or sliding speed, (2) clearance between the bearing surfaces, and (3) the bearing load in psi.

The rubbing or sliding speed of a bearing is usually measured in linear feet per minute. For example: a bearing of 5″ dia. would have a circumference of approximately 15.7″. With a journal turning at 400 rpm, the rubbing speed would be 523.5 linear ft/min. As the speed goes up the adhesive qualities of the oil must go up.

The amount of clearance necessary between a journal and the bearing surface is governed by operating speed. High-speed journals require closer bearing clearances and closer clearances require lower viscosity. Lower journal speeds allow large clearances and require higher viscosity oils.

Bearing loads are calculated in pressure per square inch of the bearing area. In general the higher the load the higher the viscosity must be. Piston pin bushings and connecting rod bearings carry high loads in diesel engines.

Basic Requirements of a Lubricant

A lubricating oil for diesel engines must accomplish the following:

1. Provide an oil film between the shafts and bearing surfaces at main, crankpin, and wristpin bearings
2. Perform the same job for the camshaft, valve gear, and various engine auxiliaries
3. Maintain an oil film between the piston and the cylinder wall
4. Aid in the dissipation of heat, and cool the piston and bearings
5. Aid in keeping the inside of the engine clean

Diesel Lubrication

It is fairly easy to find a lubricant that will perform the work outlined at the start of this chapter and to devise a satisfactory system for delivering the lubricant to the desired spots in the right amounts. The difficulty arises from the fact that the high temperatures and pressures in diesel and high-compression gas operation cause all known lubricants to deteriorate in service. The lubrication of small high-speed units is more difficult than that of large slow-speed engines.

The same kind of changes take place in each case, but in large engines they occur at a slower rate and in a much larger body of oil.

Oil Deterioration

Lubricating oils become impaired in service in two ways: (1) changes take place in the oil itself, and (2) outside sources contaminate the oil.

Changes that occur in the oil itself take several forms. High temperatures, particularly in the piston-ring area, cause oils to crack or break down in molecular structure. This produces *carbonaceous* (tarry) *materials* which tend to stick to the rings and cause other troubles.

Oxidation is another form of change. The hydrocarbons in the oil combine with oxygen in the air to produce what are called organic acids. The organic acids with low boiling points (volatile acids) are usually highly corrosive. The acids with high boiling points tend to form gums and lacquers.

These oxidation products are responsible for several of the problems encountered in diesel operation. If allowed to become concentrated, the volatile organic acids attack certain bearing metals, causing pitting and failure. Also they react with the remainer of the oil to form soft masses, or sludges. Sludges give trouble by depositing in the valve chambers, in the sumps, on the filters and strainers, and in the oil cooler.

The heavier oxidation products form hard varnish deposits on pistons, valve stems, and other metal parts. These smooth, black films develop when the oxidation products come in contact with heated metal surfaces. On valve stems, varnish may cause sticking or burning; on pistons it has a tendency to cause the rings to stick.

Outside Contamination

In the previous paragraphs you noted the lubrication problems which result from changes that take place in the oil itself due to its exposure to heat and air. Contamination of the oil from outside sources contributes to these problems and adds new ones. Among these contaminants are products of fuel combustion, such as ash, soot, unburned heavy ends of the fuel, and water. When these mix with the lubricating oil on the cylinder walls and piston, the ash, carbon, and soot help to build up piston-ring deposits. The unburned fuel may oxidize in the high-temperature zone, making a resinous gummy material that forms a binder for the soot or carbon particles.

Some of the unburned fuel may find its way past the rings. When fuel reaches the crankcase, it dilutes the lubricating oil and reduces its viscosity.

For every gallon of fuel burned, about a gallon of water results as a combustion product. This normally passes out of the engine as vapor in the exhaust, but some of it may condense, particularly on the lower and cooler parts of the cylinder liner. This water prevents the oil from properly coating the working surfaces. Also, if hydrogen sulfide or sulfur dioxide are present, because of sulfur in the fuel, they may combine with the water and cause corrosion of the metal surfaces.

Water that reaches the crankcase may have other harmful effects. It may cause emulsions (something like mayonnaise), which gather over the screens and in the circulating system. Water also tends to increase the activity of organic acids and may thus accelerate their corrosive effect on bearings and other engine parts.

The engine itself may form another source of contamination. The tiny metal particles produced by wear tend to oxidize. The resulting metallic oxides, even though present in minute proportions, act as powerful catalysts (chemical activators) and hasten the oil's deterioration.

Foreign matter from outside the engine also gets into the lubricating oil. Dirt and dust may enter the cylinders through the air-intake system, particularly if the filters are in poor condition. They may also enter the crankcase through the breather pipe if it is not protected by a filter.

Since the working oil film on cylinder walls is exceedingly thin, even the smallest dust particles have a deadly grinding action. Such abrasion may also result from small metal particles torn off the engine's wearing surfaces, such as Babbitt particles from the bearings, or bits of iron from the piston-ring edges.

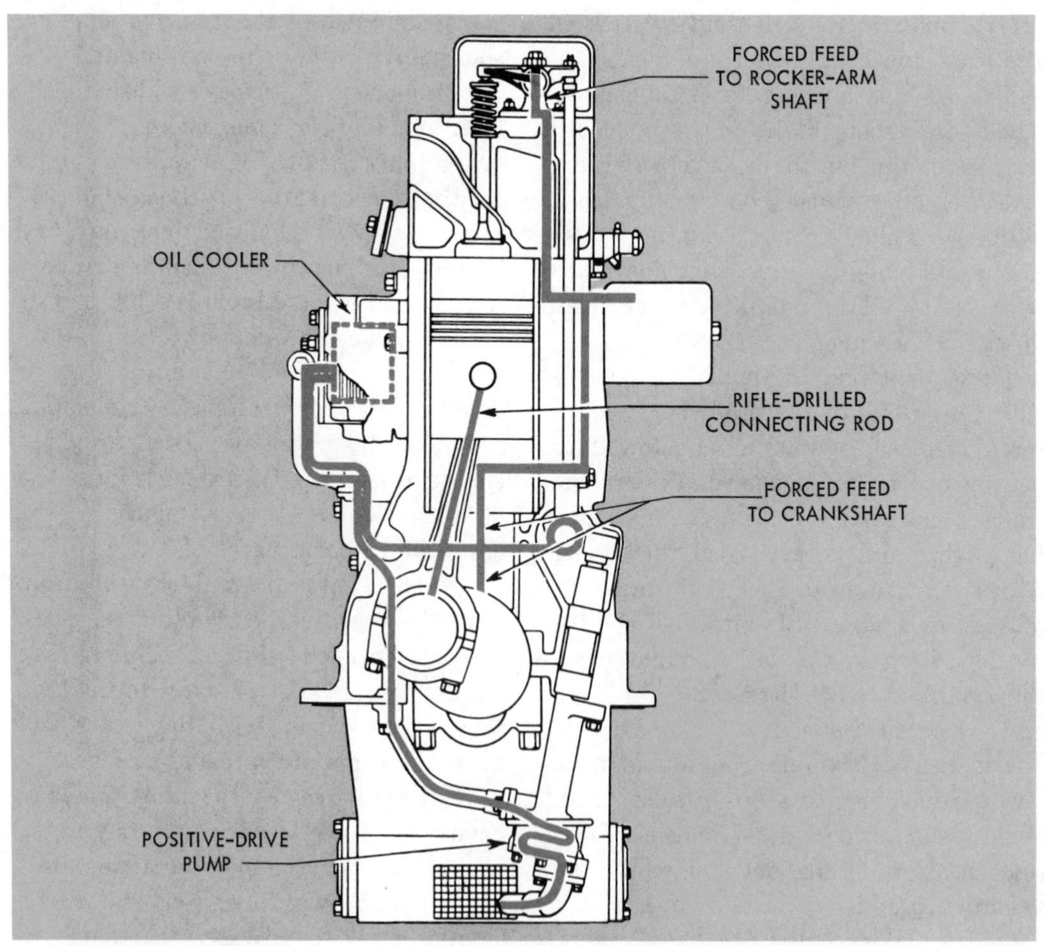

Fig. 7-3. Pressure-lubrication system of small, high-speed engine. (Power Magazine)

Solving the Lubrication Problem

After reading the previous article you realize that lubricating the diesel is a tough job. Yet it is accomplished successfully. How? For one thing, refiners are producing ever better lubricants. These oils are refined to have suitable physical characteristics and, even more important, the desired chemical characteristics.

Modern oils are made to resist oxidation. This has been accomplished to some extent by improved refining and also by adding *oxidation inhibitors* in the form of special chemical *additives*. Still other additives are used to produce so-called *detergent oils*, that is, oils which wash the metal surfaces free of deposits and keep the carbonaceous particles in such fine suspension (dispersion) in the oil that they do not deposit on the engine parts or in the filters.

The diesel operator plays his part by maintaining the engines and accessory

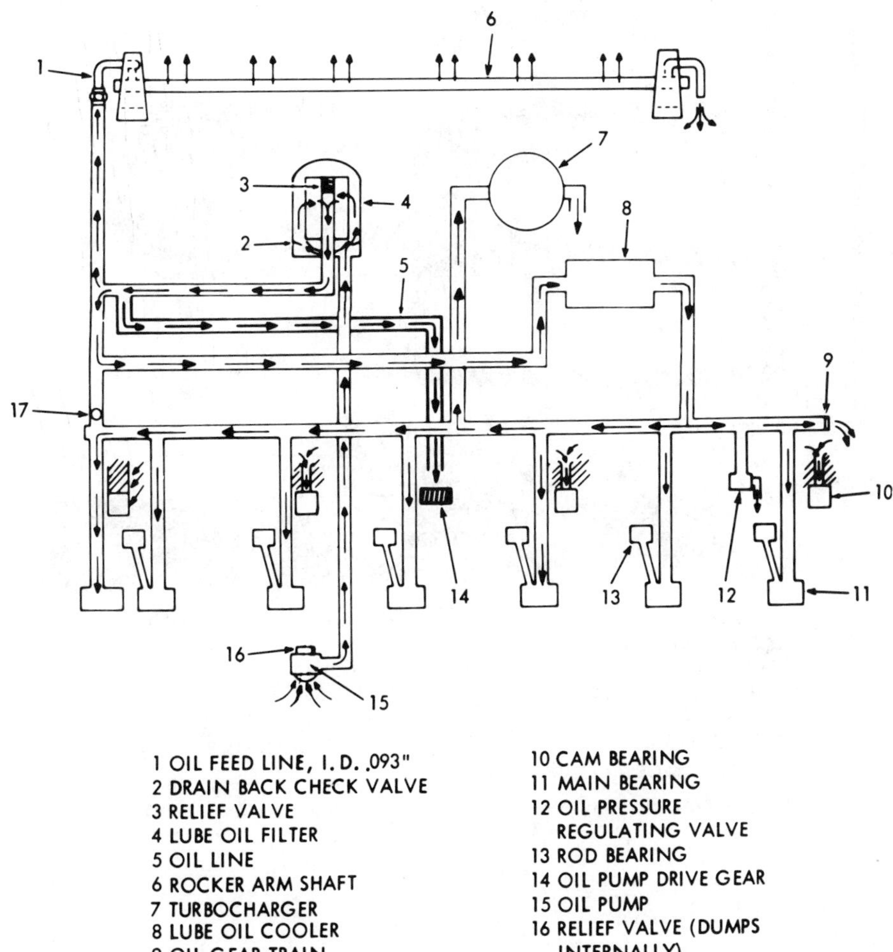

Fig. 7-4. Pressure-lubrication system of completely pressurized turbocharged engine.

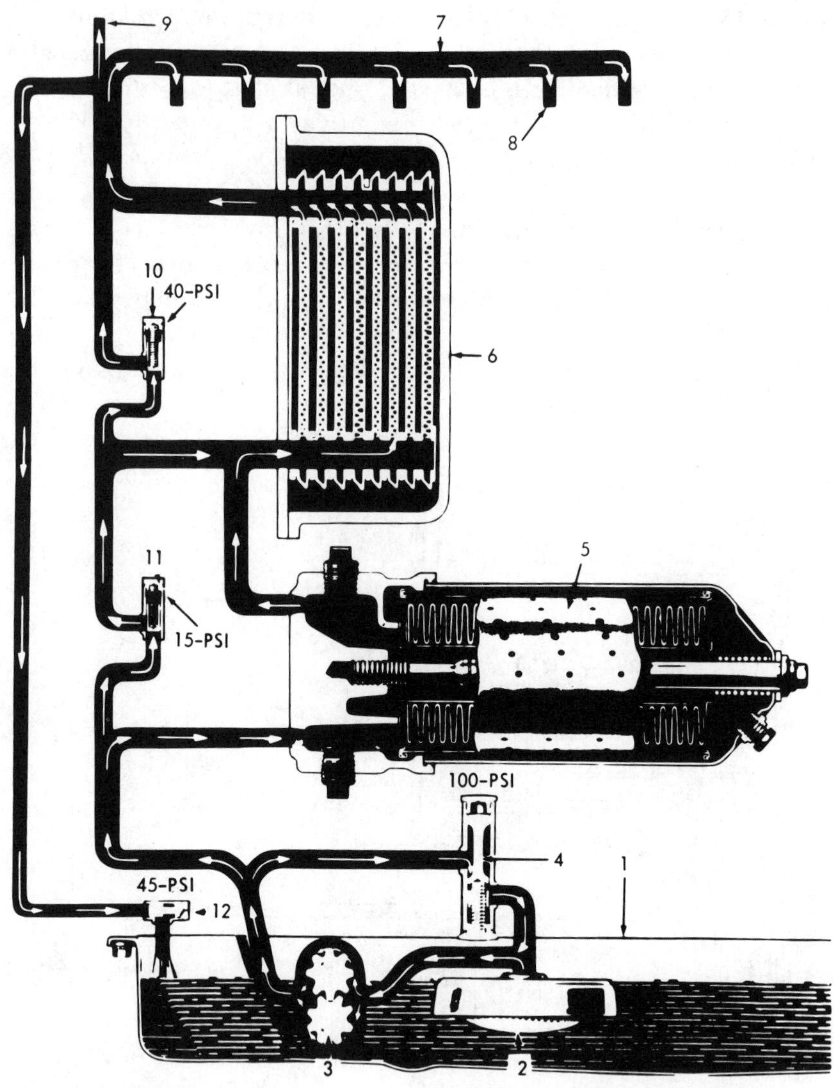

1 OIL PAN
2 SCREEN-OIL PUMP
3 OIL PUMP
4 VALVE-PRESSURE RELIEF
5 OIL FILTER
6 OIL COOLER
7 CYLINDER BLOCK OIL GALLERY
8 PASSAGE TO MAIN BEARING
9 PASSAGE TO UPPER CYLINDER BLOCK AND HEAD
10 COOLER BY-PASS VALVE
11 FILTER BY-PASS VALVE
12 PRESSURE REGULATOR VALVE

Fig. 7-5. Full-flow lubrication system for two-stroke cycle diesel engine.

equipment in good mechanical condition, thus reducing contamination caused by piston blow-by, condensation of water, and foreign matter. Also, by purifying the used oil, he gets rid of the unavoidable decomposition products and contamination, thus maintaining the oil at or near its original quality. Most additives used to improve oil performance become spent in service. Therefore the operator replaces the oil charge with fresh oil at suitable drain periods.

Diesel Engine Lubrication Systems

Most diesel engines have a *pressurized* circulation system to lubricate the main and the connecting-rod bearings, camshaft, valve gear, superchargers and auxiliary bearings. Fig. 7-3 illustrates pressure-lubrication of small high-speed diesels.

Fig. 7-4 shows a schematic diagram of a completely pressurized lubrication system on a turbo-charged diesel engine.

Fig. 7-5 illustrates a full-flow engine lubrication system on a 2-stroke cycle diesel engine. Notice the specific control of pressures throughout the system.

Finally, Fig. 7-6 shows a lubricating oil pump (gear type), with its pressure relief valve.

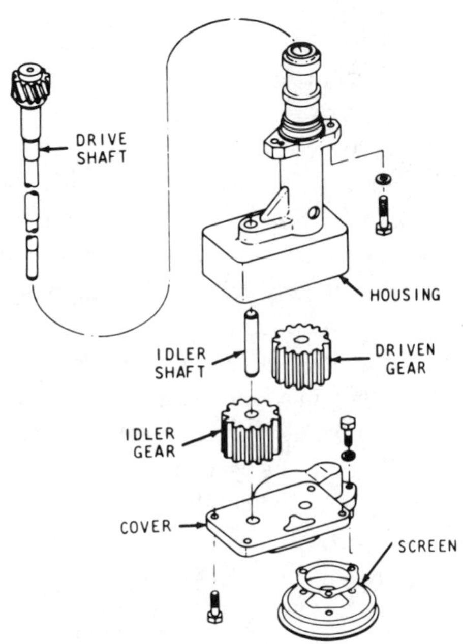

Fig. 7-6. Lubricating oil pump. (Deere and Company)

Properties of Lubricating Oils

Every lubricating oil has certain physical properties or characteristics that can be measured by laboratory tests. Such tests are widely used for two purposes: (1) to specify a particular type of oil and thus assure uniformity from one shipment to another, and (2) to give some indication of how the oil is likely to perform in an engine. However, an actual service test is the only real proof of an oil's suitability for a particular size and type of engine because no effective laboratory tests have been found for some of the important properties.

The chief properties of lubricating oils measurable by laboratory tests are:

1. Viscosity
2. Carbon residue
3. Pour point
4. Flash point
5. Water and sediment
6. Acidity or neutralization number
7. Precipitation number
8. Gravity
9. Color

Viscosity. The *viscosity* of a fluid is a measure of its tendency to hold together, as shown by its resistance to flow. Viscosity is the most important property of a lubricating oil because it determines how effectively the oil film separates the moving metal surfaces and keeps them from rubbing directly on each other. Thus, viscosity controls the load-carrying ability, affects the friction and wear of bearings, governs the sealing effect of the oil and the rate of oil consumption.

The viscosity of an oil changes with its temperature, the oil becoming thinner (lower viscosity) as the temperature rises. Therefore, the viscosity should be measured at a temperature which corresponds to the operating temperature of that particular part of the engine which the oil is to lubricate. For example, where separate oils are used to lubricate the cylinders and the bearings, the cylinder oil viscosity is usually determined at 210° F and the bearing oil at 130° F. Where the same oil is used for both cylinders and bearings, the viscosity at both temperatures is important.

The Saybolt Universal Viscosimeter is used to measure lubricating oil viscosity and gives a reading in seconds. For example, the viscosity of a particular oil might be reported as 280 Saybolt Seconds Universal (SSU) at 130°F, and 60 SSU at 210°F. This means that the viscosity at a temperature of 130°F is 280 seconds, Saybolt Universal, and decreases to 60 seconds when the temperature is raised to 210°F. Viscosity in SSU is in this case determined by heating a carefully measured amount of oil to 130°F or to 210°F, and then measuring the time in seconds required for that amount of oil to flow through a round opening of definite size.

Carbon Residue. The laboratory test for *carbon residue,* sometimes called a Conradson test, gives a clue to the amount of carbon that may be deposited in an engine. It is made by heating the oil in a vacuum until all the volatile matter has been driven off in the form of vapor. The remaining carbonaceous deposit is the carbon residue, which is expressed as a percentage of the original weight of the oil.

As a general rule, oils from any given crude type show greater residue the greater the viscosity. Napthene-base oils, for instance, show less carbon residue than paraffine-base oils. Also, the more thorough the refining, the smaller the amount of residue.

Don't assume that the laboratory test of carbon residue gives an exact measure of the amount of carbon which will be deposited in an actual engine. This depends not only on what oil, but also on the design of the engine and the conditions under which it operates, such as load, type of fuel, etc.

Neither does the carbon residue test show the nature of the carbon deposit that will form in an engine. This is quite important—some types of oils form a hard, gritty deposit in the piston ring grooves that may cause the rings to stick fast. Other oils form a soft, fluffy carbon that does not tend to stick to the rings; in fact, some of it blows out with the exhaust gases and may be noticed as sooty chunks.

Pour Point. The temperature at which the oil will barely flow under controlled test conditions is its *pour point.* Low pour point oils are needed for engines which must start in cold weather. If the pour point is too high, the oil will not reach the working surfaces when the engine starts, and wear will be excessive.

Flash Point. Heating the oil in an open cup and noting the temperature at which flashes of flame occur when a lighted taper is passed across the surface of the oil is a procedure that determines *flash point.* Practically all lubricating oils have flash points sufficiently high to eliminate fire hazard in storage. The flash test gives no information as to how the oil will perform in an engine.

Water and Sediment. In this test the oil is thinned with benzol and then whirled in a centrifuge to separate the water and sediment. New lubricating oils rarely contain any water or sediment. If present they generally have come from contamination during shipment or storage. On the other hand, oil which has been used in an engine usually picks up substantial amounts of water and sediment.

Acidity or Neutralization Number. The neutralization number of an oil is the measure of its acid content. It is determined by finding how much of a known alkaline (opposite of acid) solution is needed to neutralize the acid in the oil. In the case of new oil containing no additives, a low neutralization number, below 0.1, shows that the oil is free from the mineral acids which might remain from the refining process. But oils of the heavy duty type contain acid or alkaline additives and may show misleading neutralization numbers.

After oil has been used in an engine its neutralization number will test higher than when it was new, because service operation oxidizes the oil and produces organic acids. Some of these organic acids corrode certain bearing metals, but the neutralization number does not measure the corrosive tendency of an oil because it does not distinguish between corrosive and noncorrosive acids. Some engine instruction books advise draining the oil when its neutralization number exceeds a specified maximum, such as 0.5 to 1.5.

Precipitation Number. The volume of sediment in a used lubricating oil is measured by the *precipitation number* of the oil. The test consists of thinning the oil with petroleum naphtha and then centrifuging it to separate the sediment in the form of a precipitate. The longer an oil is used in an engine, the more sediment will accumulate and the higher its precipitation number will be, unless the engine's filter removes the sediment as fast as it forms. Some engine instruction books recommend draining the oil when its precipitation number exceeds a specified figure, such as 0.5 or 1.0.

Gravity. As in the case of fuel oils, the *gravity* of lubricating oils is expressed either as specific gravity or API gravity (American Petroleum Institute standard). The gravity of a lubricating oil has no relation to its quality as a lubricant, but is useful for weight and volume computations. Gravity is also useful, in combination with other tests, for identifying the type of oil. For example, paraffine-base oils run 8 to 10 degrees (API) higher in gravity than naphthene-base oils of equal viscosity.

Color. In the laboratory, this property of an oil is tested by comparing the oil sample in a tube of certain size with glass color standards. Color tests are used in the refinery for assuring uniformity in the production process, but the color of an oil has nothing to do with the way it will lubricate an engine.

Selection of Lubricating Oil

Having learned from the preceding article the meaning of the laboratory tests used to measure certain oil properties, you should now understand why such tests give no sure answer as to how the new oil will perform in a particular engine used in a particular kind of service. The final and only reliable method is a thorough test in the engine itself. However, once an oil has proved itself in an engine, the laboratory tests can be used to check new oil deliveries from time to time in order to reveal possible changes.

Engine builders and oil refiners have cooperated in making thorough service tests of numerous brands of oil in many types of engines. As a result, many engine builders and reputable oil refiners have prepared reference lists of suitable oils for various types of engines in the several kinds of service. Whenever possible, the engine user should select his first supply of lubricating oil from such a list, and thus avoid unnecessary and possibly costly trials. However, if such lists are not available, the engine user should inform one or more oil suppliers as to the type and size of the engine, its speed, type of lubricator, type of filter, and kind of service, so that a suitable oil can be recommended.

Oil suppliers, upon request, will furnish users with the characteristics of the selected oil. By specifying these characteristics in future orders, and by checking the properties from time to time, the user can make sure that he is receiving a uniform product.

Checking On Your Knowledge

The following questions give you the opportunity to check up on yourself. If you have read the chapter carefully, you should be able to answer the questions. If you have any difficulty, read the chapter over once more so that you have the information well in mind before you go on with your reading.

DO YOU KNOW

1. Define all types of friction, and give examples of each type.
2. Explain the difference between *cohesion* and *adhesion.*
3. Explain what happens when load is increased on a bearing.
4. The hydrodynamics oil-film and wedge produces several hundred pounds per square inch in pressure. Explain why.
5. What are three factors which govern the degree of viscosity required in a bearing?
6. List the five basic requirements of a lubricant for diesel engines.
7. How does oil deteriorate? What happens?
8. How can deterioration of diesel lubricant be reduced?
9. List and describe the nine properties of lubricating which are measurable by laboratory tests.
10. How is the best lubricating oil for a particular diesel selected?

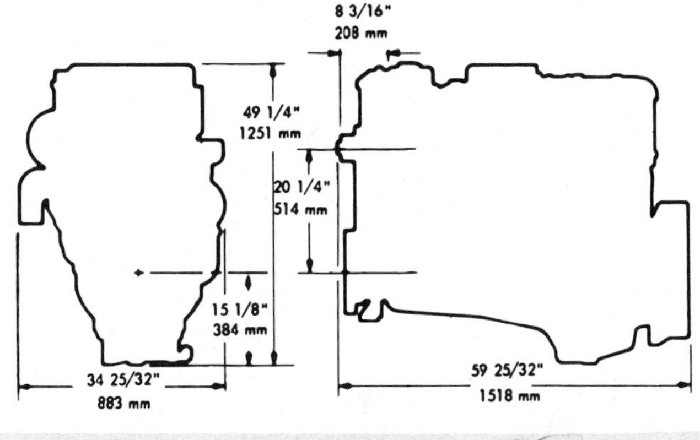

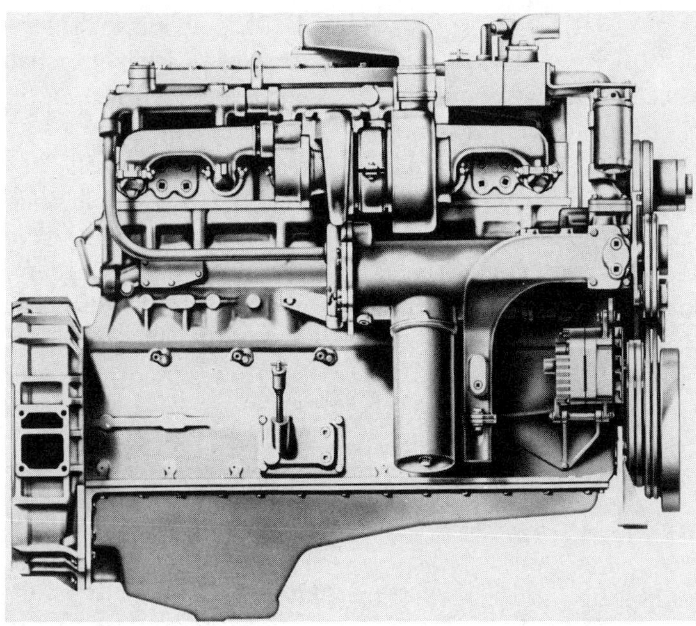

Recently available Cummins NTA-400 diesel engine. Note metric dimensions. (Cummins Engine Company, Inc.)

Basic Terms of Physics and Engineering

Chapter

8

The purpose of this chapter is to introduce you to some basic terms of physics and engineering connected with diesel engines. You have already learned what a diesel engine is, what kinds of diesel engines there are, and something about the fundamental construction of a diesel engine. You should therefore now understand the general features of a diesel.

Now it's time to pick up some numerical or quantitative knowledge about diesel performance. You learned, in Chapter 1, that some of the advantages which diesels enjoy compared to other prime movers are that diesels consume less fuel than gasoline engines and steam engines, and that diesels burn oil, which is cheaper than gasoline. But these are only general statements—if you want to know how to make the best use of diesels, you must learn something about the actual numerical values relating to them, such as the amount of power a particular diesel produces and the amount of oil it consumes to produce that power.

Beginning in this chapter, therefore, example problems and solutions are given. Many of the calculations were done as an engineer would do them—with a slide rule. It is important that you become accustomed to seeing calculations which are satisfactory for the purpose of solving such problems.

In the next few chapters you're going to learn some simple engineering as applied to diesels. Long figures in the calculations are often rounded out. We'll start by learning the meanings of some basic terms of physics and engineering which we'll often use later.

Basic Units of Measurement

The physical quantities of the universe can be expressed in three basic terms: *length, time,* and *force.* All natural phenomena that we're aware of belong to one of those three categories, and we speak of length, time, and force as *quan-*

tities that can be measured.

These three quantities are measured according to various systems which have become standardized in different parts of the world. In the United States, they are expressed in basic units of measurement as follows:

Length is expressed in feet (ft). Length may also be expressed in various other units such as inches, yards, miles, etc.

Time is expressed in seconds (sec). Time may also be expressed in minutes, hours, days, etc.

Force is expressed in pounds (lb). Force may also be expressed in ounces, tons, etc.

For several decades, though, there has been an effort to convert all users of the foot-pound-second (British) system to metric units. This is because most other countries use the meter-gram-second system unless they are purchasing imports from countries using the British system.

The British have been gradually converting to the meter-gram-second (Metric) system over the last decade. Various scientific associations in the United States and Canada have taken the lead to put all scientific and technical papers into both systems of symbols. The engineering world is following.

Eventually all automotive parts and tools will use the Metric system, so that every mechanic will need knowledge of the differences between these two systems.

Everyone knows the meaning of *length* and *time*. But just what is a *force?* In physics, a force is anything that changes or tends to change the motion of a body. If the body is at rest, anything that tends to put it in motion is a force. For instance, if an engine valve is resting on its seat, a *force* is needed from the valve lever to start moving the valve open. Likewise, if a body is already in motion, anything that tends to change either the direction or rate of its motion is a force.

Using the engine valve again as an example, the valve spring exerts a *force* on the valve to slow it down and bring it to rest as it reaches its wide-open position. A *force* should always mean the pull, pressure, rub, attraction, or repulsion of one body upon another. The most common force is the force of gravity; you'll learn more about the force of gravity later in this chapter.

Derived Units of Measurement

Derived units of measurement come from the basic units. The derived units are used to measure the following quantities encountered in engineering practice.

Area. A measure of surface is called *area,* and is expressed as the product of the length and width (two characteristic lengths) of a surface. Areas are expressed in *square* units, such as square feet (sq ft) or square inches (sq in).

The area of a circular surface, such as the top of a piston, is frequently used in diesel calculations. The area of a circular surface is 0.785 times the diameter squared. *Squared* means multiplied by itself, as: 10 × 10 = 100.

EXAMPLE 1. What is the area of a piston having a 3″ diameter?

$$\begin{aligned} \text{Area} &= 0.785 \times \text{diameter squared} \\ &= 0.785 \times (3 \text{ in} \times 3 \text{ in}) \\ &= 0.785 \times 9 \text{ sq in} \\ &= 7.065 \text{ sq in} \end{aligned}$$

Volume. A measure of space is called *volume;* it is expressed as the product of area and height (three characteristic lengths) of the space. Volumes are measured in *cubic* units, such as cubic feet (cu ft) or cubic inches (cu in). The *displacement* of a diesel engine is a volume term which is often used. The displacement of one piston of an engine is the volume of the space through which the piston travels in one stroke.

EXAMPLE 2. A piston of 3″ diameter (dia) has a stroke of 4″. What is its displacement?

$$\begin{aligned} \text{Displacement} &= \text{area} \times \text{length} \\ &= \text{area of piston} \times \\ &\qquad \text{length of stroke} \end{aligned}$$

Where:

$$\begin{aligned} \text{Area} &= 7.065 \text{ sq in} \\ \text{Length} &= 4'' \text{ stroke} \end{aligned}$$

Therefore:

$$\begin{aligned} \text{Displacement} &= 7.065 \text{ sq in} \times 4 \text{ in} \\ &= 28.26 \text{ cu in} \end{aligned}$$

[sq in × in = cubic inches (cu in)]

Rotary Motion. The movement of a point or body along a circular path is called *rotary motion.* Such is the motion of the crankpin of a diesel engine around the main bearing of the crankshaft. The position of the rotating body may be expressed by the angle through which it has rotated. Thus, in Fig. 8-1, when point *A* (center of crankpin) has moved to *B*, the position of *B* is determined by the angle of rotation *C*. Angles are measured in degrees, one complete revolution being 360 degrees.

Velocity or Speed. For purposes of this discussion, *velocity* and *speed* can be considered synonymous. Velocity or speed is a quantity that relates the basic quantities of length and time—it is the distance traveled by a moving object or a point in a unit of time, such as second (sec), minute (min), or hour (hr). Velocity is computed by dividing the distance traveled by the time used for the travel, thus:

$$\text{Velocity or speed} = \frac{\text{distance}}{\text{time}}$$

Velocity is expressed in various *derived units,* based upon the units used to express distance and time. For example, feet per minute (ft per min, or fmp), miles per hour (mi per hr, or mph).

Velocity may be *uniform* or *varying.* If the motion is constant, the velocity is uniform and the above expression will give the *actual velocity.* On the other hand, if the motion is not uniform, neither is the velocity, and the above expression will give the *average velocity.* If while driving your car at varying speeds you traveled 30 miles in one hour, the average velocity was 30 mph.

The up-and-down (reciprocating) motion of a piston in an engine cylinder is another case of motion which is not uniform. The average piston velocity is called the *piston speed,* and is always stated in fpm.

EXAMPLE 3. A piston having a 9 inch stroke makes 800 revolutions per min. What is the piston speed?

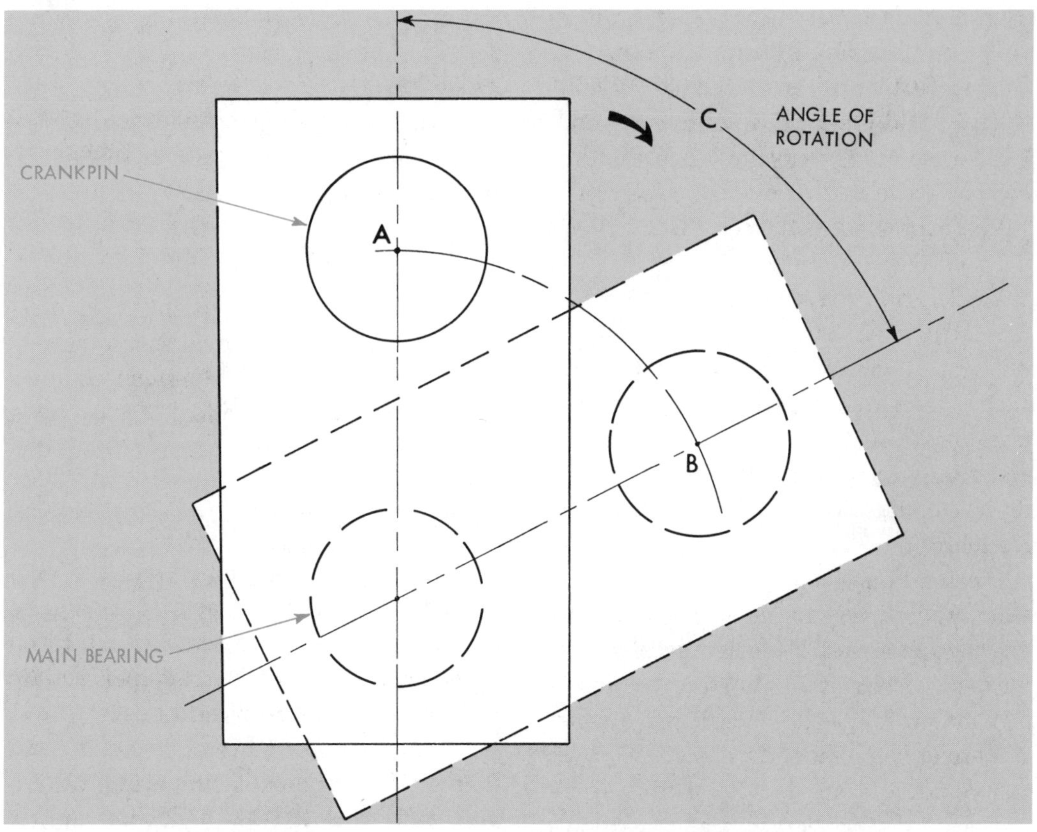

Fig. 8-1. The calculation of rotary motion is based in a diagram of the motion.

Piston speed = average piston velocity

$$= \frac{\text{distance traveled}}{\text{time to travel}}$$

$$\text{Piston speed} = \frac{9 \text{ in} \times 800 \text{ RPM} \times 2}{1 \text{ min}}$$

$$= 14{,}400 \text{ in/min (ipm)}$$

To express this in feet per minute, we divide by 12 (inches in one foot), thus:

$$\text{Piston speed (fpm)} = \frac{\text{Piston speed (ipm)}}{12}$$

$$\text{Piston speed} = \frac{14{,}400}{12} = 1{,}200 \text{ fpm}$$

Speed is also used to refer to rotary motion, such as that of an engine crankshaft. Thus, *engine speed* is said to be so many revolutions of the crankshaft per minute, and is designated as rpm.

Acceleration. The *change* in velocity of a moving body in a unit of time is *acceleration.* It is computed by dividing the *change* in velocity by the time during which this change takes place, thus:

$$\text{Acceleration} = \frac{\text{change in velocity}}{\text{time}}$$

Acceleration is expressed in various units, depending on the units used to express velocity and time. For example, if the velocity is expressed in feet per minute (ft per min), and the time in which the velocity changes is also expressed in minutes, then the acceleration will be expressed in feet per minute per minute, which in abbreviated form is ft/min/min. This is further abbreviated to ft/min^2 which is read "feet per minute squared."

Likewise, if the velocity is expressed in feet per second (ft/sec) the acceleration will be expressed in ft/sec^2 (feet per second squared).

Acceleration may be uniform or varying. If the rate of change in velocity is uniform, the acceleration is uniform, and the above expression will give the *actual acceleration.* On the other hand, if the change in velocity is not uniform, then the above computation will give the *average* acceleration.

Acceleration is called *positive* when the velocity increases, and *negative* when the velocity decreases. Negative acceleration is called *deceleration.* When an engine is brought up to speed it accelerates; when it is brought to rest it decelerates.

A common example of constant acceleration is the acceleration of the earth's force of gravity, which is 32.2 ft/sec^2. If a body is released above the surface of the earth, it will begin to fall, being attracted toward the center of the earth. As the body falls, it keeps on gaining speed (unless like a parachute it is checked by air resistance). The acceleration of gravity means that, during each second, the velocity of the falling body, disregarding the resistance of the air, will increase or accelerate by 32.2 ft/sec.

EXAMPLE 4. A body at rest (zero *inertia*) is acted upon by a force which after 5 seconds has caused it to acquire a velocity of 9000 fpm. What is the average acceleration in ft/sec^2?

First express the velocity in ft/sec.

Since 1 min = 60 sec,

$$9000 \text{ fpm} = \frac{9000}{60} = 150 \text{ ft/sec}$$

Then, as

$$\text{Average acceleration} = \frac{\text{change in velocity}}{\text{time}}$$

It can be computed that

$$\text{Average acceleration} = \frac{150}{5} = 30 \text{ ft/sec}^2$$

Pressure. Another quantity is *pressure,* which may be defined as a force acting on a unit of area. Pressure may be exerted either by a solid body or by a fluid (gas or liquid). Thus:

$$\text{Pressure} = \frac{\text{force}}{\text{area}}$$

If the force is in pounds and the area upon which the force acts is in square inches (sq in), the pressure is expressed in derived units of pounds per square inch, which is abbreviated to *psi.*

EXAMPLE 5. Compute the average pressure on the cork pads supporting an engine weighing 12,000 lb. There are six pads, each of which is in inches square.

Total area of the six pads is 6 × 10 in × 10 in = 600 sq in. The force is the weight of the engine, 12,000 lb. Therefore:

$$\text{Pressure} = \frac{\text{force}}{\text{area}} = \frac{12{,}000}{600} = 20 \text{ psi}$$

Notice that the pressure computed in the foregoing example is the *average* value, as no information was given as to how much of the total engine weight was supported on *each* cork pad.

When a force is transmitted by a fluid, either liquid or gas, the pressure between the fluid and the walls of the container will be uniform and equal in all directions, regardless of the shape of the walls. This fact makes it easy to compute the force exerted by a fluid on a certain object when the pressure of the fluid is known. Instead of writing:

$$\text{Pressure} = \frac{\text{force}}{\text{area}}$$

we turn the relationship (or formula) around to read:

$$\text{Force} = \text{pressure} \times \text{area}$$

Let's apply this relationship to the piston of a diesel engine which is being pressed upon by combustion gases. Fig. 8-2 shows a typical piston, the top of which has a dished shape. Although the area of this dished shape is greater than if the top were flat, the *effective area* on which the gas pushes in the *direction* of the *piston motion* is exactly the same as if the piston top were flat.

EXAMPLE 6. Referring to Fig. 8-2, compute the gas force acting to move the piston. The pressure of the gas is 500 psi, and the diameter of the piston is 5 inches. As the effective area of the top of the piston is equal to the area of a circle having a 5 inch diameter,

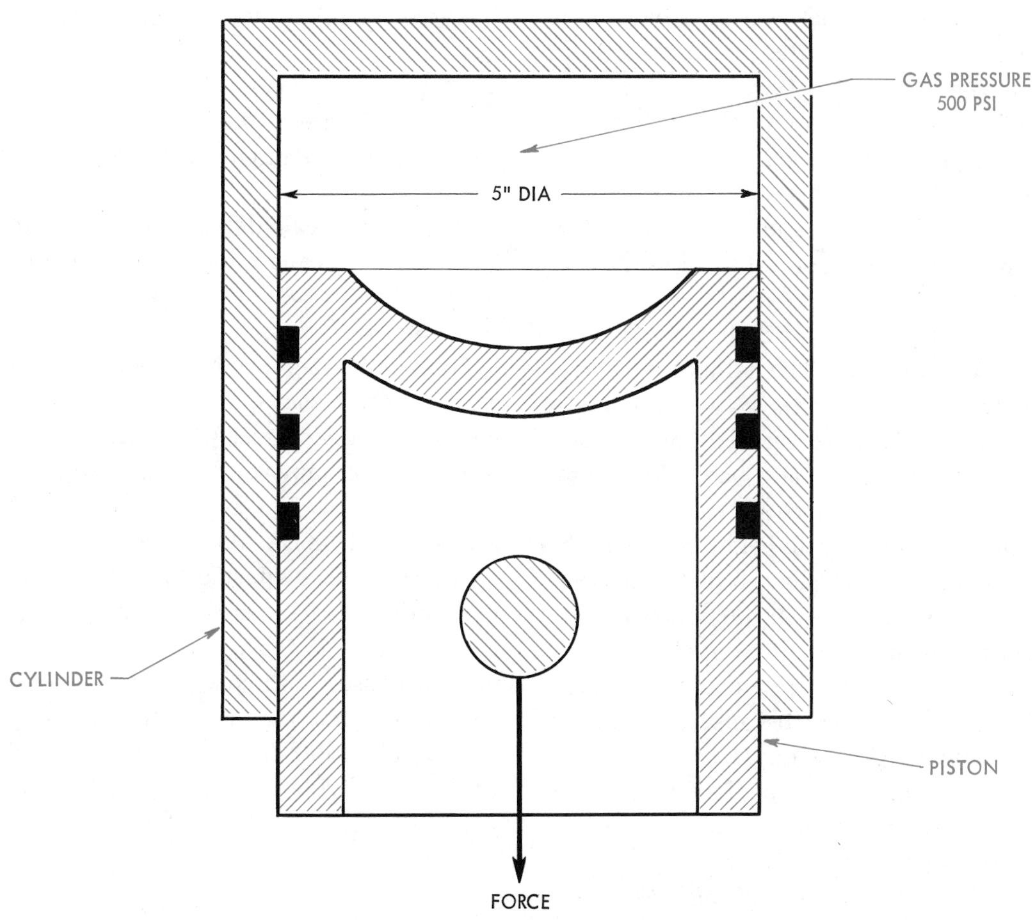

Fig. 8-2. Forces acting on engine piston.

$$\text{Area} = 0.785 \times 5^2 = 19.63 \text{ sq in}$$

Then, as

$$\text{Force} = \text{pressure} \times \text{area}$$

$$\text{Force} = 500 \times 19.63 = 9815 \text{ lb}$$

Specific Gravity. The ratio of the weight of a certain volume of liquid to the weight of an equal volume of pure water is called the *specific gravity* of the liquid. For practical use, remember that 1 cu ft of pure water weighs 62.4 lb, and 1 gallon of pure water weighs 8.34 lb.

Specific gravity of a liquid may be greater or less than that of water, that is, greater or less than one, or unity. For instance, the specific gravity of the acid solution in an ordinary storage battery is about 1.28 when the battery is fully charged.

EXAMPLE 7. Compute the weight of 1 gal of fuel oil which has a specific gravity of 0.87.

The weight of the oil will be equal to the weight of 1 gal of water times the specific gravity, or

$$8.34 \times 0.87 = 7.26 \text{ lb per gal.}$$

Temperature. The temperature of a body is a characteristic of its heat content which can be determined only by comparison with another body. When two bodies are placed in close contact, the one which is hotter will begin to heat the other and is said to have a higher temperature. For instance, if you touch the exhaust pipe of a running diesel engine, it will transmit heat to your hand; your hand feels the heat because the exhaust pipe has a higher temperature.

The temperature scale used in the United States is the *Fahrenheit scale* (F). In this scale, the two reference points are the temperature of melting ice, designated as 32°F, and the temperature of steam when the water is boiling at normal barometric pressure, designated 212°F. The distance on the scale between these two points is divided into 180 equal parts called *degrees* (deg or °). The scale is continued in both directions, above 212 degrees and below 32 degrees. Below 0°F, the temperatures are designated by a minus (−) sign.

In theoretical calculations pertaining to gases, another scale is used, called the *absolute* or *Rankine scale.* In this scale the unit (degree) is the same as in the Fahrenheit scale, but the absolute zero is placed at − 460°F. Thus, the relation between the absolute temperature and the corresponding Fahrenheit temperature is:

$$\text{Degrees absolute} = \text{degrees Fahrenheit} + 460$$

In technical calculations pertaining to gases, 68°F is called *normal* or *standard temperature.*

Mass and Weight. Mass is the quantity of matter contained in a body. It is not easy to define just what mass is, nor is a definition important for the purposes of this book. What is important is the way in which mass is *measured.* Mass is measured by the effect a force has upon the motion of the mass when the mass is free to move. When a force acts upon a mass it causes it to change its velocity a certain amount, that is, to give it a certain acceleration. The mass is equal to the force divided by the acceleration which the force produces. Thus:

$$m = \frac{F}{a} \text{ or } F = ma$$

where

$$m = \text{mass of the object}$$
$$F = \text{force in lb}$$
$$a = \text{acceleration, ft/sec}^2$$

This formula finds many applications in engineering because, if we know the mass of an object, we can compute what effect a given force will have on its motion.

The mass of an object is directly proportional to its *weight*. Let us see why. First, what does *weight* really mean? Of course, you can measure the weight of an object easily by using a spring scale, but what actually does the reading of the scale show? The weight shown by the scale is the amount of the force exerted on the mass of the object by the earth's gravity.

We learned above that $F = ma$, that is, force equals mass × acceleration. In this case F, the force (in pounds), is the *weight* of the object (in pounds). We can say, therefore, that

$$\text{Weight} = \text{mass} \times \text{acceleration of gravity}$$

We also learned that the acceleration produced by the earth's force of gravity is 32.2 ft/sec². Therefore,

$$\text{Weight} = \text{mass} \times 32.2$$

or

$$\text{Mass} = \frac{\text{weight}}{32.2}$$

where both mass and weight are expressed in lb.

You will find the factor 32.2 appearing in many engineering calculations because, although we start with the *weight* of an object, its motion is related to its *mass*. The figure 32.2 converts one into the other.

EXAMPLE 8. What is the mass of an engine piston that weighs 55 lb?

$$\text{Mass} = \frac{\text{weight}}{32.2} = \frac{55}{32.2} = 1.71 \text{ lb}$$

Work and Power

Work is done when a force moves a body through a certain distance. For instance, when you push a stalled auto from the street to the curb, work is measured by multiplying the force (F) by the distance (d) through which the body moves in the direction of the force. Thus:

$$\text{Work} = \text{force} \times \text{distance}$$

or

$$W = F \times d$$

Work is expressed in ft-lb or in-lb.

EXAMPLE 9. Find the work necessary to raise a weight of 150 lb a distance of 8 ft.

The force is equal to the weight, 150 lb. Thus:

$$\text{Work} = \text{force} \times \text{distance} = 150 \times 8 = 1200 \text{ ft-lb}$$

Power is the *rate* at which work is done. It depends upon how fast the work is done, that is, upon *time*. In other words power is the number of units of work performed in one unit of time. Thus:

$$\text{Power} = \frac{\text{work}}{\text{time}}$$

Power is expressed in various units, one of which is ft-lb/min. In engineering calculations the term *horsepower* (hp) is frequently used as a measure of power. One horsepower is 33,000 ft-lb/min or 550 ft-lb/sec.

EXAMPLE 10. Find the power required to do the work of Example 9 if the work is to be performed: (*a*) in 2 sec, or (*b*) in 8 sec.

For case (*a*):

$$\text{Power} = \frac{\text{work}}{\text{time}} = \frac{1200\ \text{ft-lb}}{2}$$
$$= 600\ \text{ft-lb sec}$$

Expressed in horsepower, this is:

$$\text{Horsepower} = \frac{600\ \text{ft-lb/sec}}{550\ \text{ft-lb/sec}} = 1.09\ \text{hp}$$

For case (*b*):

$$\text{Power} = \frac{\text{work}}{\text{time}} = \frac{1200}{8} = 150\ \text{ft-lb/sec}$$

Expressed in horsepower, this is:

$$\text{Horsepower} = \frac{150\ \text{ft-lb/sec}}{550\ \text{ft-lb/sec}} = 0.273\ \text{hp}$$

Note how *time* enters in the the problem.

Electric power is measured in units called *watts;* 1000 watts are called 1 kilowatt (kw). The relations between horsepower and kilowatts are:

$$1\ \text{hp} = 0.746\ \text{kw}$$
$$1\ \text{kw} = 1.341\ \text{hp} = 44{,}268\ \text{ft-lb}$$

EXAMPLE 11. An electric generator takes 150 kw to drive it. Compute the equivalent horsepower.

$$150\ \text{kw} \times 1.341 = 201.15\ \text{hp}$$

Energy

The energy of a body is the amount of work it can do. Energy may exist in several different forms; a body may possess energy through its *position,* its *motion,* or its *condition.*

Energy due to the *position* occupied by the body is called *potential energy.* An example is a body located at a higher level, such as water behind a dam.

When a body is in *motion* it is said to possess *kinetic energy;* for example, a ball rolling upon a level floor or the rotating flywheel of an engine.

Internal Energy. Energy due to the *condition* of a body is called *internal energy.* The energy stored *within a body* due to the forces between molecules or atoms composing it, such as in steam or gas pressure, is called that body's internal energy. *Chemical energy,* such as a fuel can release when it burns, or such as a storage battery can release when it is discharged, is also classified as internal energy.

Nuclear Energy. The energy stored within the nucleus of an atom is termed *nuclear energy.* It differs from all the other forms of energy in the fact that when nuclear energy is released, matter is actually destroyed and weight is lost. Although nuclear energy is having an ever-increasing importance in our lives, it has nothing to do with the action of diesel engines and is therefore quite outside the scope of this book.

Disregarding nuclear energy, the principle of *conservation of energy* states that energy may exist in many varied and interchangeable forms but may not be quantitatively destroyed nor created. In fact, work is merely mechanical or electrical energy in a *state of transfer* from one form to another form. The work that is done by raising a body stores potential energy in the body due to the force of gravity; work done to set a body in motion stores kinetic energy; work done in compressing a gas stores internal energy in the gas; electrical work can be transformed into mechanical work by means of an electric motor.

Heat, like work, is energy in a *state of transfer* from one body to another, due to a difference in temperature between the bodies.

Units of Energy. *Work* and *heat* are measured by separate units of energy:

1. The *foot-pound* (ft-lb) is the unit of energy based on *work* and is equivalent to the energy of a force of 1 lb acting through a distance of 1 ft.

2. The *British thermal unit* (Btu) is the unit of energy based on *heat;* it is the energy required to raise the temperature of 1 lb of pure water by 1 degree Fahrenheit at standard atmospheric pressure.

Despite the two different units, there is a fixed relation between the unit of energy based on work (ft-lb) and the unit of energy based on heat (Btu), as required by the principle of conservation of energy previously stated. The conversion factor, called the *mechanical equivalent of heat,* is:

$$1 \text{ Btu} = 778 \text{ ft-lb}$$

Engineering calculations employ two other energy units which are derived from the unit of ft-lb:

3. The *horsepower-hour* (hp-hr), which is the transfer of energy during one hour at the rate of one horsepower, that is, at the rate of 33,000 ft-lb/min, or a total of 1,980,000 ft-lb. Using the conversion factor 778, 1 hp-hr = 2544 Btu.

4. The *kilowatt-hour* (kw-hr), which is the transfer of energy during one hour at the rate of one kilowatt, that is, at the rate of 1000 watts per hour. This is equivalent to a rate of 1.341 hp, or 44,268 ft-lb/min, or a total of 2,656,100 ft-lb. Using the conversion factor 778, 1 kw-hr = 3413 Btu.

Kinetic Energy. A body's kinetic energy (the energy due to its motion) is computed by a formula that is slightly more complicated, namely:

$$\text{Kinetic energy, } KE, = \frac{1}{2} \times \frac{W}{g} \times V^2$$

where

W is the weight of the body, lb

g is the acceleration due to gravity, 32.2 ft/sec^2

V is the velocity of the body, ft/sec

If you wonder what g, the acceleration due to the earth's gravity has to do with the energy of a body in motion, recall what you learned earlier in this chapter. As we are dealing with the energy of motion, we are concerned with the *mass* of the body. But W is the *weight* of the body; therefore, to find its mass we must divide its weight by the acceleration due to gravity. Hence the expression $\frac{W}{g}$.

Suppose we want to compute the work done by kinetic energy when the velocity of a body changes. The work done equals the *change* in kinetic energy. So we use the preceding formula and compute the kinetic energy *twice*. We compute the kinetic energy of the body first at its initial velocity, V_1, and again at its final velocity, V_2. The difference between the two kinetic energies is the work done. The following example will make this clearer.

EXAMPLE 12. Find the work done by 500 lb of exhaust gases discharged upon the blades of a supercharger turbine if the initial velocity of the gases was 9000 and the final velocity was 5400 fpm.

First, the velocity must be changed to ft/sec:

$$V_1 = \frac{9000}{60} = 150 \text{ ft/sec}$$

$$V_2 = \frac{5400}{60} = 90 \text{ ft/sec}$$

Using the preceding formula, the kinetic energy of the gases before entering the turbine is:

$$KE_1 = \frac{1}{2} \times \frac{500}{32.2} \times 150^2 = 174{,}690 \text{ ft-lb}$$

and the kinetic energy of the gases after leaving the turbine is:

$$KE_2 = \frac{1}{2} \times \frac{500}{32.2} \times 90^2 = 62{,}888 \text{ ft-lb}$$

and the work done = 174,690 − 62,888 = 111,802 ft-lb.

Checking On Your Knowledge

The following questions give you the opportunity to check up on yourself. If you have read the chapter carefully, you should be able to answer the questions. If you have any difficulty, read the chapter over once more so that you have the information well in mind before you go on with your reading.

DO YOU KNOW

1. If a piston has a 6″ dia and an 8″ stroke and has six cylinders, what is the total displacement?
2. A piston having a 5″ stroke makes 2250 revolutions per minute. What is the piston speed?
3. Explain the relationship between *acceleration* and *inertia*.
4. What is *pressure?*
5. If the firing pressure in a cylinder is 900 psi, find the force in pounds on a 6″ dia piston.
6. What is specific gravity?
7. What is the relationship of mass and acceleration?
8. Explain the differences between power, force, horsepower, and torque.
9. What is the relationship of energy and work in ft-lb?
10. What is the relationship between kinetic and potential energy?

Heat and Combustion

Chapter

9

The purpose of this chapter is to make clear how heat affects the actions of gases, and how certain chemical reactions take place when fuels burn to produce heat. You'll learn how all of this is governed by simple basic laws and how easy it is to calculate the results. This will lead to a lot of practical information on the power and fuel consumption of engines.

As you learned in Chapter 1, a diesel engine produces power by burning oil in a certain manner. Burning is a way of producing heat—that's why you must start by learning something about heat.

After seeing what heat is, you'll learn how heat flows from one body to another, and how heat changes the pressure and temperature of gases such as those that do the work in an engine cylinder.

Then you'll learn some elementary chemistry so that you can understand the chemical changes that take place when fuels burn and why the heat produced is a definite quantity.

Here and there you'll pick up the meanings of some more words and phrases that are often used, such as gage and absolute pressures, air-fuel ratio, high-heat value and low-heat value.

What Heat Is

In the previous chapter, *heat* was defined as a form of energy, the energy being in a state of transfer from one body to another due to a difference in temperature between the bodies. Now let's learn some more about heat.

The energy that constitutes heat comes from the motions of the tiny particles called *molecules,* which compose all substances. Even in a piece of solid metal these molecules are continually flying back and forth in very short paths—paths that are much too short to be visible even in a high-powered microscope. If we supply heat to a body, we add energy to these flying molecules and cause them to fly faster and farther. If the body is a solid, such as the cylinder of a diesel engine, it gets hotter when we supply heat because its molecules move faster. And if we apply a thermometer, we find that its *temperature* has risen. In other words, higher temperature means higher speed of the molecules.

Solids have a fixed shape, but gases (such as air) move freely and fill their containers, no matter how large they are. Their flying molecules are continually striking the walls of the container with a force which we can measure, and which we call the *pressure* of the gas. When we apply heat to a gas, its molecules fly faster and it gets hotter. If the gas is confined in a fixed space, the heavier blows with which the molecules hit the walls of the container are shown by an increase in pressure.

This is what happens when the fuel burns in a diesel engine cylinder. The burning fuel adds heat to the gases confined in the combustion space above the piston, and increases the pressure with which the gases press upon the confining surfaces. One of these surfaces, the piston-top, is movable, consequently the increased pressure pushes the piston down on its power stroke. As the piston descends, the space occupied by the gases increases, thus the *volume* of the gases becomes greater and the gases are said to *expand.* Later in this chapter you will learn the simple laws by which the temperature, pressure, and volume of a gas are related to each other.

Heat Flow

In our study of engines, we are much concerned with the *flow* of heat from one body to another. To measure this flow, we measure the change in temperature of one of the bodies—heat is brought in if the temperature of the body rises, and taken away if the temperature falls. Our unit of heat, the Btu, is itself defined as the quantity of heat required to raise the temperature of 1 lb of water 1 degree Fahrenheit.

Substances differ as to how much heat flow is needed to raise their temperature. For instance, to increase the temperature of cast iron requires only about one-eighth the heat flow needed to raise the temperature of an equal weight of water. To take account of this fact, we say that every substances has its own *specific heat.*

Specific Heat

The *specific heat* of any substance is the *ratio* of the heat flow required to raise by one degree (1°F) the temperature of a unit weight of the substance, to the heat flow required to raise by 1°F the temperature of an equal weight of water. The specific heats of a few substances in which we are interested in this book are given in Table 9-1.

In the case of gases, specific heat varies with conditions of pressure and volume. The specific heat of pure water is 1 by definition, because 1 Btu will raise by 1°F the temperature of 1 lb of pure water. This fact permits us to say that the specific heat of a substance is equal to the heat flow, in Btu, required to raise by 1°F the temperature of 1 lb of the substance.

Denoting the specific heat by c, the heat flow, Q, required to raise the temperature of W lb of a substance from t_1 to t_2 (degrees F) is:

Heat flow = weight × specific heat × temperature change or

$$Q = Wc\ (t_2 - t_1)$$

EXAMPLE 1. Find the heat required to raise the temperature of 42 gal of lubricating oil from 95° to 185°F. The specific heat of the oil is 0.5 Btu/lb/°F, and its specific gravity is 0.92.

First, we find the weight of the oil, W. One gallon of water weighs 8.34 lb.

Therefore, 1 gal of the oil weighs 8.34 × 0.92 = 7.67 lb; and 42 gal weigh 42 × 7.67 = 322.26 lb.

Therefore, the heat required is:

$$Q = Wc\ (t_2 - t_1)$$
$$Q = 322 \times 0.5(185 - 95)$$
$$Q = 14{,}490 \text{ Btu}$$

Heat Transfer

Ordinarily heat flow, or *heat transfer*, is accomplished in three ways: *conduction*, *radiation*, and *convection*.

Conduction is heat flow by *actual contact* from a substance at a higher temperature. Conduction may take place within a body if its parts are at different temperatures, or from one body to another.

Radiation is energy transfer *through space* from a hotter body to a colder body.

TABLE 9-1 SPECIFIC HEAT OF SELECTED SUBSTANCES

SOLIDS		LIQUIDS		GASES	
Substance	Specific Heat	Substance	Specific Heat	Substance	Specific Heat at Constant Pressure
Aluminum...	0.225	Alcohol (methyl)	0.601	Air	0.237
Brass.......	0.089	Petroleum oil...	0.511	Carbon dioxide..	0.217
Cast iron....	0.125	Water.........	1.000	Water vapor (212°F).......	0.421

Convection is heat transfer accomplished by the movement of masses of matter and contained heat from one location to another. An example of convection is the movement of heated air from one part of a room to another.

Always remember the *basic principle of heat flow:* Heat can flow from one body to another only if the first body is hotter than the second.

Gases—Pressure and Volume

In our study of diesel engines, we are continually dealing with gases. The air that is compressed in the cylinders is a mixture of gases; when the oil fuel is sprayed in, the gaseous mixture burns and gets hotter; the gases then expand as they push the pistons down, and finally the spent or exhaust gases are expelled to the atmosphere. So you need to understand something about the general properties of gases.

Pressure of Gases

By their very nature, gases always have pressure. This is because their molecules are continuously flying about and pressing upon the walls of their containers, as you learned earlier in this chapter.

The pressure of a gas may be expressed not only in pounds per square inch (psi), but also in terms of the height of the column of liquid which will balance the gas pressure. The liquids generally used are water or mercury. If water is used, the relationship is that a pressure of 1 psi equals 2.309 ft of water. Let's see why this is so.

One cu ft of water weighs 62.4 lb. Imagine a cube of water having edges 1 ft long as shown in Fig. 9-1. The weight of this cube is supported by the area of its bottom surface, which is 1 ft × 1 ft, or expressed in inches, 12 in × 12 in = 144 sq in. Consequently, the weight of water supported by each square inch of the bottom surface is:

$$\frac{62.4}{144} = 0.433 \text{ psi}$$

This is the pressure per square inch of a column of water 1 ft high. If the column were only 1 inch high, the pressure would be:

$$\frac{0.433}{12} = 0.0361 \text{ psi}$$

Putting it the other way, if a column of water 1 ft high produces a pressure of 0.433 psi, the height of a column of water producing a pressure of 1 psi

$$= \frac{1 \text{ ft of water}}{0.433}$$

$$= 2.309 \text{ ft (or } 2.309 \text{ ft} \times 12 = 27.7 \text{ in)}$$

If a mercury column is used to measure a pressure, the height of the mercury column is much less than that of a water column, because mercury weighs 13.6 times as much as water. Therefore, a pressure of 1 psi

$$= \frac{27.7}{13.6}$$

$$= 2.036 \text{ inches of mercury}$$

Conversely, 1 inch of mercury =

$$= \frac{1 \text{ psi}}{2.036} = 0.491 \text{ psi}$$

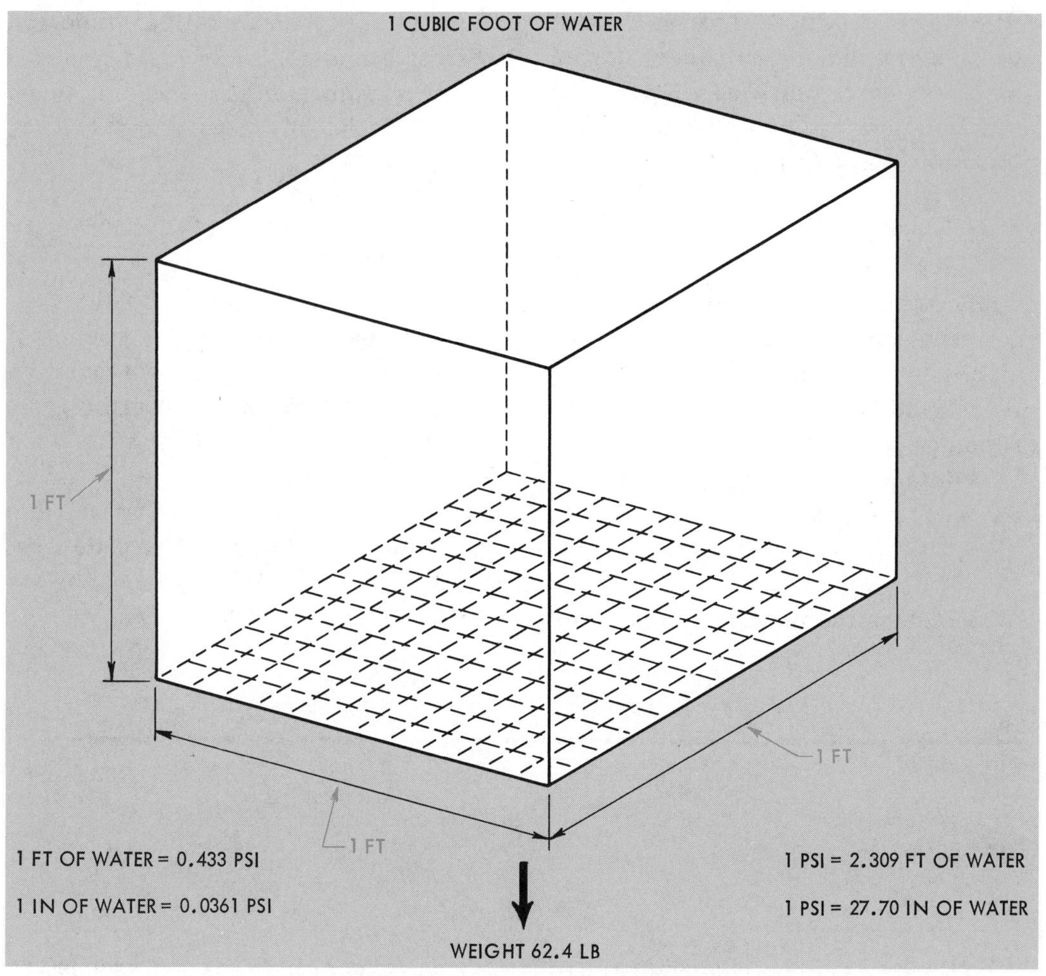

Fig. 9-1. Volume of one cubic foot of water.

EXAMPLE 2. The back pressure of the exhaust gases of a diesel engine is measured with a water gage and found to be 9 inches. What is the pressure in psi and in inches mercury?

As 1 inch of water = 0.0361 psi; then:

$$\begin{aligned}\text{Back pressure} &= 9 \text{ inches water}\\ &= 9 \times 0.0361\\ &= 0.325 \text{ psi}\end{aligned}$$

As 1 psi = 2.036 inches of mercury,

Then:

$$\begin{aligned}\text{Back pressure} &= 0.325 \text{ psi}\\ &= 0.325 \times 2.036\\ &= 0.662 \text{ inches of mercury}\end{aligned}$$

Gage and Absolute Pressures

Pressure gages measure the pressure of gases above or below the pressure of the surrounding atmospheric air, which is also called *barometric pressure.*

Pressures measured thus are called gage pressures and are sometimes termed *pounds per square inch gage* (psig). In gas calculations the *total* pressure of the gas is often needed; this can be obtained by adding to the gage pressure the barometric pressure. This total pressure is called *absolute pressure,* and is termed *pounds per square inch absolute* (psia). Ordinarily, psi is assumed to mean psig.

If absolute is designated p_a, gage pressure p_g, and barometric pressure b, the relation can be written as:

Absolute pressure = gage pressure + barometric pressure or

$$p_a = p_g + b$$

The barometric pressure b is not constant, since it changes with altitude and weather conditions. *Normal* or *standard barometric* pressure at sea level is taken as 29.92 inches of mercury, or:

$$\frac{29.92}{2.036} = 14.7 \text{ psia}$$

Volume of a Gas

The space that a gas occupies within walls confining it is called the *volume* of a gas. In the combustion space of an engine, the volume of the working gases changes when the piston moves. When the piston rises and approaches the cylinder head, the volume decreases and the gas is *compressed;* when the piston descends, the volume increases and the gas *expands.*

Gas Laws

The three measurable characteristics of any gas—pressure, volume, and temperature—are connected by a simple relation, namely:

$$pV = WRT$$

where

p is the absolute pressure in pounds per square *foot* absolute, psfa

V is the volume, cu ft

W is the weight of the gas, lb

T is the absolute temperature, °R (degrees Rankine) and

R is a constant, called the *gas constant,* the value of which is known for every gas

The value of R for air is 53.3

This fundamental gas law enables us to find the weight of a certain amount of gas if we know its pressure, volume, and temperature. It also leads to some simple rules which enable us to compute the change in one characteristic of a given amount of gas when a change occurs in one of the other characteristics.

EXAMPLE 3. Find the weight of air contained in a cylinder having a volume of 9500 cu in, at a pressure of 5 psig and a temperature 180° F. Proceeding, we set forth:

$$\text{The volume } V \text{ in cu ft} = \frac{9500}{12 \times 12 \times 12}$$

$$= \frac{9500}{1728} = 5.5 \text{ cu ft}$$

The absolute pressure p, assuming a normal barometric pressure of 14.7 psi, is $5 + 14.7 = 19.7$ psia.

The absolute temperature T is $180 + 460 = 640°R$.

Then, by relating the above data in an equation to solve for the weight, W, we proceed:

$$W = \frac{pV}{RT}$$

$$W = \frac{2840 \times 5.5}{53.3 \times 640} = 0.458 \text{ lb}$$

The following three rules come directly from the fundamental gas law:

RULE 1. At any fixed pressure, a gas expands or contracts in exact proportion to its absolute temperature.

EXAMPLE 4. An engine discharges 1200 cu ft/min of exhaust gases at 600°F into a pipe in which the gases cool (without change in pressure) to an outlet temperature of 400°F. What is the rate of flow at the pipe outlet?

The rate of flow is proportional to the volume of the gases, and the volume is proportional to the absolute temperature.

The absolute temperature of the gases entering the pipe is $600 + 460 = 1060°R$.

The absolute temperature of the gases leaving the pipe is $400 + 460 = 860°R$. Therefore:

$$\text{Rate of flow at pipe outlet} = 1200 \times \frac{860}{1060}$$

$$\text{Rate of flow at pipe outlet} = 972 \text{ cu ft/min}$$

RULE 2. At any fixed volume, the absolute pressure of a gas rises or falls in exact proportion to its absolute temperature.

EXAMPLE 5. An engine starting tank is filled with compressed air at 210 psig and 160°F, and then closed off. What will be the pressure after tank and contents cool to 60°F?

The corresponding absolute temperatures are $160 + 460 = 620°F$, and $60 + 460 = 520°F$. Proceeding, we set forth:

$$\text{Initial pressure, absolute} = 210 + 14.7 = 224.7 \text{ psia}$$

$$\text{Final pressure, absolute} = 224.7 \times \frac{520}{620} = 188.5 \text{ psia}$$

$$\text{Final pressure, gage} = 188.5 - 14.7 = 173.8 \text{ psig}$$

(Note how much starting pressure can be lost if the starting tanks are charged hot and then closed off.)

RULE 3. At any fixed temperature the volume of a given weight of gas is inversely proportional to its absolute pressure.

EXAMPLE 6. Air is compressed from 240 cu in to 15 cu in and then allowed to cool to its initial temperature. If the initial pressure was normal atmospheric, what will the final pressure be?

$$\text{Initial pressure, absolute} = 14.7 \text{ psia}$$

$$\text{Final pressure, absolute} = 14.7 \times \frac{240}{15} = 235.2 \text{ psia}$$

$$\text{Final pressure, gage} = 235.2 - 14.7 = 220.6 \text{ psig}$$

Basic Terms of Chemistry

Having learned what heat is and how heat affects the pressure and volume of a gas, it is time you learned something about *combustion,* because it is the combustion of fuel that supplies heat to the working gases inside a diesel engine cylinder.

In order to understand what we mean by combustion, let's consider some basic definitions and facts of chemistry.

Chemistry is the science that treats of the *composition* of substances, and of the transformations *in composition* which they undergo. These transformations are called *chemical reactions.*

Combustion is that kind of chemical reaction that is accompanied by release of considerable light and heat.

Burning, in the usual sense, is that kind of combustion in which a fuel chemically combines with oxygen of the air, thereby releasing considerable light and heat. In a diesel engine, it is this heat of burning that increases the temperature of gases in the cylinders.

To better grasp the idea of elements combining chemically, let's proceed further with some simple chemistry.

An *element* is a basic substance, one that cannot be subdivided into other substances. Typical elements with which we deal in diesel engines are oxygen, hydrogen, carbon, iron, and sulfur.

The smallest piece of any element is its *atom.* If you smash the atom you have no element left, nothing but the common particles of all matter such as protons, neutrons, and electrons which, in different combinations, form the atoms of all elements.

In an element the atoms are all alike. A *compound,* on the other hand, contains the atoms of two or more elements. For example, carbon dioxide is produced when the elements carbon and oxygen unite chemically in burning. It is a compound of atoms of carbon and oxygen.

A *molecule* is the grouping of two or more atoms of the same or different elements, which acts as a unit, its atoms being bound together by mutual attraction. Putting it another way, a molecule of an element or a compound is the smallest particle of matter which can maintain its separate identity.

The molecules of the common gaseous elements with which we deal in diesel engines, namely oxygen and hydrogen, consist respectively of *two* atoms of oxygen or of hydrogen.

For the sake of brevity in describing chemical substances and their reactions, every element is known by a symbol, and every compound can be designated by a combination of these symbols. In other words, we use a sort of chemical shorthand. For example, carbon is C, and a molecule of oxygen is shown as O_2, which means that the molecule of oxygen contains two atoms. When a molecule of carbon unites with a molecule of oxygen, the product is the compound carbon dioxide, shown as CO_2. (*Dioxide* means *two* atoms of oxygen.)

TABLE 9-2 ELEMENTS CONCERNED WITH COMBUSTION

Name	Symbol	Atomic Weight	Molecule	Molecular Weight
Oxygen.........	O	16	O_2	32
Hydrogen........	H	1	H_2	2
Carbon..........	C	12	C	12
Sulfur	S	32	S	32

TABLE 9-3 CONSTITUENTS OF AIR

Element	Percentage by Weight	Percentage by Volume
Oxygen	23.1	20.9
Nitrogen, argon, etc. ..	76.9	79.1
Total...............	100.0	100.0

Atomic Weight

All the atoms of any one element have the same weight (disregarding the technicality of the isotopes of nuclear science, with which this book is not concerned). The atom of oxygen, however, is much heavier than the atom of hydrogen, in fact, 16 times as heavy. Since hydrogen is the lightest element, we call the weight of a hydrogen atom unity or 1, and say that the *atomic weight* (at. wt) of any element is the weight of its atom in terms of the hydrogen atom. Table 9-2 gives the atomic weights of the elements entering into engine combustion, and also gives the symbols and molecular weights.

Air, which supplies the oxygen that takes part in engine combustion, is not a chemical compound; it is a mixture of gases which are not chemically combined with each other. Oxygen is only about a fifth of the total; the remainder is mainly nitrogen and a small fraction of rare gases. Since the nitrogen and rare gases do not take part in the combustion reactions, we call them inert and group them together in Table 9-3, which shows the proportions by weight and by volume.

Chemistry of Engine-Fuel Combustion

Fuel oil is mainly a mixture of *hydrocarbons* in liquid form. A hydrocarbon is a chemical compound composed of hydrogen and carbon. There are many kinds of hydrocarbons but, so far as we are concerned in this book, they differ chemically only in the relative proportions of hydrogen and carbon. For instance, the hydrocarbon known as ethane, $C_2 H_6$, has a molecule of two atoms of carbon combined with six atoms of hydrogen, whereas a molecule of propylene, C_3H_6, contains three atoms of carbon combined with six atoms of hydrogen.

Gaseous fuels used in gas engines and dual-fuel engines (the latter burn both liquid and gaseous fuels) likewise contain hydrocarbons in gaseous form. Some gaseous fuels also contain carbon monoxide, CO, which is a combination of carbon and oxygen.

Thus, the combustible elements of the fuels which we use in the engines covered by this book are mainly carbon and hydrogen. Another combustible but undesirable element found in fuel oils and gases, fortunately in small amounts, is sulfur.

Now let's look at the chemical reactions which govern the burning of these combustible elements when they unite with the oxygen of the air in the engine cylinder.

Chemical Reactions

First we'll take up hydrogen. When hydrogen combines with oxygen, water is produced according to the equation:

$$2H_2 + O_2 = 2H_2O$$

Stated simply, this equation reads: "Two molecules of hydrogen (a molecule of hydrogen contain *two* atoms) plus one molecule of oxygen form two molecules of water." Now let's put this equation into *weight proportions*, using the atomic weights shown in Table 9-2.

	$2H_2$	$+ O_2$	$= 2H_2O$
Wt	4	+ 32	= 32
	1 part	+ 8 parts	= 9 parts
	1 lb	+ 8 lb	= 9 lb

Thus, we find that 8 lb of oxygen are needed to burn a pound of hydrogen and that 9 lb of water are produced. Note that the atoms balance on both sides of the equation and so do the weights.

EXAMPLE 7. Find the weight of hydrogen which will unite with 24 lb of oxygen, and the weight of water produced.

From preceding equation,

1 lb hydrogen + 8 lb oxygen = 9 lb water

This problem deals with 24 lb oxygen, while the equation dealt with 8 lb. Therefore:

$$24 \div 8 = 3$$

Multiplying all parts of the equation by 3,

3 lb hydrogen + 24 lb oxygen = 27 lb water

Answer: 3 lb hydrogen, 27 lb water

How Carbon Burns

Hydrogen combines with oxygen in only one proportion, as shown above, but *carbon* may burn in two proportions, depending on whether the combustion is complete or incomplete. If plenty of oxygen is present (and this is usually the case in a diesel engine), complete combustion takes place according to the following equation and weight ratios:

	C	$+ O_2$	$= CO_2$
Wt	12 at wt	+ 32 parts	= 44 at wt
	1 part	+ 2 2/3 parts	= 3 2/3 parts
	1 lb	+ 2.67 lb	= 3.67 lb

The compound CO_2 is called *carbon dioxide.*

If not enough oxygen is present (as when too much fuel is injected into the engine cylinder), incomplete combustion of the carbon may take place, according to the following equation:

$$
\begin{array}{llll}
 & 2C & + O_2 & = 2\,CO \\
\text{Wt} & 24 \text{ at wt} & + 32 \text{ at wt} & = 56 \text{ at wt} \\
 & 1 \text{ part} & + 1\,1/3 \text{ parts} & = 2\,1/3 \text{ parts} \\
 & 1 \text{ lb} & + 1.33 \text{ lb} & = 2.33 \text{ lb}
\end{array}
$$

The compound CO is called *carbon monoxide*, which is a poisonous gas. It is always present in the exhaust gases of a *gasoline* engine because gasoline engines run on a mixture of fuel and air in which there is not enough air to permit the carbon to burn completely. As we shall see later, when carbon burns to carbon monoxide, the heat produced is only 30 percent of that produced when it burns to carbon dioxide.

Carbon monoxide, CO, which is a constituent of some gas, readily combines with oxygen to form carbon dioxide CO_2 in accordance with the equation:

$$
\begin{array}{llll}
 & 2\,CO & + O_2 & = 2CO_2 \\
\text{Wt} & 56 \text{ at wt} & + 32 \text{ at wt} & = 88 \text{ at wt} \\
 & 1 \text{ part} & + 4/7 \text{ parts} & = 1\,4/7 \text{ parts} \\
 & 1 \text{ lb} & + 0.57 \text{ lb} & = 1.57 \text{ lb}
\end{array}
$$

How Sulfur Burns

Sulfur, S, burns with oxygen to form sulfur dioxide, SO_2, as follows:

$$
\begin{array}{llll}
 & S & + O_2 & = SO_2 \\
\text{Wt} & 32 \text{ at wt} & + 32 \text{ at wt} & = 64 \text{ at wt} \\
 & 1 \text{ part} & + 1 \text{ part} & = 2 \text{ parts} \\
 & 1 \text{ lb} & + 1 \text{ lb} & = 2 \text{ lb}
\end{array}
$$

Sulfur dioxide is a gas which is corrosive in the presence of water. It is responsible for the corrosion found inside engine exhaust pipes.

Combustion Relations for Oil

As you have already learned, diesel oils are mixtures consisting almost entirely of compounds known as *hydrocarbons* because they are made up of hydrogen and carbon. Any one oil contains many different hydrocarbons, each of which is composed of hydrogen and carbon in a cetrain ratio.

Fortunately, we do not need to know how much of each particular hydrocarbon is contained in a given oil in order to work out the combustion relations for the oil as a whole. We only need to know the proportion of *all* the hydrogen to *all* the carbon in the oil, because during the process of combustion, the fuel oil particles split into their component elements, and each element combines with the oxygen of the air separately. Knowing the over-all proportions of hydrogen and oxygen, we then use the simple combustion equations for each. The following example will make this clear:

EXAMPLE 8. A certain oil contains 14 percent hydrogen and 86 percent carbon. If 120 lb of the oil are burned, how much oxygen unites with the oil and what are the weights of the products of combustion?

Beginning our calculations, we set forth that:

120 lb oil contains $120 \times 0.14\% = 16.8$ lb hydrogen

120 lb oil contains $120 \times 0.86\% = 103.2$ lb carbon

The hydrogen burns as follows:

$$\begin{aligned} 2H_2 + O_2 &= 2H_2O \\ 1\text{ lb} + 8\text{ lb} &= 9\text{ lb} \end{aligned}$$

Multiplying by 16.8 (lb hydrogen in the oil),

$$16.8\text{ lb } H_2 + 134.4\text{ lb } O_2 = 151.2\text{ lb } H_2O$$

The carbon burns as follows:

$$\begin{aligned} C \quad + O_2 \quad &= CO_2 \\ 1\text{ lb.} + 2.67\text{ lb} &= 3.67\text{ lb} \end{aligned}$$

Multiplying by 103.2 (lb carbon in the oil),

$$103.2\text{ lb C} + 275.5\text{ lb } O_2 = 378.7\text{ lb } CO_2$$

Thus,

Total O_2 uniting with oil = 134.4 + 275.5 = 409.9 lb oxygen and

Products of combustion

H_2O = 151.2 lb water

CO_2 = 378.7 lb carbon dioxide

Note that the weight of oil plus oxygen (120 + 409.9 = 529.9 lb) is equal to the weight of the products of combustion (151.2 + 378.7 = 529.9 lb). This is a useful check on the calculations.

Air-Fuel Ratio

So far, our calculations have dealt with oxygen instead of air. In engines, we are of course interested in air quantities. To change weight of oxygen into weight of air, simply remember that *air contains 23.1 percent oxygen by weight* (Table 9-3).

EXAMPLE 9. Find the weight of air which would be used to burn the 120 lb of oil in Example 8.

The weight of the oxygen was 409.9 lb. As 1 lb of air contains 0.231 lb oxygen, the weight of air containing 409.9 lb oxygen would be computed:

$$W = \frac{409.9}{0.231} = 1774\text{ lb air}$$

The preceding example shows how many pounds of air are *theoretically* needed to burn a pound of oil. Dividing 1774 (lb air) by 120 (lb fuel) we get 14.8 lb air per pound of fuel. We call this the *theoretical air-fuel ratio.*

Commercial oils are not composed solely of carbon and hydrogen as in the foregoing example, but contain a small percentage of other elements such as sulfur, nitrogen, and oxygen. This reduces slightly the amount of air required to unite with the fuel, so that the theoretical air-fuel ratio for commercial oils is about 14.5 lb air per pound of oil.

The theoretical air-fuel ratio tells us the *least* amount of air which must be supplied to an engine cylinder in order to burn a given amount of fuel. But it is impossible to make a diesel engine run satisfactorily on the *theoretical* amount of air. If only this amount of air is supplied, many particles of oxygen will not be able to participate in the process of combustion.

You'll realize how hard it is to mix all the fuel evenly with all of the air if you look at the quantities involved. A drop of liquid oil entering the combustion space of a diesel engine must be uniformly mixed with about *900 times its volume of air* in order to complete its combustion. And this mixing must be done in an exceedingly short time—only a few thousandths of a second. If only the theoretical amount of air were present, it would not be possible for each particle of fuel to find its corresponding particle of oxygen, especially since the oxygen is further diluted with the products of combustion after the burning begins. Free carbon would be released and some carbon monoxide would be formed.

Therefore, to insure complete combustion of the fuel and to avoid the heat loss due to the formation of carbon monoxide and the release of free carbon, we must always supply an excess amount of air to the cylinder. The ratio of the actual weight of air supplied to the weight of fuel injected during each power stroke, is the *actual air-fuel ratio.*

When a diesel engine is operating at a light load (and therefore a reduced amount of fuel is injected on each stroke into the normal amount of air), the actual air-fuel ratio is several times greater than the theoretical value of 14.5. As the load is increased, the air-fuel ratio decreases; but even when the engine is overloaded, the air-fuel ratio must be at least 25 to 30 percent greater than 14.5 (that is at least 18 or 19). There must be present that much of an excess of air over the minimum theoretically required for complete combustion. If we injected enough fuel to reduce the air-fuel ratio to 14.5, we would find that the engine developed *less* power instead of more.

Air-fuel ratio in diesel engines is a term that corresponds to *rich* and *lean* mixtures in gasoline engines. In gasoline engines the air-fuel ratio is determined by the adjustment of the carburetor which forms the mixture of fuel and air *before* it enters the engine cylinders.

The air-fuel ratio in gasoline engines does *not* change with the load. The carburetor is generally set to give an air-fuel ratio slightly less than the theoretical in order to obtain maximum power. This is called a slightly rich mixture because it is richer in fuel than theoretically required. If, on the other hand, the carburetor is adjusted to mix more air with a given amount of fuel, in order to avoid waste of fuel, the resulting higher air-fuel ratio is called a lean mixture.

Heat Quantities from Combustion

As you have learned in the preceding article, precise *weight* relations are involved whenever substances unite in the process of combustion. Now you'll find out that the amount of heat produced is also a fixed quantity. In other words, when a pound of a combustible unites with oxygen to form one or more compounds, the same amount of heat always appears. This is called the *heat value* and is expressed in Btu/lb. Table 9-4 shows the heat values of a few combustible elements

TABLE 9-4 HEAT VALUES OF COMBUSTIBLE ELEMENTS

Reaction	Heat Value
Hydrogen burning to water, H_2O	62,032 Btu/Lb
Carbon burning to carbon dioxide, CO_2	14,542 Btu/Lb
Carbon burning to carbon monoxide, CO	4,451 Btu/Lb
Sulfur burning to sulfur dioxide, SO_2	3,940 Btu/Lb

in which we are interested.

Oil is a mixture of combustible hydrocarbons in various proportions; the kinds of hydrocarbons and the proportions of each vary with the source of the oil and its refining. (You will learn about this in the next chapter.) Since each combustible hydrocarbon has a definite heat value, the oil mixture also has a fixed heat value. In other words, a pound of one particular oil will always produce the same amount of heat when it burns.

It would be a complicated and expensive process to find the heat value of a given oil sample by analyzing it to determine exactly which hydrocarbons it contained and the proportion of each. The practical method is to actually burn a known weight of the oil and to measure the amount of heat it produces. This is done in a laboratory instrument called a *calorimeter*. The heat produced is found by absorbing the heat in a known weight of water and measuring the rise in temperature.

High and Low-Heat Values. When oil is burned in a calorimeter, the water vapor which is formed by the union of its hydrogen with oxygen is condensed—changed from a vapor to a liquid, by the cool walls of the calorimeter. The heat measured by the calorimeter is called the *high-heat value*.

But in an actual engine, the water vapor formed by the combustion of the hydrogen does *not* condense while the engine is running; it passes out of the engine in the form of vapor in the hot exhaust gases. Thus, the high-heat value of fuels containing hydrogen includes some heat which cannot be converted into power in any internal-combustion engine. This fact has brought about the use of the term *low-heat value*. The low-heat value of a fuel is less than its high-heat value by the amount of heat which the water vapor would give up if it were condensed.

To give you an idea of the numerical values, a typical diesel oil of 28° API gravity has a high-heat value of 19,350 Btu/lb and a low-heat value of 18,190 Btu/lb (about 6 percent less). In the case of many gaseous fuels, the difference between the high and low-heat values is much more.

We will need to refer to high-heat and low-heat values in succeeding chapters. Be sure that you understand these terms.

Checking On Your Knowledge

The following questions give you the opportunity to check up on yourself. If you have read the chapter carefully, you should be able to answer the questions. If you have any difficulty, read the chapter over once more so that you have the information well in mind before you go on with your reading.

DO YOU KNOW

1. Explain basically what heat is.
2. What are the three common methods of heat transfer?
3. What does *specific heat* mean?
4. Find the heat required to raise the temperature of 20 gallons of water from 60°F to 175°F.
5. The sealed crankcase pressure of a rebuilt diesel engine is 7 inches on a water type monometer. What is the pressure in inches of mercury?
6. Explain the relationship of pressure, volume, weight, temperature and gas constant of a gas.
7. What is the difference between an element and a molecule?
8. Why is fuel oil considered a *mixture* of hydrocarbons?
9. How does air-fuel ratio relate to variations in power produced in a diesel engine?
10. How are the heat values of different fuels determined?

Chapter

10 Oil and Gaseous Fuels

The purpose of this chapter is to tell you the nature of the different kinds of oils and gases that are used as fuels in diesel, dual-fuel, and spark-ignited gas engines.

You will learn that the oils we use in engines are obtained by refining processes from crude oil produced from wells.

Diesel oils differ greatly in their properties, and therefore act differently when burned in diesel engines. In discussing each of these properties you will learn how each affects the operation of an engine. Because the ignition process is so important we'll give special attention to how oils ignite when they are injected into the engine cylinder.

You will also learn about dual-fuel engines, which burn mostly gas and a little oil, and about spark-ignited gas engines, which burn gas only. These engines have become popular because of the availability of natural gas at low cost. Since they can also burn many other kinds of gases, we'll discuss the various commercial gaseous fuels and see how they differ from each other.

All of this will lead you smoothly into the next chapter, which is on engine power and fuel consumption.

In this chapter you'll learn the meaning of some more common terms such as: oils of paraffin base, mixed base and asphalt base; cracked oils, cetane number, Saybolt viscosity and fuel knock.

What Oil Is

Petroleum or crude oil is usually a dark brown liquid; it is a mixture of a large number of compounds. The main chemical elements which form these compounds are hydrogen and carbon, which we call hydrocarbons, as you learned in Chapter 9. The amount of hydrogen present in the compounds varies from about 11 to 15 percent, by weight, the balance being carbon.

Some of these hydrocarbons, when separated from the crude oil, are gases at ordinary temperatures. These can be stored under pressure in steel cylinders and used as fuel for household purposes, such as propane. Other compounds, when separated from the crude oil, are solids at ordinary temperatures; for example, asphaltum used as a road surfacing material, and paraffin wax, used for sealing purposes. Most of the compounds are liquids that vary considerably in composition and properties. Crude oil also contains varying amounts of impurities, such as sulfur, oxygen, nitrogen, water, salt, sand, and clay.

Crude petroleum is found in pockets under the earth's surface in many parts of the world. The crude oil found in a certain locality normally exhibits some properties which distinguish it from crude oils found in other places. In the continental United States there are three main oil-producing areas; the characteristics of petroleum from these areas differ considerably.

Eastern crude oils (sometimes called Pennsylvania oils) consist chiefly of compounds containing much light oil or gasoline and considerable amounts of paraffin wax. They are often termed *paraffin-base oils*. They contain very little asphaltic material and sulfur. The proportion of hydrogen in products obtained from this crude oil is usually higher than in similar products from crudes from other fields.

Mid-continent crudes vary widely. In general, however, they contain a slightly lower amount of paraffin compounds and more asphaltic materials. They are often termed *mixed-base oils*. In some cases these crudes contain so much sulfur that they must be specially treated to remove it.

Western or California crudes contain large amounts of asphaltic (naphthenic) materials and relatively high percentages of sulfur. They are termed *asphalt-base* or *naphthene-base* oils.

Although large producing areas exist in other parts of the United States, and in the rest of the world, their crudes generally have characteristics that classify them with one of the three main types.

How Oil Is Refined

Although crude oils are sometimes used as diesel fuels in their original state as they come from the wells, most crude oils are put through a refining process whereby (*a*) impurities are removed, (*b*) the oil is separated into several component parts or *fractions*, and (*c*) some of the fractions are re-formed or converted to become other products. An oil refinery is an intricate chemical factory which utilizes its equipment in various ways, depending upon the type of crude oil and the products desired.

Distillation

The first major step in refining, called *fractional distillation*, separates the crude oil into fractions from which commercially useful products can be made. Such separation is possible because compounds of petroleum boil (vaporize) at different temperatures.

A simple distillation process, using a *tank still*, works as follows: Crude oil contained in a closed vessel is heated by a coil through which steam or hot gases are circulated. At first, the low-boiling compounds are driven off the crude oil in vapor form. The vapors are taken away by a pipe connected to the top of the vessel, are condensed by passing through a cooling coil, and are collected as liquids in a tank. The temperature of the crude oil is kept constant.

After all compounds which boil below this temperature are driven off, the temperature of the crude oil is increased. The vapors distilled at the higher temperature are collected in another tank after condensing. Again the process is repeated. The three main products which are thus obtained, in the order of their increasing boiling temperature, called the *boiling point*, are: (1) gasoline, (2) kerosene, and (3) fuel oil. Lubricating oils are distilled off later or are left in the unvaporized residue.

The simple *batch process* of separation, using the tank still as described above, has been superseded by the continuous flow *pipe still*, which is more efficient, faster, and more flexible.

Here the crude oil is heated in pipes, and the resulting mixture of hot vapors and liquid is discharged into a fractioning or bubble tower. The vapors rise through the tower, where the sorting process is conducted in one continuous operation. Since the vapors cool as they rise, the heavier fractions (which have the higher boiling points) condense in the lower part of the tower. The lighter fractions, which rise higher before they condense, become liquid in the upper part of the tower. The tower is filled with numerous liquid-collecting plates and draw-off pipes whereby the various fractions are drawn off at different levels.

Conversion

The second major step of refining is the *conversion process*, which serves the purpose of increasing the output of motor gasoline. Here the ratio of carbon to hydrogen is changed, and the size and arrangement of the hydrocarbon molecules are altered. Undesired molecules are forced, by an increase in temperature or by a chemical reagent called a *catalyst*, to convert themselves into more useful types of molecules.

Conversion processing began with the invention of *thermal cracking*. This process utilizes a combination of heat and pressure to cause chemical rearrangement, or cracking, of the heavier hydrocarbons, to increase the output of lighter products. Diesel fuels produced as a result of thermal cracking generally are more difficult to ignite than those produced by distillation only.

A highly important advance in refining occurred with the development of the *catalytic cracking* process. A catalyst is any material which promotes chemical changes in other compounds without being itself changed or destroyed. The *Houdry catalytic* cracking process uses a fixed bed, while the later *fluid catalytic* cracking process is a continuous one that uses a powdered catalyst which flows like a liquid. Diesel fuels produced as a result of catalytic cracking generally have good ignition qualities.

Still other conversion processes are employed in modern refineries to increase the quantity or quality of motor gasoline. One of these is *polymerization,* which combines small molecules into larger ones —an action opposite to that of cracking. Using controlled heat and pressure with a catalyst, it converts refinery gases into liquids suitable for blending into motor gasoline.

Properties of Diesel Fuels

Most diesel fuels are made from the fuel oil fraction. This same fraction is also widely used as a household furnace oil as well as for cracking stock for making gasoline as just described. Because of its many uses, the fuel oil fraction is valuable and commands a higher price than the heavier fractions and residue oils. Consequently, cheaper mixtures composed of fuel oil and the heavier fractions are widely used in large diesel engines that can burn the heavier oils satisfactorily.

The properties or *characteristics* of the fuel have considerable influence on the performance and reliability of a diesel engine. These properties can be measured by laboratory tests which will give some indication of the way the fuel will perform in practice, although there is no real substitute for an actual engine test.

The chief fuel properties affecting engine operation are:

1. Ignition quality
2. Heating value
3. Volatility
4. Flash point
5. Pour point
6. Carbon residue
7. Viscosity
8. Sulfur content
9. Water and sediment content
10. Ash content
11. API gravity
12. Specific gravity

Ignition Quality. The ability of a diesel fuel to ignite by itself, or *self-ignite,* under the conditions existing in an engine cylinder, is called the fuel's *ignition quality.* A fuel with a good ignition quality is one which will self-ignite at a low temperature. Such fuel performs better in an engine; it makes starting easier, produces less smoke and reduces fuel knock. Ignition quality is one of the most important properties of a diesel fuel; high-speed engines, particularly, require fuels of good ignition quality.

Ignition quality is expressed by an index called the *cetane number.* As a diesel-fuel characteristic, cetane number corresponds in a way to *octane number* of gasoline. Because of its importance we'll discuss the whole topic of ignition and cetane number in more detail in the next section.

Heating Value. An oil's *heating value* is important because it enables us to find how much heat energy is supplied to an engine and thus how well the engine converts heat energy into work. Heating value can be found by actual test in a calorimeter. But you have just learned that heating value is closely related to gravity. Since calorimeter tests are costly, it is common practice to assume the heating value of a given oil to be the same as that previously found by testing some other oil of the same gravity. Table 10-1 shows the heating value, expressed as

TABLE 10-1 GRAVITY, WEIGHT, AND HEATING VALUE OF FUEL OILS (AT 60° F)

GRAVITY, DEGREES API	SPECIFIC GRAVITY	POUNDS PER GALLON	HIGH-HEAT VALUE	
			Btu per Lb	Btu per Gal
10	1.0000	8.33	18,540	154,620
11	0.9930	8.27	18,590	153,740
12	0.9861	8.22	18,640	153,220
13	0.9792	8.16	18,690	152,510
14	0.9725	8.10	18,740	151,790
15	0.9659	8.05	18,790	151,260
16	0.9593	7.99	18,840	150,530
17	0.9529	7.94	18,890	149,980
18	0.9465	7.89	18,930	149,360
19	0.9402	7.83	18,980	148,610
20	0.9340	7.78	19,020	147,980
21	0.9279	7.73	19,060	147,330
22	0.9218	7.68	19,110	146,760
23	0.9159	7.63	19,150	146,110
24	0.9100	7.58	19,190	145,460
25	0.9042	7.53	19,230	144,800
26	0.8984	7.49	19,270	144,330
27	0.8927	7.44	19,310	143,670
28	0.8871	7.39	19,350	142,990
29	0.8816	7.35	19,380	142,440
30	0.8762	7.30	19,420	141,770
31	0.8708	7.26	19,450	141,210
32	0.8654	7.21	19,490	140,520
33	0.8602	7.17	19,520	139,960
34	0.8550	7.12	19,560	139,270
35	0.8498	7.08	19,590	138,690
36	0.8448	7.04	19,620	138,120
37	0.8398	7.00	19,650	137,550
38	0.8348	6.96	19,680	136,970
39	0.8299	6.92	19,720	136,460
40	0.8251	6.87	19,750	135,680
41	0.8203	6.83	19,780	135,090
42	0.8156	6.79	19,810	134,510

high-heat value, which can be expected in oils of various gravities. These are average figures, derived from numerous tests.

Volatility. The readiness with which a liquid changes to a vapor is known as the *volatility* of the liquid. In the case of diesel fuel, volatility is indicated by the *90 percent distillation temperature.* This is the temperature at which 90 percent of a sample of fuel has distilled off. The lower this temperature, the higher the volatility of the fuel. In small diesel engines a higher fuel volatility is needed than in larger engines in order to obtain a low fuel consumption, low exhaust temperature, and minimum smoke.

Flash Point. An oil's *flash point* is the lowest temperature at which it will give off inflammable vapors in sufficient quantity to flash or momentarily ignite when brought in contact with a flame. It is an index of fire hazard; a fuel oil having an excessively low flash point is dangerous in storage and handling. Minimum flash points are fixed by law in some states and also by insurance regulations.

Don't make the mistake of assuming that flash point indicates how an oil will ignite in a diesel engine cylinder. That depends upon the ignition quality of the oil. In fact gasoline, which has a very low flash point, is a very poor diesel fuel because of its poor ignition quality.

Pour Point. The temperature at which an oil solidifies or congeals is its *pour point.* It indicates the suitability of the oil for cold weather operation. A high pour point indicates that in cold weather the oil will not flow readily through the fuel system and will not produce a good spray when injected into the engine.

Carbon Residue. After all volatile matter in a sample of oil has been evaporated off by heating in a closed container in the absence of air, that which remains is *carbon residue.* This test is a measure of the amount of heavy components which, instead of evaporating when heated, remain behind and form coke (a form of carbon). Thus the carbon residue of an oil indicates its tendency to form carbon deposits on engine parts. The permissible amount of carbon residue depends largely on engine size and speed, being more for large slow-speed engines than for small high-speed engines.

Viscosity. A measure of a fluid's internal friction or resistance to flow is called its *viscosity.* Viscosity is expressed by the number of seconds required for a certain volume of liquid at some standard temperature to flow out through an orifice or hole of a certain small diameter. The longer time it takes, the higher is the viscosity of the fluid.

The instrument used in the United States for determining viscosity of oils is the *Saybolt Viscosimeter* with the *Universal* orifice, and the test results it gives are designated as S.S.U., meaning *seconds, Saybolt Universal.*

As regards diesel engines, viscosity indicates the ability of the fuel to flow in the fuel system and also determines the lubricating value needed to lubricate the fuel pump and injector. Since lubrication of the pump and injection nozzle depends entirely on the fuel oil, its viscosity must not be too low. Viscosity also affects the pattern of the oil spray in the engine cylinder. Low viscosity oils tend to give finely divided, short sprays, while high viscosity oils produce coarser atomization and longer sprays. You will learn more about this in the chapter on fuel injection.

Sulfur Content. *Sulfur* in the fuel burns in the engine cylinder along with the rest of the fuel and forms gases which become corrosive liquids when they react with water. Since water vapor is always present in the products of combustion, sulfur can cause corrosion of the cooled cylinder walls, especially when the engine operates at a low load and the cylinder wall temperature drops low enough to condense some of the water vapor. Similar corrosion from sulfur is frequently found in the exhaust systems of diesel engines.

Water and Sediment Content. *Water* and *sediment* are generally measured together by a centrifuge which separates these heavier materials from the remainder of the oil. Water and sediment may cause corrosion or wear of the fuel pump or injector. Excessive water may also cause irregular combustion.

Ash Content. A fuel's *ash content* is determined by burning a known quantity of the oil and weighing the incombustible ash which remains. The ash usually consists of such impurities as sand and rust, which are extremely abrasive (like sandpaper). Ash content must therefore be low to insure against excessive engine wear. Refined oils generally have a low ash content.

Specific Gravity. The weight of an oil compared to the weight of an equal volume of water is the *specific gravity* of the oil. It is only a rough indication of the grade of the oil. An oil of high specific gravity is termed a heavy oil; of low specific gravity, a light oil. Heavier oils generally contain a greater proportion of heavier hydrocarbons. Such oils tend to flow less readily, to produce poorer sprays and to burn less completely. However, there are many exceptions—a particular heavy oil may be a much better diesel fuel than some other light oil. Putting it another way, oils of the same gravity can differ widely as to viscosity and ignition quality, the two most important diesel fuel properties.

Specific gravity is, however, a necessary test because it is closely related to the heating value of the oil, and it is much easier to measure the specific gravity of an oil than to measure its heating value. High specific gravity of an oil means low heating value per pound. This relation is true because oil is composed mainly of hydrogen and carbon, and because carbon has a much lower heating value and higher atomic weight than hydrogen (as you learned in Chapter 9). Thus the proportions of carbon and hydrogen affect both the specific gravity and the heating value of the oil.

To make this clear, let's compare a heavy oil (high specific gravity) with a light oil (low specific gravity). The reason why the heavy oil is heavier than the light oil is that it contains more carbon and less hydrogen. But because it has more carbon and less hydrogen, the heavy oil must have less heating value per pound. So heavy oils have lower heating value on a *weight* basis. The numerical values are given in Table 10-1.

Diesel fuels are always sold on a *volume* basis (gallons or barrels), whereas their heat value depends on their weight. To convert volume to weight we need to know specific gravity.

API Gravity. The commercial term devised to express specific gravity in the more convenient form of degrees rather than a decimal ratio is *API Gravity*. (API stands for American Petroleum Institute.) The API Gravity Scale is generally

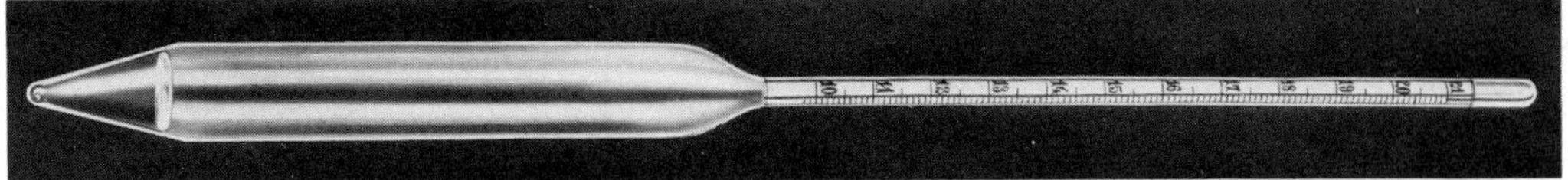

Fig. 10-1. Hydrometer showing API degrees. (Weston Electrical Instrument Corp.)

used in the United States. Gravity in degrees API is related to specific gravity by the following equation:

$$\text{Gravity, API deg} = \frac{141.5}{\text{specific gravity}} - 131.5$$

Table 10-1, which covers the range of diesel fuels, gives the numerical relations between gravity in API deg, specific gravity, and pounds per gallon. Note that the API gravity of water is 10. Most oils are lighter than water and therefore have an API gravity exceeding 10.

To measure the gravity of an oil we use a simple instrument called a hydrometer, illustrated in Fig. 10-1. This consists of a weighted bulb with a narrow stem. When put into the oil it will sink to a depth which depends on the gravity of the oil—the heavier the oil, the less it will sink. The stem is graduated in specific gravity, or API degrees; the graduation found at the surface of the oil shows its gravity.

Ignition

In diesel engines, the fuel is ignited and the combustion started by the heat of compression alone. The several steps are as follows: The fuel enters the cylinder as a fine mist or fog produced by the injection nozzle. It finds the cylinder filled with almost pure air whose temperature has been raised by compression to about 1000°F. The particles of fuel pick up heat from the air, begin to turn into vapor, and soon some vapor particles ignite. This igniting develops additional heat and helps to ignite other vapor particles.

Ignition Quality of a Fuel

A fuel is said to have high ignition quality when it will self-ignite at a low temperature. Thus, fuel of high ignition quality requires but little time to vaporize and attain combustion within a diesel engine. Ignition quality affects engine performance in the following ways:

Ignition Delay. The time which it takes to heat the fuel particles, turn them into vapor, and bring about combustion is called *ignition delay* or *ignition lag.* The higher the ignition quality of a fuel, the shorter the ignition delay. Characteristics of the engine, particularly speed and compression ratio, also affect ignition delay.

In present high-speed engines, ignition delay is only one or two thousandths of a second (0.001–0.002 sec). Ignition delay decreases when an *engine runs faster,* because the more vigorous mixing of fuel and air at the higher speed makes the fuel particles heat up faster. Ignition delay also decreases when the *compression pressure is raised,* because the higher pressure makes the air hotter and thus heats the fuel particles quicker.

Engine Starting. When an engine is being cranked while being started, the injected fuel will ignite and the engine will start if the temperature of the air in the engine cylinder is above the self-ignition temperature of the fuel. The temperature in the engine cylinder depends on the temperature of the outside air—the cooler the outside air, the lower will be the temperature after compression. Consequently, a diesel engine will start in colder weather with an oil of better ignition quality, because the oil will self-ignite at the lower temperature prevailing in the cylinder.

Exhaust Smoke. If the self-ignition temperature of the fuel is not sufficiently low, part of the fuel will fail to burn completely and the resulting soot (unburned carbon) will appear as smoke in the exhaust gases. The smoking will be worst at light loads when the engine temperatures are low.

Fuel Knock. A sharp sound within the cylinder, which is produced by too sudden or too great an increase in pressure when the fuel ignites, is called fuel knock or combustion knock. Fuel knock is caused by the rapid burning of the charge of fuel which accumulated during the ignition delay period, between the time the first of the fuel was injected and the time when the ignition started.

The better the ignition quality of the fuel, the lower will be its self-ignition temperature and the sooner will it ignite. The sooner it ignites, the shorter will be the delay period. The shorter the delay period, the less the quantity of fuel that will have accumulated, and consequently the less fuel will be available to burn rapidly and thus cause fuel knock.

Cetane Number

The term cetane number is used to express the ignition quality of a particular fuel. The cetane number of the fuel is the percentage of cetane in a mixture of two hydrocarbons that has the same ignition quality (self-ignition temperature) as the fuel being tested. The two hydrocarbons are *cetane,* which has an excellent ignition quality, and *alpha-methyl-naphthalene,* which has an exceedingly poor ignition quality. The scale runs from 0 to 100, pure alpha-methyl-naphthalene corresponding to 0, and pure cetane to 100. Thus a fuel having a cetane number of 42 would have the same ignition quality as a mixture composed of 42 percent cetane and 58 percent of the other chemical.

The cetane number of a fuel sample is determined by burning it in a special single-cylinder test engine which has an adjustable compression ratio. The test procedure is based on the fact that in a given engine, running at a fixed speed, two fuels of the same ignition quality will show the same ignition delay at the same compression ratio. The delay period is measured from the moment when the fuel injector opens until the pressure starts to rise in the cylinder.

Using the test fuel, the compression ratio is raised until a standard ignition delay period is reached. Similar tests are then made with differing mixtures of the reference hydrocarbons in order to find the mixture which shows the same delay period for the same compression ratio as was determined for the fuel under test. The percentage of cetane in this mixture is the cetane number of the fuel under test. In other words, the cetane number of the test fuel is the percentage of cetane in the cetane mixture that requires the same compression ratio as the test fuel to produce the standard delay period.

Dual-Fuel Engines Burn Gas and Oil

The dual-fuel engine was developed from the diesel engine to take advantage of the relatively cheap natural gas which pipelines were distributing all over the United States. It is called dual-fuel because it burns both gas and oil.

The dual-fuel engine is a highly efficient engine which closely resembles the diesel engine in several respects. First, it uses the same high compression that a diesel engine does. Second, it operates with much excess air, its mixture being quite lean compared to that used in a gas or gasoline engine. Putting it another way, the air-fuel ratio is high, like that of a diesel engine.

The dual-fuel engine, like the gas or gasoline engine, mixes the fuel and air *before* compression. But despite the high compression to which the mixture is subjected, the mixture is so *lean* that it does not self-ignite in the short time available. A third similarity between dual-fuel and diesel engines is in the method of ignition. No spark plug is used in the duel-fuel engine; instead, the gaseous mixture is ignited by injecting *pilot oil* into the heated mixture. The pilot oil ignites first, in the same way as the oil injected into a diesel engine cylinder; the burning oil then ignites the gaseous mixture.

Spark-ignited gas engines also burn natural gas and other gaseous fuels, but instead of injecting a small charge of oil to ignite the gaseous mixture, they use an electric spark to do the job.

In the next section you will learn something about the gaseous fuels which these engines burn.

Gaseous Fuels

Dual-fuel and spark-ignited engines run successfully on many different kinds of gaseous fuel, so let's first see the general picture of gaseous fuels. We can divide such fuels into three broad classes: (1) natural gas, (2) manufactured gases, and (3) by-product gases. These classes differ not only in the properties of the gas but, importantly, in commercial availability.

Natural Gas

The ideal fuel is *natural* gas. It is found in commercial quantities in more than thirty states, and huge networks of pipelines distribute it to some part of nearly every state.

The origin of natural gas is not known, but it is frequently found associated with oil, and the two fuels are believed to have a common source. Natural gas has neither color nor odor. Its composition varies with the source, but methane (CH_4) is always a major constituent, running from 75 to 95 percent. Most natural gas contains some ethane (C_2H_6) and a small amount of nitrogen. Gas from some areas contains sulfur, usually in the form of hydrogen sulfide (H_2S), which is highly corrosive; such gas is often called sour gas.

The heating value of natural gas, depending on its source, is about 975 to 1200 Btu/cu ft, high-heat value. The low-heat value ranges from about 875 to 1075 Btu/cu ft. Natural gas is usually sold by the cu ft, but may be sold by the *therm*, which is 100,000 Btu.

A high percentage of dual-fuel and spark ignited engines use natural gas.

By-Product Gases

Gases produced in the course of manufacturing other substances are termed *by-product gases*. For instance, the well-known bottled fuels, *butane* and *propane*, are essentially by-products of the manufacture of natural gasoline and of certain oil refinery operations. Both have great heating value (about 2500 Btu/cu ft for propane and 3200 for butane), and are easily liquefied at low pressure. Being by-products, they are relatively cheap.

It has become common practice to utilize the gas produced when sewage sludge is decomposed in sewage disposal plants. This *sewage gas* is often used as fuel in dual-fuel or spark-ignited engines which supply some or all of the power required in the operation of the disposal plant. Sewage gas runs about two-thirds methane and one-third carbon dioxide. Usually some hydrogen sulfide is present. Heat value is about 650 Btu/cu ft high-heat.

Checking On Your Knowledge

The following questions give you the opportunity to check up on yourself. If you have read the chapter carefully, you should be able to answer the questions. If you have any difficulty, read the chapter over once more so that you have the information well in mind before you go on with your reading.

DO YOU KNOW

1. What are the differences between paraffinic, asphaltic and naphthenic based crude oils?
2. What is involved in fractional distillation of crude oil?
3. What is *thermal cracking?*
4. What is *catalytic cracking?*
5. What is *polymerization?*
6. What are the chief fuel properties affecting diesel engine operation and explain the principle of each property as it relates to diesel engine operation.
7. What is the difference between *cetane* number and *octane* number?
8. What are the advantages and disadvantages of gaseous type fuels as compared to liquid type fuels?

Engine Power and Fuel Consumption

Chapter

11

In the previous chapter you learned the theory of heat and combustion. The purpose of this chapter is to make clear how the amount of power which an engine produces and the amount of fuel it burns are related to the heat energy of the fuel.

You'll trace the steps which the heat energy takes as it turns into mechanical energy at the engine shaft, and why no engine can convert *all* of the heat energy of its fuel into actual power output.

You will discover what we mean by efficiency and why an engine gets more work out of its fuel if it has high compression in its cylinders.

But this chapter will do more than discuss general principles. All along we'll work out some simple methods of calculation that you can use in your own daily work, such as how to compute the fuel consumption of an engine.

While reading this chapter you'll become acquainted with some more terms you'll want to understand, such as torque, brake mean effective pressure, and thermal efficiency.

Indicated Power

The term *indicated power* is used to express the power produced *inside* the engine cylinders; it is the first step in changing the heat energy of the fuel into mechanical energy. The reason we call it indicated power is that we can measure it with an *indicator*. This is an instrument which draws lines that show the gas pressure existing inside the cylinder at each point of the piston's stroke. The pictures which the indicator draws are called *indicator cards* and look like Fig. 11-1. Indicator cards are drawn to scales. By measuring the indicator card one can calculate the

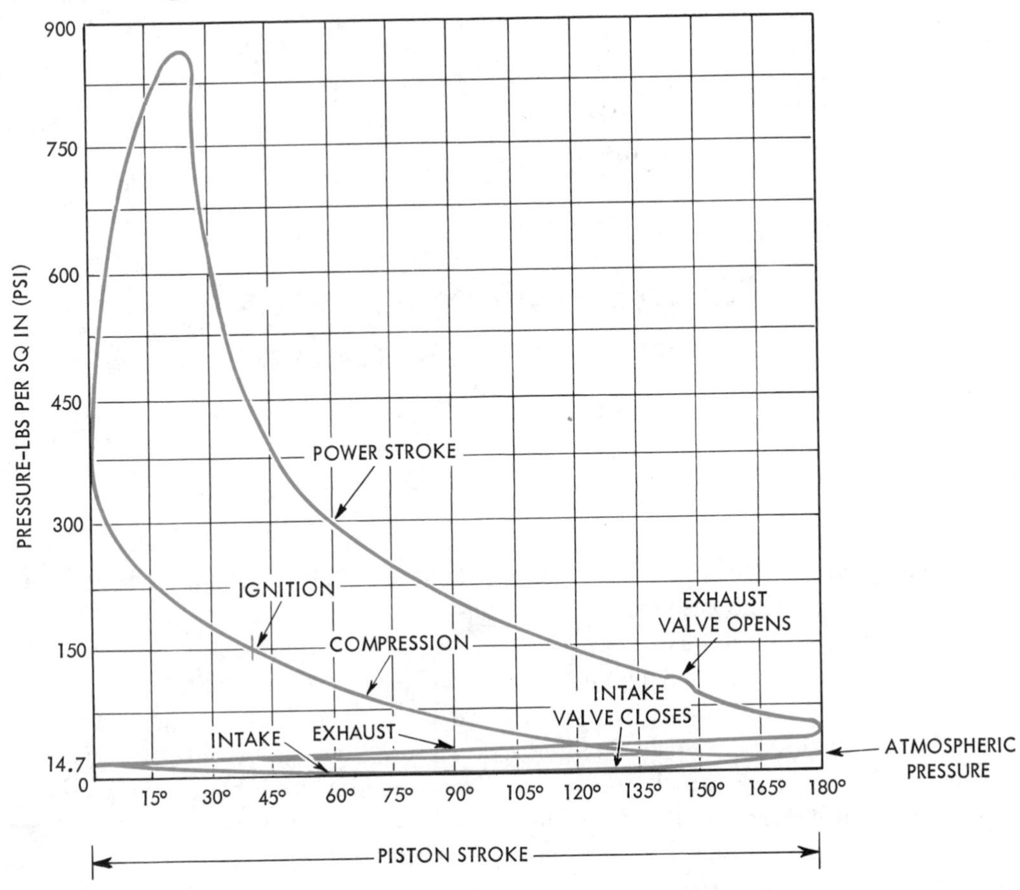

Fig. 11-1. Typical indicator card for four-cycle engine.

amount of mechanical energy which existed in the gases within the cylinder. This is the indicated power.

Indicator cards are not often taken for the purpose of measuring the indicated power of diesel engines in actual use, except on very large, slow-speed engines such as those used on motor ships.

In practical work you are more interested in the actual power available at the engine's crankshaft, which we call *brake power* and which we'll discuss later. Nevertheless, you should know what indicated power is, because scientific tests with indicators have greatly improved the performance of diesel engines by showing the effect of changes in design of combustion space, fuel injectors, etc. In other words, the larger the amount of indicated power we produce by burning a given amount of fuel, the more effectively we have converted the heat or thermal energy of the fuel into mechanical work.

Brake Horsepower

The net power taken from an engine is called the power output or brake power. It is more usually expressed as *brake horsepower* (bhp). The word brake comes from the fact that the power output of an engine can be measured by absorbing the power with a brake.

To illustrate the principle of measuring brakepower, Fig. 11-2 shows a simplified diagram of a brake called a prony brake.

A sophisticated application of the prony brake is being used by one engine test equipment manufacturer that produced a model which will connect to the tractor power-take-off shaft. This power absorption device looks and operates similar to a drum-type brake on an automobile.

Another manufacturer utilizes this power absorption device in an automotive chassis dynamometer.

Chapter 17 includes much more details regarding power measuring devices.

The principle of the prony brake (Fig. 11-2) is to absorb the power output of the engine in the friction of a series of wood blocks, *A*, squeezed against the rim of a wheel (often the flywheel) fastened to the engine's shaft. The belt of wood blocks is attached to a lever arm, *L*, the end of which presses on the platform of a weighing scale, *W*.

Before the engine is tested, the wood blocks are loosened from the wheel by backing off the adjusting nuts, *B*. The dead weight of the lever arm is then weighed on the scale; call this weight W_0.

The engine is then brought up to speed and the friction belt is applied by tightening the hexnuts, *B*. The friction between belt and wheel increases, causing the lever arm to press down on the scale platform with a certain force; the amount of this force can be found by balancing the scale. Call this measured force W_t. Then the net force *F* which the engine is producing is equal to the scale reading when the engine is running, less the dead weight of the lever, that is,

$$F = W_t - W_o$$

The amount of this force, in pounds, is one of the test figures we need to calculate horsepower.

The other figure we need is the distance through which the force travels in one minute. To determine this, we measure the engine speed and also the length of the lever arm, *L*. The effective length or radius of the lever arm is the distance *R*, which is the distance from the center-line of the wheel to the point where the lever arm presses on the scale. (To improve accuracy, a knife edge, *E*, is used where the lever arm presses on the scale.) The distance through which the force at the end of the scale arm acts in one revolution of the wheel is the circumference of a circle whose radius is the length of the lever arm.

You will recall that the circumference of a circle is 3.14 times its diameter or 6.28 times its radius; consequently, the distance through which the force acts *in one*

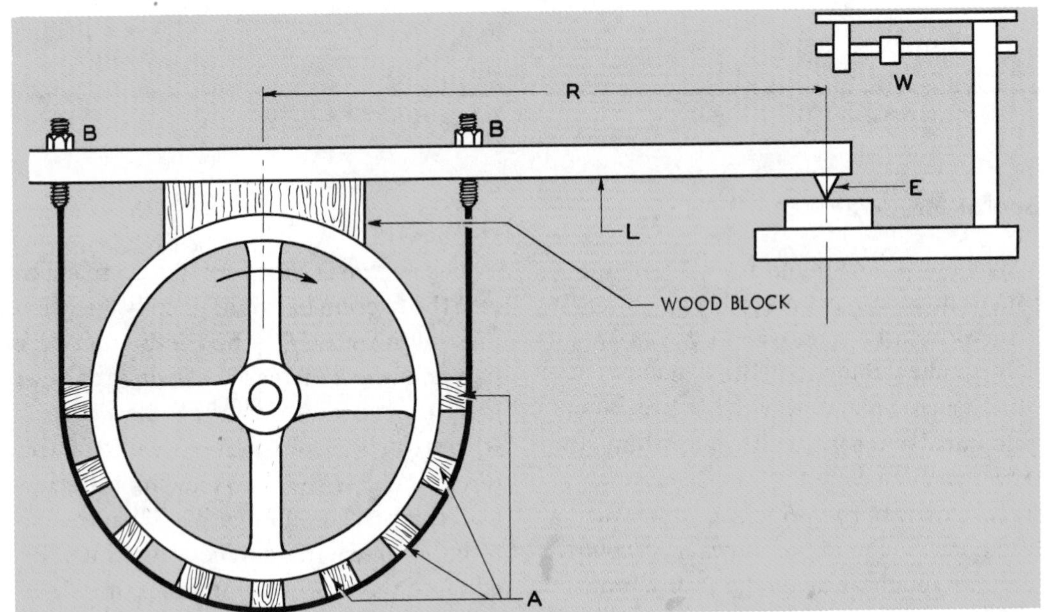

Fig. 11-2. Diagram of prony brake.

revolution is 6.28 × *R*. In *one minute* the force acts through a distance equal to the circumference multiplied by the number of complete turns it makes in a minute. So, if the speed of the engine (and wheel) is found to be *N* rpm, the distance the force acts in a minute is equal to the circumference × *N*, or 6.28 *RN*.

Now we have all we need to calculate the brake horsepower of the engine. Remembering that we have one horsepower when the force (lb) times the distance (ft) through which it acts in one minute equals 33,000, we find:

Brake horsepower =

$$\frac{\text{net force on scale} \times \text{distance covered in one minute}}{33{,}000}$$

or

$$\text{Brake horsepower} = \frac{F \times 6.28\,RN}{33{,}000}$$

$$= \frac{FRN}{5250}$$

where

R = effective length of brake lever, ft
N = speed of engine, rpm
F = net force measured on scale, lb

EXAMPLE 1. Find the brake horsepower developed if an engine runs at 1150 rpm and the test scale reading is 151 lb. The prony brake used has a lever arm with an effective length of 4 ft and a dead weight of 22 lb.

First, we find the net force:

$$F = W_t - W_o$$
$$F = 151 - 22 = 129 \text{ lb}$$

Then, as

$R = 4$ ft
$N = 1150$ rpm

we proceed:

$$\text{Brake horsepower} = \frac{FRN}{5250}$$

$$= \frac{129 \times 4 \times 1150}{5250} = 113 \text{ bhp}$$

Power output can be measured in other ways than by a prony brake. One method that is often used is to connect the engine to an electric generator and then measure either the twisting force or torque required to turn the generator or the amount of electric energy delivered by the generator. (You'll learn more about torque later in this chapter.)

The brake horsepower of an engine is always less than the indicated horsepower. The indicated horsepower is the power developed by the gases within the cylinders, but before this gas power appears at the engine crankshaft as actual output it must pass through the moving parts of the engine. These parts, such as pistons and bearings all consume a certain amount of power because of their *friction*. Friction is the loss of power which occurs when one surface rubs on another; the *friction horsepower* of an engine is the horsepower required to overcome the friction of its moving parts, including any auxiliary equipment driven directly by the engine. The actual or brake horsepower of an engine is therefore equal to its indicated horsepower less its friction horsepower (fhp).

Torque

Earlier in this book you learned that the crankshaft of an engine is forced to turn by the pushes given to it by the cranks to which the pistons are attached. This turning or twisting force is called the engine *torque*. After torque has been applied to a crankshaft by its cranks, the shaft can in turn apply torque to whatever the engine drives, such as an electric generator or the drive shaft of a vehicle. Thus torque is directly related to the power of an engine; if we know an engine's torque and speed we can directly calculate its power. On the other hand, if we know an engine's power and speed, we can calculate how much turning effort the crankshaft will apply to the engine's load.

Torque is expressed quite simply. You merely multiply the turning force by the length of the lever on which it acts at a right angle. Fig. 11-3 shows a diagram of one crank of an engine. If a force F acts on the crankpin at a right angle to the line joining the centers of the crankpin and the main journal, the torque T tending to turn the crankshaft is equal to $F \times R$, where R is the distance between the centers of the crankpin and the main journal. The distance R is called the *lever arm*. If the force is expressed in pounds and the lever arm in feet, then the torque will be expressed in pound-feet (lb-ft). Torque is also expressed sometimes in pound-inches (lb-in).

$$T = F \times R \text{ (lever arm)}$$
$$T \text{ (lb-ft)} = F \text{ (lb)} \times R \text{ (ft)}$$

Notice that torque is always a product of the amount of force and the length of lever arm. Consequently, for a given torque, we can have numerous combinations of amount of force and length of

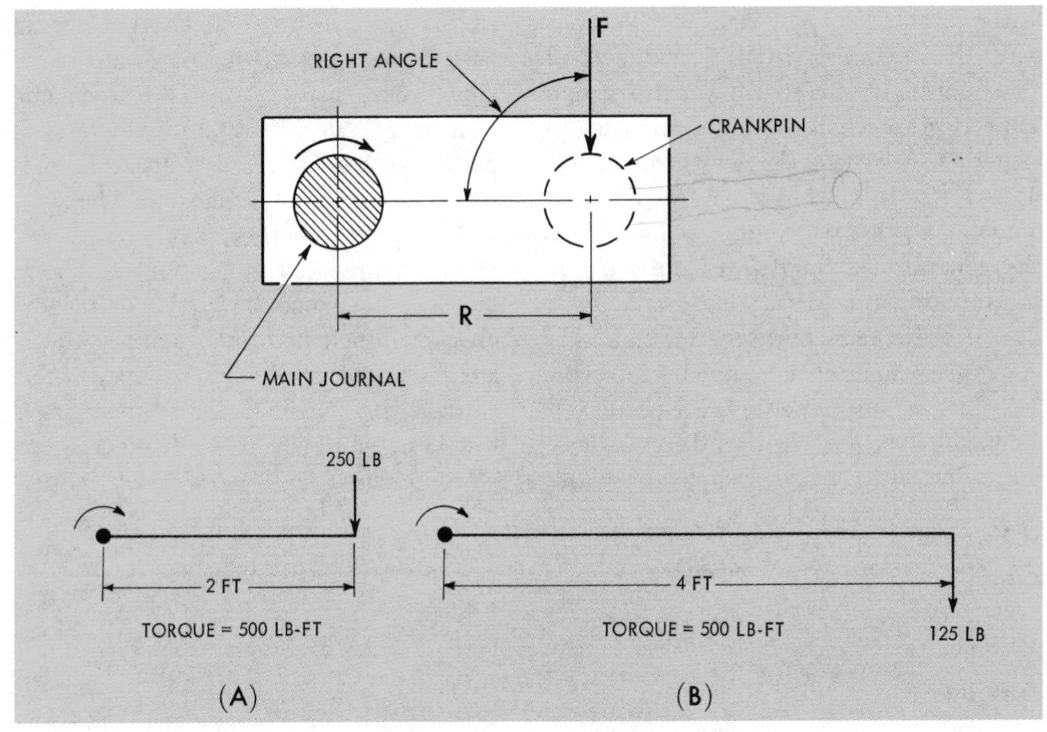

Fig. 11-3. Engine torque.

lever arm. For example, suppose the force on the crankpin in Fig. 11-3A is 250 lb and the lever arm (which is the length of the crank) is 2 ft. Then the torque rotating the crankshaft is 500 lb-ft. If the crankshaft is connected to a load having a lever arm of 4 ft, then the force which will be applied to the load is $\frac{500}{4} = 125$ lb, as shown in Fig. 11-3B.

Now let's find the relation between torque and power. You've just learned that torque $T = FR$. You also learned in the previous section that brake horsepower $= \frac{FRN}{5250}$. Substituting T for FR in this equation we get:

$$\text{Brake horsepower, bhp,} = \frac{TN}{5250}$$

where

$T =$ torque in lb-ft
$N =$ speed in rpm

This equation, rearranged, shows also that:

$$\text{Torque, } T, = \frac{5250 \times \text{bhp}}{N}$$

Example 2. What torque is produced by an engine which develops 62 bhp at 850 rpm?

By writing terms in an equation and substituting, we find:

$$T = \frac{5250 \times \text{bhp}}{N} = \frac{5250 \times 62}{850}$$
$$= 383 \text{ lb-ft}$$

Brake Mean Effective Pressure

The actual output of an engine, its brake horsepower, can be conceived as being related to the number of cylinders, the diameter and stroke of the pistons, the engine speed, and one other factor called the *brake mean effective pressure*, which we abbreviate to *bmep.* Brake mean effective pressure is a useful concept that helps us judge the safe power output of an engine, so let's try to picture what it means. Suppose that instead of the gas pressure on the piston changing all the time, the gas-pressure remained fixed at the *average* pressure.

Then let's assume that this average gas pressure is reduced by the amount needed to overcome the friction losses; the result is what we call the brake mean effective pressure. Looking at it another way, we can picture the brake mean effective pressure as the average pressure which, if imposed on the pistons of a frictionless engine (of the same dimensions and speed), would produce the same power output as the actual engine.

Brake mean effective pressure cannot be measured directly, but it can easily be calculated by means of the brake horsepower equation if we know the horsepower output of an engine, its speed and the diameter and stroke of its pistons. When we discuss engine rating in the next chapter, you'll see how useful the term mean effective pressure is.

A simplified formula for calculating brake mean effective pressure for common four-cycle engines is:

$$\text{bmep} = 1{,}008{,}000 \frac{\text{bhp}}{D^2LMN}$$

in which

bmep = brake mean effective pressure, psi
bhp = brake horsepower
D = diameter of each piston, inches
L = stroke of each piston, inches
M = number of engine cylinders
N = engine speed, rpm

If the engine is two-cycle, divide the answer by two.

EXAMPLE 3. Compute the brake mean effective pressure of a 6-cylinder, four-cycle diesel engine, 5¾″ bore, 8″ stroke, which develops 120 bhp at 1000 rpm.

Expressing terms in an equation, we set forth:

$$\text{bmep} = 1{,}008{,}000 \frac{\text{bhp}}{D^2LMN}$$

where

bhp = 120
$D = 5\frac{3}{4}$
$L = 8$
$M = 6$
$N = 1000$

Substituting, we have:

$$\text{bmep} = 1{,}008{,}000 \frac{120}{5\frac{3}{4} \times 5\frac{3}{4} \times 8 \times 6 \times 1000}$$

$$\text{bmep} = 76.2 \text{ psi}$$

Relation between Bmep and Torque

If you happen to know the engine torque instead of its horsepower, you can calculate the brake mean effective pressure by the following simplified formula applicable to common four-cycle engines:

$$\text{bmep} = 192 \frac{\text{torque}}{D^2LM}$$

in which torque is expressed in lb-ft, and the other terms have the same meaning as before. For a two-cycle engine, divide the answer by two.

Note that you can calculate bmep from the torque *without knowing the engine speed.*

EXAMPLE 4. Compute the brake mean effective pressure of a 4-cylinder, two-cycle engine, 4¼″ bore, 5″ stroke, when it develops a torque of 350 lb-ft.

Setting forth the values given as terms of an equation, we have:

$$\text{bmep for a four-cycle engine} = 192 \frac{\text{torque}}{D^2LM}$$

where

$$\text{torque} = 350$$
$$D = 4\tfrac{1}{4}$$
$$L = 5$$
$$M = 4$$

Substituting, we have:

$$\text{bmep, if a four-cycle engine,} = 192 \frac{350}{4\tfrac{1}{4} \times 4\tfrac{1}{4} \times 5 \times 4} \quad \text{bmep} = 186$$

and

$$\text{bmep for this two-cycle engine} = \frac{186}{2}$$

$$= 93 \text{ psi}$$

Efficiency and Fuel Consumption

In engineering, *efficiency* means the ratio of the output to the input. For example, if we are concerned with the efficiency with which a diesel engine converts the heat energy of its fuel into mechanical power, the output is the work delivered and the input is the heat supplied in the fuel. There are various kinds of efficiencies, as you will see a little later.

Efficiencies are usually expressed as percentages and are always less than 100 percent. The difference between the efficiency and 100 percent is the percent loss incurred during the process concerned.

Some of the efficiencies that enter into diesel engine performance are: (1) mechanical efficiency, (2) indicated thermal efficiency, (3) brake thermal efficiency, and (4) volumetric or charge efficiency.

Mechanical Efficiency

You have learned the meanings of the terms *indicated horsepower* and *brake horsepower*. The first term refers to the power developed by the gases inside the engine cylinder, while the second term represents the actual power output of the engine. In transmitting the gaseous power to the output end of the crankshaft, several *mechanical losses* occur due to friction caused by the rubbing of the pistons and piston rings against the cylinder walls, friction in the various bearings, power absorbed by the valve mechanisms, fuel pumps and injectors, and power absorbed by various auxiliaries, such as the pumps for lubricating oil and water, and the blowers for scavenging or supercharging. These mechanical losses are quite substantial; their amount is shown by the *mechanical efficiency*, which is the ratio of the brake horsepower to the indicated horsepower, expressed as a percentage. Thus:

$$\text{Mechanical efficiency} = \frac{\text{brake horsepower (bhp)}}{\text{indicated horsepower (ihp)}} \times 100$$

The mechanical losses are sometimes expressed by the term *friction horsepower*. Thus:

$$\text{Friction horsepower (fhp)} = \text{ihp} - \text{bhp}$$

In the case of small engines, for which it is difficult to measure indicated horsepower, the friction horsepower is often determined by direct measurement. The engine, with its fuel shut off, is rotated at the desired speed by an electric motor, and the power required to rotate it is measured. We can then obtain the indicated horsepower from the relation:

$$\text{Indicated housepower (ihp)} = \text{bhp} + \text{fhp}$$

EXAMPLE 5. Find the mechanical efficiency of an engine which delivers an output of 57 bhp at a certain speed and which requires 19 hp to rotate it, without fuel, at the same speed.

First find indicated horsepower by substituting given values for terms in the relation ihp = bhp + fhp:

$$\text{ihp} = 57 + 19 = 76$$

Then compute according to equation:

$$\text{Mechanical efficiency} = \frac{\text{bhp}}{\text{ihp}} \times 100 = \frac{57}{76} \times 100 = 75 \text{ percent}$$

Mechanical efficiency depends upon the construction of the engine, workmanship of the various engine parts, and the operating conditions, such as the amount of load carried, temperature of the cooling water, and so on.

Two-cycle engines, generally speaking, have a slightly lower mechanical efficiency than four-cycle engines, due to the power absorbed by the scavenging pump or blower, which is partly offset by its freedom from the friction losses during the exhaust and intake strokes of a four-cycle engine.

An engine having a better designed lubricating system, better workmanship and better alignment of its moving parts, has smaller friction losses and therefore a higher mechanical efficiency. An engine operating at part-load has a lower mechanical efficiency than at full output because most mechanical losses remain nearly constant regardless of the load, and therefore, when the load decreases, the mechanical losses become *relatively* greater.

EXAMPLE 6. Find the mechanical efficiency of the engine discussed in Example 5 when it delivers half the bhp output. Assume that the mechanical losses remain constant.

First we set forth:

$$\text{bhp} = \frac{57}{2} = 28.5$$

and

$$\text{ihp} = \text{bhp} + \text{fhp} = 28.5 + 19 = 47.5$$

Therefore,

$$\text{Mechanical efficiency} = \frac{\text{bhp}}{\text{ihp}} \times 100 = \frac{28.5}{47.5} \times 100 = 60 \text{ percent}$$

Mechanical efficiency of most diesel engines, at full load, is 70 to 85 percent.

Indicated Thermal Efficiency

The ratio of the work done by the gases in the cylinder to the heat energy, or *thermal* energy, of the fuel supplied in the same time is called *indicated thermal efficiency*. Since the work done by the gases is called *indicated* work, the thermal efficiency thus determined is termed *indicated thermal efficiency*. We can say, therefore, that:

$$\text{Indicated thermal efficiency} = \frac{\text{indicated work}}{\text{heat input}} \times 100$$

Indicated thermal efficiency is a term employed in research and laboratory work, but is of little practical value to owners and operators of diesel engines. It leads us, however, to another kind of thermal efficiency which is useful to the reader of this book and which we will next discuss.

Brake Thermal Efficiency

What is known as *brake thermal efficiency* differs from indicated thermal efficiency in the fact that, instead of considering the work done by the gases in the cylinder, we use the actual power output of the engine, that is, its useful work in brake horsepower. It is expressed as follows:

$$\text{Brake thermal efficiency} = \frac{\text{actual output}}{\text{heat input}} \times 100$$

Thus, the brake thermal efficiency tells us how effectively an engine converts the heat energy in its fuel into actual (net) power at the shaft. Brake thermal efficiency takes account of *all* the losses in the engine—the thermal losses when the heat energy of the fuel is converted into temperature and pressure of the gases within the cylinder, and the mechanical losses when the gaseous work (indicated power) is converted into useful output at the engine shaft. Brake thermal efficiency is therefore equal to indicated thermal efficiency multiplied by mechanical efficiency. For this reason it is sometimes called *overall efficiency*.

Fuel Consumption. Before making use of the preceding expression for brake thermal efficiency you must take care to express both the actual output and the heat input *in the same units*. The heat input is most readily expressed in Btu, because we find the heat input by measuring the amount of fuel which the engine consumes in a certain length of time. This is called the *fuel consumption*.

We must also know the heating value of the fuel. In the case of oil the fuel consumption is usually measured in pounds, and the heating value in Btu/lb. In the case of the gaseous fuels used in dual-fuel or gas engines we measure the fuel consumption in cubic feet, and the heating value in Btu/cu ft. In either case, the heat input in a given time equals the fuel consumption multiplied by the heating value of the fuel.

In order to express the actual output in the same unit, Btu, use the conversion factor you learned when you studied energy. This factor is: 1 hp-hr = 2544 Btu. Thus, if an engine develops, say, 50 bhp for two hours, its power output is $50 \times 2 = 100$ bhp-hr, which can be expressed as $100 \times 2544 = 254{,}400$ Btu. If in the *same length of time*, two hours, the fuel consumed by the engine contained 800,000 Btu, the brake thermal efficiency would be:

$$\frac{254{,}400}{800{,}000} \times 100 = 31.8 \text{ percent}$$

EXAMPLE 7. Find the brake thermal efficiency of an engine which consumes 16.2 lb of fuel in 20 minutes while developing a power output of 126 bhp. The fuel has a heating value of 19,300 Btu/lb.

During the 20-minute period of the test, the engine develops 126 bhp for a period of $\frac{1}{3}$ hour, or $126 \times \frac{1}{3} = 42$ bhp-hr. In terms of Btu, this is $42 \times 2544 = 106{,}900$ Btu.

During the same period the heat input = lb of oil consumed × heating value per lb = $16.2 \times 19{,}300 = 312{,}700$ Btu.

Therefore:

$$\text{Brake thermal efficiency} = \frac{\text{actual output}}{\text{heat output}} \times 100$$

Substituting, we have:

$$\text{Brake thermal efficiency} = \frac{106{,}900}{312{,}700} \times 100 = 34.2 \text{ percent}$$

Test results and guarantees of diesel engines are usually expressed as pounds of fuel per brake horsepower-hour (lb/bhp-hr). Because the correct name of this term, *specific fuel consumption rate*, is so long, it is usually abbreviated to *fuel consumption.* Thus we say that the fuel consumption of a certain diesel engine is 0.42 lb/bhp-hr.

Similarly, the fuel consumption in the case of gaseous fuel is usually expressed either as cubic feet of gas/bhp-hr, or as Btu/bhp-hr.

When the fuel consumption of an engine is stated in terms of the amount of fuel consumed per brake horsepower-hour, you can calculate the brake thermal efficiency quite easily, as the following example shows.

EXAMPLE 8. Find the brake thermal efficiency of an engine whose fuel consumption is 0.44 lb/bhp-hr, the fuel having a heating value of 19,500 Btu/lb.

For each bhp-hr, the actual output is 2544 Btu. For each bhp-hr, the heat input is $0.44 \times 19{,}500 = 8580$ Btu. Therefore:

$$\text{Brake thermal efficiency} = \frac{\text{actual output}}{\text{heat input}} \times 100$$

Substituting:

$$\text{Brake thermal efficiency} = \frac{2544}{8580} \times 100 = 29.6 \text{ percent}$$

Reversing this calculation, we can estimate how much fuel an engine will consume if we know its brake thermal efficiency. The expression becomes:

$$\text{Heat input} = \frac{\text{actual output}}{\text{brake thermal efficiency}} \times 100$$

EXAMPLE 9. A certain type of diesel engine is known to have a brake thermal efficiency of 38 percent. Estimate its fuel consumption when using fuel of 18-deg API gravity.

We want to find the fuel consumption per *bhp-hr;* therefore, we calculate the heat input needed for an output of one bhp-hr. We know that one bhp-hr equals 2544 Btu per hour; therefore, the amount of actual output, in Btu, is 2544.

$$\text{Heat input} = \frac{\text{actual output}}{\text{brake thermal efficiency}} \times 100$$

Substituting:

$$\text{Heat input} = \frac{2544}{38} \times 100$$

Heat input = 6695 Btu/bhp-hr

From Table 11-1 we find that the heating value of oil of 18-deg API gravity is 18,930 Btu/lb. The amount of this oil which is needed to give a heat input of 6695 Btu is $\frac{6695}{18,930} = 0.35$ lb. This weight of oil will produce one bhp-hr in the engine having a brake thermal efficiency of 38 percent. In summary:

Fuel consumption = lb of oil/bhp-hr
Fuel consumption = 0.35 lb/bhp-hr

Volumetric Efficiency

The *volumetric efficiency* of a four-cycle engine (unsupercharged) is the ratio of the actual volume of air taken into the engine cylinder during the intake or suction stroke, to the volume of the piston's movement. The latter volume is equal to the piston area multiplied by the piston stroke, and is called piston *displacement.* The actual volume of air taken in is stated in terms of standard temperature and pressure, namely, 60°F and 14.7 psi (Chapters 8 and 9). Thus: Volumetric efficiency (percent) =

$$\frac{\text{volume admitted (std. temp \& pressure)}}{\text{piston displacement}} \times 100$$

Volumetric efficiency shows the amount of air actually taken in as compared with the maximum possible amount of air represented by the piston displacement.

How Volumetric Efficiency Affects Engine Power

You learned, in Chapter 6, that a given volume of gas at a certain pressure and temperature has a certain weight. Therefore, volumetric efficiency tells us the *weight* of air actually in the cylinder compared to the weight, at standard temperature and pressure, of the volume of air equal to the piston displacement.

The weight of air taken into the cylinder is less than the weight of the piston displacement at standard temperature and pressure because the air in the cylinder is always hotter and its pressure is always lower than the standard values of 60°F and 14.7 psi. The temperature of the fresh air drawn into the cylinder is raised, (1) due to mixing with the hot gases left in the clearance space from the previous cycle and, (2) due to contact with the hot surfaces of cylinder walls, piston crown, and cylinder head. The pressure in the cylinder at the end of the intake stroke is lower than the atmospheric pressure outside because of the resistance to air flow through the intake valve (port) and intake passages.

Engine power capacity is directly related to volumetric efficiency—the lower the volumetric efficiency, the less power an engine can develop. This is because with lower volumetric efficiency we have less weight of air in the cylinder, which in turn means that less fuel can be burned and less power can be developed. Thus, the power capacity of an engine can be increased if its volumetric efficiency is improved by taking air from a point in the engine room where the air is cooler or by improving the valve timing so as to increase the pressure at the end of the intake stroke.

In present four-cycle engines, volumetric efficiency is on the order of 80 to 87 percent, the higher figure being reached at slower engine speed.

For two-cycle engines the concept of volumetric efficiency does not apply. Instead, the term *scavenge efficiency* is used, which shows how thoroughly the burned gases are removed and the cylinder filled with fresh air. Scavenge efficiency depends greatly on the arrangement of the exhaust and scavenge air ports and valves. As in the case of four-cycle engines, the power capacity can be increased by taking the air for the scavenge blower at a point where the air is cooler.

In a supercharged four-cycle engine (you learned in Chapter 1 what supercharging is) the intake air is not drawn in by the piston but is forced in by a separate supercharging blower. Consequently the amount of air in the cylinder is not directly related to the piston displacement, and the term volumetric efficiency does not apply. In most supercharged engines the amount of air in the cylinder is considerably more than that corresponding to the piston displacement.

Effect of Compression Ratio on Thermal Efficiency

You have learned that diesel engines have high efficiency (low fuel consumption) because the diesel principle permits the use of high compression ratios. This high compression ratio is accomplished by reducing the clearance space, which is the space left above the piston when it has reached the top of its stroke; this space is also called the *clearance volume*. The *compression ratio* is the ratio between the total volume above the piston when it is at the bottom of its stroke, and the clearance volume. Thus, in Fig. 11-4, A is the total volume, B is the clearance volume, and $\frac{A}{B}$ is the compression ratio.

Now let's see why we get higher efficiency when we increase the compression ratio. The real reason is that when we increase the compression ratio, we likewise increase the *expansion ratio*, which is the ratio of the total volume at the end of the expansion or power stroke to the volume at the beginning of the expansion stroke.

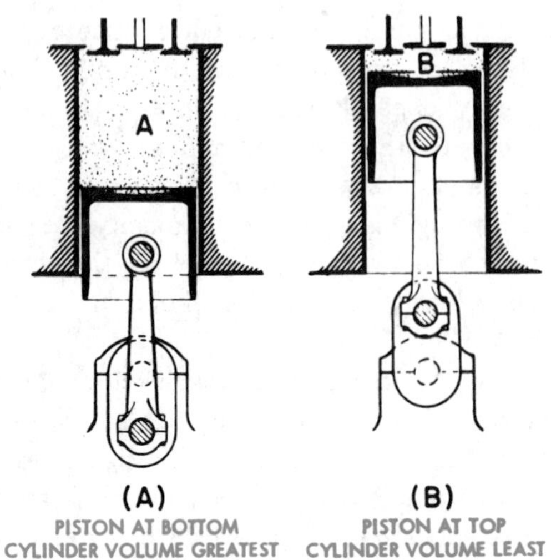

Fig. 11-4. Compression ratio.

Both the compression ratio and the expansion ratio depend upon the clearance volume—when we reduce the clearance volume, we increase both the compression ratio and the expansion ratio.

It is the increase in expansion ratio that gives us better efficiency. The more we expand a gas, the more heat we take out of it; in other words, the more of its heat energy (Btu) is transformed into mechanical energy (ft-lb).

Theory indicates and tests prove that the drop in temperature of the gases (that is, the loss in their heat energy) in a diesel engine during the power stroke becomes greater if the compression ratio is increased. Therefore, the improved efficiency obtained with higher compression is not due to compression itself, but to the fact that higher compression results in more complete expansion, which in turn converts more of the thermal energy of the combustion products into mechanical work.

To give you an idea of the numerical relation between compression ratio and thermal efficiency, Table 11-1 shows the calculated figures for various compression ratios. (For simplicity, the figures were calculated for an ideal engine without friction and other losses.)

Note from Table 11-1 that the gain in thermal efficiency is quite rapid at first, but that the *rate* of gain falls off as the compression ratio becomes higher. This is one important reason why it does not pay to use excessively high compression ratios.

TABLE 11-1 RELATION OF COMPRESSION RATIO TO INDICATED THERMAL EFFICIENCY OF AN IDEAL DIESEL ENGINE

Compression Ratio	Percent Thermal Efficiency
3	34.3
4	41.0
6	48.9
8	53.5
10	56.9
12	59.4
14	61.2
16	62.8
18	63.9

Where the Lost Heat Goes

Few diesel engines convert as much as half the heating value of their fuel into gaseous energy in the cylinder (indicated power). What becomes of the remainder of the heat energy? Part of it is picked up by the surrounding metal surfaces of the cylinder, cylinder head, and piston. After traveling through these engine parts, it either heats the water which is passing through the cooling jackets or is radiated to the surrounding air. The rest of the heat remains in the gases after they have completed their expansion during the power stroke, and passes out of the engine in the exhaust gases.

This lost heat is important for several reasons.

1. Some of it can be recovered and utilized for heating purposes.

2. The heat which enters the metal walls must not be excessive; otherwise the engine parts may burn or crack.

3. The heat in the exhaust gases tends to burn the exhaust valves of a four-cycle engine or the exhaust ports of a two-cycle engine.

Table 11-2 shows the distribution of heat energy in a typical diesel engine when running at full load. Engines of different designs and sizes will show considerable variation.

Note the large amount of heat which finds its way out of the engine in the jacket water and in the exhaust gases.

TABLE 11-2 HEAT DISTRIBUTION IN TYPICAL DIESEL ENGINE AT FULL LOAD

Where the Heat Goes	Btu per Bhp-Hr	Percent of Heat in Fuel
Brake work	2545	33
Friction	680	9
Heating of jacket water	2100	27
Heat in exhaust gases	2200	28
Radiation	255	3
Total	7780	100

Checking On Your Knowledge

1. What is an indicator card and how is it used to find engine efficiency?

2. What is the relationship between *brake horsepower* and *indicated horsepower?*

3. Compare the importance of *brake horsepower* and *torque* in diesel engine performance specifications.

4. What is *brake mean effective pressure* and of what significance is it to engine power?

5. What is the relationship of *brake mean effective pressure* and *torque?*

6. Compare *mechanical efficiency, indicated thermal efficiency, brake thermal efficiency,* and *volumetric efficiency.*

7. Why does *compression ratio* affect *thermal efficiency?*

8. Explain the various heat losses in a diesel engine.

Chapter

12 Engine Rating and Performance

The purpose of this chapter is to explain what an engine's capacity or *rating* is, and then to discuss the various factors that limit (*a*) the amount of load which an engine can safely carry, and (*b*) the highest speed at which it should run.

You'll find out why the power output of an engine is limited by the heating of its parts and by incomplete burning of the fuel, which shows up as smoke. In the previous chapter you learned what brake mean effective pressure is; in this chapter you'll see how *bmep* is used as a guide in checking an engine's power capacity.

You'll discover that the highest speed at which an engine can run safely depends upon its lubrication and upon the forces needed to start and stop its pistons as they make their up and down strokes. These inertial forces have an important effect on engine performance; we'll therefore give them considerable attention.

Next you'll see how ratings are actually applied, that is, the various ways in which engine manufacturers specify the ratings of their engines, and how they guarantee the fuel consumption.

Engines lose part of their power capacity when they run at elevations above sea level, where the air is thinner. They also lose power when the air they breathe gets warmer. This chapter will tell you why this is so and how much the capacity is reduced.

All of this information has great practical value. It may, for example, keep you from getting into trouble trying to make a 100-horsepower engine do a 150-horsepower job!

Engine Rating

The *rating* of a given diesel engine or, more precisely, its *capacity rating*, is the amount of brake horsepower which the engine builder states that the engine will deliver at a given speed. In Chapter 11 you learned that the brake horsepower of an engine depends upon the engine's dimensions, its speed, and its brake mean effective pressure. Therefore, for an engine of known dimensions and speed, the builder's rating will be directly proportional to its *brake mean effective pressure.* As the rating goes up, the bmep rises in exact proportion. The following example will refresh your memory.

EXAMPLE 1. Find the rated brake mean effective pressure of a five-cylinder, four-cycle diesel engine, 8″ bore, 10½″ stroke, which is rated 150 bhp at 600 rpm.

Using the formula shown in Chapter 11:

$$\text{bmep} = 1{,}008{,}000 \, \frac{\text{bhp}}{D^2LMN}$$

Where

$$\begin{aligned} \text{bhp} &= 150 \\ D &= 8 \\ L &= 10\tfrac{1}{2} \\ M &= 5 \\ N &= 600 \end{aligned}$$

And substituting:

$$\text{bmep} = 1{,}008{,}000 \times \frac{150}{8 \times 8 \times 10\frac{1}{2} \times 5 \times 600} = 75 \text{ psi}$$

Naturally, engine builders like to rate their engines as high as they can. But there is a limit beyond which it is not wise to go; the bmep of an engine for a certain horsepower and speed is a clue as to how closely the engine has approached that limit. Let's look at the factors which control engine power rating.

Combustion and Cooling Limit Power Rating

The horsepower rating of a given diesel engine, running at a specified speed, is limited mainly by its combustion and cooling characteristics. To get more power, more fuel must be burned. The more fuel is burned, the hotter will be the combustion gases. Meanwhile, the amount of air available to burn the fuel remains fixed, because an engine of a certain size, running at a certain speed, draws in practically the same amount of air regardless of the amount of oil injected into the cylinders (unless the engine is supercharged). Therefore, if we inject more and more

TABLE 12-1 APPROXIMATE BRAKE MEAN EFFECTIVE PRESSURES OF DIESEL ENGINES

Type of Engine	Approximate Bmep Lb / Sq In
4-cycle, medium-speed, continuous rating	75–85
4-cycle, supercharged, continuous rating	120–205
4-cycle, supercharged, intermittent rating	up to 300
2-cycle, various scavenging methods cont. rating	30–75
2-cycle, supercharged, continuous rating	100–130
2-cycle, supercharged, intermittent rating	up to 150

oil, we reach a point where the combustion is no longer complete, part of the fuel is wasted, and the exhaust grows smoky. Also the cooling jackets become unable to cope with the excessive amount of heat produced in the cylinders, whereupon some of the engine parts overheat and fail.

For a given design, the engine builder therefore rates his engine on the basis of a bmep that will insure satisfactory continuous operation. Table 12-1 gives an idea of the bmep found in diesel engines as now built.

Lubrication and Inertia Limit Speed Rating

Engine rating, in terms of brake horsepower, depends not only on bmep but also on speed, as you have seen. Speed, too, has its limits. If the engine speed is increased too much, the pistons rub so fast on the cylinders that it becomes impossible to maintain a lubricating oil film between the rubbing surfaces, whereupon *scoring* or *seizure* is likely to occur. Also, as the speed goes up, the *inertial forces* of the pistons and connecting rods increase rapidly and may become great enough to damage the engine bearings or stretch the connecting rod bolts. Let's spend a few moments on inertia and its effects.

Inertia and Its Effects

Inertia is the resistance of a body to a change of motion. It is the tendency of an object to remain at rest if it is stationary, or to continue to move if it is moving. Thus, a force must be applied *to* an object to start it moving or to make it move faster. Likewise, a force is developed *by* an object when its motion is slowed down or when it is brought to rest.

Inertia comes into play whenever pistons move. The piston is at rest for an instant at each end of its stroke. It has been moving in one direction and is about to move in the other; at the moment the

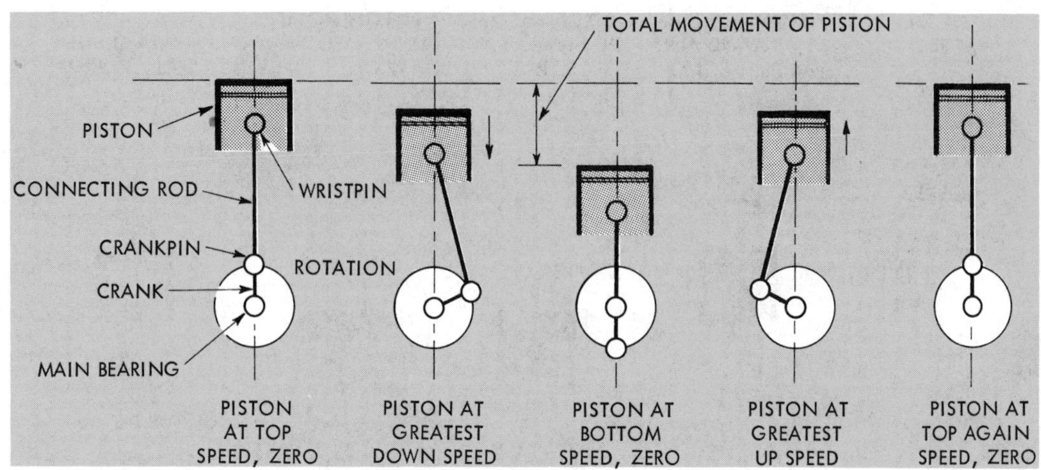

Fig. 12-1. How speed of piston changes.

direction of motion changes, the piston is at rest. The piston's motion is controlled by the crankshaft through the connecting rod. Since the crankshaft rotates at an almost constant speed; the resulting effect on the piston is to make the piston move faster and faster (accelerate) until it gets to nearly the middle of its stroke, where the piston moves at the same speed as the crankpin; at this point the piston starts slowing down, and moves slower and slower (decelerates) until it comes to rest at the other end of its stroke. Study Fig. 12-1 and you will see why this is so.

The force which must be applied to the piston to make it accelerate and the force which is developed by the piston when it decelerates is the piston's *inertial force.* Fig. 12-2 will help you to picture how the inertial force changes when the piston of a four-cycle engine moves into various positions. In Fig. 12-2A, the piston is at rest momentarily at the top of its stroke; there is no inertial force. In Fig. 12-2B the piston's downward speed is increasing. The force required to overcome its inertia comes, on the power stroke, from the gas pressure on the piston top. (But in very high-speed engines the gas pressure is not sufficient of itself to start the piston moving, and the connecting rod must actually *pull down* on the piston to help the gas pressure get started.) In Fig. 12-2C the piston's downward speed is decreasing; consequently the piston's inertial force is *pushing* on the connecting rod in addition to the gas pressure.

In Fig. 12-2D is shown the beginning of the upward, exhaust stroke, when the piston is accelerating upwards. Now the connecting rod is *pushing* the piston, in order to overcome its inertia and move it faster. In Fig. 12-2E is shown the second half of the exhaust stroke when the piston is coming to rest and is *pulling* on the connecting rod. Similar actions take place on the next two strokes, namely, the downward suction stroke and the upward compression stroke. The inertial forces are the same as on the preceding downward and upward strokes; only the gas pressures are different.

You have just seen that sometimes the connecting rod is *pushing* the piston and

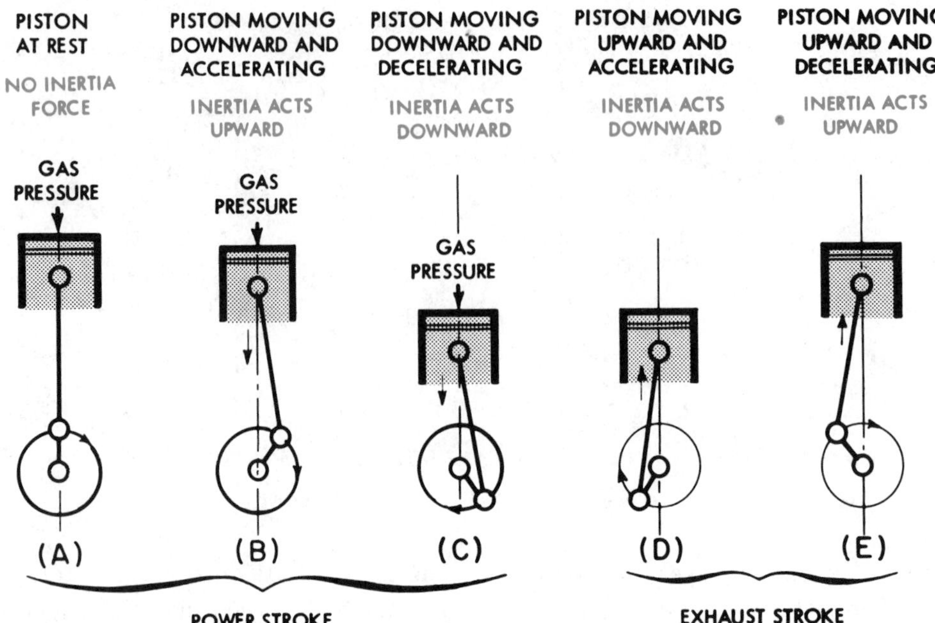

Fig. 12-2. Changes in inertial forces.

at other times it is *pulling* the piston. When the rod is pushing the piston it must be stiff enough to resist bending. When the rod is pulling the piston, Fig. 12-3 shows what happens. Fig. 12-3 is the same as Fig. 12-2E, except that it is enlarged to show the bearings in more detail, as well as the crankpin bolts which enclose the crankpin bearing.

The gas pressure on top of the piston is only the back pressure of the exhaust gases, which is insignificant. Consequently, the forces come solely from the inertial force which the rising piston develops when it is forced to slow down and come to rest at the top of its stroke. Notice how the inertial force of the piston travels through the *upper* half of the wristpin bearing and then through the connecting rod. Next it *pulls* on the crankpin bolts and presses on the *lower* half of the crankpin bearing. Finally it passes through the crank and presses on the *upper* half of the main bearing.

The reason why we have examined in detail what happens near the end of the exhaust stroke (Fig. 12-3.) is that this is the part of the cycle when the inertial forces exert their greatest pull on the crankpin bolts. Because of space limitations these bolts cannot be made as large as we would like. It is important that the inertial forces should not overstrain these bolts.

Inertial forces increase with the square of the engine speed. That is, the inertial force will be increased four times if the engine speed is doubled. Therefore, inertial forces and resulting loads on bearings will soon become excessive if engine speed is increased beyond that for which the engine was designed. Consequently, an increase of 25 percent in engine speed would not only increase the *rubbing*

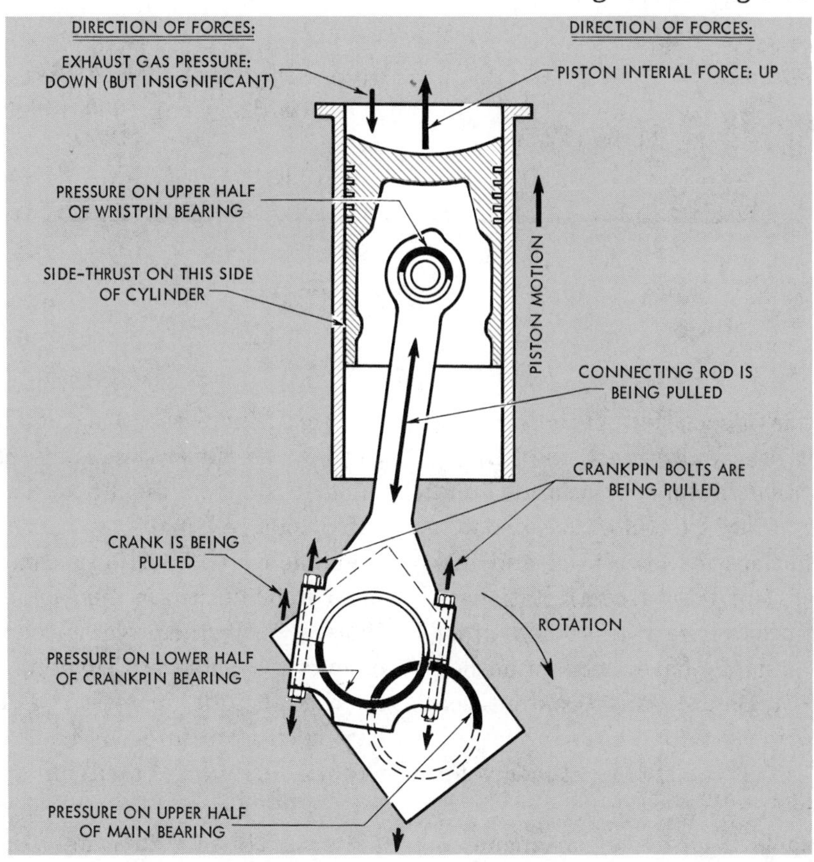

Fig. 12-3. Forces on engine parts caused by piston inertia at end of exhaust stroke.

speed of pistons and bearings 25 percent but, it would increase the inertial forces about 56 percent (1.25 squared equals 1.56) and thus increase by 56 percent the *pressures* on the bearings and the pull on the crankpin bolts.

Piston Speed

You will now realize that the highest speed in rpm at which an engine can be run safely depends directly on the speed with which its pistons travel. That speed is not constant, since the pistons start and stop on every stroke. For rating purposes, therefore, we use the *average* or *mean* piston speed. This is the constant speed with which the piston would have to move to travel the same distance (in the same time) as it travels with the actual speed. This average piston speed, which we simply call *piston speed,* is easily calculated from the engine speed and the length of the piston stroke by the following relation:

$$\text{Piston speed (ft/min)} = \frac{\text{engine rpm} \times \text{piston stroke (in)}}{6}$$

Example 2. Find the piston speed of a $4\frac{1}{2}'' \times 5\frac{1}{2}''$ ("bore × "stroke) engine running at 1200 rpm.

Setting forth terms in an equation, we have:

Piston speed =

$$\frac{1200 \times 5\frac{1}{2}}{6} = 1100 \text{ ft/min}$$

In present diesel engines, the rated piston speed varies from 1000 to about 2000 ft/min.

Standard Ratings

The trade association of builders of diesel engines, known as the Diesel Engine Manufacturers Association, and often called DEMA for short, has adopted standard ratings for low speed and medium speed (up to 750 rpm) stationary diesel engines. These ratings and other DEMA standards have been published by DEMA in Chicago in a book entitled *Standard Practices for Stationary Diesel Engines*. Part of the DEMA standard for power rating follows:

"The sea level rating of an engine is the net brake horsepower that the engine will deliver continuously when in good operating condition and located at an altitude of not over 1500 feet above sea level, with atmospheric temperature not over 90°F and barometric pressure not less than 28.25 inches of mercury. Engine manufacturers offer engines with sufficiently conservative sea level ratings so that the engines will be capable of delivering an output of 10 percent in excess of full load rating, with safe operating temperatures, for 2 continuous hours, but not to exceed a total of 2 hours, out of any 24 consecutive hours of operation."

Ratings at Higher Altitudes

You learned in Chapter 1 that the pressure of the earth's atmosphere decreases as the altitude or height above the earth increases. Thus, at higher altitudes the air is less dense and contains a smaller weight of oxygen per cubic foot. A smaller weight of oxygen available for combustion means that less fuel can be burned in the engine cylinder. Consequently, the power which any diesel engine is capable of delivering decreases as the altitude increases. The normal barometric pressures to be expected at various altitudes, based on a temperature of 90°F, are shown in Fig. 12-4.

Using these relations, DEMA has adopted a standard practice for reducing the power ratings of non-supercharged (naturally aspirated) engines. Fig. 12-5 shows the standard rating at various altitudes of such engines, expressed in percent of sea level rating. Note that the standard sea level rating applies for all altitudes up to 1500 ft above sea level. This is simply a practical provision to make it unnecessary to correct engine ratings for the numerous installations at low altitudes.

Ratings at Elevated Temperatures

As quoted above, DEMA power ratings at sea level and at altitudes are based on atmospheric temperatures at the engine air intake of not more than 90°F. The power which an engine can deliver decreases as the intake air temperature in-

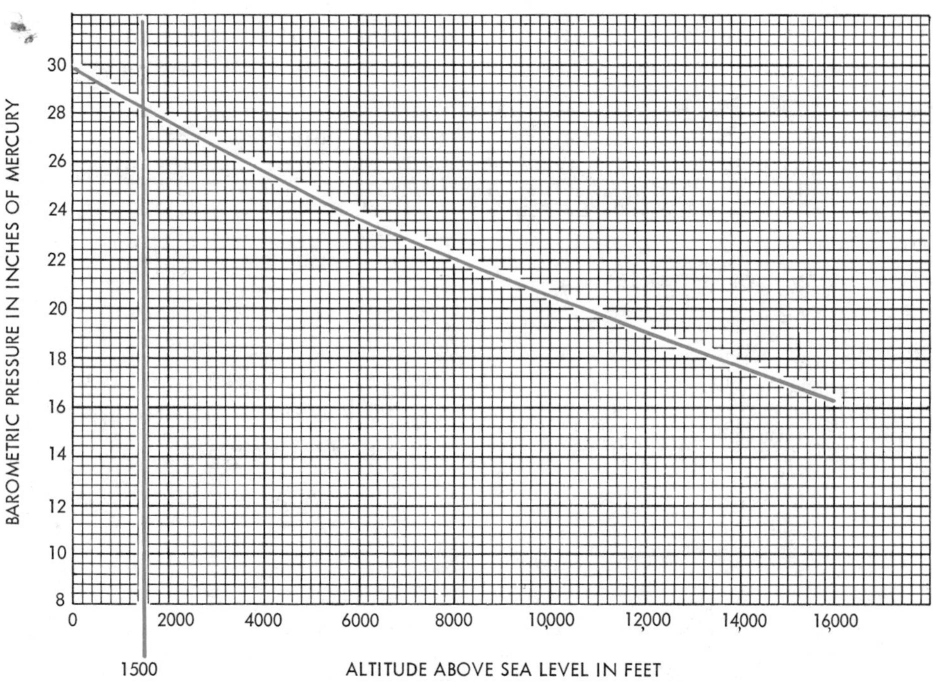

Fig. 12-4. Barometric pressure at various altitudes based on temperature at 90°F. (Diesel Manufacturers' Assn.)

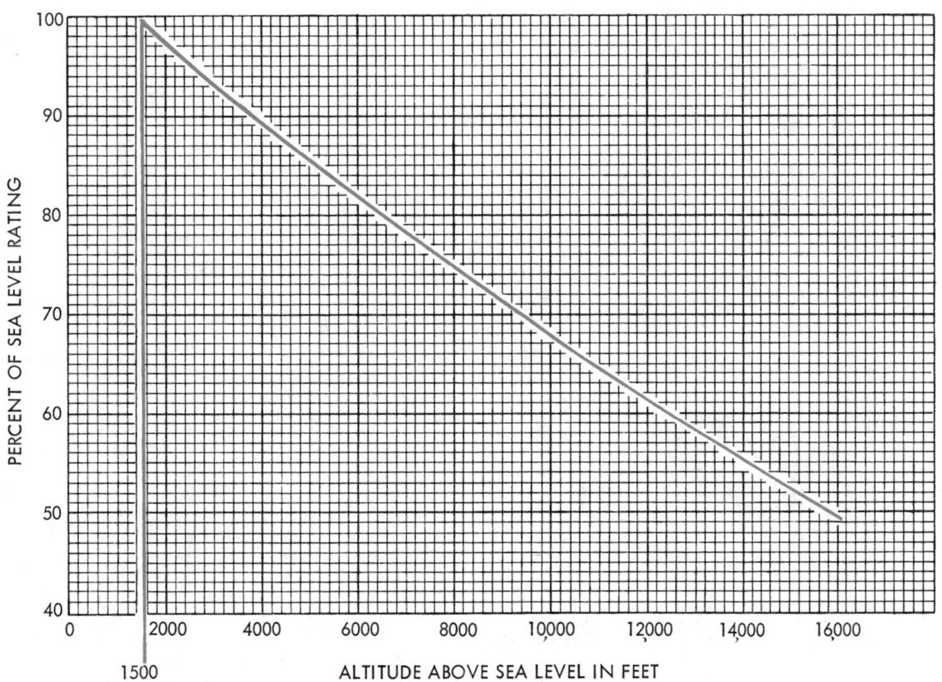

Fig. 12-5. Altitude ratings for non-supercharged diesel engines expressed in percent of sea level brake horsepower ratings. (Diesel Engine Manufacturers' Assn.)

creases. The reason is that the warmer air is less dense and therefore contains less oxygen to burn the fuel in the engine cylinder. Since few localities (in the U.S.) exceed 90°F air temperature for more than a few hours on a few days in the year, DEMA practice is to make engine ratings apply to temperatures up to 90°F, and to reduce the rating only for higher temperatures. In such case the rating is reduced as shown in Fig. 12-6 (in addition to any reduction due to altitude).

Effects of Supercharging and Inlet Air Cooling on Derating

Superchargers and intake air coolers recover part of the loss in engine capacity which would result from operation at high altitudes or at elevated atmospheric temperatures. The percent of capacity so recovered varies greatly with different designs and operating conditions. For this reason no general formula has been agreed upon for making the derating calculations for engines equipped with su-

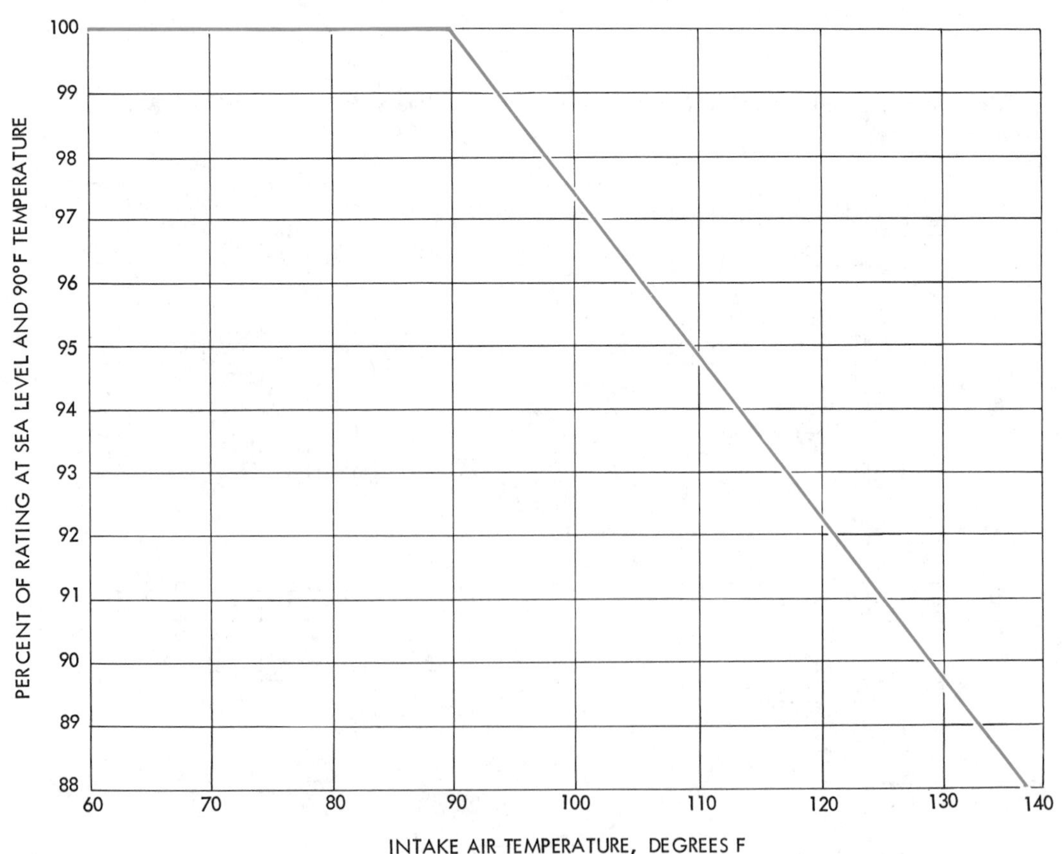

Fig. 12-6. Elevated temperature ratings for non-supercharged engines having no intake air-cooling equipment, expressed in percent of normal rating. (Diesel Engine Manufacturers' Assn.)

perchargers or intake air coolers. Consequently, Figs. 12-5 and 12-6 are applicable only to non-supercharged engines having no intake air cooling equipment; that is, normally operated engines.

Other Power Ratings

Many builders of high-speed diesel engines have not adopted the DEMA standards for power rating. It is common practice for such manufacturers to state two or more ratings for an engine in accordance with the kind of service the engine is to perform. For example, one important manufacturer uses three different kinds of ratings, namely: maximum, intermittent, and continuous. Since the engines are used at several speeds, these ratings are shown in the form of curves on a power chart. Fig. 12-7 is such a chart for a six-cylinder, 5.4″ bore, 6.5″ stroke, su-

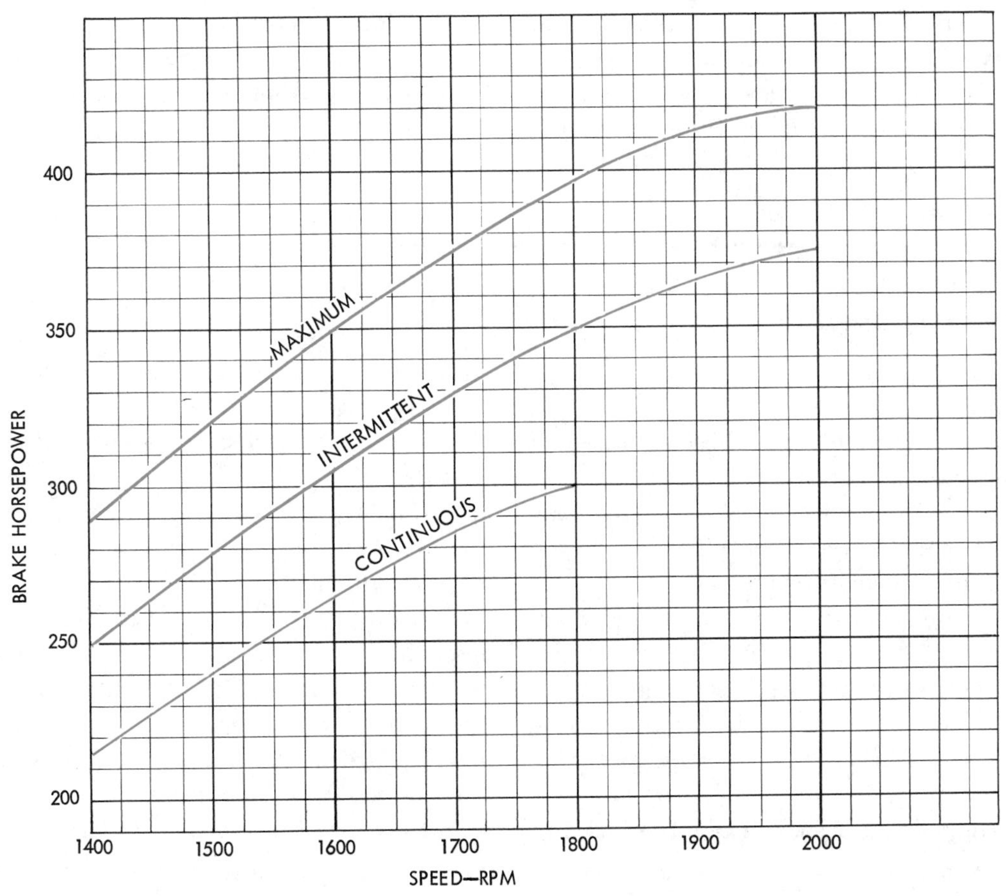

Fig. 12-7. Power rating chart for six-cylinder, 5.4″ bore, 6.5″ stroke, supercharged four-cycle engine. (Caterpillar Tractor Co.)

percharged four-cycle engine.

The *maximum rating*, in the manufacturer's language, is "the horsepower and speed capability of the engine which can be utilized for about five minutes duration, followed by reasonable periods of operation at reduced output. A plus or minus factor of 5 percent is applied to all maximum ratings to allow for production tolerances."

The *intermittent rating* is "the horsepower and speed capability of the engine which can be utilized for about one hour, followed by about an hour of operation at or below the continuous rating." If engines are operated continuously at the intermittent rating, more maintenance will be needed.

The *continuous rating* is the horsepower and speed capability of the engine which can be utilized without interruption or load cycling. This can extend for months or years of operation if the engine is equipped for non-stop lube oil and filter changes.

The foregoing output figures are based on sea level barometric pressure and 60°F temperature.

Using Fig. 12-7, we find that at a speed of 1800 rpm this engine is capable of the following outputs:

Maximum rating ....397 bhp
Intermittent rating ......350 bhp
Continuous rating ....300 bhp

The foregoing shows you how necessary it is to select the *kind* of rating that applies to the *kind* of service which the engine is to perform.

Fuel Consumption

Diesel Engines

Fuel consumption of diesel engines, as you learned in Chapter 11 is always expressed in terms of pounds of fuel per unit of net power output. The net power output is usually expressed in terms of brake horsepower-hours (bhp-hr) and the corresponding fuel consumption as lb/bhp-hr. In the case of diesel engines driving electric generators, the net power output is sometimes expressed in electrical terms, namely, kilowatt-hours, and the corresponding fuel consumption as lb/kwhr.

Guarantees of fuel consumption are made for the engine delivering full rated load, three-quarter load and half-load, when operating at rated rpm.

In accordance with DEMA standards and the "Power Test Code" of the American Society of Mechanical Engineers, fuel consumption guarantees for diesel engines are based on fuel oil having a high-heat value of 19,350 Btu/lb (commercially the heating value of oil is always understood to be high-heat value, unless otherwise stated). If the oil used has a different heat value, the test results are corrected in proportion to the heat value. If the test was made with fuel below standard heat value, the corrected fuel consumption will be less than the test figure. See the following example.

EXAMPLE 3. Tests of a diesel engine using oil of 18,900 high-heat value show a fuel consumption of 0.46 lb/bhp-hr. Correct the fuel consumption to a basis of fuel of 19,350 Btu heating value.

Setting forth terms in an equation and substituting, we have:

$$\text{Corrected fuel consumption} = \frac{\text{heat value of test fuel}}{\text{heat value of reference fuel}} \times \text{actual fuel consumption}$$

$$\text{Corrected fuel consumption} = \frac{18{,}900}{19{,}350} \times 0.46 = 0.449 \text{ lb/bhp-hr}$$

Fig. 12-8 shows curves giving the fuel consumption at 1600, 1800, and 2000 rpm of the diesel engine whose power output was shown in Fig. 12-7. Such curves are useful in finding the fuel consumed by an engine when it is running at partial loads as well as at full load.

For example, the curve for 1800 rpm shows that when the engine is delivering its continuous rating of 300 bhp at 1800 rpm, its fuel consumption (point *A* on the curve) is 0.39 lb/bhp-hr, or 0.39 × 300 = 117 lb/hr. If the engine is called upon to deliver only half of its continuous capacity, or 150 bhp, its fuel consumption (point *B* on the curve) is 0.432 lb/bhp-hr, or 432 × 150 = 64.8 lb/hr. Note that, although the engine uses less fuel at the reduced load, the fuel consumption *per unit of power output* is about 11 percent more.

Dual-Fuel Engines

For dual-fuel engines, which burn gaseous fuel and also a small amount of pilot oil, the gas consumption is expressed and guaranteed as *heat consumption,* in terms of Btu, low-heat value, per net bhp-hr. Although cu ft of gas are measured when testing, and the consumption could therefore be expressed in the form

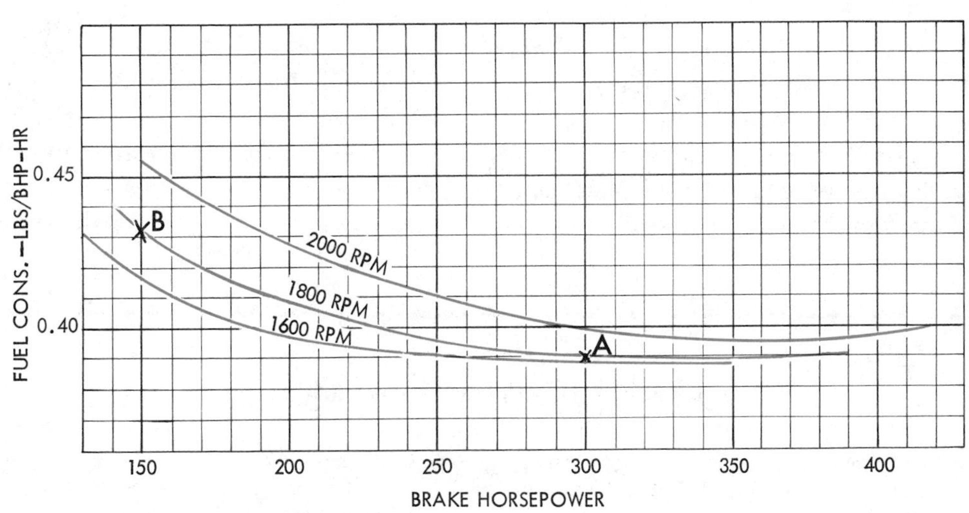

Fig. 12-8. Power rating chart for same engine at higher rpm. (Caterpillar Tractor Co.)

of cu ft of gas per bhp-hr, this term has fallen into disfavor because gas fuels vary widely in heating value, far more, in fact, than do fuel oils. Therefore, *cu ft gas/bhp-hr* would mean little unless we knew the heating value of the gas used in the test.

The reason we express the Btu in terms of low-heat value instead of high-heat value is that, for gaseous fuels, the percentage difference between the low and high-heat values may range from zero to more than 15 percent, because of varying hydrogen content. The engine can utilize only the low-heat value, because the heat in the water vapor produced by burning the hydrogen passes out with the exhaust gases. Consequently, fuel guarantees for engines burning gaseous fuels are always made on the basis of the low-heat value of the gas.

The heat value of the pilot oil consumed by a dual-fuel engine must be added to the heat value of the gaseous fuel in order to obtain the total heat consumption. Logically, the low-heat value of the pilot oil should be used. However, it is commercial practice to use the high-heat value of the pilot oil, because the high-heat value is always employed for oil consumed by diesel engines. The error thus introduced is commercially acceptable because the difference between the high and low-heat values of an oil is only about 6 percent, and the pilot oil contributes less than 10 percent of the total heat. Thus the error is less than 0.6 percent.

For the same reason, we generally assume the heat value of the pilot oil to be 19,350 Btu/lb (which is the basic figure used for diesel engines). This is the high-heat value of 28° API gravity oil and comes close to the actual heat value of any oil suitable for use as pilot oil.

EXAMPLE 4. Express the heat consumption of a dual-fuel engine which consumes 6.8 cu ft of natural gas and 0.021 lb of pilot oil per bhp-hr. The natural gas has a low-heat value of 950 Btu/cu ft. Assume the heat value of the pilot oil is 19,350 Btu/lb.

Heat consumption of gaseous fuel, low-heat value basis,

$$= 6.8 \times 950 = 6460 \text{ Btu/bhp-hr}$$

Heat consumption of pilot oil

$$= 0.021 \times 19{,}350 = 406 \text{ Btu/bhp-hr}$$

Heat consumption, low-heat value basis, total: 6866 Btu/bhp-hr

Gas engine fuel consumption, like that of dual-fuel engines, is expressed and guaranteed as *heat consumption*, in terms of Btu, low-heat value, per net bhp-hr. (There is no pilot oil to be considered.)

Checking On Your Knowledge

The following questions give you the opportunity to check up on yourself. If you have read the chapter carefully, you should be able to answer the questions. If you have any difficulty, read the chapter over once more so that you have the information well in mind before you go on with your reading.

DO YOU KNOW

1. How does combustion and cooling limit power rating of a diesel engine?
2. Piston speed varies throughout each stroke. What significance is this to design of diesel engines? What position in the stroke is the piston moving at the greatest speed?
3. How does altitude affect diesel engine ratings?
4. How does temperature affect diesel ratings?
5. Compare supercharging with and without inlet air cooling and explain.
6. How does diesel engine fuel consumption relate to load and explain the relationship?
7. How does diesel engine fuel consumption relate to speed and explain the relationship?

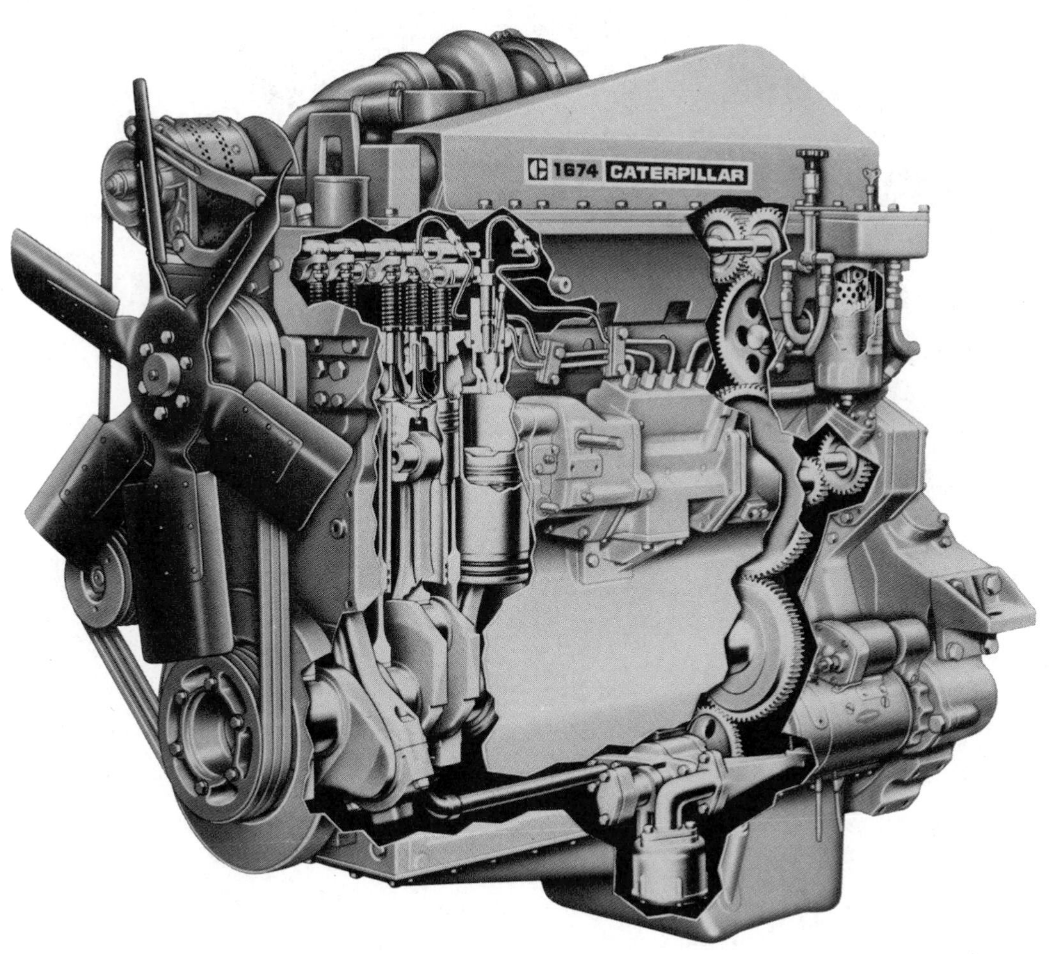

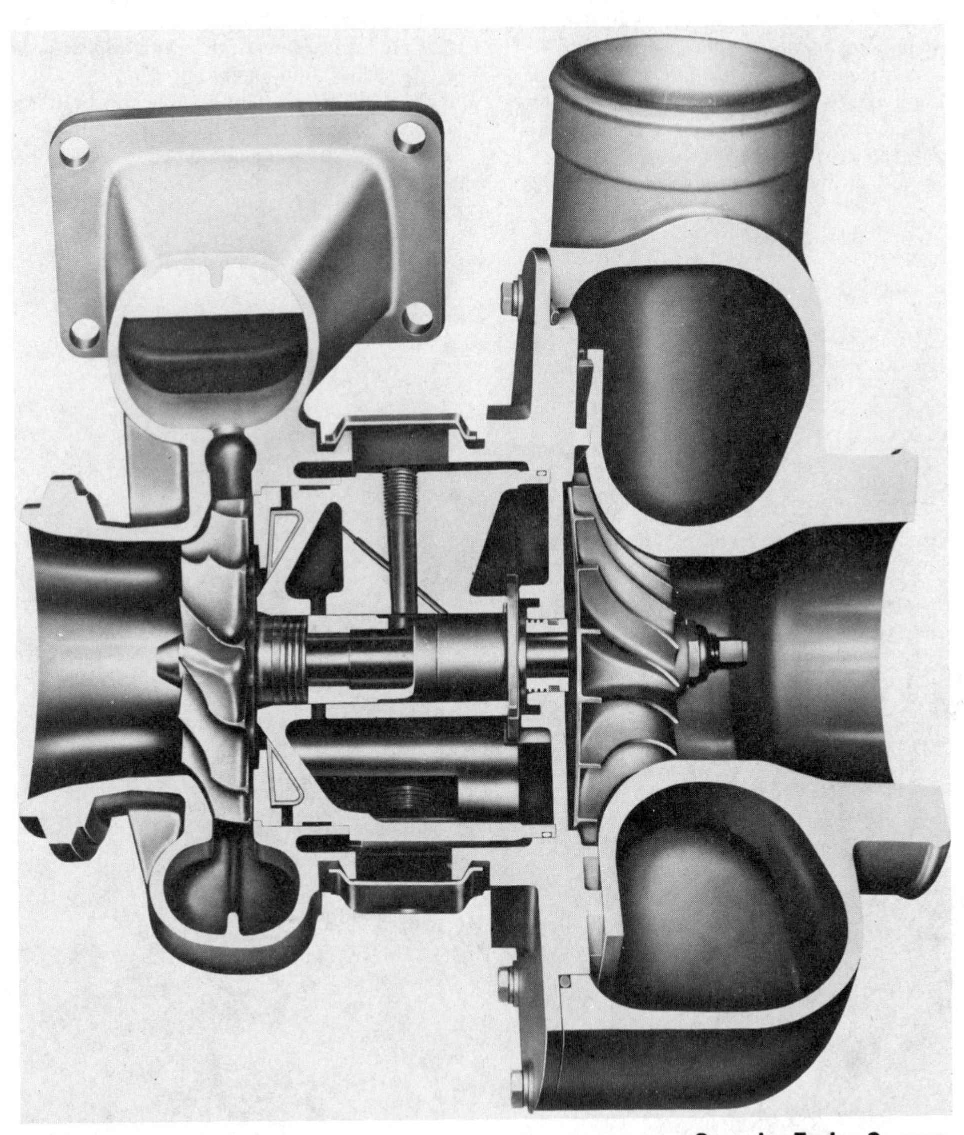

Cummins Engine Company

Intake and Exhaust Systems–Scavenging and Supercharging

Chapter **13**

Having examined in previous chapters the major mechanical features of diesel engines, you are now ready to study the methods of charging the cylinder with air and removing the products of combustion —the burned gases. In four-cycle engines the pumping strokes of the piston provide the forces to fill and discharge the combustion chamber so that the only additional part needed is valve gear to control the movement of the gases. This will be considered first.

In two-cycle engines, there must be some means of forcing fresh air in to push out the exhaust gas and charge the cylinder. This is the job of the *scavenging* system. You will see how crankcase scavenging is accomplished and will look at the construction of various types of scavenging pumps and blowers.

We can charge the cylinder with more air if we *supercharge* it, so you will see how superchargers work. Many engines use *turbochargers* which are driven by the engine's exhaust gases. You will see the many ways in which turbochargers are applied, both to four-cycle and to two-cycle engines.

Intake and Exhaust Systems

The purpose of the intake system is to supply air (oxygen) for burning the fuel in the combustion chamber. It includes the air filter, inlet manifold, inlet port, and the intake valve. The purpose of the exhaust system is to scavenge the exhaust gases efficiently from the combustion chamber and conduct them to the outside air with a minimum of noise.

The exhaust system consists of exhaust valves, exhaust manifold, muffler, and exhaust piping. If the engine is equipped with a supercharger its compressor is part of the intake system. If the engine is equipped with a turbocharger, its compressor is part of the intake system and its turbine is part of the exhaust system.

Valve Gear

In four-cycle engines, separate piston strokes fill and discharge the cylinder, so

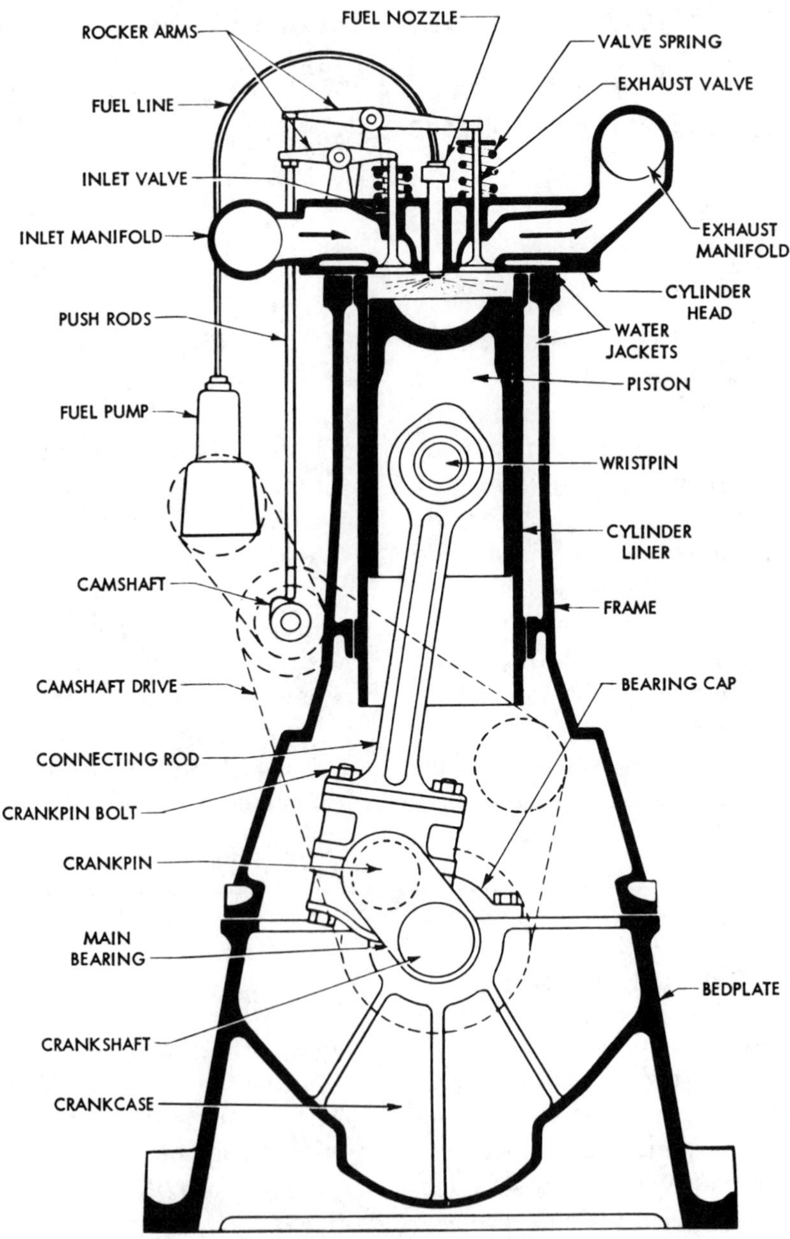

Fig. 13-1. Arrangement of four-cycle diesel engines.

that the only additional parts needed for a basic engine are valves and valve-actuating gear to control the flow of air and exhaust gas. Since these are important to the basic understanding of intake and exhaust systems, they will be considered first.

Valves

The action of the various parts of a valve gear of a four-cycle engine, as indicated in Fig. 13-1, is as follows: the crankshaft drives the camshaft by chain or gearing; the cam lifts the push rod, thus transmitting cam action to the rocker arm; the latter changes the upward motion of the push rod to a downward motion of the valve. The valve is open. As the smaller portion of the cam moves under the push rod, the valve spring starts to push the valve to its seat, closing it.

Valve Requirements

Getting the fresh air into and the exhaust gases out of the engine cylinders requires power, referred to as the *pumping loss*. In order to reduce the back-pressure during the exhaust process, the exhaust valve openings are made as large as practical. This is particularly important in the case of two-cycle engines, since the entire exhaust process occurs in a small fraction of the piston stroke, and scavening must be accomplished entirely by the pressure of the fresh-air charge. For these reasons, two-cycle engines which use exhaust valves generally employ two or four valves per cylinder.

In the case of a four-cycle engine, the exhaust valve size is not so critical because the exhaust gases are forced out by the piston during an entire stroke. The inlet valve opening is more important since all the intake air enters the cylinder through it. Restrictions to the air flow not only increase the pumping loss but also reduce the volumetric efficiency, as seen in Chapter 12. Most moderate-speed four-cycle engines have one inlet and one exhaust valve per cylinder, usually of the same size.

Air-flow restriction becomes more pronounced as engine speeds are increased, due to the higher velocity of gas flow. For this reason the power of every high-speed engine reaches a peak, beyond which an increase in speed results in a drop in power output. Because of these conditions, intake valves of high-speed four-cycle engines are made as large as possible and are sometimes as much as 25 to 35 percent larger than the exhaust valves.

You will find that some four-cycle engines use two inlet and two exhaust valves per cylinder, as in Fig. 13-2. The advantages of this construction are: (1) a larger valve area within the available space in the cylinder head; (2) less valve life is needed because the valves are smaller, therefore, valves can be opened and closed more rapidly; and (3) cooler operation because the heat picked up by the valve head has a shorter distance to travel to the valve seat where it is carried off. However, doubling the number of valves increases the complexity of the valve mechanism.

Valve Construction

Intake and exhaust valves used on diesel and dual-fuel engines are all of the mushroom-shaped poppet valve type with which you are familiar because of its use in automotive engines. The seating edge of the valve head is generally beveled at a 45° angle.

Fig. 13-2. Cylinder head using four valves per cylinder. (Allis-Chalmers, Engine Div.)

Inlet valves offer no serious problems as to suitable material because, although they are exposed to the heat of combustion, they are well cooled by the current of air which flows over them when they are open. Ordinary carbon or low-alloy steel serves the purpose. But exhaust valves run much hotter than inlet valves because, instead of being cooled by low-temperature air, they are swept by hot exhaust gas. Exhaust valves are usually made of silicon-chromium steel (silchrome) or steel alloys containing a high percentage of nickel and chromium.

Fig. 13-3 shows the simplest valve design; the valve head and stem are in one piece, with the stem riding in a removable guide. Springs (often only one) hold the valve firmly against its seat; a rocker arm above (not shown) pushes down on the valve stem and forces the valve open against the spring pressure. The valve seats directly in the cylinder head. The head of the valve is cooled by conduction of heat to the seat surface and to the stem guide; therefore designers make every effort to bring the cooling water passages as close to these points as possible.

The valve itself, and the seat, must resist severe mechanical wear and in the case of exhaust valves, high temperatures as well. For these reasons, valves are

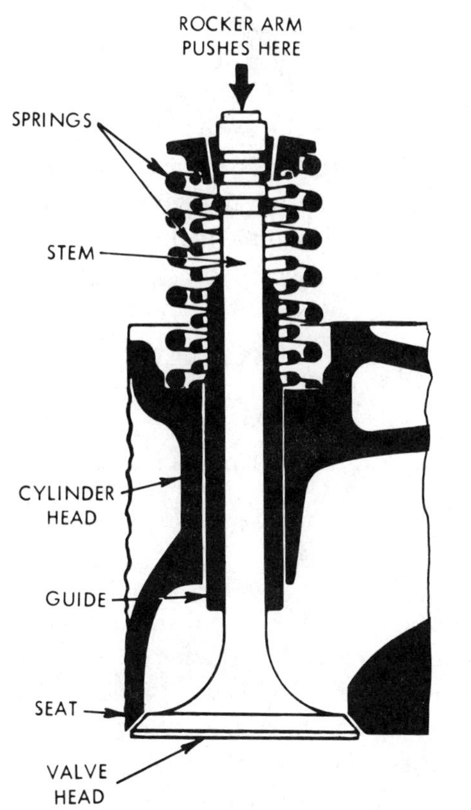

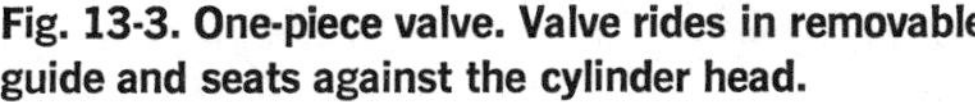

Fig. 13-3. One-piece valve. Valve rides in removable guide and seats against the cylinder head.

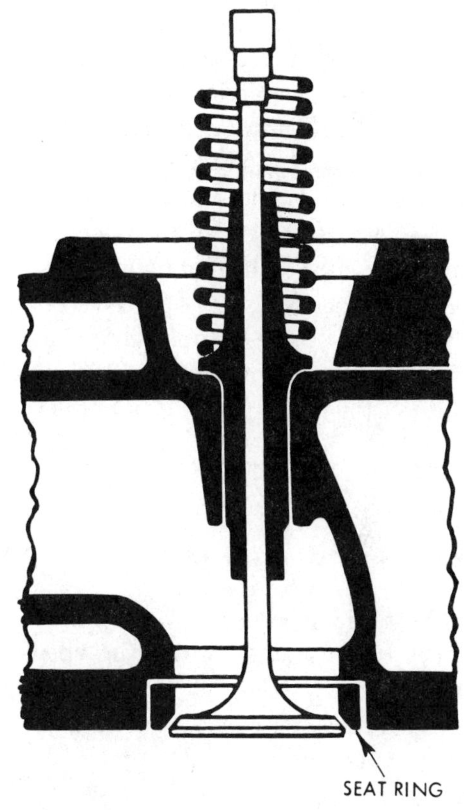

Fig. 13-4. Seat-ring of wear-resisting material. Ring is shown inserted in cylinder head.

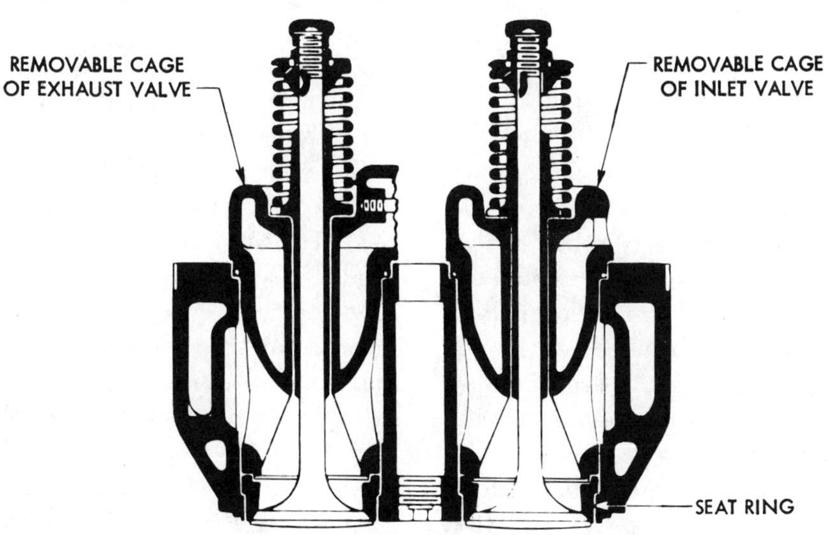

Fig. 13-5. Removable cages. Cages carry seat rings, stem guides, and spring.

often made in two parts, with a steel stem welded to a heat-resisting alloy head. It is also common practice to insert a seat ring of wear-resisting material, as shown in Fig. 13-4. In some cases, the seating surfaces of valve and seat are hard-faced with a cobalt-chromium-tungsten alloy, such as stellite, which remains hard at high temperatures.

Valves for large engines may be assembled in cages, as in Fig. 13-5, which shows the whole cylinder head with inlet valve and exhaust valve and also the recess for the fuel-injection valve between them. Each valve cage, which is removable from the cylinder head as a unit, carries, in addition to the valve itself, the seat ring, stem guide and spring. The exhaust-valve cage shown is water-cooled. The cage is usually made of cast iron. When the seat ring wears, it may be renewed at low cost.

Valve-Actuating Gear

The basic job of the valve-actuating gear is to cause and control the opening and closing of the inlet and exhaust valves. It may also actuate the fuel-injection valves and the fuel pumps, or the air-starting valves.

In most engines, this gear consists of rocker arms which actuate the valves, pushrods which connect the rocker arms and the cams on the camshaft, and a drive connecting the camshaft to the crankshaft.

Camshaft Drives

In four-cycle engines, the camshaft speed must be exactly one-half the crankshaft speed, so that the camshaft makes one complete revolution while the crankshaft makes two. In two-cycle engines, the camshaft speed must be exactly the same as the crankshaft speed. Because these speed relations must be exact, the connecting drive must be positive; this requires the use of gears or chains. Many drive arrangements are used; Fig. 13-6 shows six possible layouts.

The drive arrangement used for any particular engine depends to a large extent on where the camshaft is located, and on whether an auxiliary camshaft (for fuel pumps, etc.) or a power take-off shaft is included. The camshaft may be located low, near the crankshaft, using long push rods; it may be located on the cylinder block, using short push rods; or at the cylinder head level, without push rods.

For the sake of good appearance and cleanliness, the camshaft and push rods are often enclosed completely.

Gears for camshaft drives must be accurately cut and heat-treated to resist wear. Helical teeth (teeth placed at an angle) are frequently used in place of spur teeth (teeth placed straight) for greater quietness and more even transmission of power. A fiber or other nonmetallic gear is sometimes introduced into the train of gears for the same reason. Chain drives are of the roller or silent type. Provision is also made for adjustments necessitated by wear.

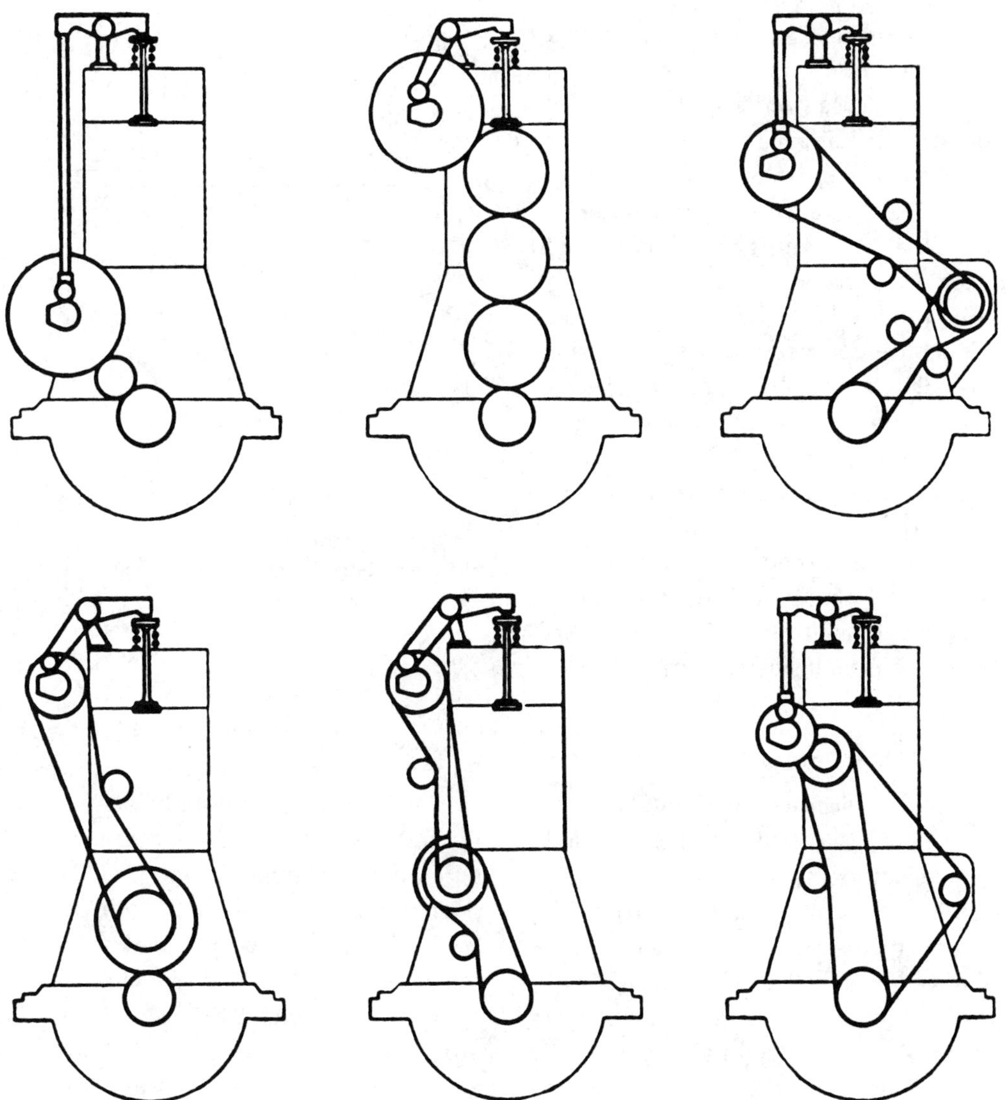

Fig. 13-6. Possible camshaft drives.

Camshafts

The camshaft in four-cycle engines carries the cams for actuating the inlet and exhaust valves. In valve-scavenged two-cycle engines, it carries the cams for the exhaust valves. In addition, the camshaft may carry cams for fuel-spray valves, fuel injection pumps, or air-starting valves. Some engines have two camshafts, one of which handles the inlet and exhaust valves, and the other the fuel pumps and other auxiliaries.

The camshaft may be constructed in several ways. It may be forged in one piece, including the cams themselves, as integral cams. Or the camshaft may consist of a steel shaft with separate forged-steel or cast-iron cams keyed on. Another construction, used on large engines, is to make up the camshaft in sections, with cams either integral or separate. In the sectional design, each section handles one cylinder, or a group of cylinders, and enough sections are bolted together to handle the whole engine.

To insure good support, the camshaft is usually carried by a series of *camshaft bearings*, one bearing being located between each pair of cylinders. The bearings may be either plain bushings or split sleeves. If plain bushings are used, their bores are larger than the cams, so that the camshaft may be withdrawn endwise. If split bushings are used, the camshaft may be removed sidewise from the engine.

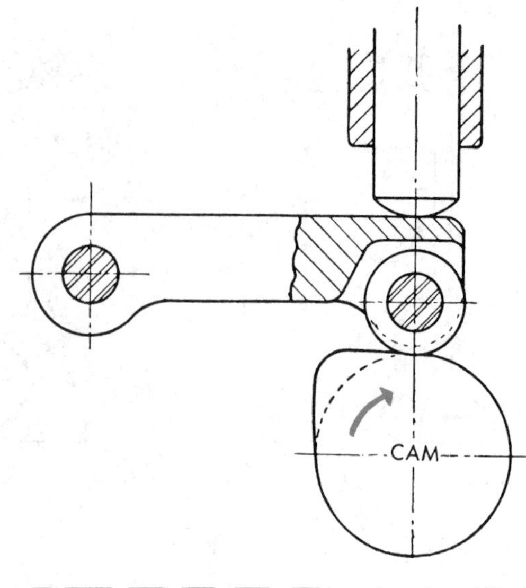

Fig. 13-7. Hinged cam-follower.

Push Rods and Rockers

To obtain stiffness without unnecessary weight, the push rods are usually hollow, they are tubes rather than rods. In the simplest arrangement (often used on small high-speed engines), the lower end of the push rod carries a head or follower of flat or mushroom shape which rides on the cam. A rounded head at the upper end fits into a cup on one end of the valve rocker arm. In many engines, side-thrust on the push rod is avoided by using a hinged follower which rests on the cam and transmits the cam action to the push rod. The follower usually carries a roller which runs on the cam and thus reduces friction. See Fig. 13-7.

Rocker arms swing on a steel fulcrum pin or pivot, resting in a bronze bushing or sometimes a needle bearing. The rocker arm may contact the end of the valve stem by means of a roller, but some form of set screw or tappet is more usual. The set screw is not only simpler and lighter than the roller but also permits adjusting the tappet clearance (lash) so that the valve closes firmly on its seat. If no adjustment is provided where the rocker arm and the valve stem meet, it must be provided on the push-rod side in order to take care of the changes in length required to compensate for wear.

In many cases the fulcrum bearing of the rocker arm and all contact points of the push rod and valve stem are pressure-lubricated.

Automatic Valve-lash Adjusters are used on some engines to avoid the necessity of a clearance otherwise needed in the valve gear to allow for expansion due to temperature changes. They also eliminate the need for manual adjustment in order to take care of the wear at various points of the valve gear. Automatic ad-

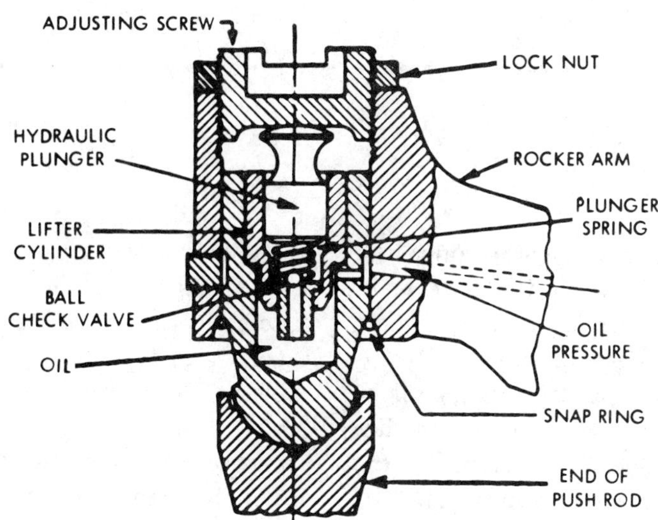

Fig. 13-8. Hydraulic lash adjuster.

justers may be either mechanical or hydraulic. The mechanical type uses a cam (generally located at the end of the rocker arm over the valve stem) and a spring which turns with the cam so as to take up the clearance when the valve is on its seat.

Fig. 13-8 shows a hydraulic lash adjuster built into the end of the rocker arm above the push rod. It consists of a small lifter cylinder containing a plunger, a spring, and a ball check valve. The plunger rests against the upper part of the rocker arm, while the spring pushes the cylinder downward toward the push rod.

In operation, oil under pressure from the lubricating oil system enters the lifter cylinder past the ball check valve and is trapped under the plunger, which has previously taken up the clearance. When the rocker arm moves downward to open the valve, the trapped oil transmits its force through the cylinder to the push rod. If the valve stem expands, there is sufficient leakage of oil past the plunger to permit the lifter cylinder to rise slowly so that there is no danger of holding the valve open.

Valve springs serve to close the valves. Valve springs used on diesel and gas-burning engines are made of highly tempered round steel wire, wound in a spiral coil. Only a small portion of the spring force is needed to keep the valve tight on its seat.

The principal duty of the spring is to provide sufficient force, while the valve is being lifted, to overcome the inertial forces of the valve gear caused by the rapid motion. By overcoming these inertial forces, the spring keeps the push rod in contact with the cam to prevent the bouncing that would otherwise occur. In order to give the spring sufficient force for this purpose, it is always compressed when it is installed (it is further compressed when the valve is opened).

Valve Timing

In an actual four-cycle engine, the dividing points between the four main events (*suction, compression, expansion, exhaust*) do not come at the very beginning and end of the corresponding strokes. See Fig. 13-9. The differences are smaller in low-speed engines and increase as the engine speed increases. Valves must be opened gradually to avoid excessive inertial forces and must be lifted well off their seats before they open effectively. The gases themselves have inertia, which means that it takes time to start them moving. Once moving, they tend to continue to move.

These conditions must all be allowed for in the valve timing of an actual engine. Consequently the intake valve is

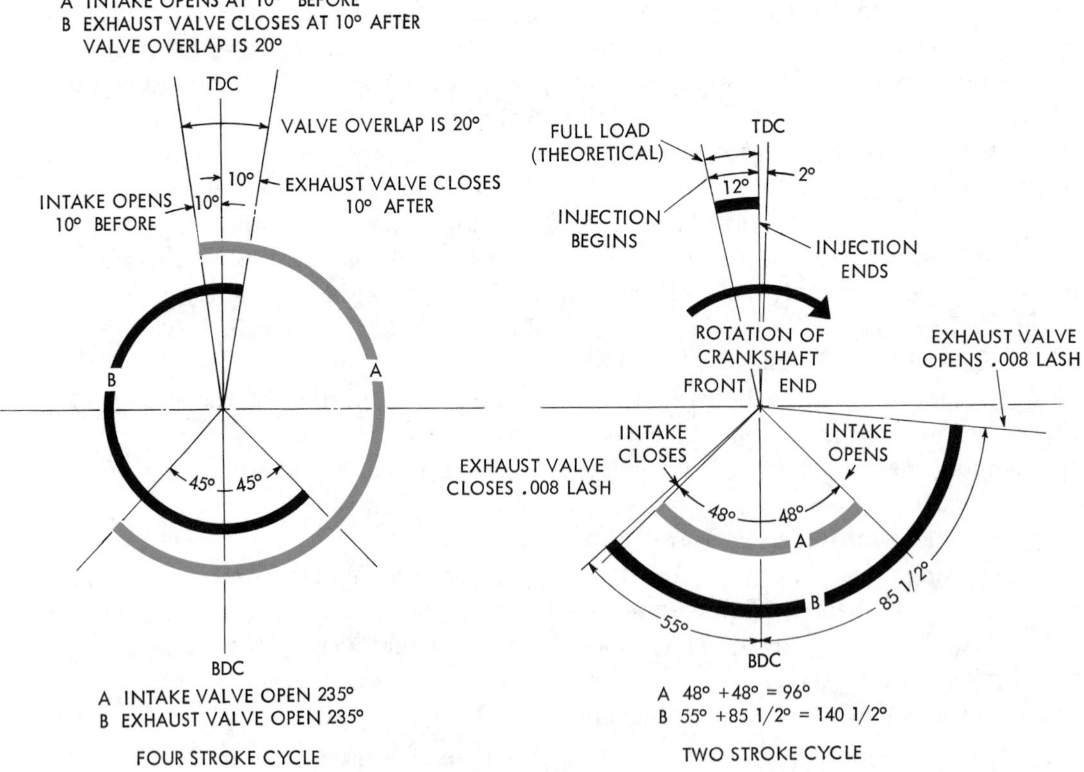

Fig. 13-9. Typical crankshaft-valve timing diagram of four-cycle and two-cycle diesel engines.

opened *before* top center by 10 to 25 degrees of crank angle; it is closed from 25 to 45 degrees *after* bottom center. In order to release the exhaust gases in proper time, the exhaust valve begins to open 30 to 60 degrees *before* bottom center (on the expansion stroke) and closes 10 to 20 degrees *after* the top center.

Note that when the piston is near the top center at the end of the exhaust stroke (and the beginning of the suction stroke), *both* valves are open during a considerable period. The period that the valves are open together is called *valve overlap*. Its purpose is to get more burned gas out and more fresh air in.

The best valve timing for a particular engine depends upon such factors as valve lift, shape of the cam, and speed of the engine. The proper timing is found by actual trial at the factory and is given in the manufacturer's instruction book. Note that an increase in clearance will delay, or retard the opening of a valve and will speed up, or advance its closing, thus shortening the time the valve is open. A decrease of the clearance will act the other way. An excessive decrease of the clearance may prevent the valve from seating firmly and thus cause loss of power, burning of the valve seat, or other problems.

Exhaust valves of two-cycle engines open 25 to 35 degrees before bottom center.

Before we get into scavenging and supercharging, let's consider the rest of the intake and exhaust systems.

Air-Intake System

The purpose of the air-intake system is to provide the air required for the combustion of the fuel. In the cases of two-cycle engines and supercharged engines, additional air is needed to scavenge the cylinders of the spent gases. Essentially, the air-intake system consists of piping from the source of fresh air to: (1) the air-intake manifold of a four-cycle engine, (2) the scavenging pump inlet of a two-cycle engine, or (3) the supercharger inlet of a supercharged engine. The system usually includes a cleaner to remove dust and other harmful particles and often a silencer to subdue annoying noises from air pulsations.

Piping

In order for an engine to deliver its full power and burn its fuel efficiently, it needs a full supply of air. Therefore, intake piping should be as short and as straight as possible, and bends should have a long sweep in order to reduce the resistance to air flow.

Air Cleaners or Filters

Air-cleaning devices have come into general use because they prevent the internal engine surfaces from being damaged by abrasive material which the air might carry in accidentally. They greatly extend the life of cylinder liners and piston rings by cleaning the air of most of the dust which it normally holds.

All atmospheric air is polluted to some extent with sand-like particles which increase the rate of wear. During the severe dust storms which occur occasionally in the desert section of the western United States, the air pollution may be as much as several pounds per 1000 cu ft of air.

But even in clear-air localities where the air carries less than 0.2 grains (7000 grains equal one pound) of dust per 1000 cu ft, air cleaners reduce wear and ring-sticking. This is true because an engine gulps in a lot of air—from 2.2 to 5.1 cu ft/min/rated bhp, depending on the type of engine.

EXAMPLE: How much dust enters the cylinders of a 300-bhp, two-cycle, crankcase-scavenged engine in one year during which the engine runs 250 days of 9 hours each? The average dust content of the air (an industrial district) is 1.2 grains per 1000 cu ft. The air rate of the engine (displacement) is 3.9 cu ft/min/rated bhp.

As the amount of air taken into engine per hour = 300 bhp × 3.9 cu ft/min × 60 min/hr — 70,200 cu ft/hr, we can compute:

$$\text{Dust entering engine per hour} = \frac{1.2 \times 70{,}200}{1000} = 84.2 \text{ grains.}$$

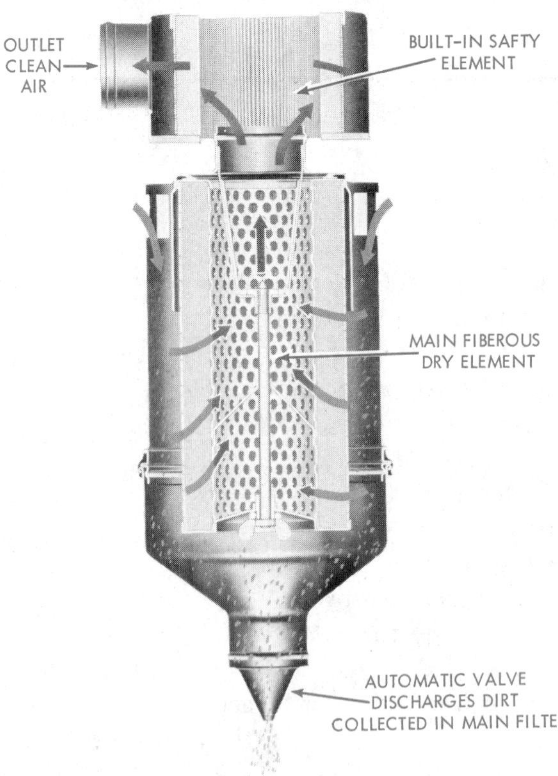

Fig. 13-10. Dry type air cleaner with automatic dirt discharge valve. (Deere & Company)

Therefore

Dust entering engine in 1 year =

$84.2 \times 9 \times 250 = 189{,}450$ grains

Since 7000 grains make one lb,

Dust entering engine in 1 year =

$$\frac{189{,}450}{7000} = 27 \text{ lb}$$

The most common types of air cleaners are: (1) dry-type filters, (2) viscous-impingement filters, (3) oil-bath filters, and (4) water sprays.

Dry-Type Filters. These clean the air by passing it through a fibrous dry element. The type of air cleaner in Fig. 13-10 uses a built-in safety element in case the main element should get damaged in cleaning. Note that the valve at the bottom of this element discharges most dirt automatically. It is in effect a primary and secondary filter operation.

Oil Bath Filters—Light Duty. The light duty oil bath air cleaner in Fig. 13-11 consists of a metal wood cleaning element supported inside a housing beneath which is contained a bath of oil. The lower portion of the housing incorporates a chamber which serves as a silencer.

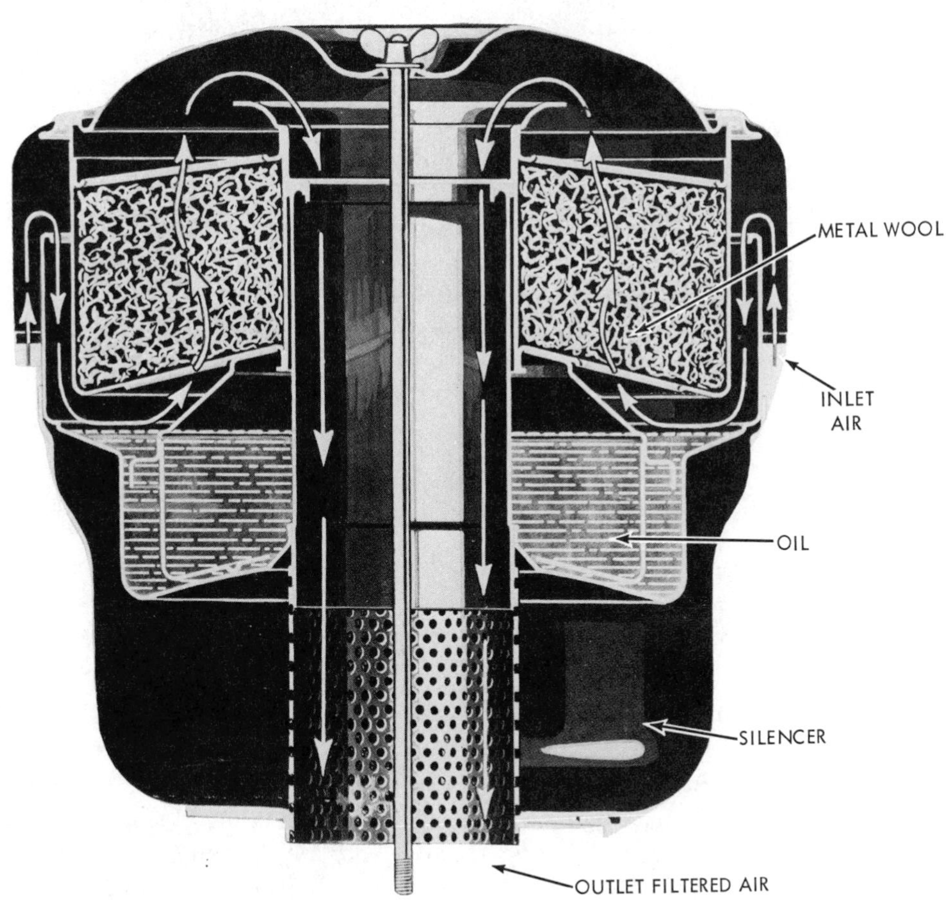

Fig. 13-11. Light-duty oil bath air filter. (General Motors Corporation)

Fig. 13-12. Heavy-duty oil-bath air cleaner. (General Motors Corporation)

Air drawn into the cleaner by the blower must change direction over the top of the oil bath. The heavy dirt is trapped in the oil. The lighter particles of dirt are trapped in the metal mesh. The mesh is continuously covered with a fine mist of oil continuously deposited by the fast-moving air.

The heavy-duty oil bath type cleaner, Fig. 13-12, operates the same way as the light-duty one, but is easier to clean and has a higher stack of metal mesh to allow for higher volumes of air. It has a side mounting.

Inlet Manifolds

A diesel engine inlet manifold cross-sectional area can be made considerably larger than a gasoline engine design. Because the diesel engine manifold handles only air without fuel, it is not necessary to maintain the high velocity. The slower velocity of air flow due to a larger inlet manifold will reduce the restriction of flow to a minimum.

The diesel inlet air should be kept as cool as possible. This will allow a greater charge of air to enter the combustion chamber. Intake manifold coolers are often used, especially with turbo-charged engines. This principle will be investigated later in this chapter.

Air cleaner size on a diesel should be larger than a gasoline engine of the same size, since the diesel engine operates at full air flow in contrast to a gasoline engine which very seldom operates at full throttle.

In a diesel engine the air filter restriction should be kept below 7 inches of pressure, as measured on a water manometer.

Scavenging

You learned in Chapter 2 that in two-cycle engines the pressure of the incoming air charge is used to scavenge or push out the exhaust gases during a short period of time when the piston is near the bottom of its stroke. There are three basic methods of supplying the incoming air charge at the low pressure (2 to 5 psi) needed to accomplish this. In (1) *crankcase scavenging*, the crankcase and underside of the piston act as an air compressor. (2) *Power-piston scavenging* is somewhat similar, except that a separate chamber is provided in the space under the piston. The downstroke of the piston compresses air which has been previously drawn into the chamber on the upstroke. The most common method, however, is (3) *pump* or *blower-scavenging*, in which scavenging air is supplied by a pump or blower, driven either directly from the engine or by an independent source of power (usually an electric motor).

In crankcase-scavenging engines, partitions between the cylinders divide the crankcase into separate compartments, as seen in a large engine, Fig. 13-13. Smaller two-stroke gasoline engines on lawn mowers use the principle today. The slight vacuum created by the piston upstroke draws air in through filters and through automatic ring valves in the partition walls. The downstroke of the piston compresses the air slightly. Near the end of the downstroke, the piston uncovers the exhaust ports in the cylinder wall and

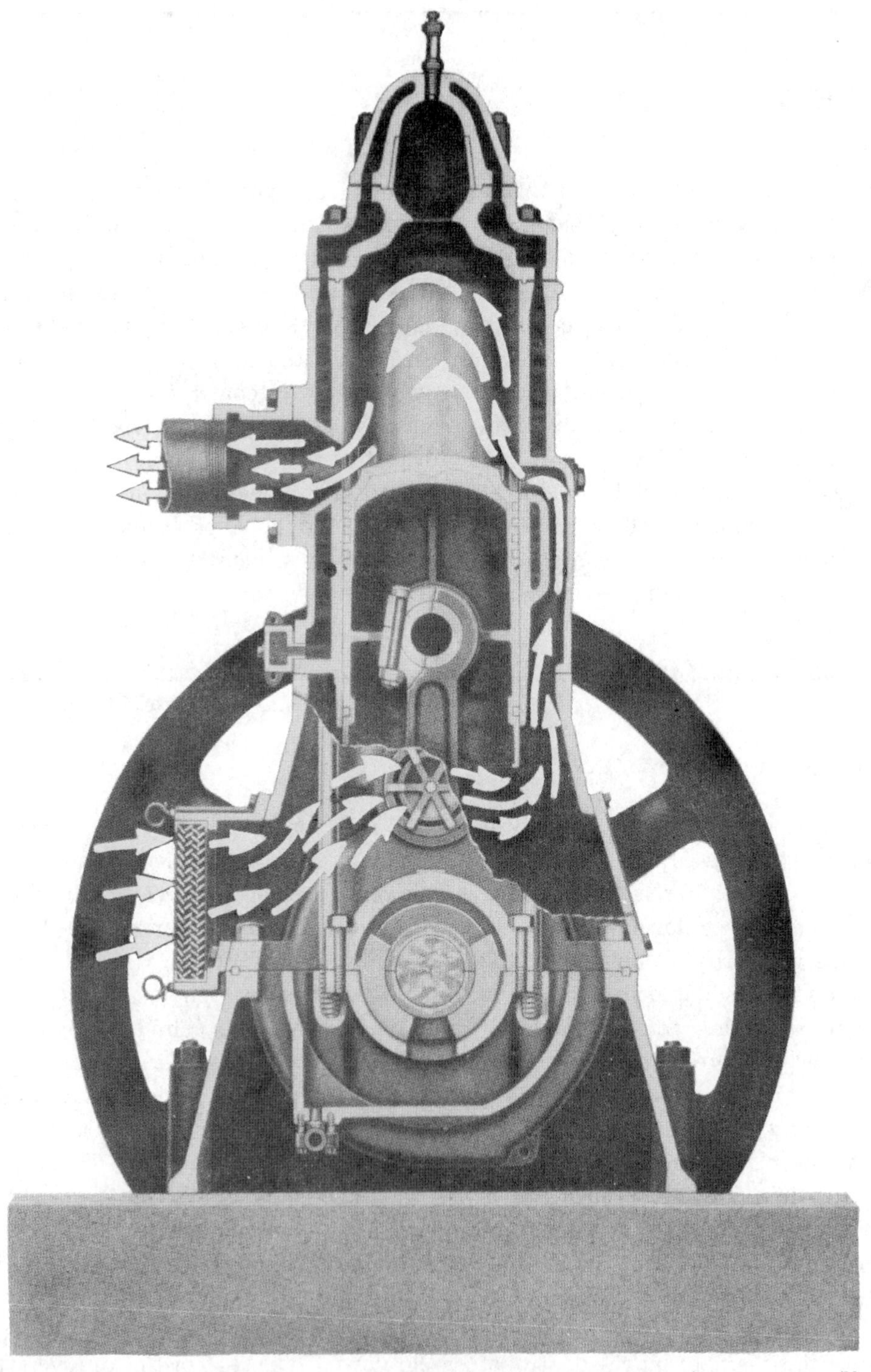

Fig. 13-13. Crankcase-scavenging engine. The underside of piston is used as air compressor. (Venn-Severin Engine Works, Inc.)

then the scavenging ports. Air from the crankcase then flows into the cylinder, displacing the burned gas and charging the cylinder for the next cycle.

Ports and Valves

Two-cycle engines may be classified according to the way the air enters the cylinder and the exhaust leaves. The most common arrangements are port-scavenging and valve-scavenging.

In *port-scavenging* engines the air enters and the gas leaves by ports which the piston uncovers, as in Fig. 13-13. Fig. 13-14 shows another port-scavenged engine, in which the scavenging air is supplied by a blower. Here the exhaust ports are placed as close as possible to the bottom of the cylinder, and their upper edges are the same level as the upper edges of the air ports. Thus the descending piston starts to uncover both sets of ports simultaneously.

The automatic air valves serve an important purpose. They keep the burned gas from flowing out through the air ports until the cylinder pressure has fallen below the scavenging air pressure, some of

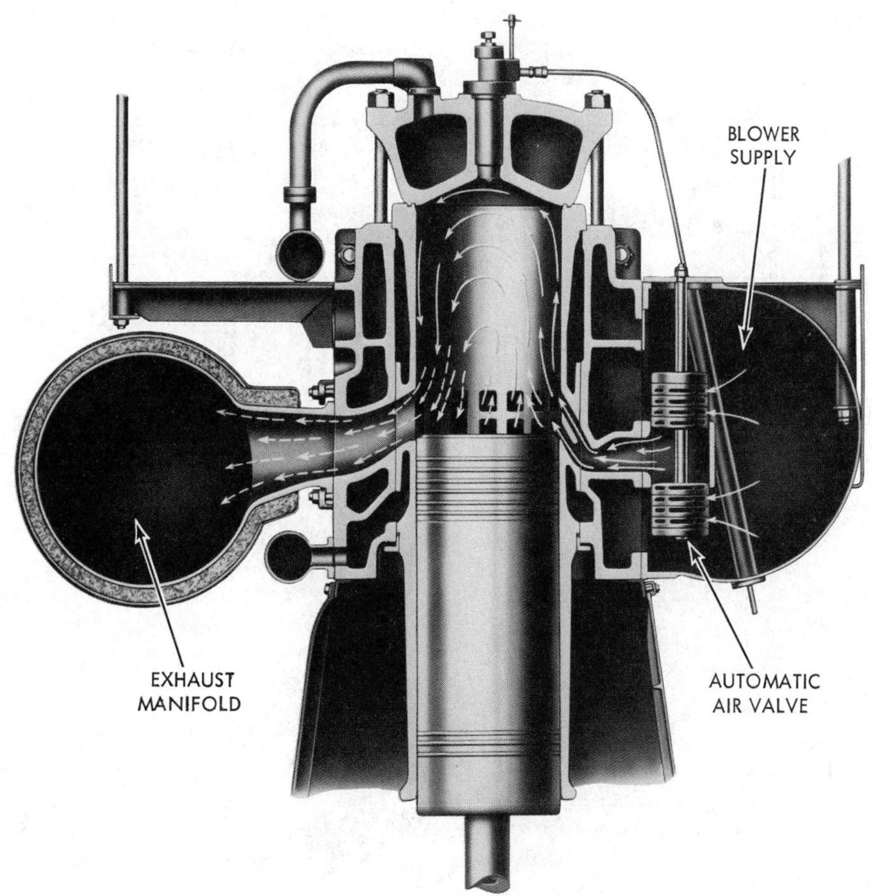

Fig. 13-14. Port-scavenged engine. Automatic valves are used at air ports. (Nordberg Manufacturing Co.)

the burned gas having been blown out through the exhaust ports. Then the air valves open and scavenging air flows in to expel the remaining burned gas.

The air valves remain open while the piston moves to bottom center, reverses its travel and moves upward, or until the compression pressure in the cylinder exceeds the pressure in the air manifold. In this way the cylinder is filled, before compression begins, with air at a slight pressure above atmospheric. In other words, there is a slight supercharge. Note the upward direction of the air ports—this is to cause the scavenging air to sweep upward in order to clear out the burned gas more effectively.

The opposed-piston engine uses a form of port-scavenging shown schematically in Fig. 13-15. The lower piston controls the exhaust ports, the upper one the inlet ports. In order to permit the burned gas to discharge its pressure before admitting the scavenge air, the crank of the lower crankshaft is advanced about 12 degrees in respect to the crank of the upper

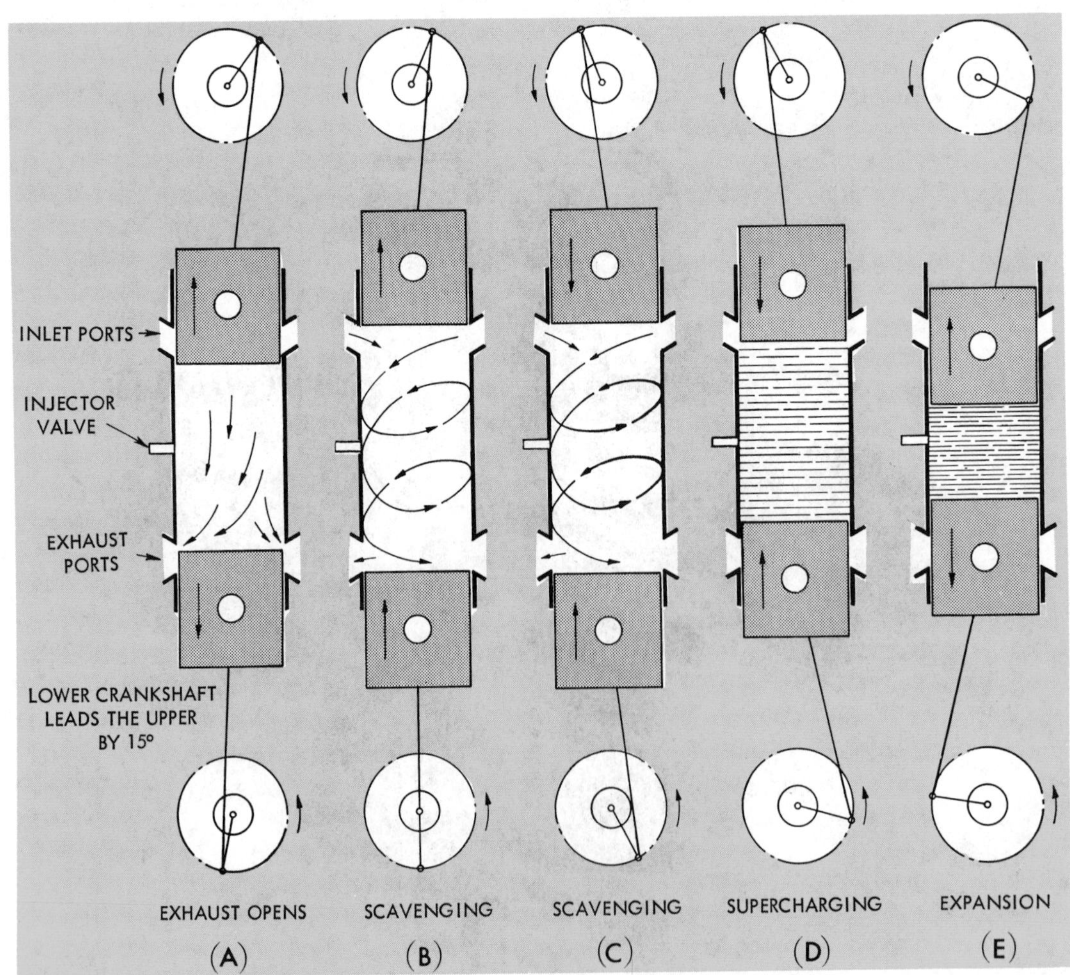

Fig. 13-15. Scavenging process in opposed-piston engine.

crankshaft, so that the lower crankshaft leads the upper one by 12 degrees.

This causes the exhaust ports to uncover ahead of the inlet ports, as shown in Fig. 13-15A. When the cylinder pressure is sufficiently lowered, the inlet ports are uncovered, Fig. 13-15B and scavenging begins to take place, as indicated by the arrows. Both ports remain open for a time, with scavenging continuing, as shown in Fig. 13-15C. The exhaust ports close before the inlet ports do, and the cylinder pressure builds up to the scavenge pressure by the time the inlet ports are covered and compression begins, Fig. 13-15D.

Slightly before the pistons come closest together, fuel is injected, ignited, and burned, causing the pistons to separate on the expansion stroke, Fig. 13-15E. A cross-section of an opposed piston engine appears in Chapter 4.

Since both the air and the burned gas flow in the same direction, this is called a *uniflow* system. It causes a minimum of turbulence and improves the effective-

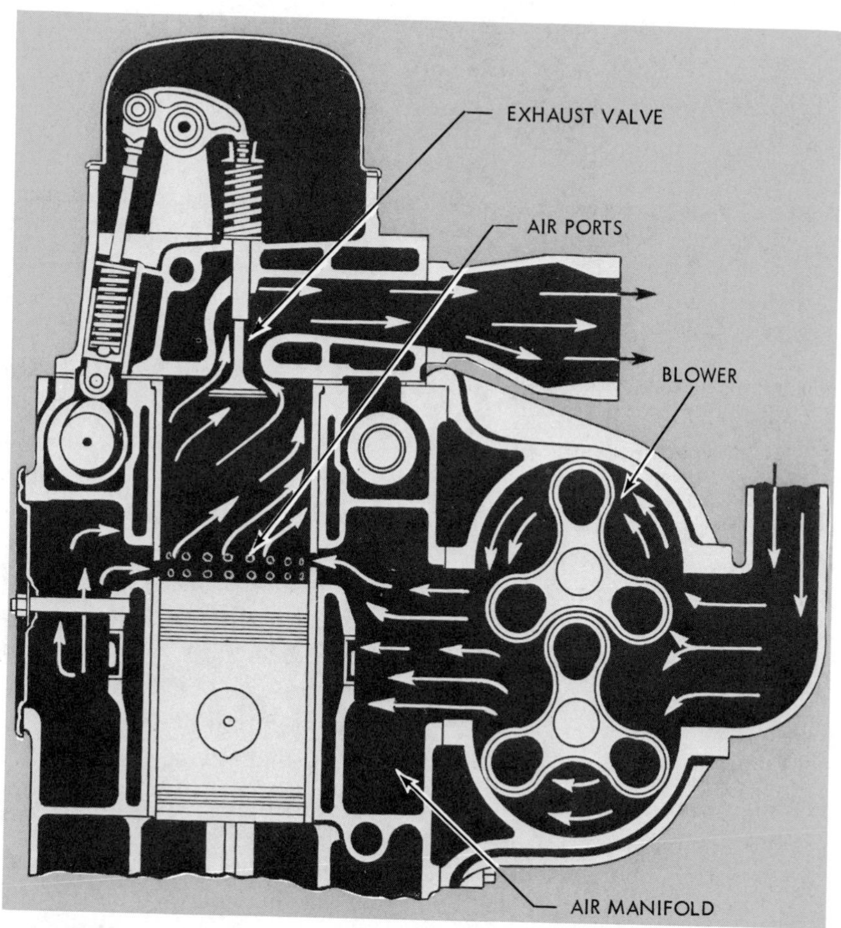

Fig. 13-16. Valve-scavenged in-line engine using blower. (General Motors Corporation)

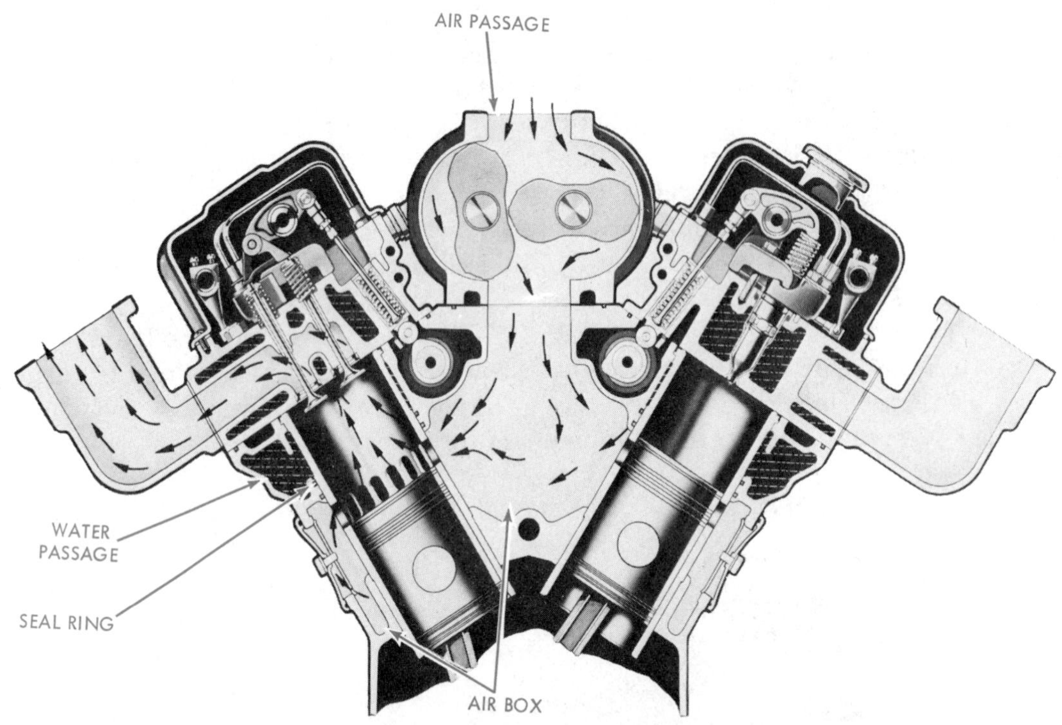

Fig. 13-17. Valve-scavenged V-type engine using blower. (General Motors Corporation)

ness of the scavenging action.

In *valve-scavenged* engines, Fig. 13-16 and 13-17, the blower forces air into the manifold and into the cylinder when the piston uncovers the ports, which are arranged around the entire circumference of the cylinder. The scavenging air flows upward or downward and out through the exhaust valves in the cylinder head. This engine, too, employs uniflow-scavenging.

Scavenging Pumps

Reciprocating pumps are frequently used to supply scavenging air. The pumps may be single-acting, but are usually double-acting. One scavenging cylinder usually supplies two or more cylinders. Fig. 13-18 shows a typical design using a vertical pump mounted at one end of the engine and driven from an extension of the crankshaft. The pump shown is double-acting, with one piston.

To save space, two pistons are occasionally used, one above the other, on the same piston rod. Valves are usually of the automatic ring or feather type, which consists of light steel rings or strips which are lifted by pressure and returned to their seats by spring action.

Another reciprocating pump principle is demonstrated in Fig. 13-19. Here the pump cylinders are at right angles to the power cylinders, with pistons driven from the same crankpins.

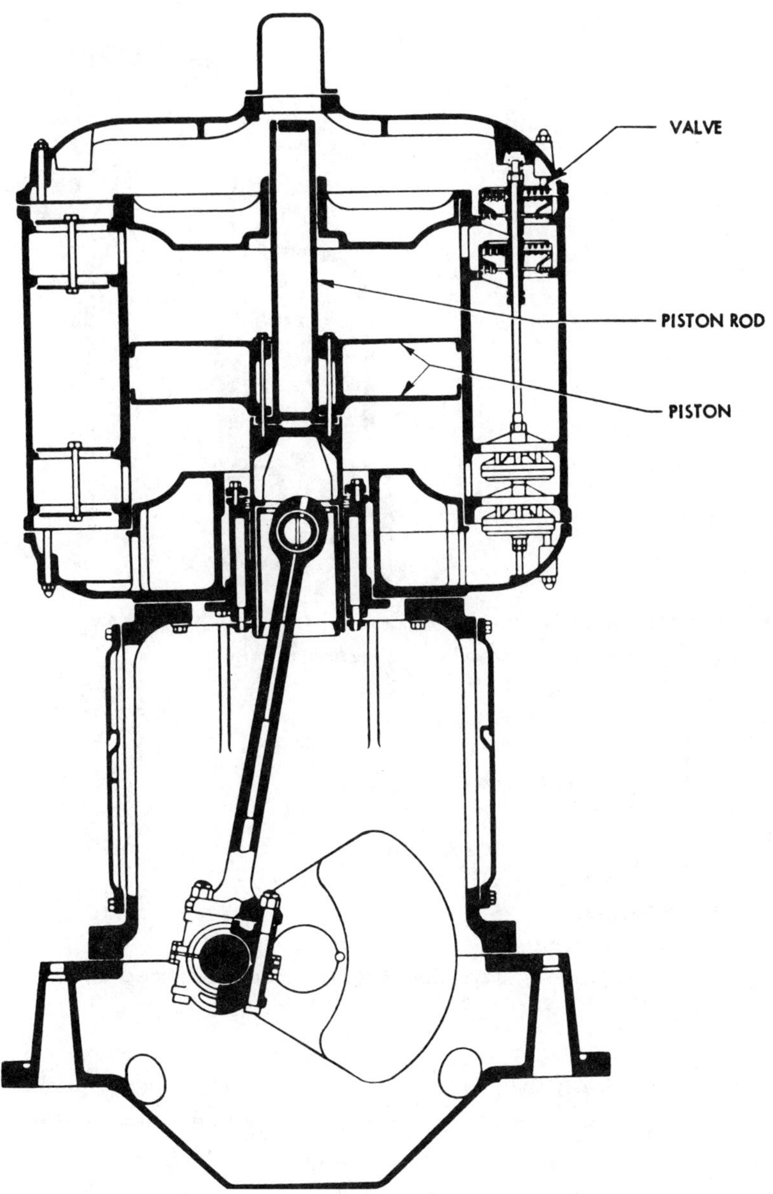

Fig. 13-18. Reciprocating scavenging pump with single piston that is double-acting.

Scavenging Blowers

Positive-displacement rotary blowers and centrifugal blowers, described in the section on Superchargers, are also widely used to supply scavenge air. The positive-displacement type works efficiently at one to three times engine speed and can be conveniently driven from the engine by

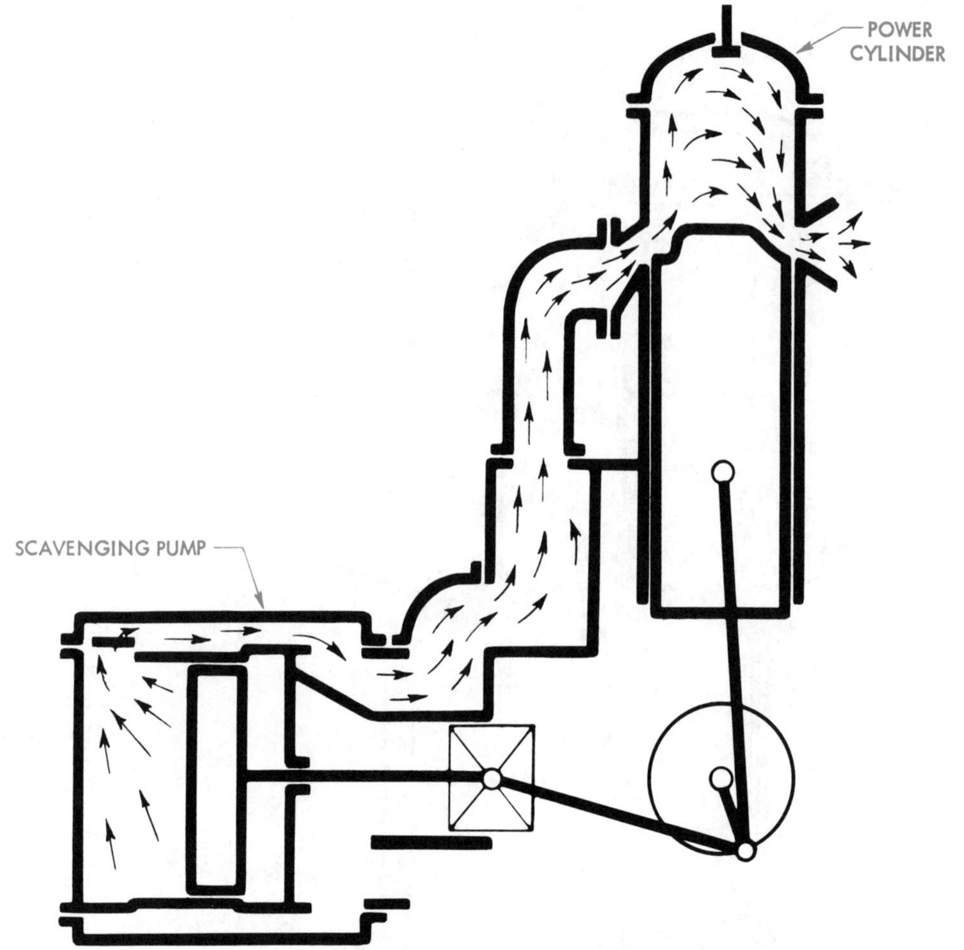

Fig. 13-19. Reciprocating scavenging pump at right angle to power cylinder.

gears or chain. Where the blower is attached to the engine, the positive-displacement blower is the preferred type. The centrifugal blower, when used for scavenging, lends itself better to electric motor drive.

Supercharging

The purpose of supercharging is to cram more air into the cylinder so that more fuel can be burned and the engine output boosted. This is accomplished by forcing air into the cylinder of an engine at a pressure substantially above atmos-

Fig. 13-20. Positive-displacement rotary blower. The impellers have two lobes each. (Roots-Connersville Blower Div., Dresser Industries, Inc.)

pheric (at least 6 psi above).

Supercharging should not be confused with scavenging. The scavenging process in two-cycle engines use a swish of air to push out the spent gases and replace them with fresh air at approximately atmospheric pressure. But supercharging, whether in two-cycle or four-cycle engines, goes a step further and packs the cylinder with still more fresh air.

The increase in output obtained from a supercharged engine depends upon various factors, the most important of which is the supercharge pressure used. Boosts of the continuous power rating usually run from 40 to 100 percent, but much higher boosts have been reached in special designs. Supercharged engines usually show better fuel economy than unsupercharged units, because the power gained by supercharging increases faster than the friction losses and also because the high-velocity air charge causes better mixing of the fuel and air.

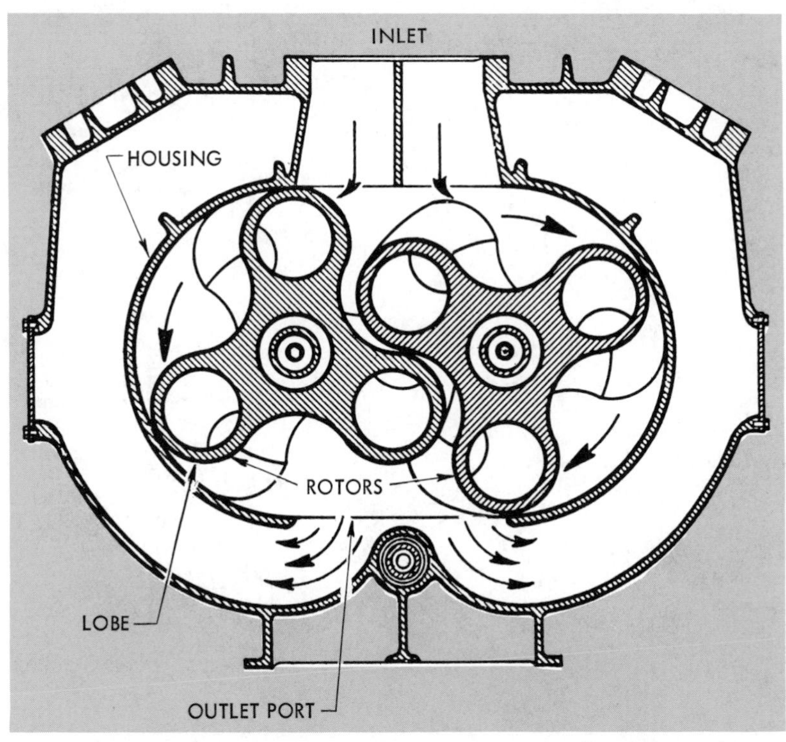

Fig. 13-21. Roots-type blower with three-lobed impellers.

For small engines the relatively high cost of a supercharger may not justify its use, but most high-capacity four-cycle engines are now supercharged because of the lower overall cost per horsepower, reduced space, lower weight and better fuel efficiency.

Superchargers

Supercharged air is provided either by *positive displacement rotary blowers* or by *centrifugal blowers* (rarely by reciprocating piston pumps). These may be driven (1) by the engine itself, (2) from a separate power source, such as an electric motor, or (3) from an exhaust-gas turbine.

A positive-displacement rotary blower of the long used Roots type consists of an oblong housing with flat end-plates, inside of which two accurately machined rotors, called impellers, rotate in oppo-

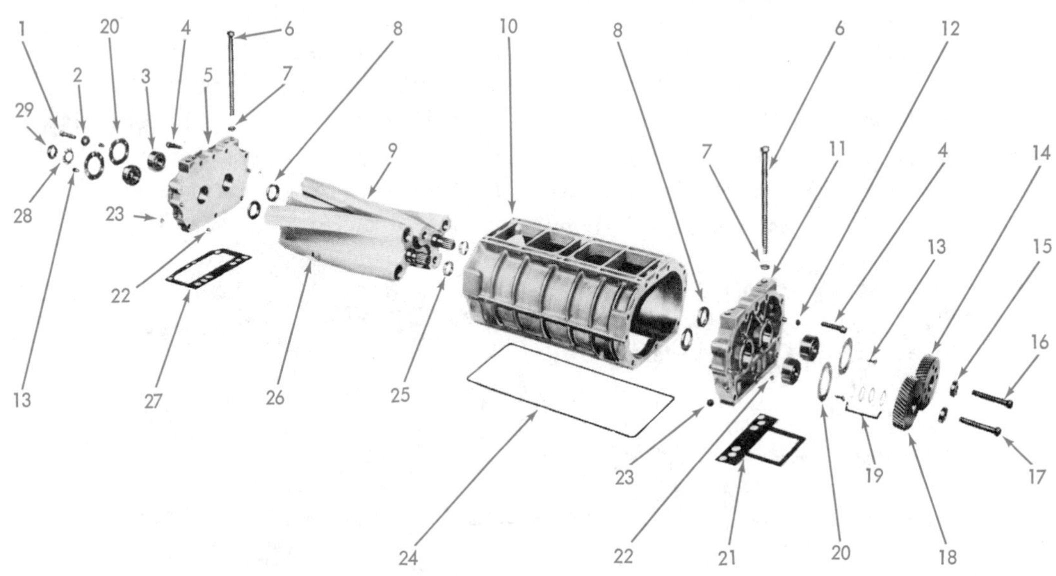

1 BOLT
2 WASHER
3 BEARING–ROTOR FRONT BALL
4 BOLT SOCKET HEAD
5 PLATE FRONT END
6 BOLT–BLOWER TO BLOCK
7 WASHER–SPECIAL
8 OIL SEAL–ROTOR SHAFT
9 ROTOR ASSEMBLY R–H HELIX
10 HOUSING BLOWER
11 PLATE–REAR END
12 BEARING–ROTOR REAR ROLLER
13 BOLT BEARING RETAINER
14 GEAR–ROTOR DRIVE L–H HELIX
15 PILOT–ROTOR GEAR
16 BOLT– DRIVE GEAR
17 BOLT– DRIVEN GEAR
18 GEAR–ROTOR DRIVE, R–H HELIX
19 SHIMS
20 RETAINER–BEARING
21 GASKET
22 STRAINER END PLATE OIL
23 PLUG–PIPE
24 GASKET
25 SLEEVE–OIL SEAL
26 ROTOR ASSEMBLY L–H HELIX
27 GASKET
28 LOCK WASHER
29 NUT

Fig. 13-22. Exploded view of three-lobed V-8 blower. (General Motors Corporation)

site directions, Fig. 13-20. The impellers are mounted on two parallel shafts and have lobes, which mesh closely with each other (like gear teeth) when they rotate.

Impellers may carry two or three lobes each; the blower shown in Fig. 13-21 has three lobes. As the rotors turn, the entering air is first trapped between the housing and the valley between two adjacent lobes and then carried to the outlet port where it is forced out at sufficient pressure to overcome that in the engine air receiver.

The two-lobed impeller gives maximum

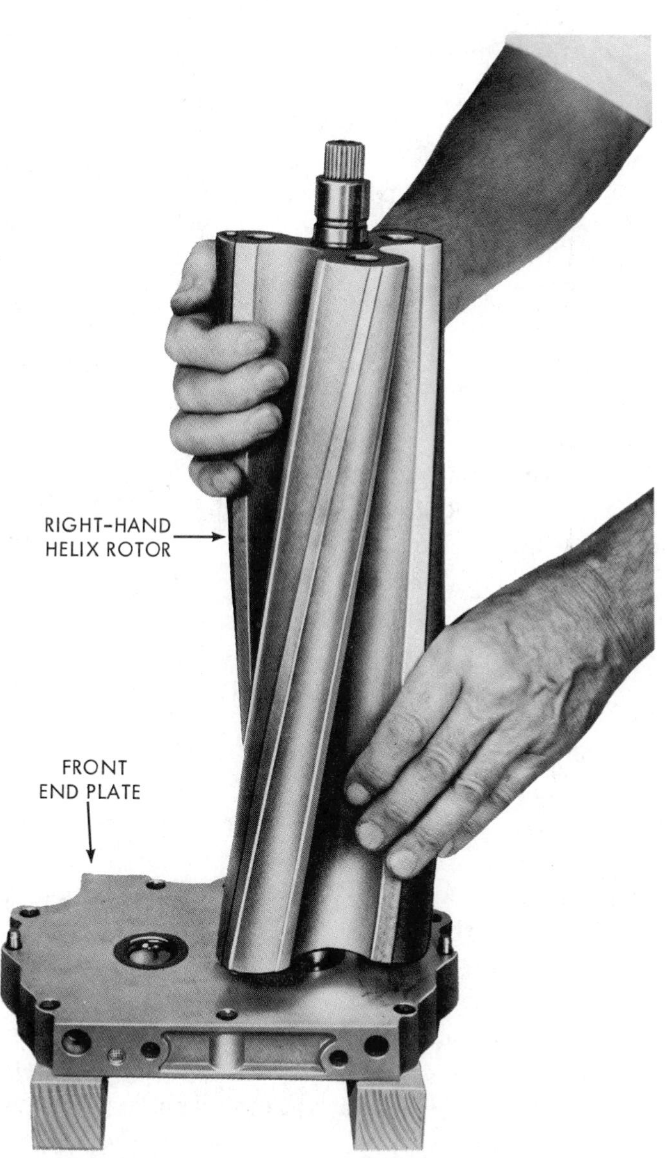

Fig. 13-23. Installing blower rotor in front end plate. (General Motors Corporation)

capacity from a given size of housing and also consumes the least amount of power. However, its delivery pulsates (throbs) and it tends to be noisy. For these reasons, three lobes are often used, and also lobes of spiral shape.

Fig. 13-22 illustrates a three-rotor blower used on a V-8 two-stroke Detroit diesel engine, Model 53.

The V-8 General Motors blower illustrated in Figs. 13-22, 13-23, and 13-24 is a precision made piece of equipment which is driven approximately twice the engine speed. Two timing gears located on the drive-end of the rotor shafts space the rotor lobes with a close tolerance. As the lobes of the rotors do not touch at any time, no lubrication is required. To adjust the timing and get the required clearance between the lobes, shims are placed behind the timing gears.

While rotor lobe clearance can be adjusted, gear backlash cannot be corrected. The gears must be replaced when they have worn to a point where the backlash exceeds 0.004 inches.

The blower bearings, timing gears, and governor drive mechanisms are pressure-lubricated by means of oil passages to the top deck of the cylinder block which lead from the main oil galleries to an oil passage in each blower end plate.

A centrifugal blower uses a high-speed wheel provided with vanes and enclosed in a carefully shaped casing or volute. Air entering the wheel near the hub is thrown outward by centrifugal force. (Centrifugal force is the force which tends to move a rotating body outwards from the center of rotation. It is proportional to the weight of the body, to its distance from the center of rotation, and to the square of its rotating speed.) The volute is so shaped that the high velocity of the air is transformed into pressure. Centrifugal blowers run at relatively high speeds (upward of 4000-5000 rpm) and are therefore driven more readily from a separate motor or a turbine than from the engine.

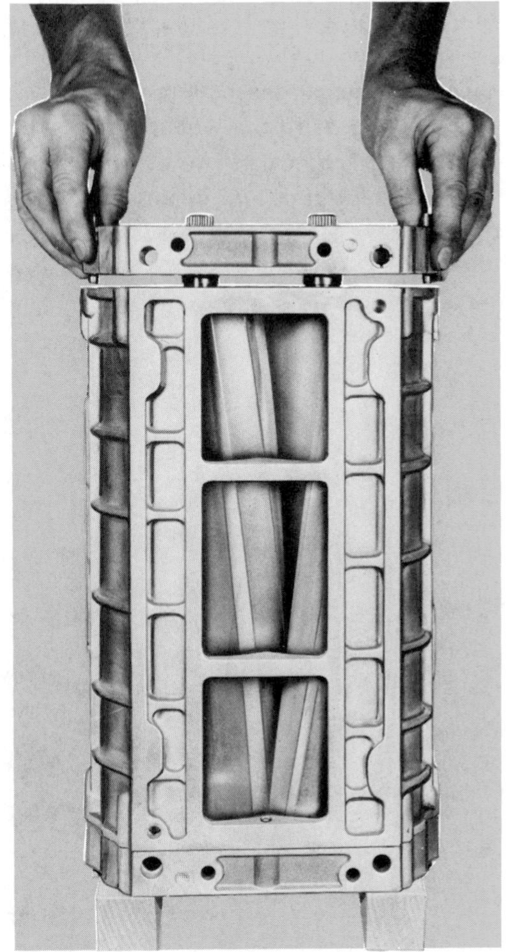

Fig. 13-24. Installing rear end plate on blower, rotors, and housing.

The positive-displacement rotary blower finds particular favor in applications where the blower is driven by an engine which is required to deliver full torque over a wide speed range (as in automotive service). Its advantage over an engine-driven centrifugal blower in such applications is that the positive-displacement blower delivers practically the same amount of air per revolution regardless of speed or working pressure. A centrifugal blower, on the other hand, develops a pressure proportional to the square of the speed; hence a centrifugal blower driven by the engine itself (through gears, etc.) would supply little air when the engine speed is reduced.

Turbochargers

Turbochargers are centrifugal blowers driven by exhaust gas turbines. They are widely used to supply supercharge air because they run largely on power that would otherwise be wasted. The hot exhaust gas from a diesel or gas engine contains considerable energy; the gas turbine can recover some of this energy and use it to drive the centrifugal blower. Fig. 13-25 shows a turbine and blower wheel mounted on a common shaft in a single casing. Fig. 13-26 shows the two elements in separate casings.

In all four-cycle engines and in most two-cycle engines, the turbocharger is mechanically independent of the engine which it serves. Its only connection to the engine are the exhaust pipe to the turbine and the air pipe from the blower to the intake system of the engine.

The turbocharger schematic in Fig. 13-27 shows the flow of air through the compressor, through the intake valve and into the combustion chamber. The exhaust gases leave the combustion chamber through the exhaust valve, hit the turbine wheel and are exhausted to the outside air. Because of the high temperature, the turbine parts are made of heat-resisting material and water jackets are sometimes provided.

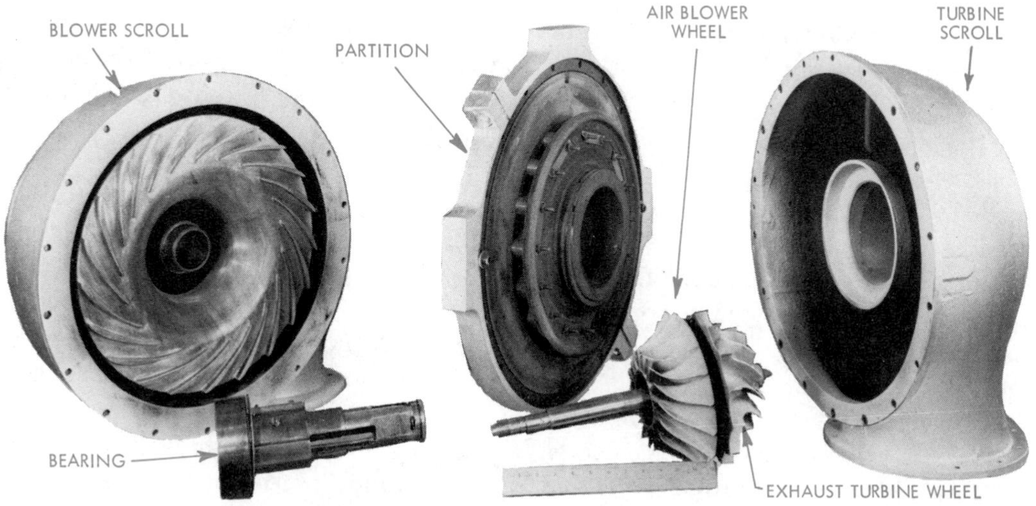

Fig. 13-25. Turbocharger with wheels close together. (DeLaval Turbine, Inc.)

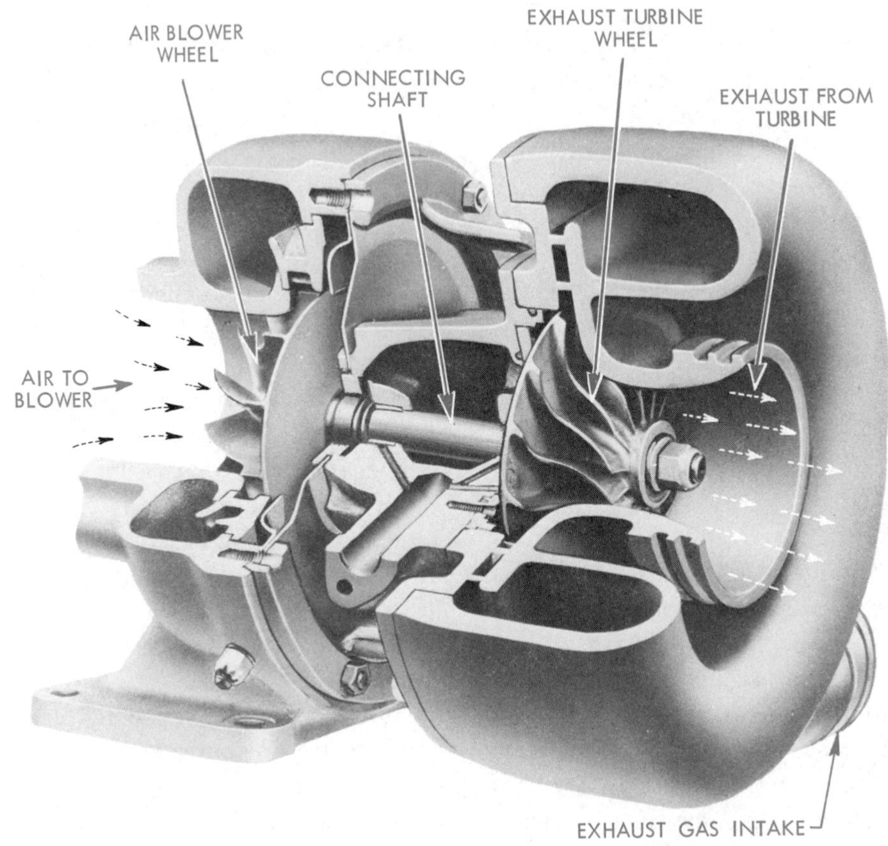

Fig. 13-26. Turbocharger with wheels separated. (Air Research Industrial Div., Garrett Corporation)

Turboblowers run at high speeds, from about 10,000 to over 100,000 rpm, depending on capacity and blower pressure. No controls are needed with diesel engines; speed, air quantity, and charging pressure follow engine load and speed changes. Turbochargers for gas engines require controls in order to maintain the desired fuel-air ratio for various operating conditions.

A turbocharger warms the air it handles, because of the compression. For this reason, an air cooler, or intercooler, is often used between the turbocharger and the engine. By reducing the temperature of the air entering the engine, the intercooler increases the density of the air and thus introduces a greater weight of it into the cylinder, increasing the supercharging effect.

The total inlet system illustrated in Fig. 13-28 is used on John Deere tractors, Models 4620 and 7020/7520 four-wheel drive models. Air can be traced through the pre-cleaner, the aspirator that expels collected dirt through the exhaust, and the path of air through the exhaust, as well as the path of air through the turbocharger and intercooler.

The intake manifold intercooler in Fig.

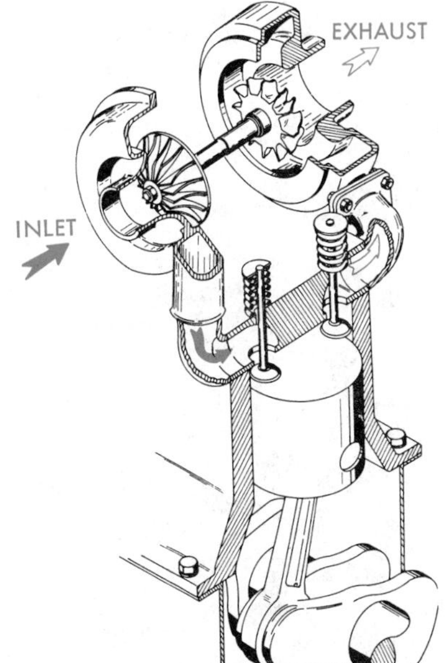

Fig. 13-27. Turbocharger operation diagram. (Allis-Chalmers)

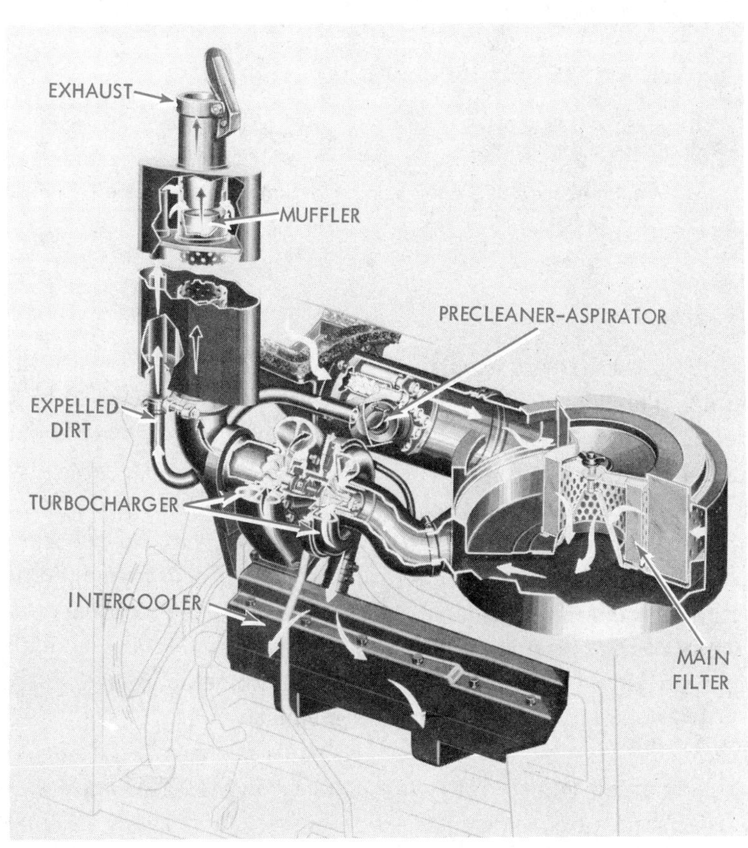

Fig. 13-28. Path of air through the air cleaner, turbocharger, and intercooler. (Deere & Company)

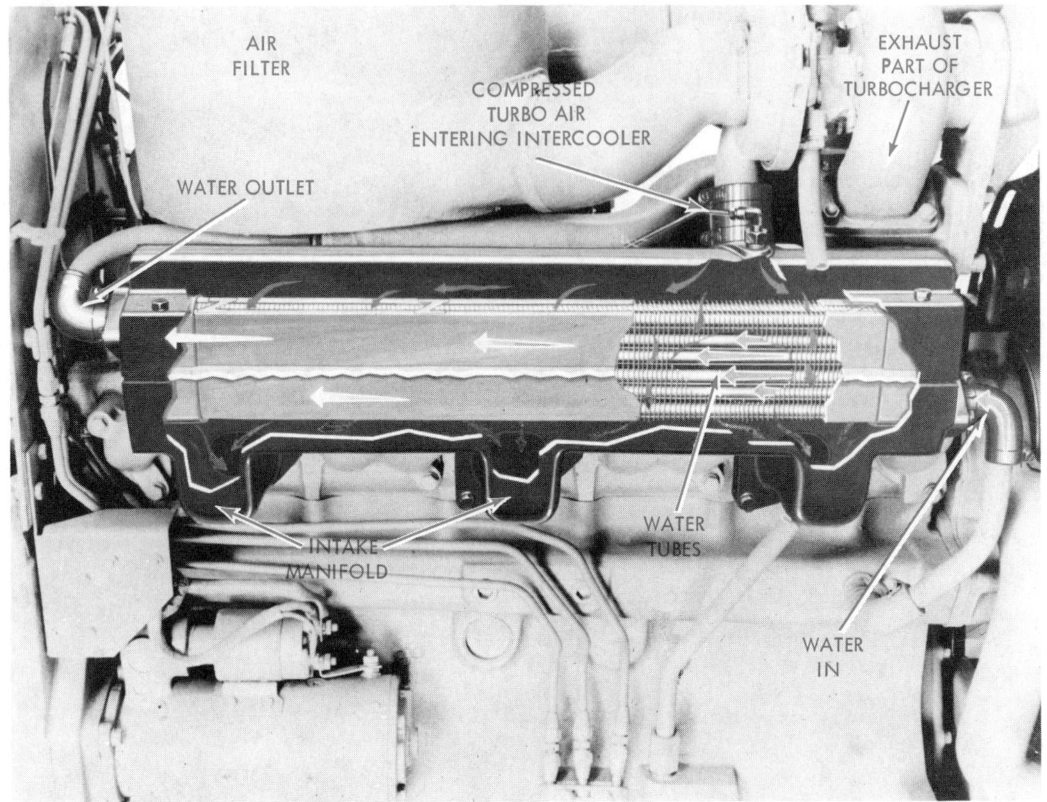

Fig. 13-29. A cutaway view of the intake manifold intercooler. (Deere & Company)

13-29 is presently a John Deere exclusive in farm tractors. This device takes air, heated by compression in the turbocharger, reduces the temperature and volume so that more air can enter the combustion chamber to burn the fuel more completely. It also provides some cooling to valves, combustion chamber walls, and the exhaust vanes of the turbo to prolong the life of these parts.

Supercharging Systems

When a supercharger is fitted to an engine, various changes are made in the engine itself:

1. The *valve overlap* in a normal four-cycle engine is about 30 to 40 degrees. In supercharged engines this is *increased to about 130 to 160 degrees* by opening the inlet valve earlier and closing the exhaust valve later. During the period that both valves are open, the supercharge air effectively rids the combustion chamber of burned gases. It also provides some cooling effect, which particularly benefits the exhaust valves.

2. It is customary to *increase the clearance volume*. This reduces the compression ratio and keeps the maximum or firing pressure down near that of an un-

supercharged engine.

3. Since the basic purpose of the supercharging is to increase the air change so that more fuel may be burned, the rate of fuel injection must be increased. This may mean *larger fuel-pump plungers*, etc.

4. The *exhaust manifolds* of non-supercharged engines are usually water-cooled. However, cooling the exhaust gases reduces their energy and is therefore undesirable in a turbocharged engine. For this reason, the *exhaust manifolds* of turbocharged engines are *insulated* to preserve the heat energy (and also to increase the operator's comfort).

5. Since high-compression gas engines are sensitive to a change in the fuel-air ratio, the amount of air delivered by the turbocharger must be adjusted to suit the prevailing engine load and speed, and also the atmospheric temperature. All large engines have *automatic control systems* to make these adjustments.

6. Two turbocharging methods for four-cycle engines are in use: the *pulse* or Buchi system, and the *constant-pressure* system. The pulse system (which was developed first) makes direct use of the high velocity of the surge of gas which bursts from each engine cylinder when its exhaust valve opens. These pulses flow through carefully sized exhaust manifolds to the turbine wheel of the turbocharger. To prevent the exhaust pulse of one cylinder blowing back into another cylinder which is being scavenged, two or more separate exhaust manifolds are used on engines with many cylinders. (As many as eight manifolds have been used on a large sixteen-cylinder engine).

In the constant-pressure turbocharging system the exhaust pulses are smoothed out in a large common exhaust manifold, so that the pressure of the exhaust gases at the turbine wheel is nearly constant. This system became practicable when design improvements increased the overall efficiency of turbochargers. The improved efficiency increases the air blower pressure relative to the exhaust back-pressure. At the engine cylinder the large pressure difference between the incoming air charge and the departing exhaust gas scavenges the cylinder thoroughly and provides a high supercharge.

7. Early closing of the inlet valve improves the supercharged engine's efficiency. It is ordinarily necessary to increase the clearance volume in order to reduce the compression ratio and thus keep the firing pressure from exceeding that of the engine. But with constant-pressure turbocharging there is a way to keep the same clearance volume and thus obtain maximum fuel efficiency while still producing far more power than in an unsupercharged engine.

In this scheme, the inlet valve cam is altered so that the inlet valve closes early on the piston's inlet stroke, *after* the highly supercharged air has entered the cylinder. As the piston continues to descend, the trapped air expands and its pressure falls. At the end of the inlet stroke, the compression starts from a much lower pressure than the supercharge pressure. This makes it possible to use the full compression ratio and thus obtain correspondingly good fuel efficiency. However, since less air enters the cylinder because of the early closing of the inlet valve, the engine cannot develop as much power.

What all this means in figures is shown in Table 13-1, comparing two large four-cycle engines of the same cylinder size,

TABLE 13-1 EFFECTS OF EARLY CLOSING OF INLET VALVE

	Pulse Turbocharging	Constant-Pressure Turbocharging with Early Inlet Valve Closing
Turbocharger Blower Pressure, psi	7	15
Compression Ratio	8.5 to 1	13 to 1
Final Compression Pressure, psi	450	450
Firing Pressure, psi	1050	1050
Fuel Efficiency, Btu/bhp/hr	6900	6300
Rated Power, Bmep	154	136

one fitted with pulse turbocharging, the other with constant-pressure turbocharging and early inlet valve closing.

Note that the engine with the early inlet valve closing gains 13 percent in fuel efficiency while sacrificing 12 percent in power, compared to the engine with pulse turbocharging, but still develops 70 percent more power than an unsupercharged engine.

8. *Variable early closing of inlet valve* is another possibility. The Nordberg

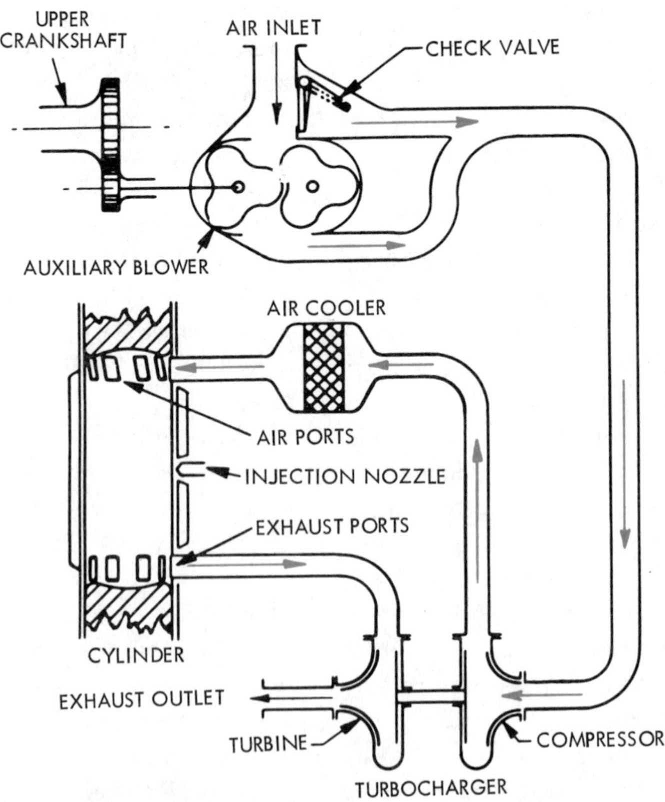

Fig. 13-30. Turbocharging system for two-cycle engine, using mechanically-driven blower preceding turbocharger. (Colt Industries, Fairbanks Morse Motor & Generator Operations)

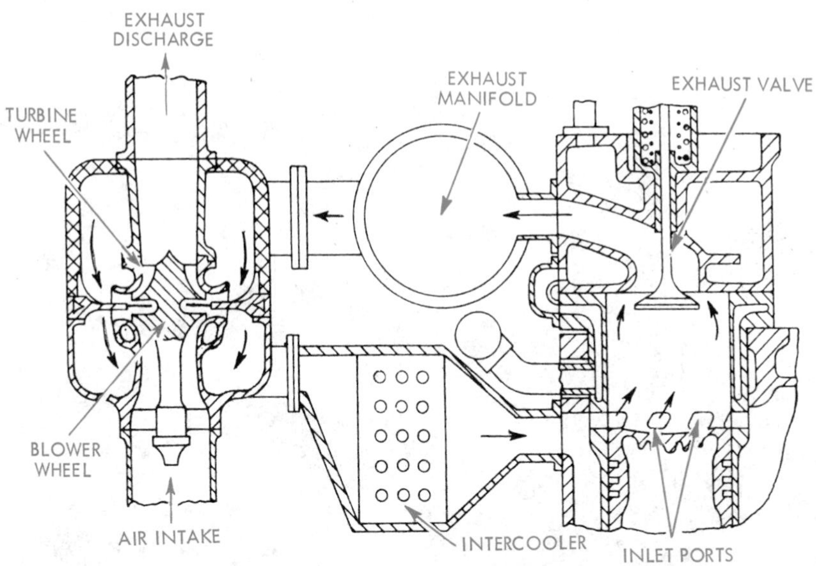

Fig. 13-31. Turbocharging system for two-cycle engine. No scavenging blower is used.

Superairthermal engine, for instance, employs early closing of the inlet valve and also *varies the closing time* automatically to maintain nearly the same compression and firing pressures regardless of load variations or ambient temperature changes. This feature results in higher pressures and temperatures at light loads, which in turn facilitates starting, and improves the light-load fuel economy.

9. Turbochargers may be fitted to two-cycle engines in several ways. The turbocharger may be used *in addition to the regular scavenging system.* In this case the atmospheric air drawn into the scavenging pump or blower is compressed to scavenging pressure in the normal manner and then passes to the turbocharger where it is raised to supercharge pressure. Fig. 13-30 shows this system as used on Fairbanks-Morse engines.

At light load, when there is little energy available to drive the turbocharger, the mechanically-driven blower alone puts the scavenging air into the cylinders. As engine load increases, the turbocharger speeds up and sucks in so much air that its inlet pressure drops to atmospheric, causing the blower check valve to open. At this engine power the blower becomes unloaded (saving engine power) and the turbocharger enters the load range where it alone does both the scavenging and supercharging.

10. The turbocharger may be used *without any other scavenging system,* Fig. 13-31. Under the most favorable conditions the engine starting air contains sufficient energy to start the turbocharger and also to supply enough combustion air to fire the fuel. Usual practice today is to equip the turbocharger with some special device to supply additional scavenging air while the engine is being started and while it is running at light load.

11. *Mechanical drive,* which was used

Fig. 13-32. Supercharged two-cycle engine with disconecting mechanical drive. (Clark Bros. Co.)

by Clark Brothers on the first American supercharged two-cycle engine, is still one of the schemes employed. Referring to Fig. 13-32, the whole supercharger (gas turbine and air blower) can be driven from the engine's crankshaft by a chain and gears, the drive passing through a disconnecting coupling. At starting and at light-load the supercharger is driven mechanically. When the load increases sufficiently, the drive coupling disconnects automatically; the supercharger then operates on exhaust gas only.

12. *Jet air starting* is another scheme; it may be used when high-pressure air is available. The compressed air is blown through nozzles into the turbocharger to get it going. The jets may be arranged to act on either the gas turbine wheel or on the air blower wheel. If the latter, the jet air passes into the engine and directly assists the scavenging process.

In some gas engines the spark is retarded automatically while the engine is starting. This causes the turbocharger to accelerate faster because retarding the spark increases the exhaust gas temperature and thus feeds more energy to the gas turbine wheel.

Checking On Your Knowledge

The following questions give you the opportunity to check up on yourself. If you have read the chapter carefully, you should be able to answer the questions. If you have any difficulty, read the chapter over once more so that you have the information well in mind before you go on with your reading.

DO YOU KNOW

1. How is scavenging accomplished in a four-cycle and a two-cycle diesel engine?
2. What is the valve gear in a four-cycle and two-cycle engine?
3. Why is the size of the intake valve so critical and not the exhaust valve on a four-cycle engine?
4. Explain why the materials used in exhaust valves are different from intake.
5. Explain why hard seats are used in exhaust valves.
6. Diagram six typical camshaft drive layouts.
7. Explain the methods of lubricating various valve-actuating systems.
8. How do hydraulic valve lifts operate?
9. Explain why valves are timed to open and close at unequal times during a complete engine cycle.
10. What is the purpose of valve overlap?
11. Explain the advantages and disadvantages of dry and wet type air cleaners.
12. What are the changes made in an engine design when a supercharger is fitted to the engine?

Chapter

14 Injecting Fuel

This chapter will acquaint you with the most important part of the diesel engine—its fuel-injection system. The major working parts which you have previously studied serve to bring in the air needed for combustion and to deliver mechanical power derived from the hot gas. This chapter, on injecting the fuel, and the next chapter, on burning it, will show you how these vital operations are performed.

You will start by learning what is required of a fuel-injection system. Next, you'll see how these requirements are met in the several kinds of injection systems in common use. Among these are multiple-plunger, distributor, and unit-injector types. Simple diagrams and pictures with explanations will help you understand how each type functions.

You will also want to know what the actual designs look like, therefore an examination of the construction and operation of some typical designs will be undertaken.

In this chapter you will also learn what is meant by metering and timing the fuel injection; spray valves with *differential area; pintle nozzles* and other terms often often used.

Fuel-Injection System

The fuel-injection system, in delivering the fuel to the combustion chamber, must do the following:

1. Meter, or measure, the correct quantity of fuel to be injected.
2. Time the fuel injection.
3. Control the rate of fuel injection.
4. Atomize, or break up, the fuel into fine particles.
5. Properly distribute the fuel in the combustion chamber.

Most diesels built today use some form

of *mechanical* or *solid injection.* Pumps and spray valves form the basic elements of all solid-injection systems. These may be combined in many ways but there are three main classes: (1) multiple plunger, (2) distributor systems and (3) unit injectors.

Metering. Accurate metering or measuring of the fuel means that, for the same fuel-control setting, exactly the same quantity of fuel must be delivered to each cylinder for each power stroke. Only by doing this can the engine operate at uniform speed with a uniform power output from each cylinder.

Timing. Proper timing, which means starting the fuel injection at the required moment in the cycle, is essential in order to obtain the maximum power from the fuel-air mixture, and thus insure fuel economy and clean burning. If the fuel is injected too *early* in the cycle, ignition may be delayed because the temperature of the air at this point is not high enough. In the meantime, unburned fuel accumulates in the cylinder; when it does ignite, it produces excessive pressure and noise.

On the other hand, if the fuel is injected too *late* in the cycle, the fuel will not all be burned until the piston has traveled well past top center. This reduces the amount of expansion of the burned gas, and in extreme cases some of the fuel may still be burning when the exhaust valve or port opens. As a result of late timing, the engine does not develop its full power, the fuel consumption is high, and the exhaust is smoky and hotter than normal.

Rate Control. The rate of fuel injection is important for the same reason that correct timing is important. If the start of injection is correct but the rate of injection is too fast, the results will be similar to an excessively early injection. If the rate of injection is too slow, the results will be similar to an excessively late injection.

Atomization. Fuel must be atomized to suit the type of combustion chamber in use. (We'll discuss combustion chambers in the next chapter.) Some chambers require very fine fuel particles; others can operate with coarser fuel particles. Proper atomization hastens the starting of the burning process by causing the fuel particles to give off vapor sooner and by exposing a greater surface of fuel particles to the oxygen with which the fuel can combine.

Distribution. Proper distribution of the fuel must be obtained in order that the fuel will penetrate to all parts of the combustion chamber where oxygen is available for combustion. If the fuel is not well distributed, some of the available oxygen will not be utilized, and the power output of the engine will be low.

Multiple Plunger Injection System

In a multiple plunger injection system, there is a separate pumping unit for each cylinder. In this system (known also as a jerk-pump system), the bulk of the job is carried out by the pump itself, which raises pressure, meters the charge, and times the injection.

An individual pump serves each engine cylinder, connecting directly with an injection nozzle containing a spring-loaded check valve. Details of the various designs of pumps and nozzles employed in this widely-used system are given later.

The requirements which a pump of this type must fulfill, as to both metering and timing, are such that they can be met only by a precision device as is indicated in the following paragraphs.

Pump Metering. The volume of the fuel injected, at full load, is only about 1/20,000 of the displacement of the engine cylinder. When the engine is idling, the ratio is only about 1/100,000. This means that, when a small engine (around 50 cu in displacement/cylinder) is idling, the fuel-injection pump must accurately meter the minute quantity of oil represented by a single round drop less than 1/10 of an inch in diameter.

Pump Timing. To gain an understanding of the extreme accuracy required of timing an engine, it takes only 1/300 of a second for an engine running at 1000 rpm to complete the entire injection period (assuming that the period covers 20 degrees of crankshaft rotation.)

The start of injection should not be permitted to vary more than one degree of crankshaft rotation; this means holding it within limits of about 1/6000th of a second!

Pump Injection Pressures. These minute quantities must be injected within these close time limits at pressures of 2500 to 3000 psi, and in some cases much higher.

Fit of Pump Plungers. All injection fuel pumps of the jerk-pump type have plungers fitted closely to the pump cylinders by lapping. Lapping means finishing hardened surfaces by working them against the surfaces of laps with an exceedingly fine abrasive material.

From a number of plungers and cylinders which have been lapped truly round but which differ slightly in diameter, the plungers are matched to their cylinders by selection. In this way a fit is obtained with a clearance less than 0.0001″ (one-ten-thousandth of an inch). Such a fit gives very little leakage, even with the high pressures used, and no packing of any kind is needed. Due to this method of mating a plunger to its cylinder, these parts are not interchangeable. If a plunger or cylinder is worn or damaged, *both* pieces must be replaced.

Principles of Operation of APE Pump

The American Bosch fuel injection pump in Fig. 14-1 is a constant stroke, lapped-plunger type, operated by self-contained cam-and-tappet arrangement. The purpose is to deliver an accurately measured quantity of fuel oil under high pressure to the spray nozzle through which the fuel is injected into the cylinder of the engine. This fuel must be delivered at a definite point in the engine cycle and within a limited number of degrees of engine crankshaft rotation.

The pump consists chiefly of a housing, the pumping element or elements—each comprised essentially of the plunger and barrel, the plunger return spring with its seats, and the geared control sleeve engaging with the control rack. The housing also contains the fuel sump, the delivery valve assembly, the delivery valve holder and the union nut for connection of the high pressure discharge tubing.

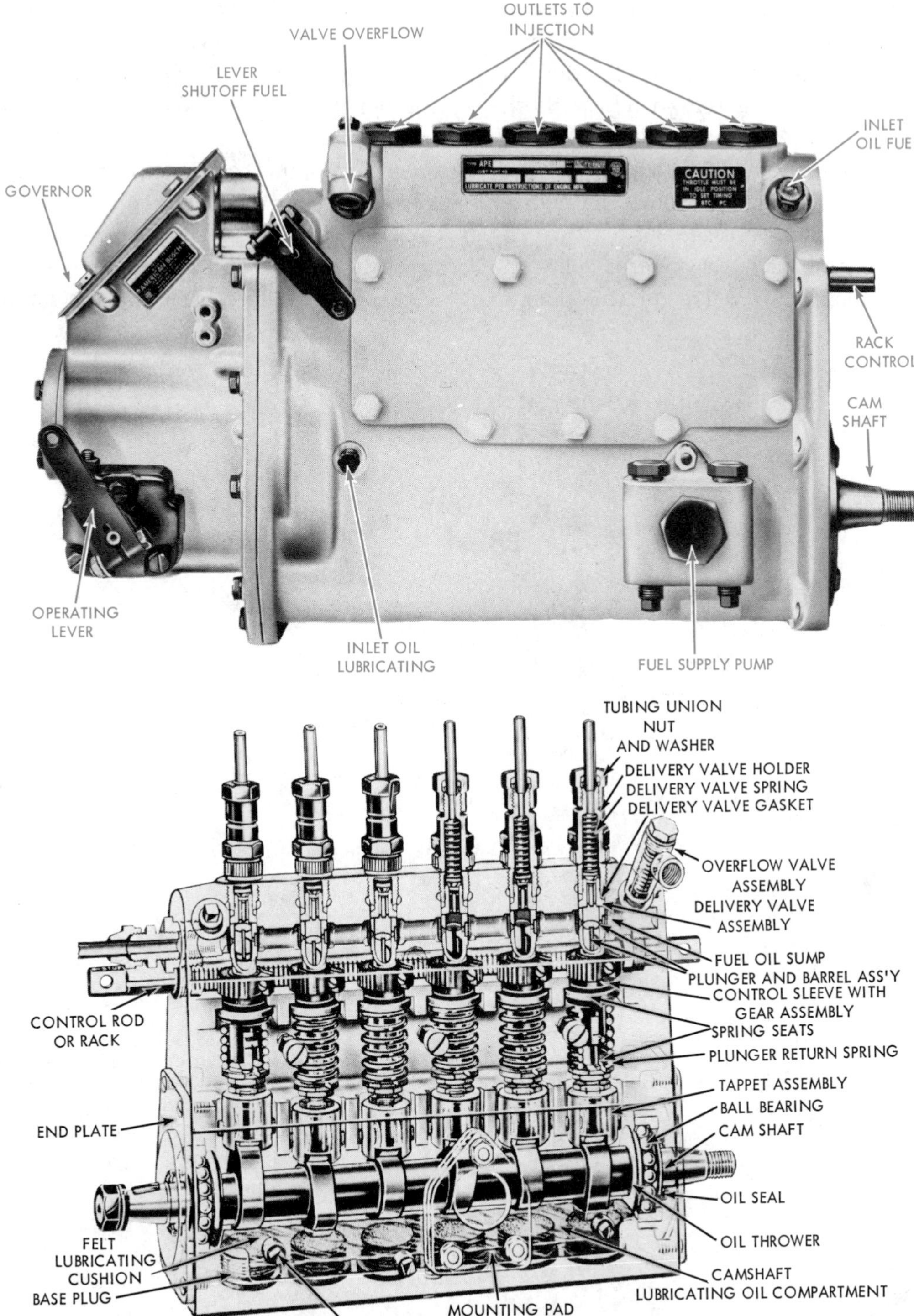

Fig. 14-1. Multiple plunger APE pump. (American Bosch AMBAC Industries, Inc.)

Pumping Principles. Fuel enters the pump from the supply system through the inlet connection and fills the fuel sump surrounding the barrel. With the plunger at the bottom of its stroke, Fig. 14-2A, the fuel flows through the barrel ports, filling the space above the plunger and also the vertical slot cut in the plunger and the cut-away area below the plunger helix (scroll). As the plunger moves upward, the barrel ports are closed, Fig. 14-2B. As it continues to move upward, Fig. 14-2C, the fuel is discharged through the delivery valve into the high pressure line.

Delivery of fuel ceases when the plunger helix passes and opens the spill port, Fig. 14-2D, and the delivery valve returns to its seat. During the remainder of the stroke, fuel is spilled back into the sump.

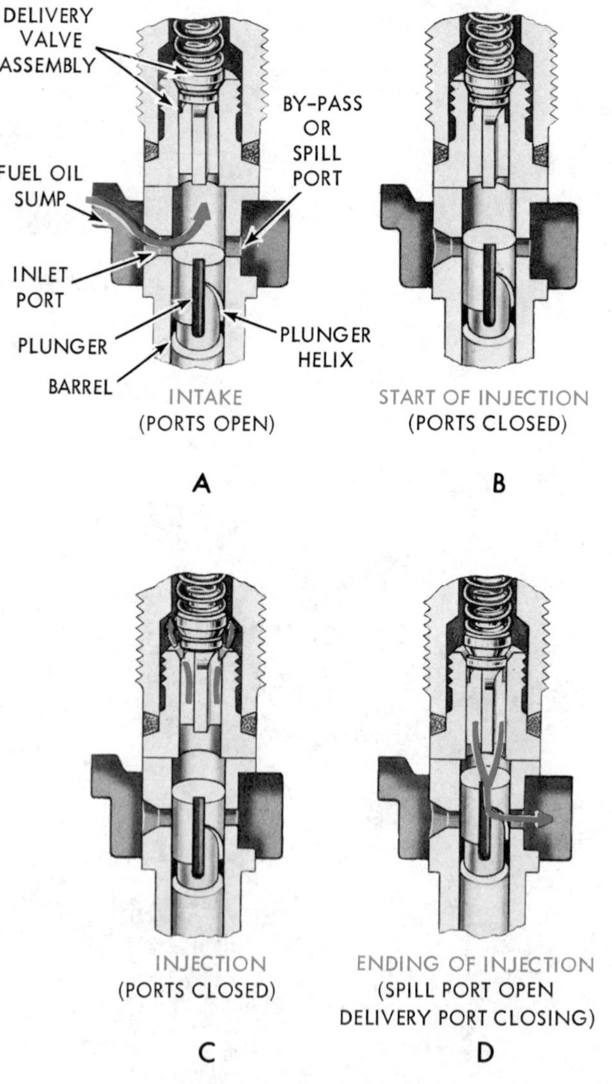

Fig. 14-2. Pumping principles of APE pump. (American Bosch AMBAC Industries, Inc.)

This termination of fuel delivery controls the quantity of fuel delivered per stroke.

Metering Principles. Fig. 14-3 top illustrates the plunger rotated in the position of maximum fuel delivery where the effective part of the stroke is relatively long before opening of the spill port. Fig. 14-3 bottom illustrates a normal delivery with a much shorter effective stroke. The plunger in Fig. 14-3 right has been rotated so that the vertical slot is in line with the spill port and the entire stroke is non-effective; there is no delivery. The amount of fuel delivered is controlled by rotating the plunger, thus changing the length of its effective pumping stroke.

Plunger Types. Plunger types include two different types of helices—a lower helix and an upper helix, as illustrated in Fig. 14-4.

For the lower helix type plunger, the effective stroke injection always begins at the same time, regardless of where the plunger is rotated (except at no delivery position) because the top end that closes the ports is horizontal or flat. Plungers of this type are used in injection pumps marked *Timed for Port Closing.*

For the upper helix type plunger the beginning of the effective stroke varies as the plunger is rotated because the top edge which closes ports is sloping. Plungers of this type are used in pumps marked *Timed for Port Opening.*

The plunger is rotated by means of the control sleeve Fig. 14-5, the lower end of which is slotted to engage the flanges of the plunger. The upper end of the control sleeve is provided with a gear segment engaging the control rack. Movement of the control rack, either manually or by governor action, rotates the plunger and varies the quantity of fuel delivered by the pump.

Delivery Valve. The delivery valve, Fig. 14-6 top left, prevents excessive draining of fuel from the discharge line. As the plunger helix uncovers the spill port, there is a sudden drop in pressure below the valve and the valve closes.

As it snaps into its seat the delivery valve also performs the important function of greatly reducing the pressure in the high pressure tubing, thus lessening the possibility of secondary injection or after-dribble at the spray nozzle. This is accomplished by an accurately-fitted relief or displacement piston located at the upper end of the valve stem below the seat.

As the lower edge of the piston passes into the valve body, Fig. 14-6 top right, all fuel in the line is trapped. Therefore, further movement of the valve to its seat, 14-6 bottom, increases the volume of space available for the existing fuel in the line and thereby greatly reduces its pressure.

Timing. Timing of the pump to the engine must be done carefully. All fuel injection pumps of the APE type are usually provided with timing reference markings on the camshaft extension and end plates, Fig. 14-7 top.

The longitudinal marking on the camshaft is properly transferred to the pump half of the drive coupling, whether this is of American Bosch or the engine manufacturer's make, as shown in Fig. 14-7 bottom.

When the reference mark of the camshaft extension or coupling half registers with lines marked R or L, *right* and *left hand rotation respectively,* it indicates that the upward movement of the plunger, in pump cylinder number one, has just

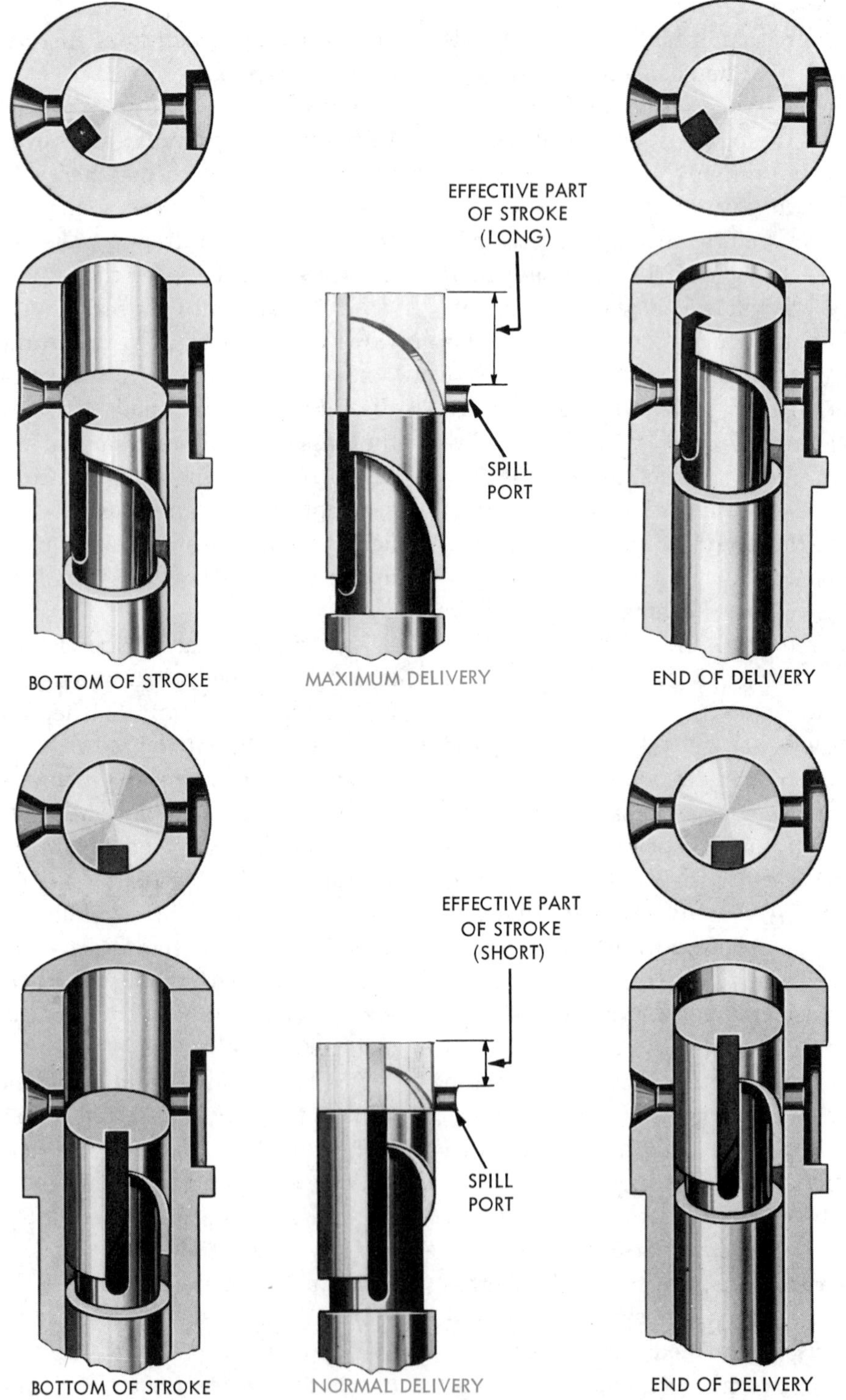

Fig. 14-3. Metering principles—maximum, normal, and (Continued on page 247) no delivery. (American Bosch AMBAC Industries, Inc.)

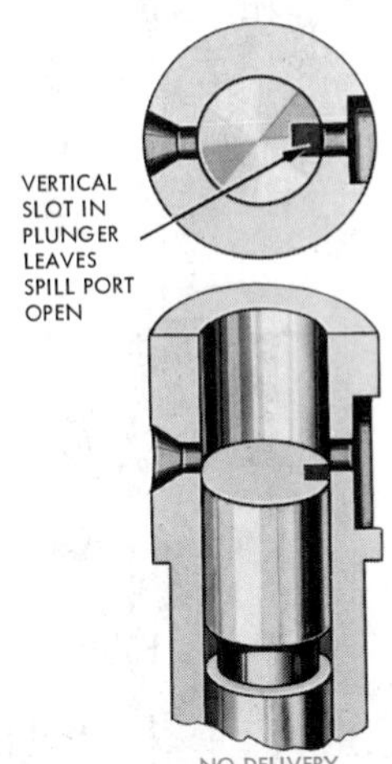

Fig. 14-3. (Cont'd) Metering principles—no delivery. (American Bosch AMBAC Industries, Inc.)

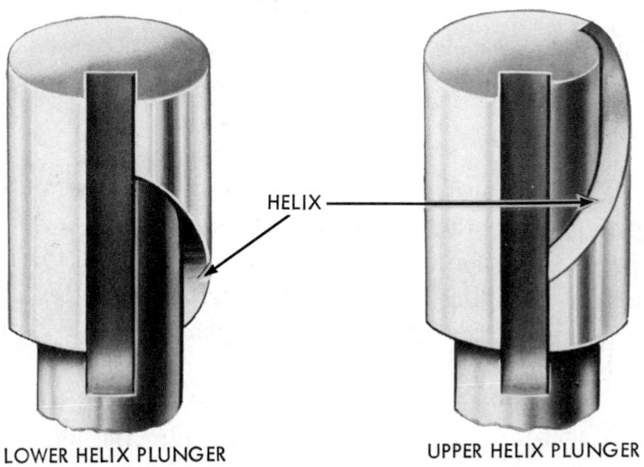

Fig. 14-4. Plunger types—lower and upper helix. (American Bosch AMBAC Industries, Inc.)

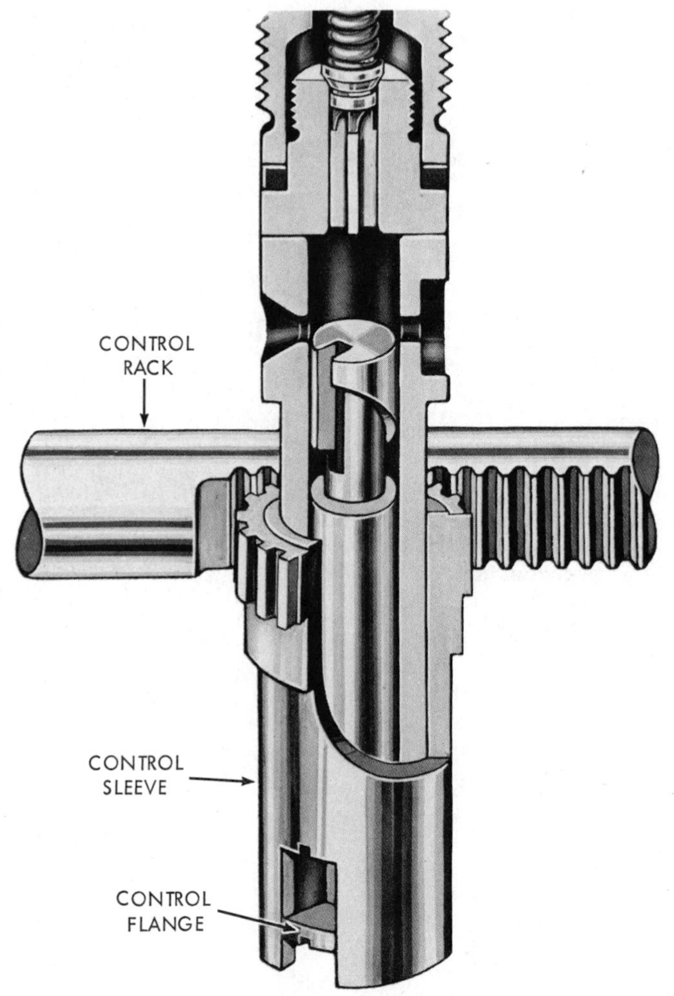

Fig. 14-5. Plunger rotation mechanism. (American Bosch AMBAC Industries, Inc.)

reached its timed position, either port-closing or port-opening, depending upon the particular engine installation.

Installation. Before installing the injection pump on the engine, consult the engine manufacturer's instructions covering its installation and timing and follow those instructions implicitly to assure the most satisfactory engine performance. The following instructions are general and for guidance only.

When attaching the injection pump to the engine, the latter's piston of number one cylinder should be brought near the top center position on the compression stroke by turning the crankshaft in the running direction of the engine. The flywheels of practically all engines are pro-

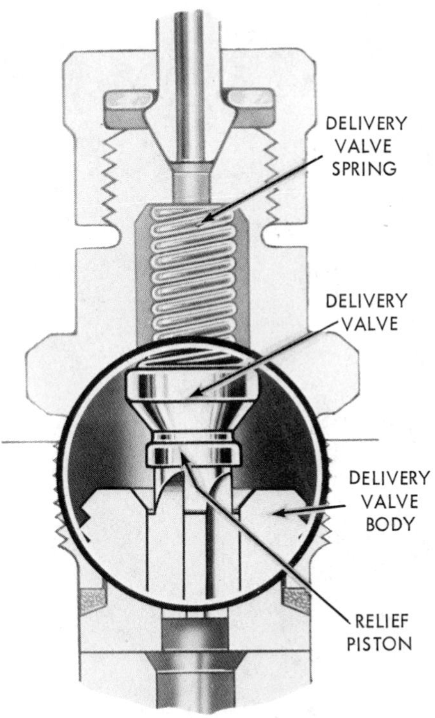

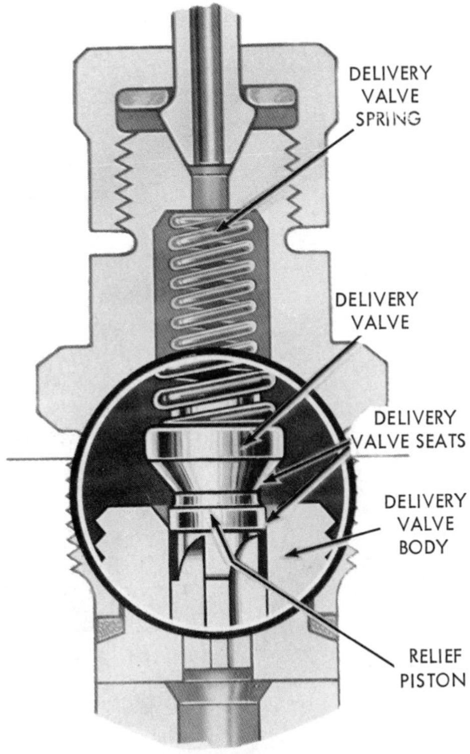

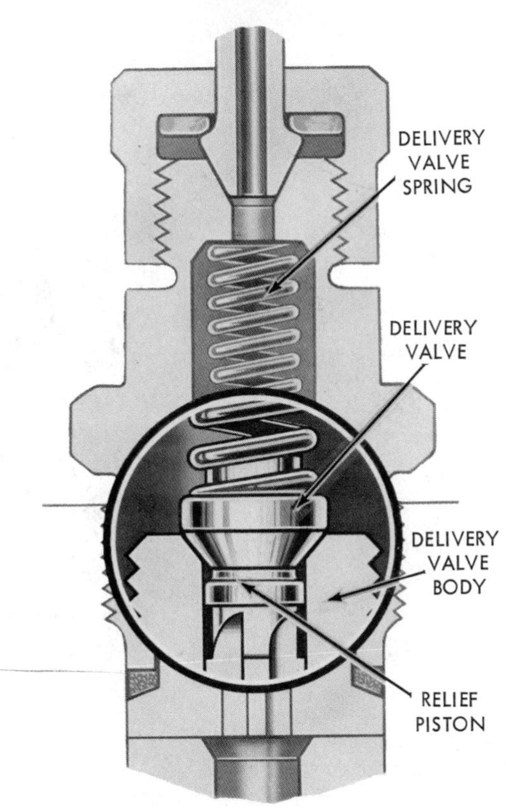

Fig. 14-6. Delivery valve in three positions. (American Bosch AMBAC Industries, Inc.)

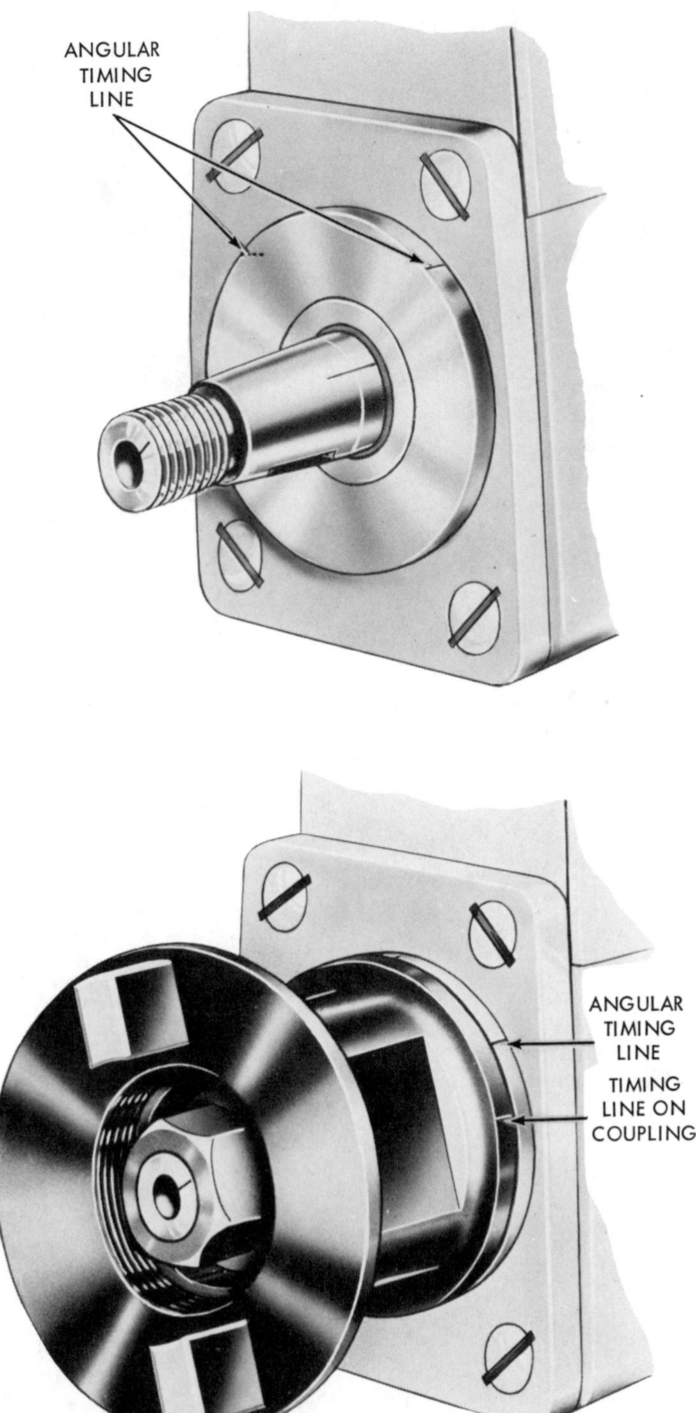

Fig. 14-7. Timing marks on APE pump and engine coupling adaptation to pump timing marks. (American Bosch AMBAC Industries, Inc.)

vided with markings denoting the fuel injection pump timing to facilitate the proper piston position for number one cylinder ahead of top center to which the injection pump must be set.

The injection pump is then mounted on the engine and the correlation of the two coupling halves arranged in such a manner as to have the proper reference marking alignment between the timing markings on the pump half of the coupling and the respective marking on the pump and plate.

Fuel Supply Pumps — Low Pressure. The fuel supply pump draws fuel oil from the fuel tank and pumps it through the fuel filtering system into the fuel injection pump.

The American Bosch supply pump is mounted directly on the housing of the injection pump and is driven by the injection pump's camshaft. It is a variable self-regulating, plunger-type pump which will build pressure up to only a pre-determined point.

The operation of the fuel supply pump is as follows, Fig. 14-8. As the injection pump cam allows the plunger to be forced by its spring toward the camshaft, the suction effect created opens the inlet valve and permits the fuel to enter the plunger spring chamber, Fig. 14-9A.

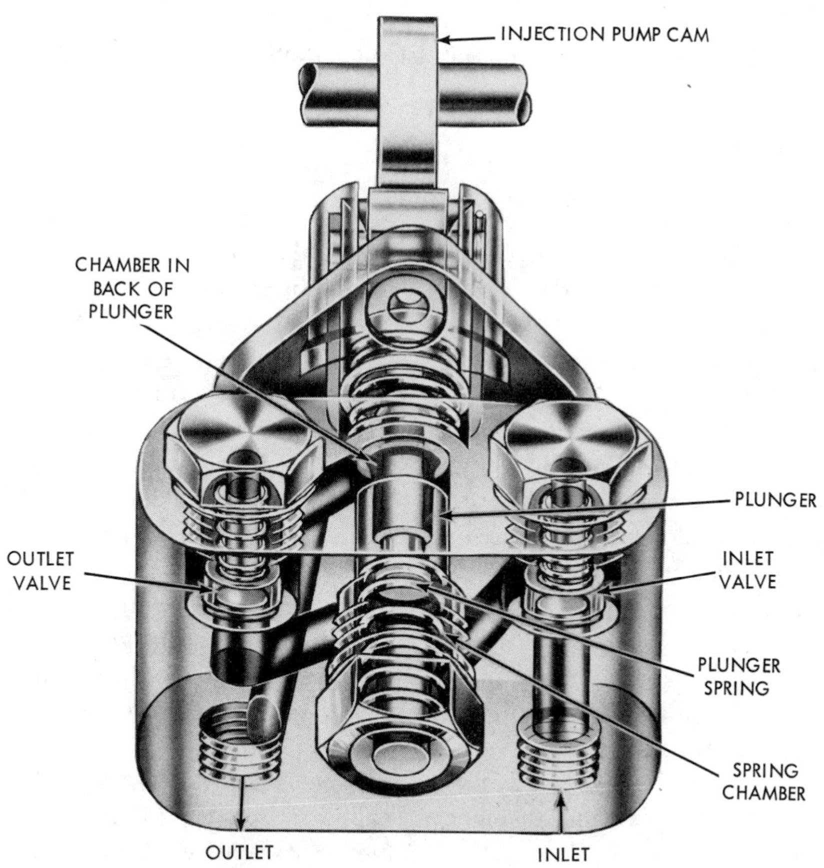

Fig. 14-8. Fuel supply—low pressure. (American Bosch AMBAC Industries, Inc.)

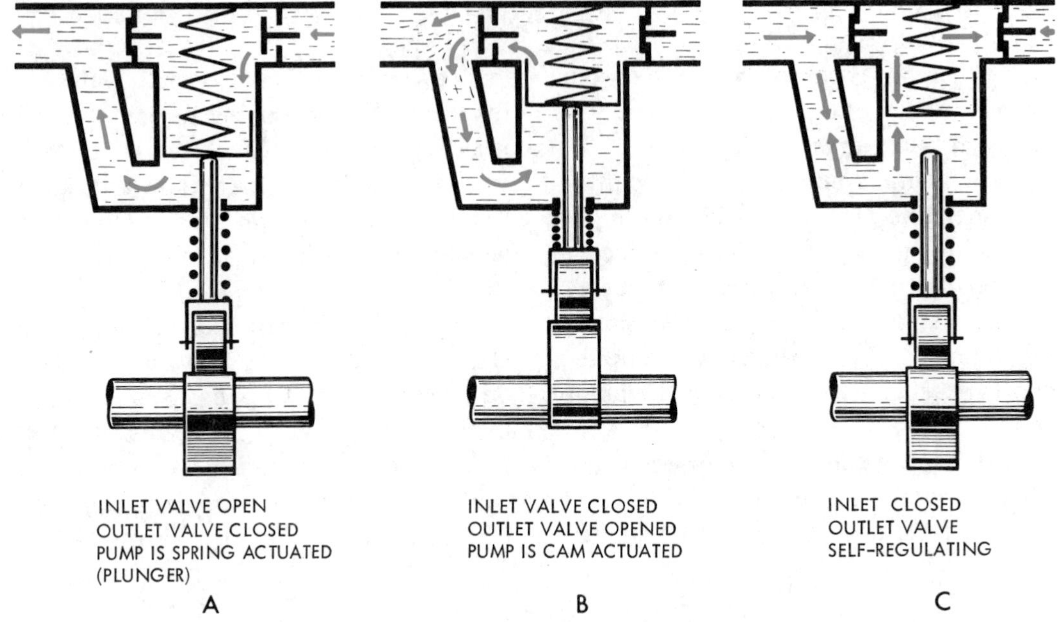

Fig. 14-9. Fuel supply pump schematic. (American Bosch AMBAC Industries, Inc.)

Then, as the cam lobe drives the plunger against its spring the fuel is forced by the plunger through the outlet valve, around into the chamber created in back of the plunger by its forward movement, as in Fig. 14-9B.

As the injection pump cam continues to rotate, it allows the plunger spring, which is now under compression, to press the plunger backward again, thus forcing the fuel oil behind the plunger out into the fuel line leading to the filters and injection pump, as in Fig. 14-9C. At the same time the plunger is again creating a suction effect which allows additional fuel oil to flow through the inlet valve into the spring chamber, as in Fig. 14-9A again.

This pumping action continues as long as the fuel is being used by the injection pump fast enough to keep the supply pressure from rising to the point where it equals the force exerted by the spring of the plunger. Then the pressure between the supply pump and the injection pump holds the plunger stationary against the spring and away from the cam. This prevents further pumping action until the pressure drops enough to permit the plunger to resume operation. This entire cycle is automatic and continues as long as the engine is running.

It is preferable to employ a fuel supply system of the return-flow type, in which the pressure is limited by a spring-loaded overflow valve, usually placed on the injection pump opposite its fuel inlet connection. The overflow valve, *usually adjusted to about 6 to 15 psi*, permits the fuel to pass entirely through the injection pump sump and back into the supply tank. Thus any air or gas which may have entered the system will be carried away. An accumulation of air or gas within the injection pump sump can easily become troublesome.

Distribution Type Fuel Injection Pump

The distributor-type fuel injection pumps are used on many of the light and medium-sized diesel engines at the present time. This type of pump is popular because of its quick response, lower cost, light weight, and compactness, simplicity of design and its ease of adaptability to the small high-speed diesel engines.

Roosa Master DC Pump

The Roosa Master DC pump, a fuel injection pump, Fig. 14-10, is the heavy duty model in a series of several models produced by the company. It is adaptable to light and medium-sized diesel engines of from two to eight cylinders. It is described as a twin cylinder, opposed plunger, inlet-metering distributor-type injection pump.

For construction details, the cutaway view, Fig. 14-11, identifies the main components. The main rotating members of the pump identified in Fig. 14-12 include: (1) the *drive shaft,* (2) *distributor rotor* (containing the plungers and mounting the governor) and (3) the *transfer pump.*

Operation. Referring to Fig. 14-12, the *drive shaft* engages the distributor rotor which fits inside the hydraulic head. The *distributor rotor* has two cylinder boxes, each containing two plungers. The plungers are forced toward each other simultaneously by an internal cam ring which is held stationary. (Fig. 14-14 shows the cam and the plungers being forced together in the pumping stroke). There are as many lobes on the cam as there are engine cylinders to be served.

The *transfer* (low pressure) *pump* was shown in Fig. 14-11. On the opposite end of the rotor is a positive displace-

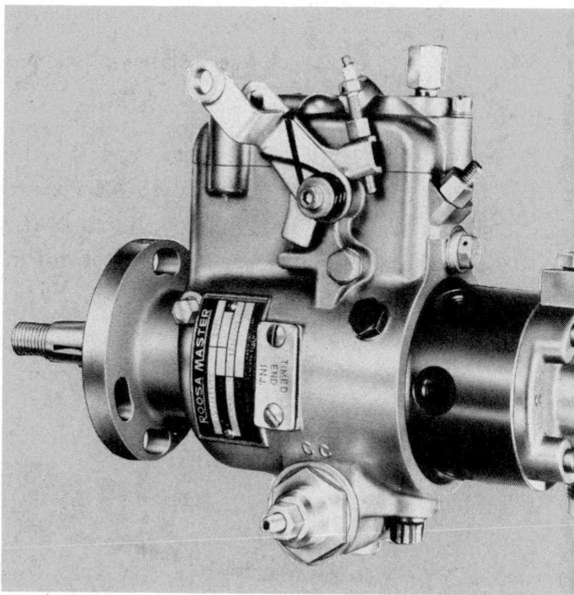

Fig. 14-10. Roosa Master DC pump. (Stanadyne/ Hartford Division)

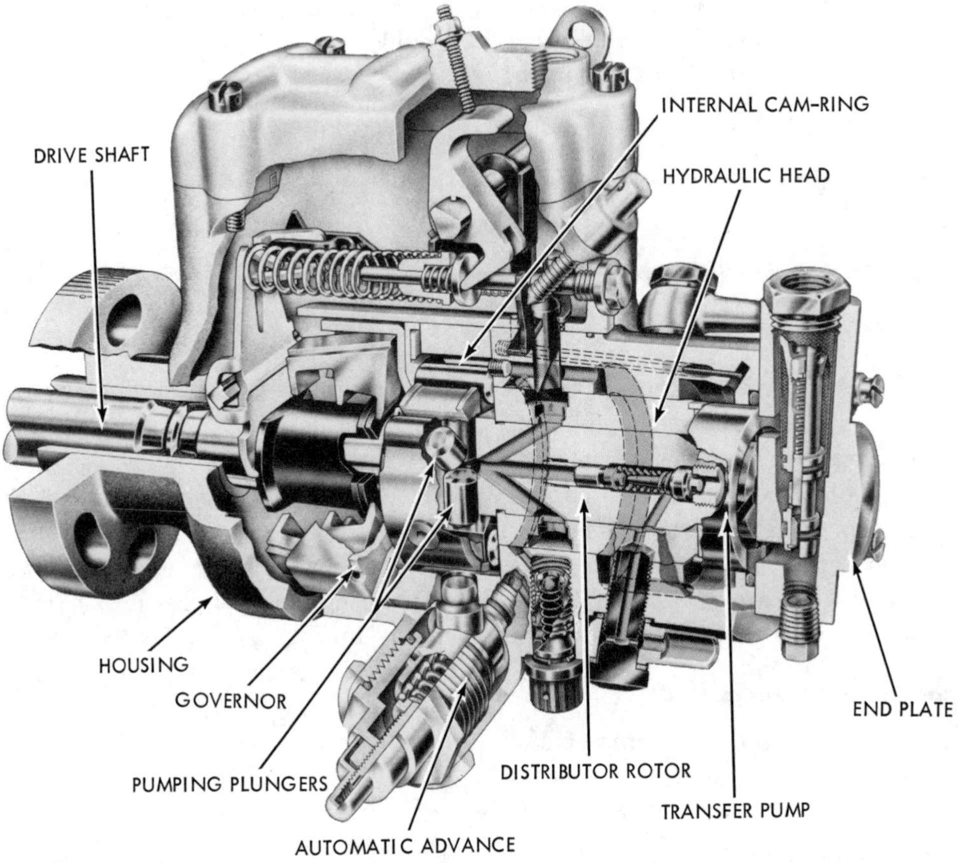

Fig. 14-11. Roosa Master DC pump components. (Stanadyne/Hartford Division)

ment vane type and is covered by the *end plate*. The *transfer pump blades* are shown in Fig. 14-12. They fit into the slots on the end of the distributor rotor.

In Fig. 14-12, one of two angled passages can be seen on the distributor rotor. These are used for charging the pumping plungers. To the right of the charging

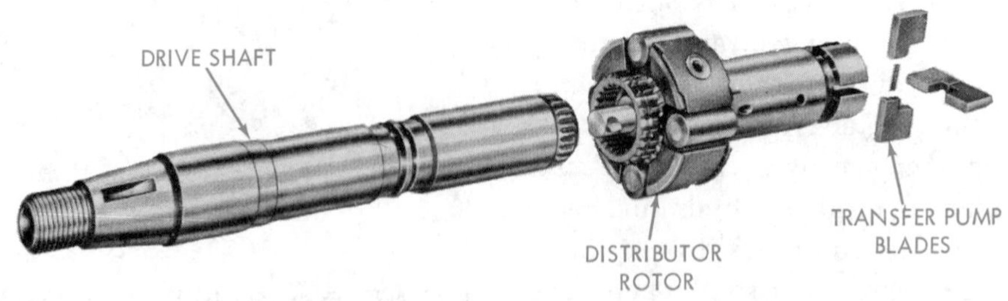

Fig. 14-12. Rotating members. (Stanadyne/Hartford Division)

passage is a discharging hole leading from a centrally drilled passage in the rotor.

The *hydraulic head* in Fig. 14-11 fits snugly over the small end of the distributor rotor and as the rotor revolves inside the hydraulic head, the discharge passage in the rotor indexes with appropriate passages in the hydraulic head to lead to the injector nozzles which are connected to each engine cylinder.

The *mechanical governor* (also in Fig. 14-11) regulates the speed of the engine within close tolerances. The action of the flyweights is transferred through a sleeve to the *governor arm* which pivots and controls the amount of fuel flow by turning the *metering valve* through a linkage.

The metering valve can be closed off by an independently operated shut-off lever.

Fuel Cycle. By following the fuel through one complete cycle, the operating principles of this pump should be quite clear.

In Fig. 14-13, fuel is drawn from the supply tank into the low-pressure transfer pump through the *inlet strainer*. The *transfer pump* has a greater capacity than

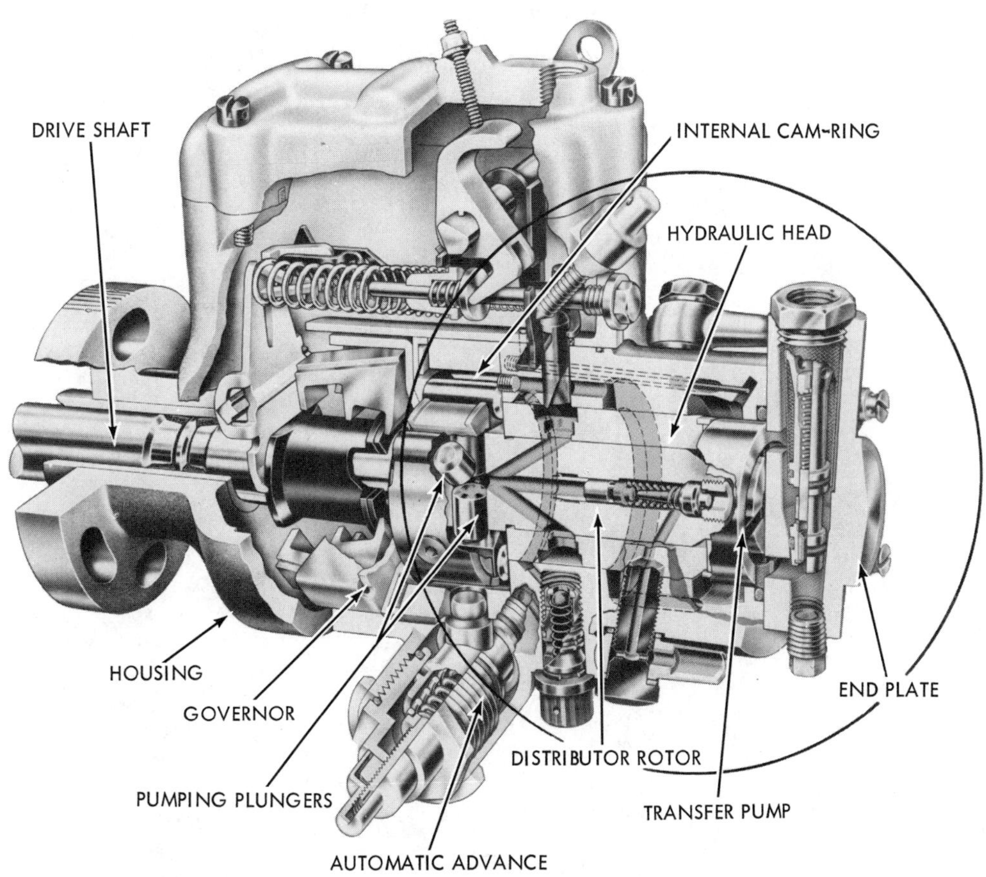

Fig. 14-13. Roosa Master DC pump operation. (Stanadyne/Hartford Division)

the maximum injection requirements, therefore a large percentage of the fuel is by-passed through the *regulating valve*, and back to the inlet side of the pump. Flow and pressure increases with increased rpm of the transfer pump. Fuel under transfer pump pressure flows through the *drilled passage* in the hydraulic head and into the *annulus*.

The fuel then flows around the annulus to the top where it goes through a *connecting passage* to the *metering valve.* The metering valve, which rotates in response to the governor, regulates the flow into the *charging ring* which includes the charging ports.

When the distributor rotor turns, the two *inlet passages* register with the two charging ports in the hydraulic head, thereby allowing fuel to flow into the pumping cylinders. As the rotor continues to turn, the inlet passages move out of registry and the single discharge port in the rotor lines up with one of the head outlets.

Charging Cycle. In Fig. 14-14, as the distributor rotor revolves the angled inlet passages in the rotor line up with opposite charging ports of the charging ring. Fuel under low pressure, controlled by the opening of the metering valve, flows into the pumping cylinders, forcing all four plungers apart. The plungers move outward a distance in proportion to the amount of fuel required and regulated by the metering valve.

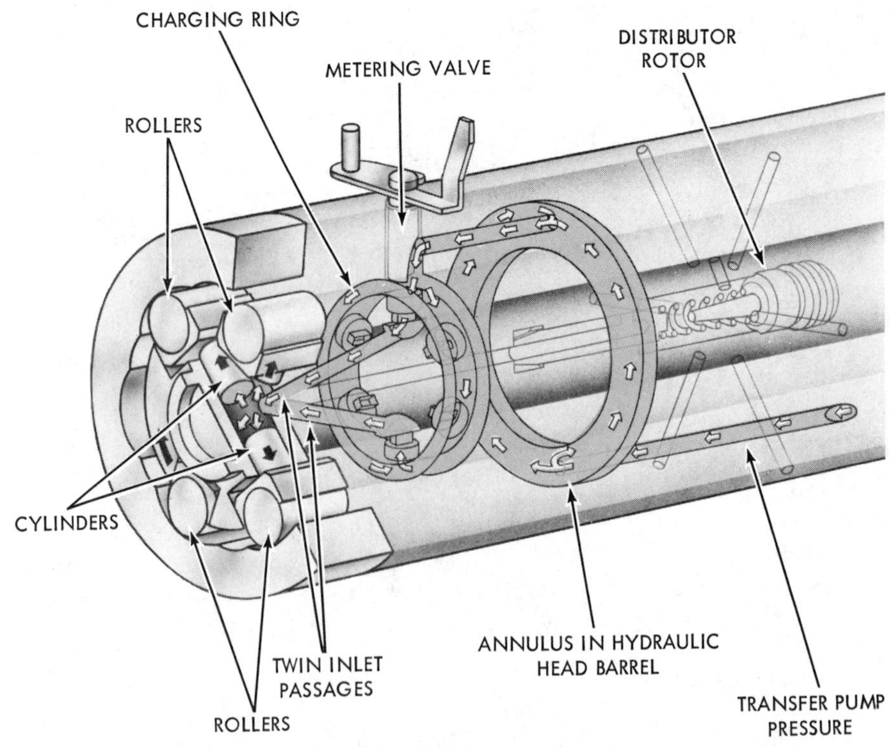

Fig. 14-14. Roosa Master DC pump-charging cycle. (Stanadyne/Hartford Division)

If the engine is idling, the plungers move out very little. Maximum plunger travel and fuel delivery is limited by adjustable leaf springs which contact the edge of the roller shoes. (Fig. 14-12 shows the springs and rollers on the large end of the rotor.) Only when the engine is operating at full power and the metering valve is wide open will the plungers move to the most outward position.

Discharge Cycle. Compare the relative positions of the plungers in relation to their position on the cam in Figs. 14-14 and 14-15. As the rotor continues to turn from the charge position, the inlet passages move out of registry with the charging ports and move into registry with one of the outlet ports.

At this time both sets of rollers contact the cam lobes and the plungers are forced together. The fuel is forced through the passage along the axis of the distributor rotor and out to the injection line. Delivery of the fuel continues until the rollers pass the high point on the cam and the plungers are allowed to move outward. At this time pressure in the axial passage drops and this allows the injection nozzle to close.

Delivery Valve Function. *Retraction* of the injection line is the most significant function of the delivery valve. It is accomplished by rapidly decreasing the injection line pressure, after injection to a pre-determined point lower than that of the injection nozzle opening pressure.

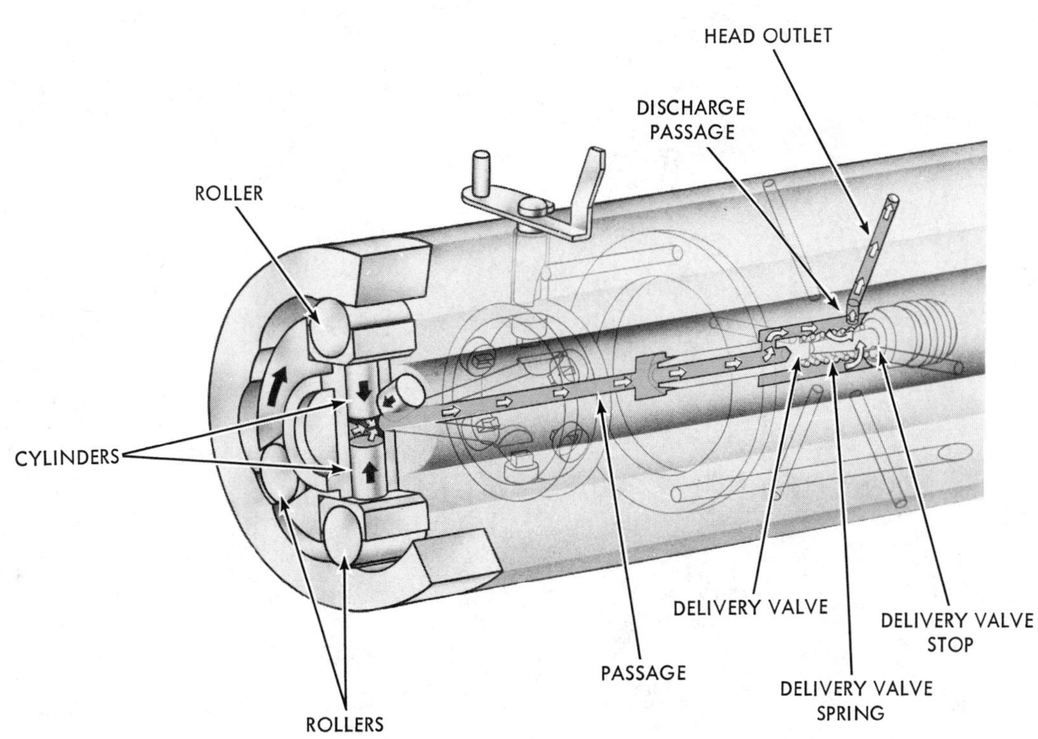

Fig. 14-15. Roosa Master DC pump-discharging cycle. (Stanadyne/Hartford Division)

This reduction in pressure causes the nozzle valve to return rapidly to its seat, achieving sharp delivery cut-off and preventing dribble of fuel into the combustion chamber.

The delivery valve, Fig. 14-16, is located and operates in a bore in the center of the distributor rotor. It is of simple construction in that it requires no seat—only a shoulder to limit travel. Sealing is accomplished by the close fit in the bore. At the start of injection, the fuel pressure moves the delivery valve slightly out of its bore and adds the volume of its *displacement section* to the enlarged cavity of the rotor occupied by the delivery valve spring. This displaces an equal volume of fuel in the spring cavity before delivery through the valve ports begin.

At the end of the injection, the pressure on the plunger side of the delivery valve is quickly reduced by allowing the plunger cam rollers to drop into a retraction step on the cam lobes. The retraction

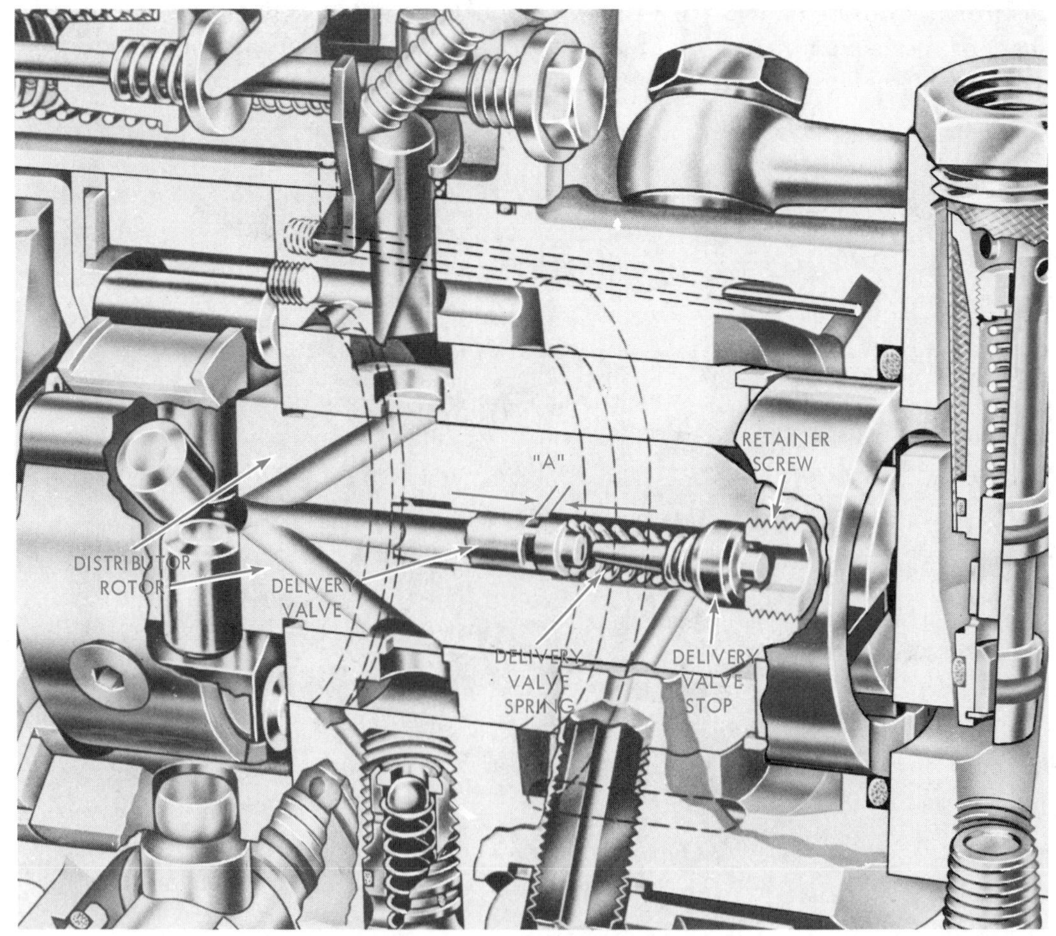

Fig. 14-16. Roosa Master DC pump—delivery valve and transfer valve. (Stanadyne/Hartford Division)

step on the cam can be seen in Fig. 14-15.

As the pressure drops, the delivery valve spring forces the delivery valve to its closed position. This removes its *displacement section* from the spring cavity, Fig. 14-16. Fuel then rushes back out of the injection line to fill the volume of the retreating delivery valve. Following this, the rotor discharge port goes out of registry and the remaining injection line pressure is trapped.

Fuel Transfer Pump and Governor Relationship. The fuel transfer pump shown at the extreme right side of Fig. 14-16 includes a pressure regulating system which will vary the pressure and fuel flow in respect to the requirements of the pump.

Mechanical Governor. In the centrifugal governor, Fig. 14-17, the movement of the flyweights against the governor thrust-sleeve rotates the metering valve. This rotation varies the registry of the metering valve slot with the passage to the rotor, thus controlling the flow to the engine.

This type of governor derives its energy from the centrifugal action of the flyweights pivoting on their outer edge in the retainer. Centrifugal force tips them

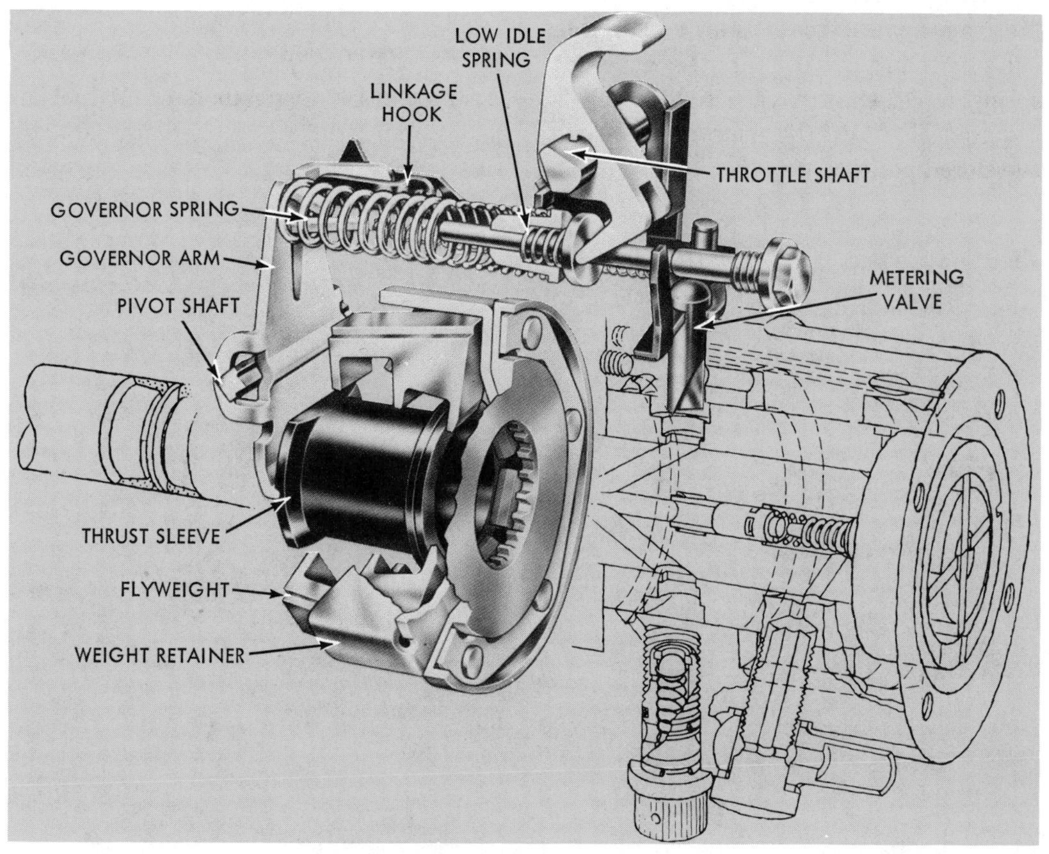

Fig. 14-17. Roosa Master—centrifugal governor. (Stanadyne/Hartford Division)

outward, moving the governor thrust-sleeve against the governor arm. The governor arm pivots on the knife edge of the pivot shaft, and is connected through a simple positive linkage to the metering valve.

The force on the governor arm caused by the centrifugal action of the flyweights is balanced by the compression type governor spring. The spring is manually controlled by the throttle shaft linkage in regulating engine speed. A light idle spring is provided for more sensitive regulation at the low speed range. The limits of throttle travel are set by adjusting screws for proper idling and and high idle positions.

A light tension spring allows the stopping mechanism to close the metering valve without overcoming the governor spring force. Only a very light force is required to rotate the metering valve to the closed position.

Return Oil Circuit. In addition to delivering fuel for the engine injection nozzles, the transfer pump supplies low-pressure fuel oil to lubricate the entire pump.

Fig. 14-18 shows how transfer pump pressure is delivered to the cavity in the hydraulic head. The upper half of this cavity is connected with a vent passage which is located behind the metering valve bore and connects with a short vertical passage to the governor linkage compartment. If air enters the suction side of the transfer pump, it passes to the air vent cavity and then to the vent passage. Air and a small amount of fuel then flow from the housing to the fuel tank via the return line.

Roosa Master DM Pump

The Roosa Master model DM fuel injection pump incorporates many new design features. See Fig. 14-19.

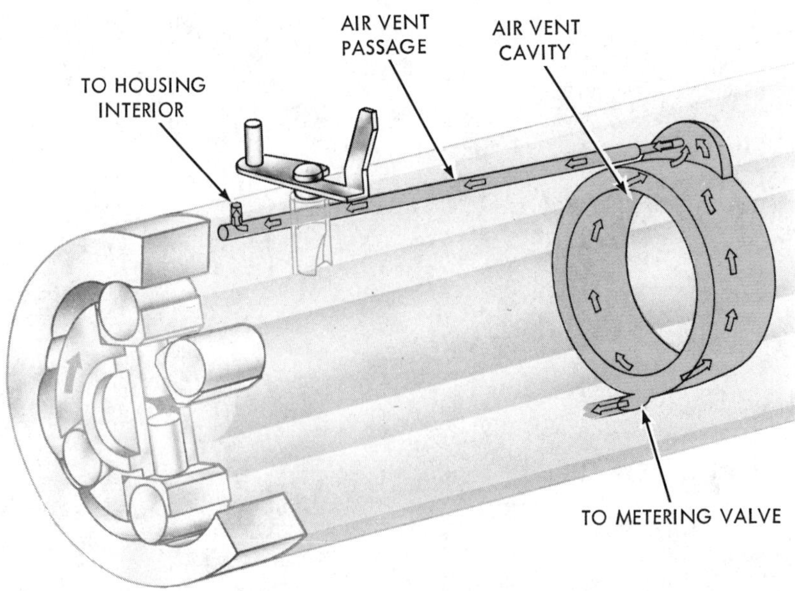

Fig. 14-18. Roosa Master DC return fuel circuit. (Stanadyne/Hartford Division)

The new design was aimed at the new, small, high speed and high output diesel engines. Fuel costs, quieter operation and lower exhaust emissions were also important considerations in the design of the DM pump.

Some of the changes in this model, over the Roosa Master DC injection pump, were a heavier housing and mounting flange, increased diameter drive shaft, supported with a heavy ball bearing instead of the sleeve bearing on the former models. Changes in the transfer pump, improved governor and automatic advance were design improvements demanded by emission standards and small, high speed engines.

PSJ American Bosch Injection Pump

The American Bosch PSJ fuel injection pump, Fig. 14-20 top is a compact, high-speed (capable of 3600 rpm), variable timing, governor-controlled, flange-mounted, single-plunger injection unit designed for use on small and medium-sized diesel engines.

The head core, Fig. 14-20 bottom, has a central bore to which the plunger is fitted; and it is also counterbored and threaded at the upper end for the delivery valve. Drilled discharge ducts (four, six, or eight, depending on the number of cylinders) extend symmetrically from the plunger bore. They end in discharge fittings that are assembled into the top of

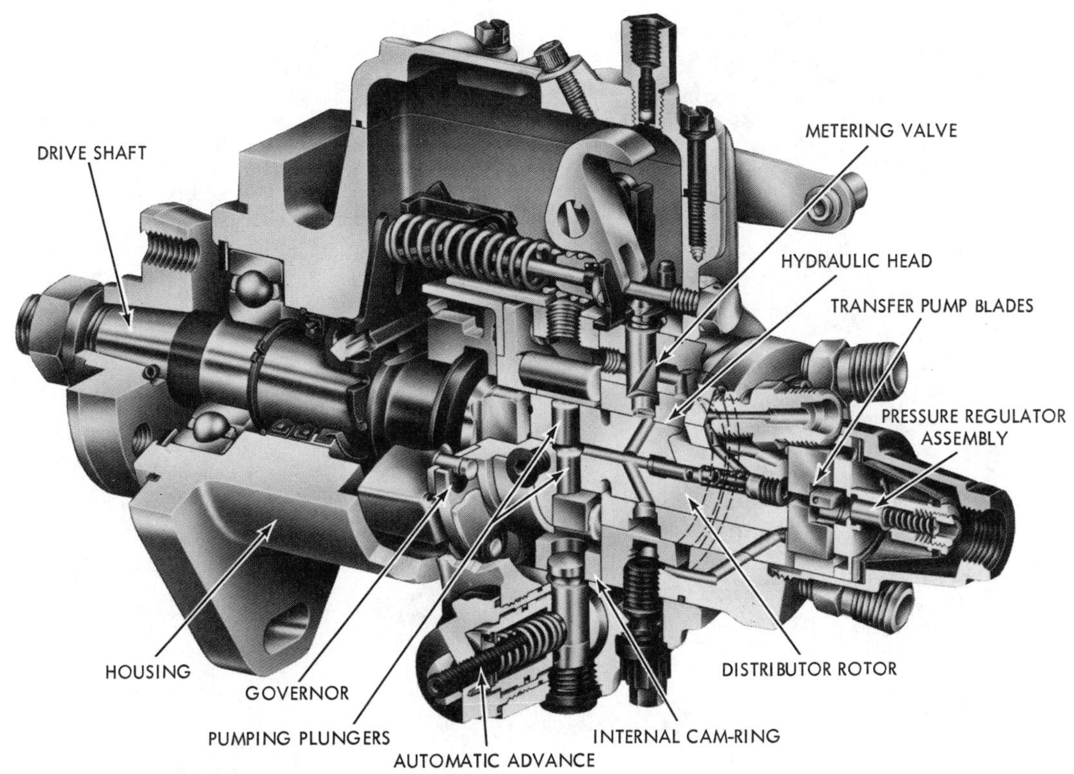

Fig. 14-19. Roosa Master DM injection pump components. (Stanadyne/Hartford Division)

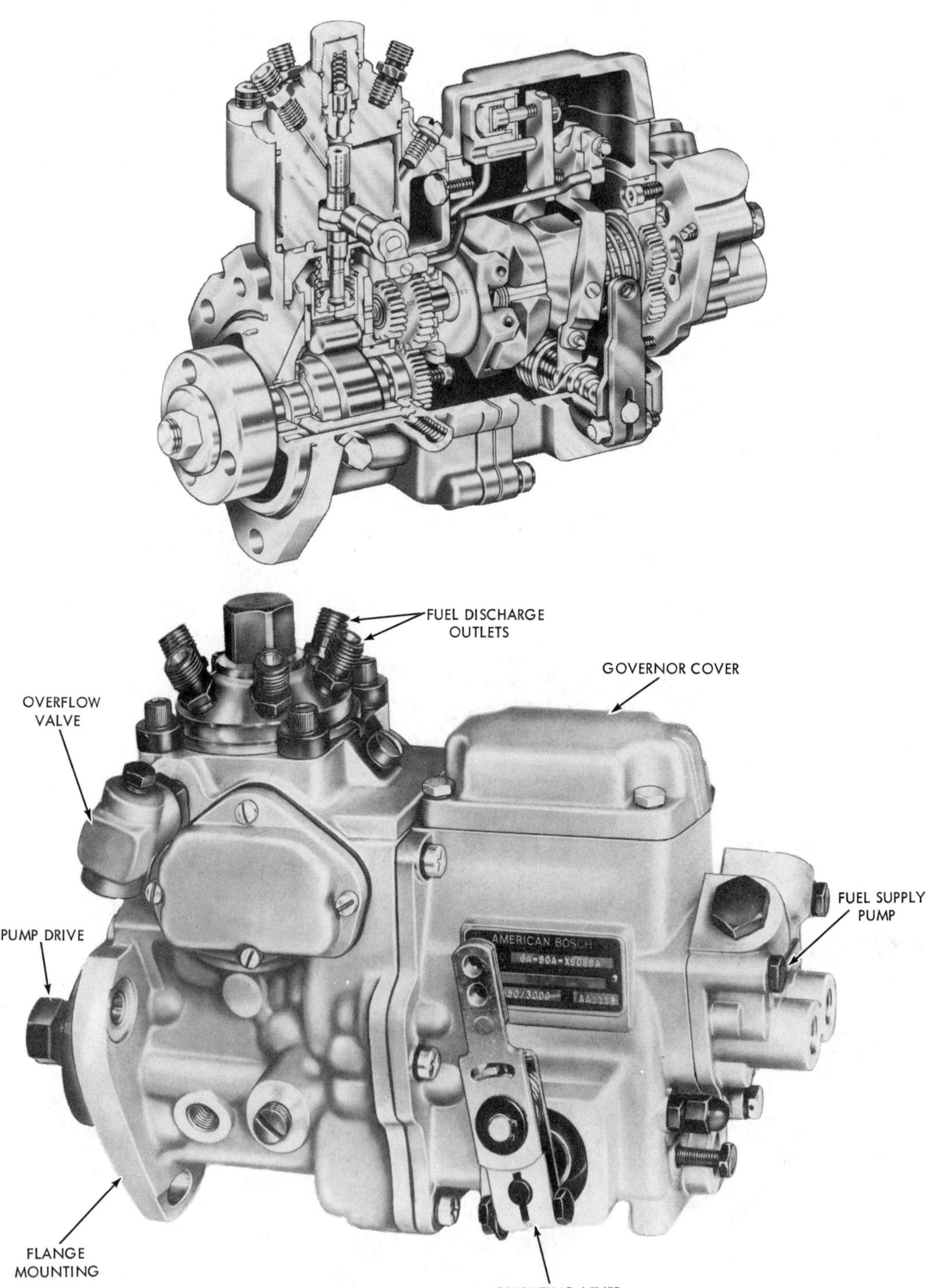

Fig. 14-20. Late Model PSJ-6A pump—single plunger type and cutaway of earlier PSJ model. (American Bosch AMBAC Industries, Inc.)

the core. A downwardly-inclined duct (Fig. 14-23) from the delivery valve bore intersects a mating duct that leads to the distributing annulus in the plunger.

The plunger makes one complete revolution for every two revolutions of the pump camshaft which is operating at engine crankshaft speed, Fig. 14-21. The plunger rotates continuously while moving vertically through the pumping cycle. Therefore, on an eight-cylinder engine, the four lobe cam actuates the plunger eight times for every two revolutions of the pump camshaft. Similarly, on a six-cylinder engine the three-lobe cam actuates the plunger six times for every two revolutions of the pump camshaft.

Fig. 14-22 shows an exploded view of the parts concerned with the pumping stroke. By comparing Figs. 14-20, 14-21 and 14-22, a clearer understanding of the relationship of parts and the pumping

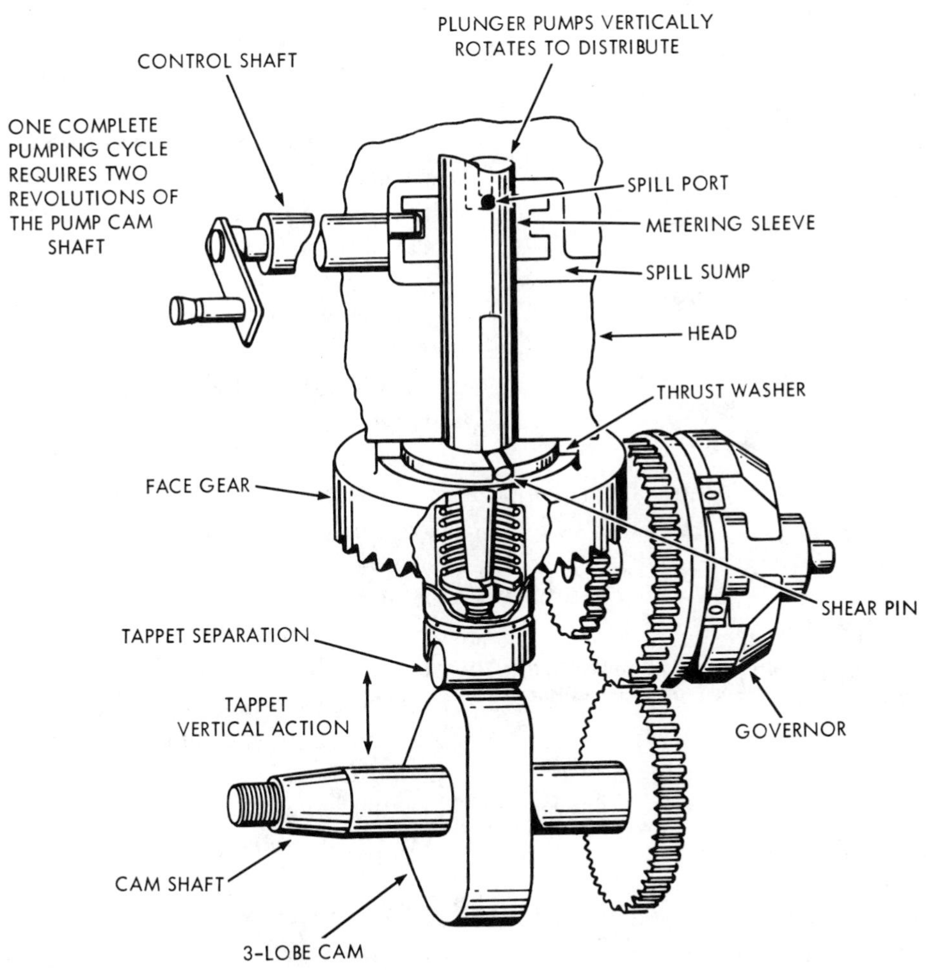

Fig. 14-21. Operational diagram for PSJ pump. (American Bosch AMBAC Industries, Inc.)

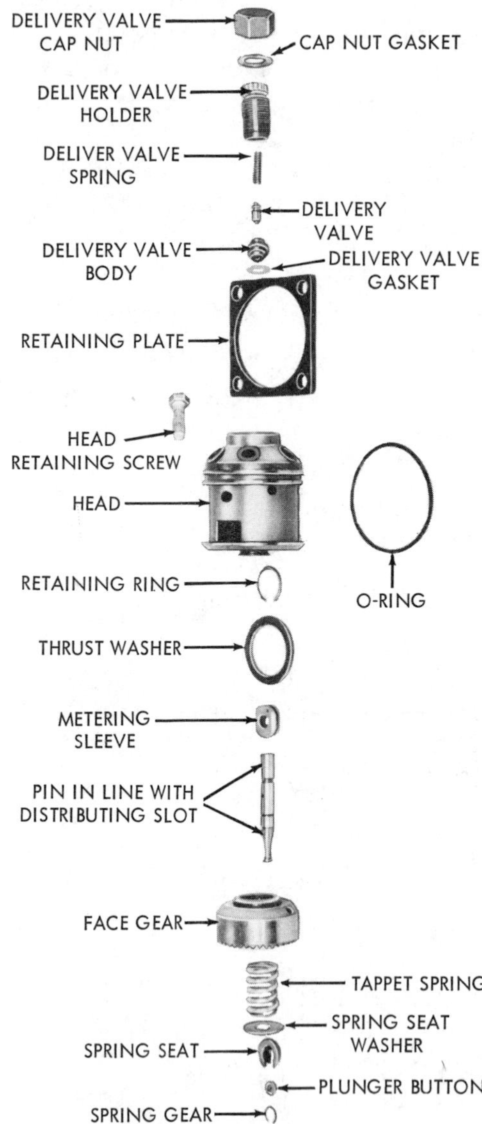

Fig. 14-22. Exploded view of PSJ hydraulic head and plunger assembly. (American Bosch AMBAC Industries, Inc.)

principle can be obtained.

Fuel Pumping Principles. Fuel enters the pump from the supply system through the pump housing inlet connection. This action fills the sump area and that portion of the head cavity between the top of the plunger and the bottom of the delivery valve when the plunger is at the bottom of the stroke. The plunger has previously been brought to this low position by the plunger return spring. The tappet is held at its lowest position by the plunger return spring during the filling sequence, Fig. 14-23A.

As the rotating plunger moves upward under cam action, it closes the two horizontal scallops which contain the inlet ports. This traps and compresses the fuel and builds up pressure until the spring-loaded delivery valve is opened, Fig. 14-23B.

As the plunger continues its upward stroke, the fuel is forced through the delivery valve and is conveyed through the intersecting duct to the annulus and the distributing slot in the plunger. The vertical distributing slot on the plunger connects to the outlet duct which is then registering as the plunger rotates, Fig. 14-23C.

The rotary and vertical motions of the plunger are so phased in relation to the outlet ducts that the vertical distributing slot overlaps only one outlet duct during each effective portion of each stroke. After sufficient upward movement of the plunger, its metering port passes the edge of the metering sleeve. The fuel under pressure then escapes down the vertical hole in the center of the plunger and into the sump surrounding the metering sleeve, which is at supply pressure, Fig. 14-23D.

With the collapse of the pressure beneath it, the delivery valve then closes, during which action the piston portion of the valve blocks the passage before the valve reaches its seat and performs its

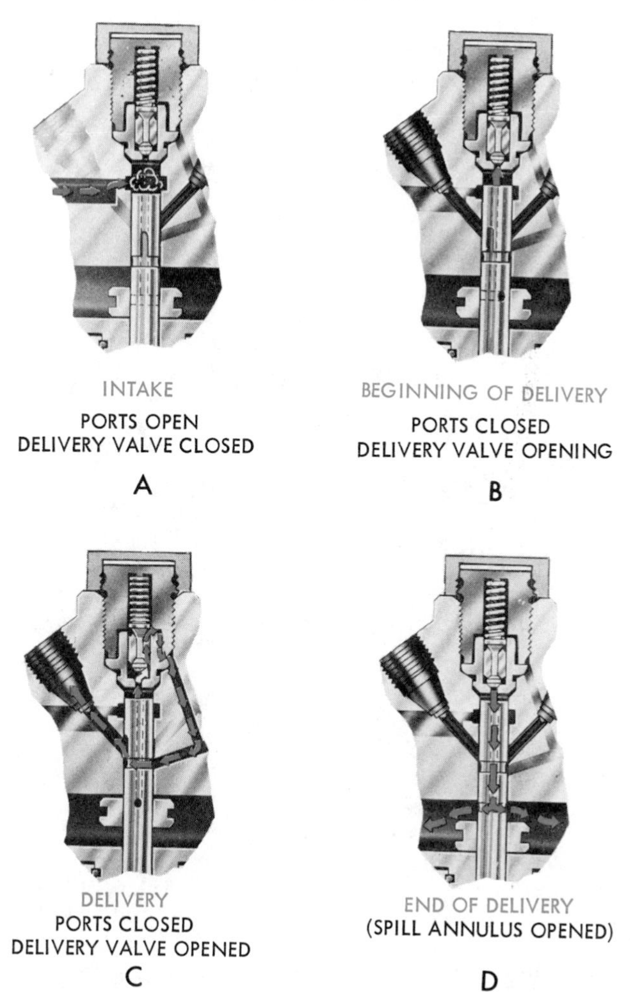

Fig. 14-23. Fuel pumping principle. (American Bosch AMBAC Industries, Inc.)

function of reducing the residual pressure in the discharge system. This completes the pumping cycle.

Fuel Metering Principles. The quality of fuel delivered per stroke is controlled by the position of the metering sleeve in relation to the fixed port closing position (the point at which the top of the plunger covers the horizontal scallops). See Fig. 14-24 top. As the horizontal metering hole in the plunger breaks over the top edge of the metering sleeve, pumping pressure is relieved down through the center hole in the plunger and out the sump surrounding the metering sleeve. Thus fuel delivery terminates despite the continued upward movement of the plunger.

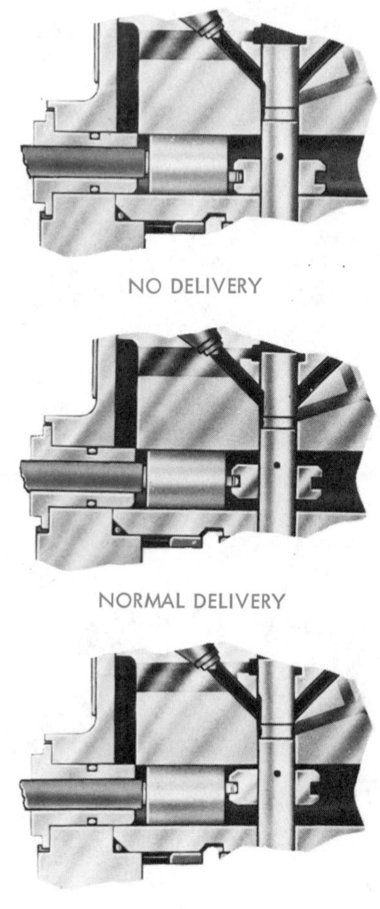

Fig. 14-24. Fuel metering principle. (American Bosch AMBAC Industries, Inc.)

If the metering sleeve is moved to mid-position, as in Fig. 14-24 middle, the hole in the plunger is uncovered later in the stroke by the sleeve, hence the effective stroke of the plunger is longer and fuel is delivered. If the metering sleeve is raised further, as in Fig. 14-24 bottom, the horizontal hole remains covered by the sleeve until late in the plunger stroke, thereby increasing fuel delivery. Upward movement of the metering sleeve increases and downward decreases the quantity of fuel pumped per stroke.

Unit Injectors

The unit injector combines the high pressure injection pump and the injector nozzle into one unit. There are some advantages to this type over the more common types of separate units. This eliminates the necessity of high pressure lines from the injection pump to the injection nozzle. Also, the individual unit injectors can be replaced independently.

Unit injectors must operate off a special cam for each injector to be adjusted to perform like all the others, an added step in tune-up procedures.

General Motors Type

The design of the unit injector combines the pump, fuel valve, and nozzle for each cylinder in one unit. Fig. 14-25 is typical of a General Motors unit injector. The pump is a jerk type, with port controlled by a recess in the plunger which is bounded by two helical (spiral form) edges. An exploded view of the injector is inverted and shown in Fig. 14-26.

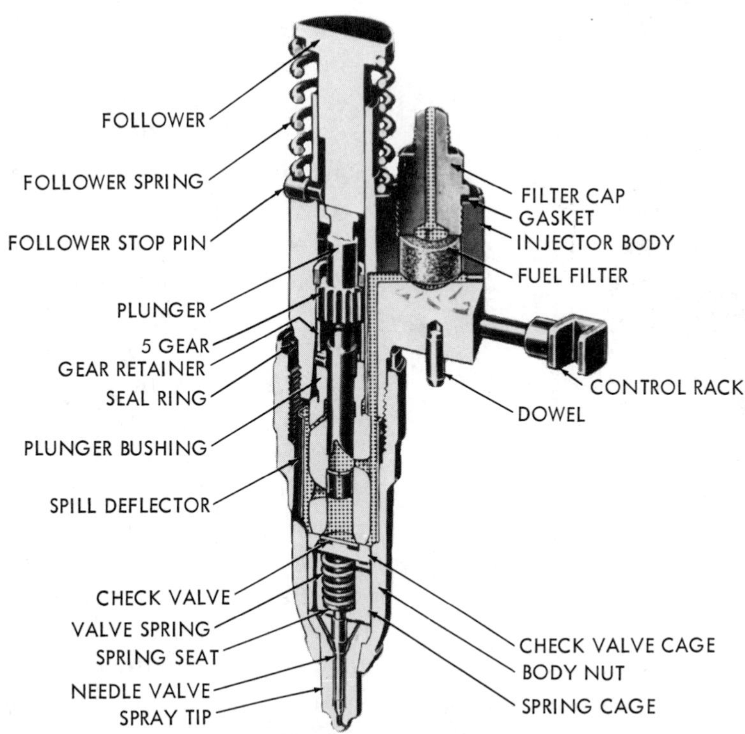

Fig. 14-25. Fuel injector assembly shows fuel flow. (General Motors Corp.)

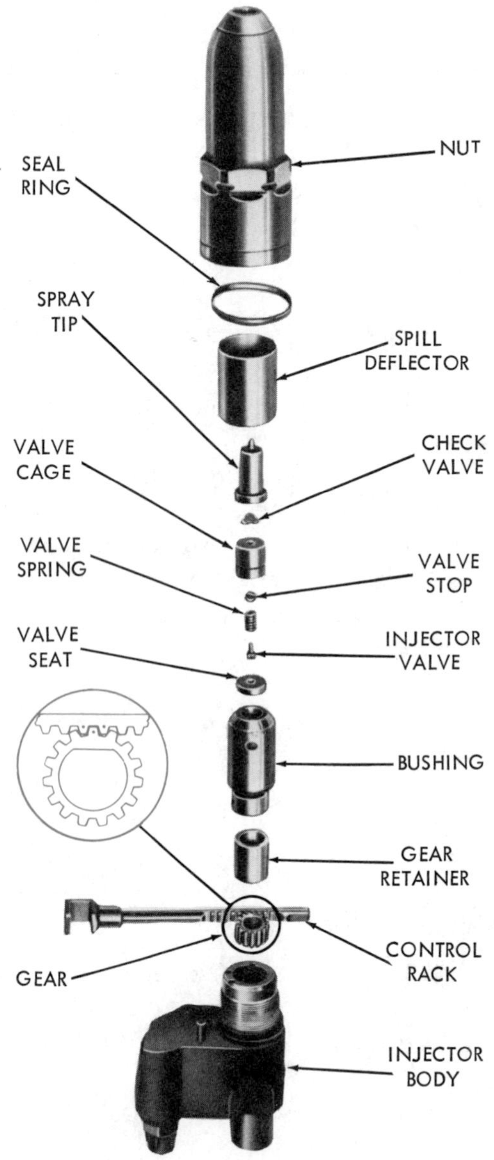

Fig. 14-26. Injector rack, gear, and spray tip details. (General Motors Corp.)

Operation. Fuel oil, supplied to the injector from the low-pressure fuel pump, enters the drop-forged steel body at the top through the filter cap. After passing through the filter element in the inlet passage, the fuel oil fills the supply chamber between the bushing and the spill deflector. The plunger moves up and down in this bushing, the bore of which is connected to the fuel supply by two funnel-shaped parts.

Motion of the injector rocker arm is transmitted to the plunger by the follower which bears against the plunger spring. The follower stop-pin positions the follower in the body.

The plunger can be rotated, in operation, around its axis by a rack-and-pinion gear arrangement, Fig. 14-27. The fuel is metered by an upper helix and a lower helix which is machined into the lower end of the plunger. The relation of these helixes to the two ports changes with the rotation of the plunger. See Fig. 14-28.

As the plunger moves down, fuel in the high-pressure cylinder is first displaced through the ports back into the supply chamber, until the lower edge of the plunger closes the lower port. The remaining fuel oil is then forced up through the center passage in the plunger into the recess between the upper helix and the lower cut-off. From there it can still flow back to the supply chamber, until the helix closes the upper port.

Rotation of the plunger, by changing the position of the helix, retards or advances the closing of the ports and the beginning and ending of the injection period. At the same time it increases or decreases the amount of fuel which remains under the plunger for injection into the cylinder in the quantity desired.

Injector Mounting. In an in-line engine, the injectors are mounted in the cylinder head, with their spray tips projecting slightly below the top of the in-

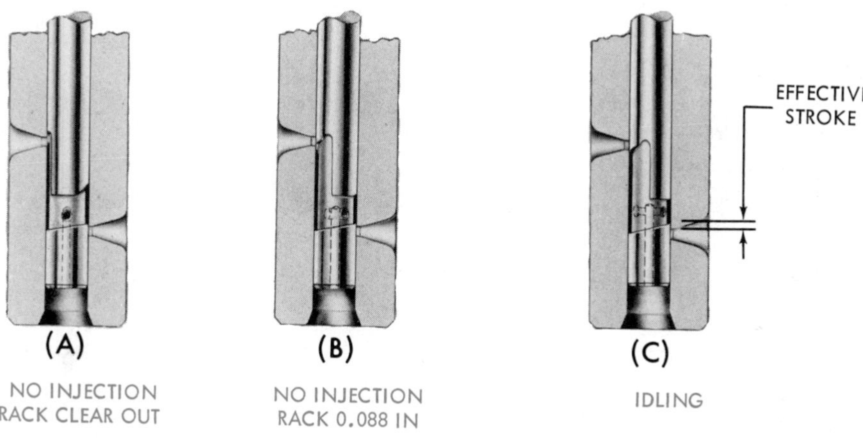

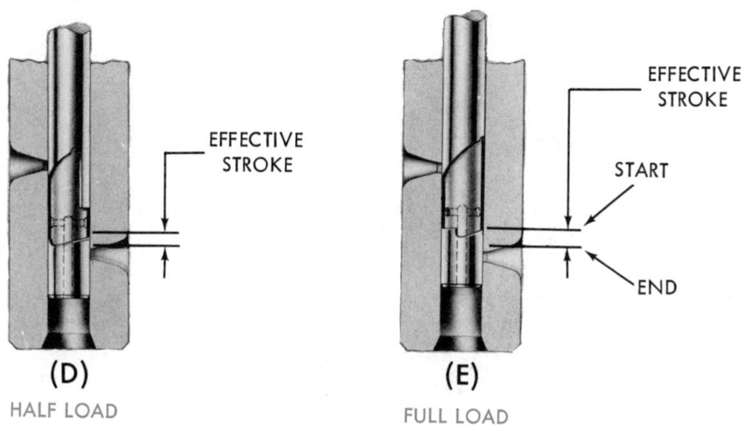

VARIED DEGREES OF INJECTION FROM NO INJECTION TO FULL LOAD PRODUCED BY ROTATING PLUNGER WITH CONTROL RACK

Fig. 14-27. Fuel metering produced by rotating plunger. (General Motors Corp.)

side surface of the combustion chambers as illustrated in Fig. 14-29. A clamp, bolted to the cylinder head and fitting into a machined recess in each side of the injector body, holds the injector in place in a water-cooled copper tube which passes through the cylinder head.

In the in-line engine and in the V-type engine, a dowel pin in the injector body registers with a hole in the cylinder head for accurately locating the injector assembly, Fig. 14-30.

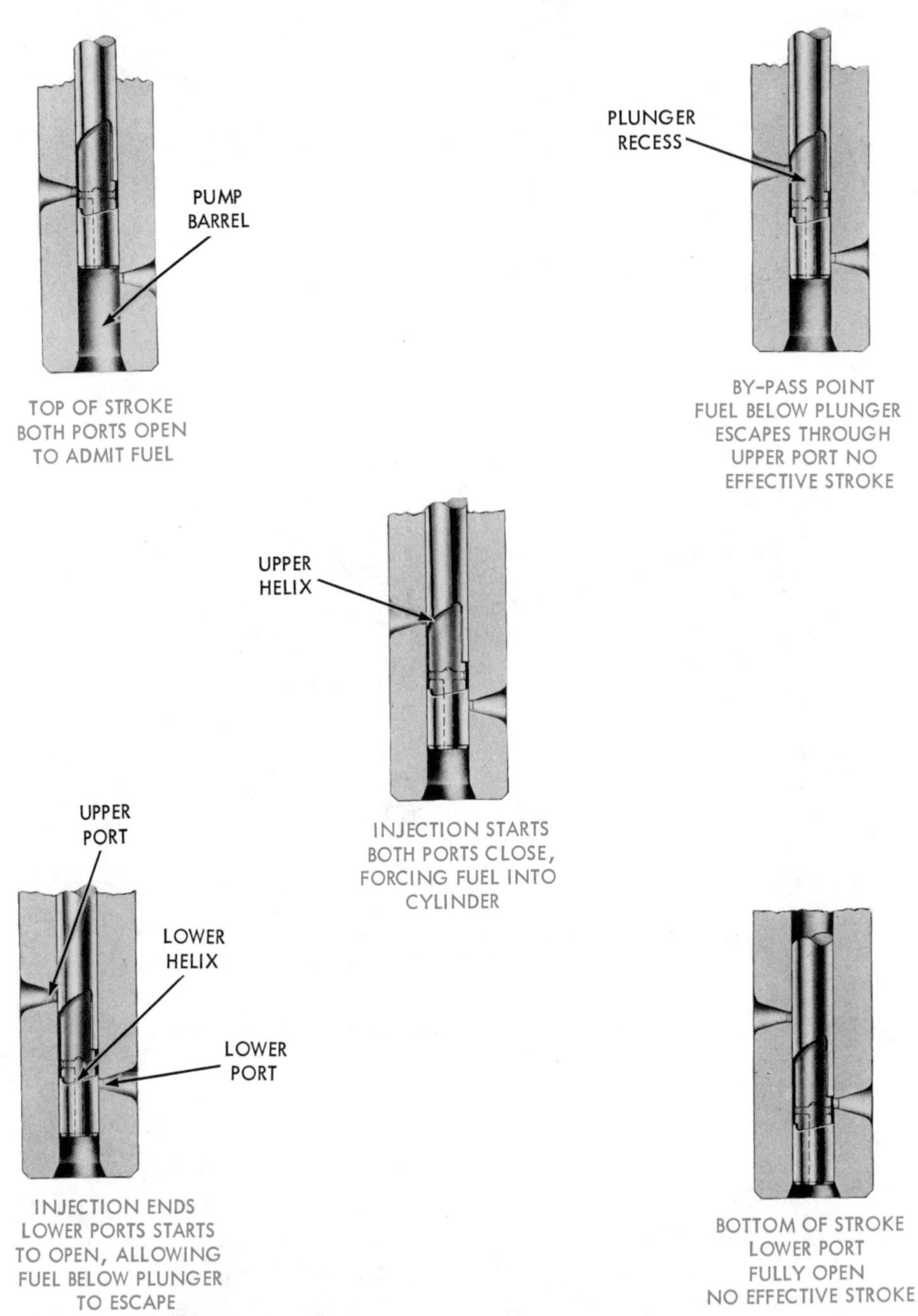

Fig. 14-28. Injector operation—vertical travel of plunger. (General Motors Corp.)

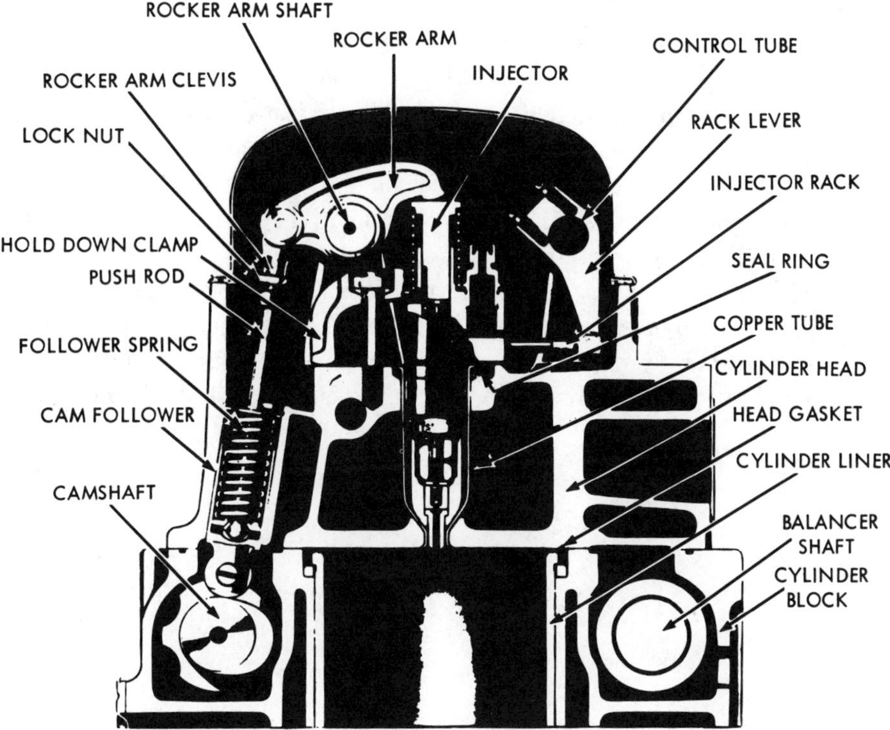

Fig. 14-29. Fuel injector and operating mechanism installed. (General Motors Corp.)

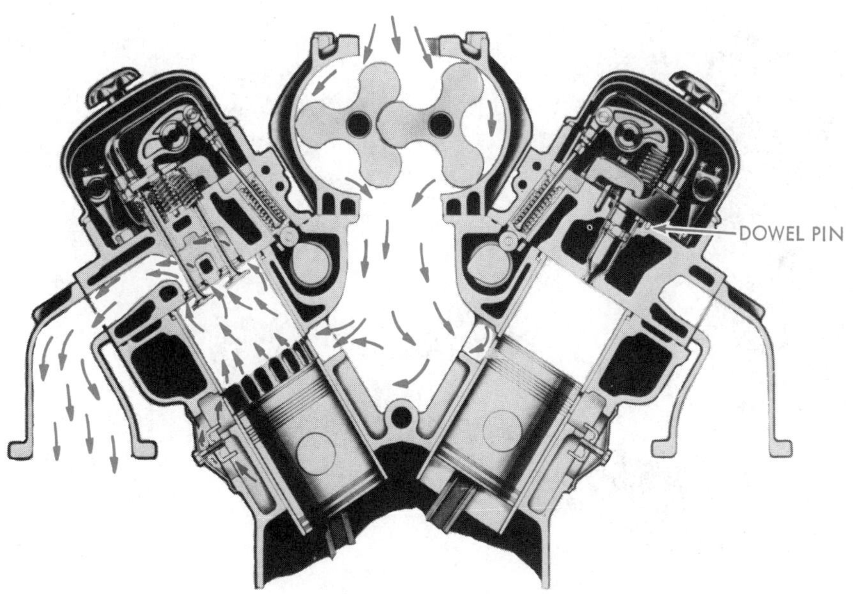

Fig. 14-30. Unit injector installed in right bank of engine. (General Motors Corp.)

Cummins PT Design

This unit-injector system works on the *pressure-time* (*PT*) principle. It consists of a gear-type fuel pump which delivers the fuel at moderate pressure (about 250 psi at rated speed) through a restricting throttle to the governor and then to a manifold which feeds cam-operated injectors in the cylinder heads. The governor, Fig. 14-31, includes a bypass valve controlled by the governor flyweights. This regulates the pressure of the fuel supplied to the injectors. Fig. 14-32 shows a cross-section of a PT fuel pump with its major parts.

The injector, Fig. 14-33, raises the pressure to that required to produce a good spray and also times the start of injection. The injector action is pictured in four stages in Fig. 14-34, which shows the nozzle tip in larger scale. In the first stage (A) the injector plunger moves up, uncovering the supply hole. This allows fuel to circulate through the injector and

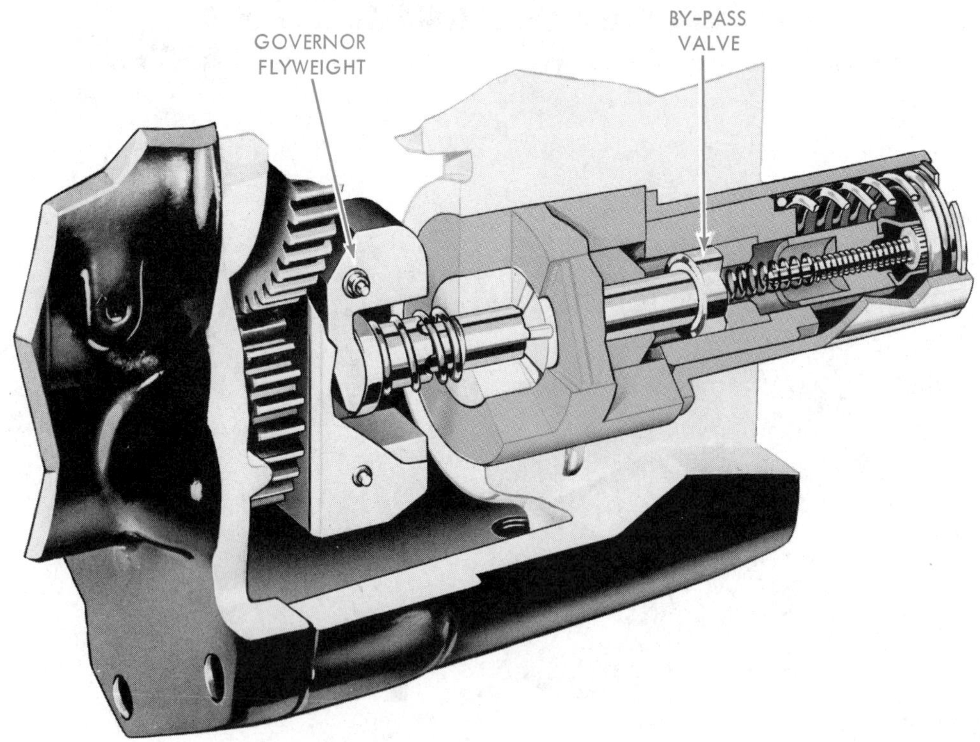

Fig. 14-31. Cummins governor and by-pass valve. (Cummins Engine Co.)

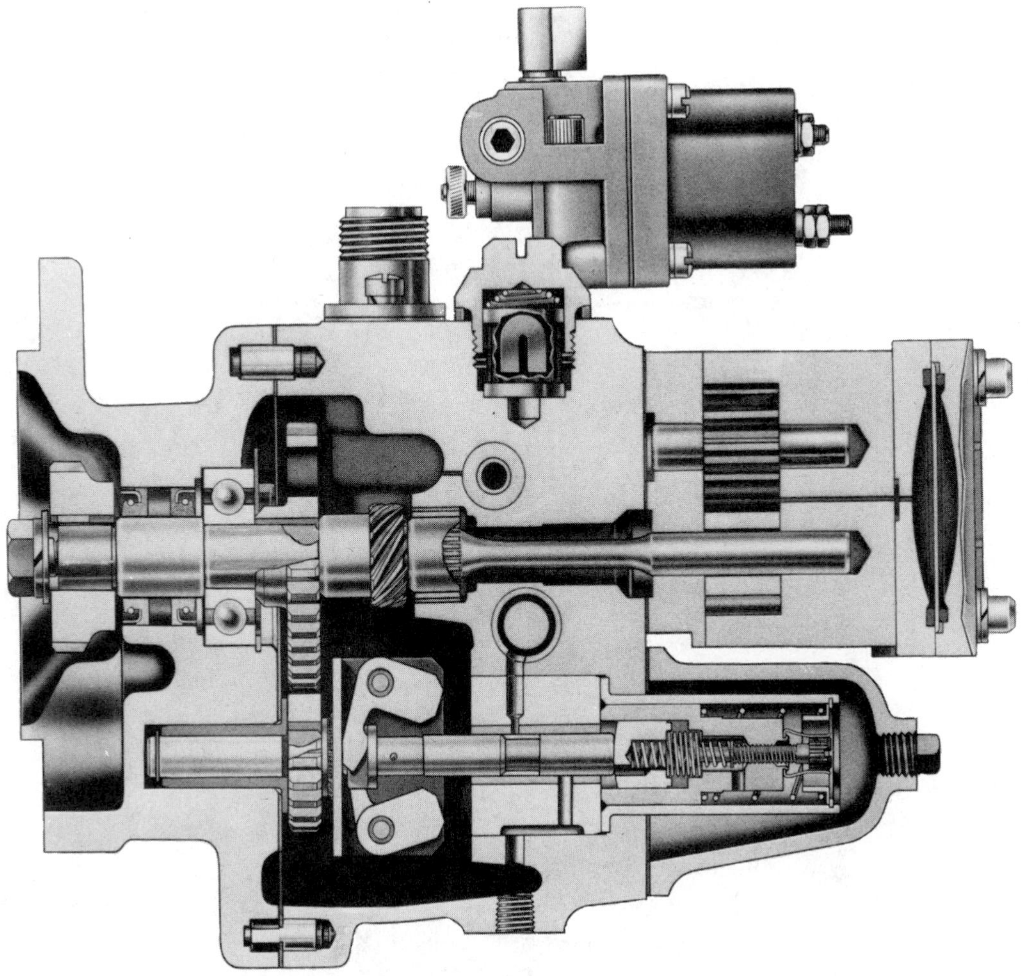

Fig. 14-32. Cummins PT fuel pump cross-section. (Cummins Engine Co.)

out of the drain at the left. Most of the fuel delivered to the injector returns to the fuel tank, thus cooling the injector and warming the fuel in the supply tank.

In stage B, Fig. 14-34 at the top of the upstroke, the plunger has uncovered the metering orifice, allowing fuel to enter the injector cup. The length of time this orifice is uncovered and the amount of pressure on the fuel determine the quantity of fuel entering the cup and later injected.

Stage C, Fig. 14-34 shows the plunger moving down and injecting fuel. First it closes the metering orifice, then it forces the trapped fuel at high pressure through holes in the tip of the injector cup and into the cylinder as a fine spray.

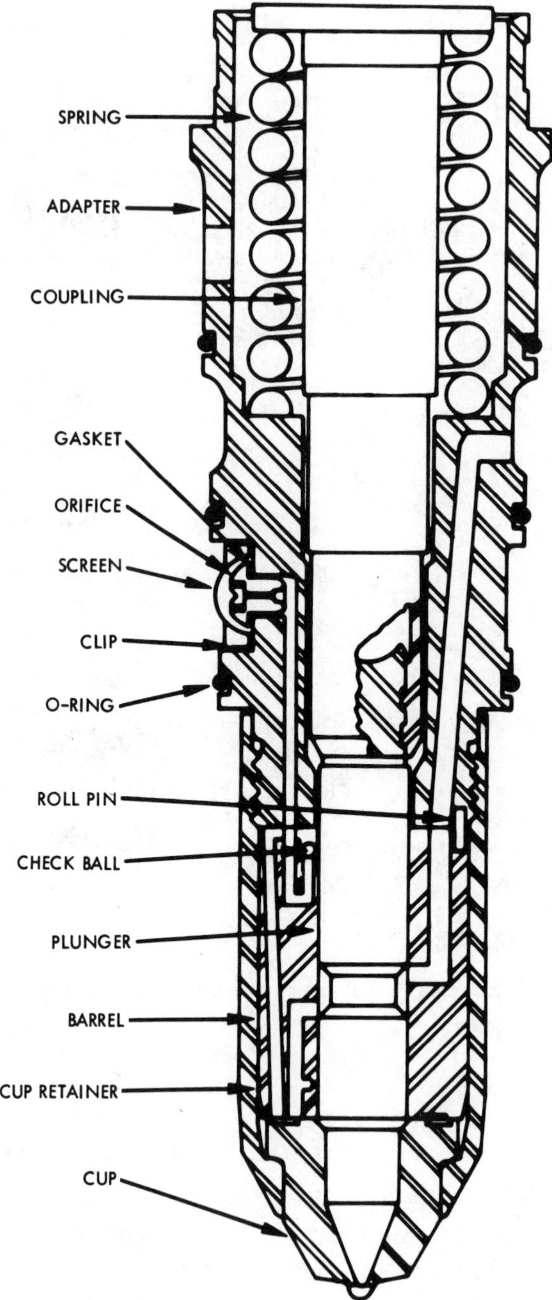

Fig. 14-33. Cummins injector. (Cummins Engine Co.)

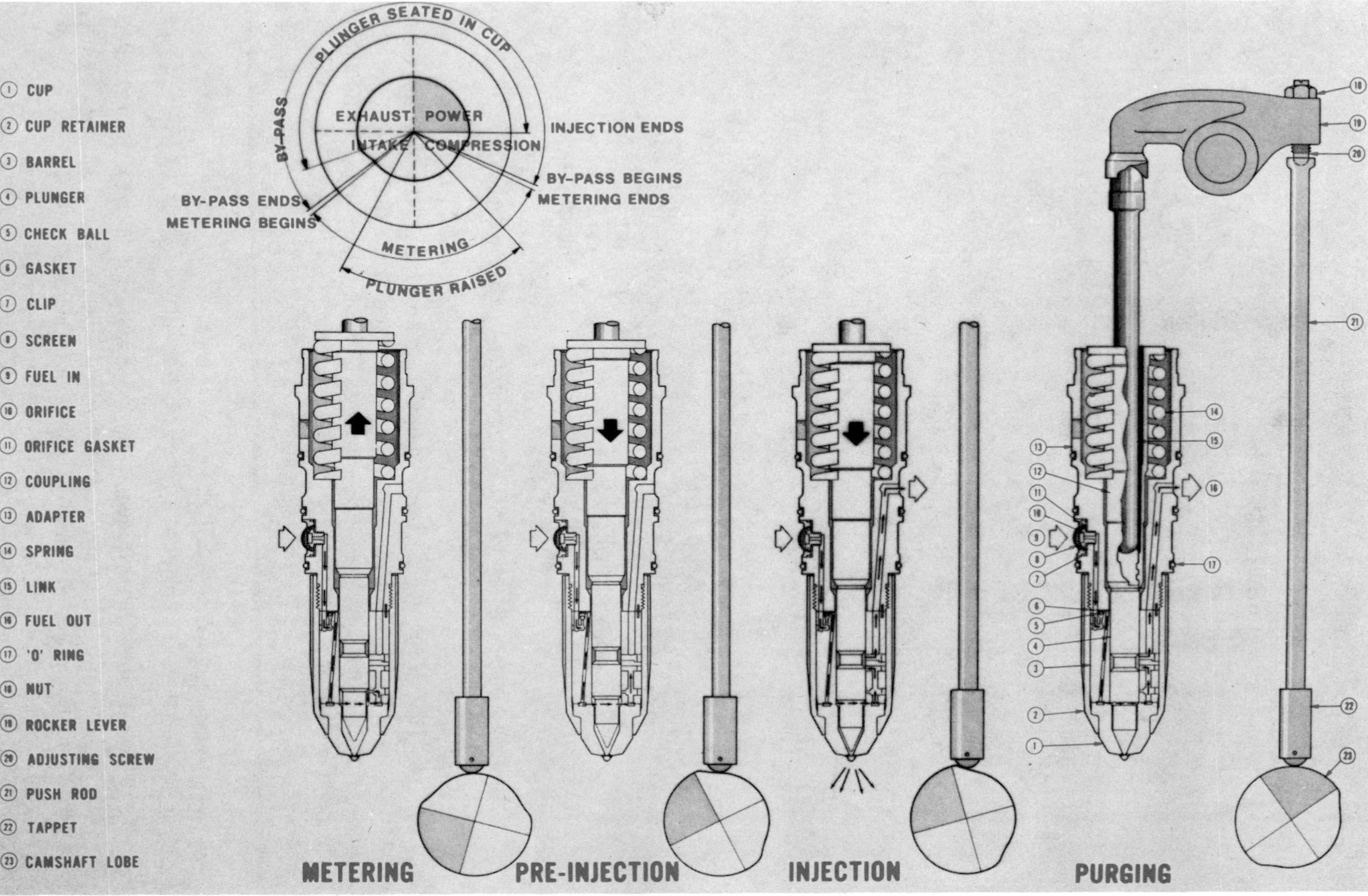

Fig. 14-34. Cummins fuel injector cycle. (Cummins Engine Co.)

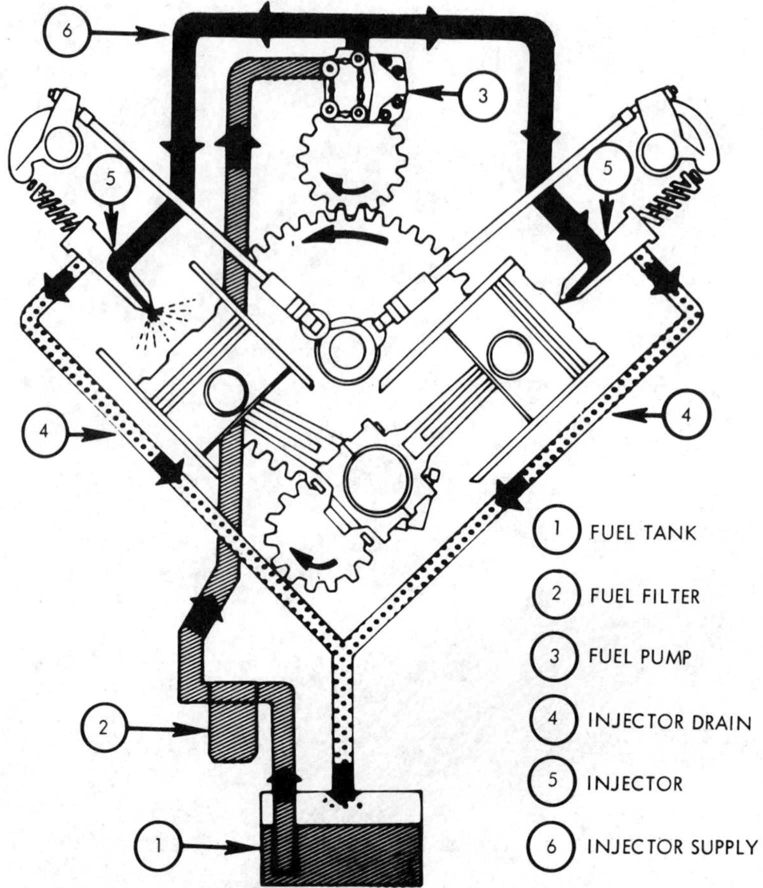

Fig. 14-35. Fuel flow through V-6 and V-8 type engines. (Cummins Engine Co.)

Stage D, Fig. 14-34 is the idle period, during the power and exhaust strokes, when the tapered end of the plunger remains in the injector cup seat. A close fit between plunger and cup assures complete injection of the fuel.

Fig. 14-35 shows the typical fuel flow through a Cummins V-6 or V-8 engine.

Injectors and Nozzles

Injection nozzles fit into the cylinder head and are hydraulically operated by fuel delivered from the injection pump. There are many types of injection nozzles; however most operate on the principle of hydraulically lifting a spring loaded valve and allowing the pressurized fuel to spray out through one or more orifices (holes) in the combustion chamber.

The three common types of nozzle tips

are single hole, multiple hole, and pintle type single hole nozzles. A conical jet (like an ordinary garden spray) having an angle of about 4 to 15 degrees is produced by a single-hole tip.

Manufacturing inaccuracies tend to cause uneven sprays when the angle is large; this limits the practical spray angle. Single-hole nozzles are used in engines where combustion-chamber shape creates turbulence (see next chapter) and where there is thus less need for a finely atomized and widespread spray. Such nozzles also offer the advantage of a fairly large opening even in small high-speed engines, which reduces the risk of clogging.

Hole Type Nozzles. The hole type nozzle consists of a body and a valve with the valve loaded by a spindle and a spring. The spring is to provide the specified nozzle opening pressure. As the plunger in the injection pump delivers fuel to the nozzle, the pressure in the line increases to a value sufficient to lift the nozzle valve against the spring pressure imposed upon it. The spill annulus in the plunger is uncovered by the metering sleeve at one point in the stroke of the injection pump plunger. This causes the valve to close rapidly, terminating fuel injection for that particular stroke.

Multiple Hole Nozzles. These nozzles, Fig. 14-36 right, find most use in open

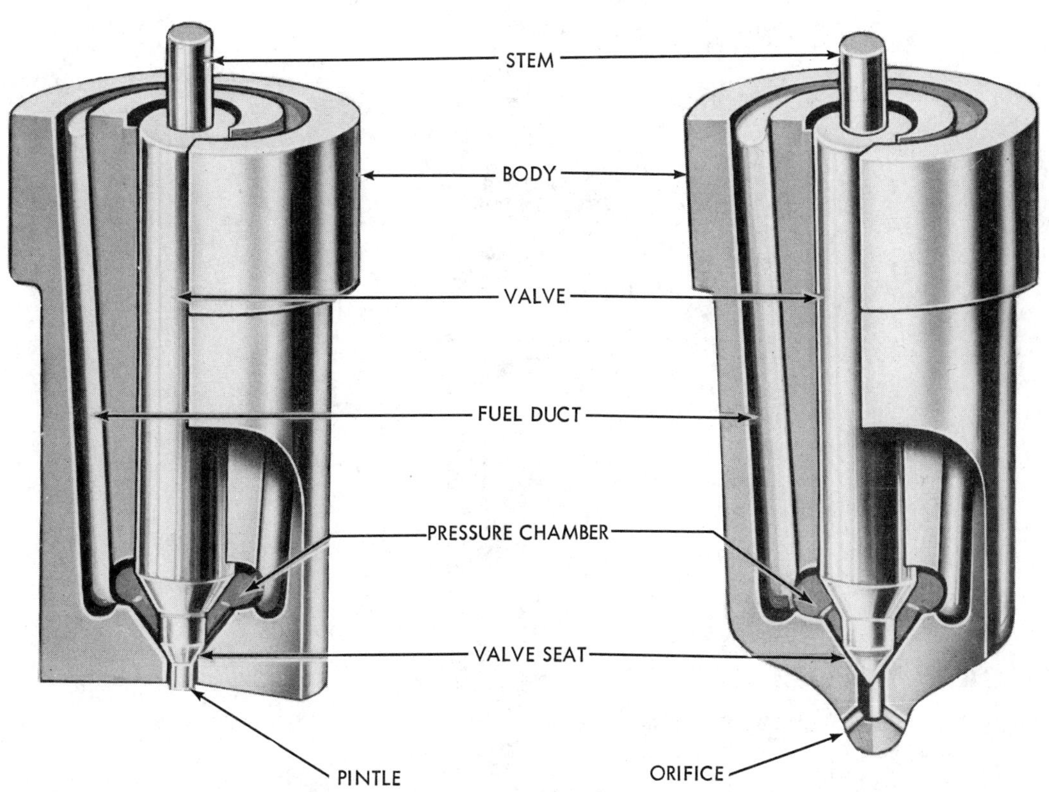

Fig. 14-36. Hole type and pintle type nozzle tips. (American Bosch AMBAC Industries, Inc.)

combustion chambers, where it is necessary to distribute the fuel spray to all parts of a wide shallow chamber. The more spray openings, the smaller each becomes and the more it is necessary that the fuel be clean. Spray openings or orifices are from 0.006″ up to about 0.033″ in diameter, and their number may vary from three to as many as eighteen for large-bore engines.

Pintle Nozzles. Nozzles of this type, Fig. 14-36 left, are fitted with valves whose ends extend into a shank or pin called a *pintle*. The shape of the pintle is made according to the spray pattern desired.

The pintle reaches into the nozzle orifice to form an annular space through which the fuel passes. By suitably shaping the pin, the designer obtains either a hollow cylindrical jet of high penetrating power, or a hollow cone-shaped spray with an angle up to 60 degrees. The pintle type nozzle works uniformly and accurately. In addition, the motion of the pintle tends to avoid formation of carbon crust on the tip of the nozzle.

Fit of Nozzle Valves. Needle valves

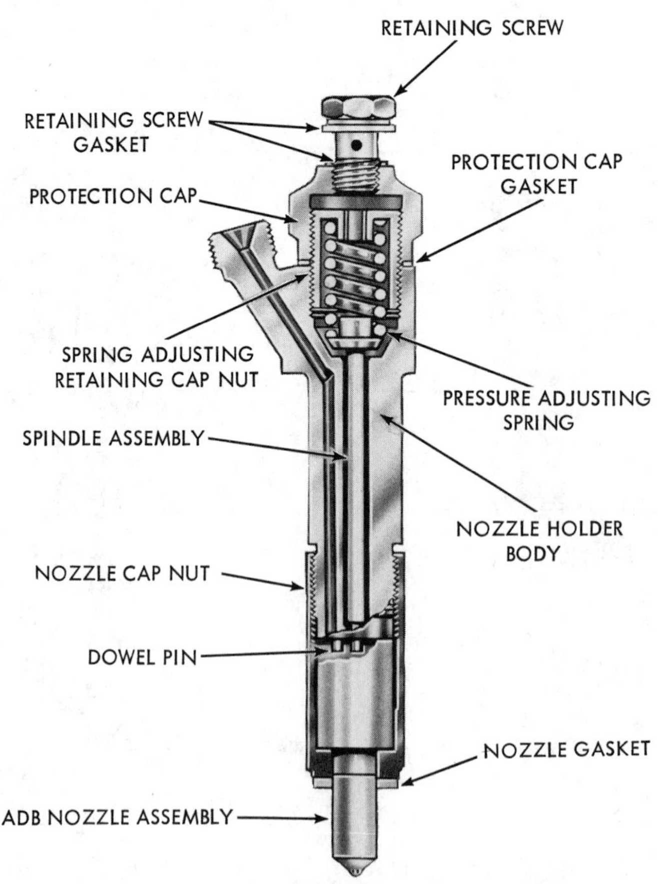

Fig. 14-37. ADB and AKF type nozzle and holder assembly. (American Bosch AMBAC Industries, Inc.)

and the nozzle bodies which guide them are usually made of alloy steel, heat-treated to reduce wear. Valves and guides are lapped together to form a precisely mated assembly, thus cannot be interchanged.

American Bosch Injectors

The American Bosch injector nozzle and holder in Fig. 14-37 includes a forged body, incorporating a spindle and spring and a high-pressure duct. The spindle bore and spring chamber act as a passage for leak-off fuel that by-passes and lubricates the nozzle valve.

The nozzle consists of a body and valve with the latter loaded by the spindle and spring. The spring is set to provide the specified nozzle opening pressure. As the plunger in the injection pump delivers fuel to the nozzle, the pressure in the line increases to a value sufficient to lift the nozzle valve against the spring pressure imposed upon it. This occurs at a point in the injection-pump plunger stroke where the spill annulus on the plunger is uncovered by the metering sleeve, causing the delivery valve to close rapidly. The consequent reduction in line pressure enables the spring in the nozzle holder to seat the valve quickly. This terminates injection of the fuel into the combustion chamber for that particular plunger stroke.

Roosa Master Pencil Type Injector

Fig. 14-38 shows a cutaway drawing of the internal parts of the Roosa Master pencil type injector. Unlike most other injectors, this type is a simplified design which can be produced on a low cost, mass production basis.

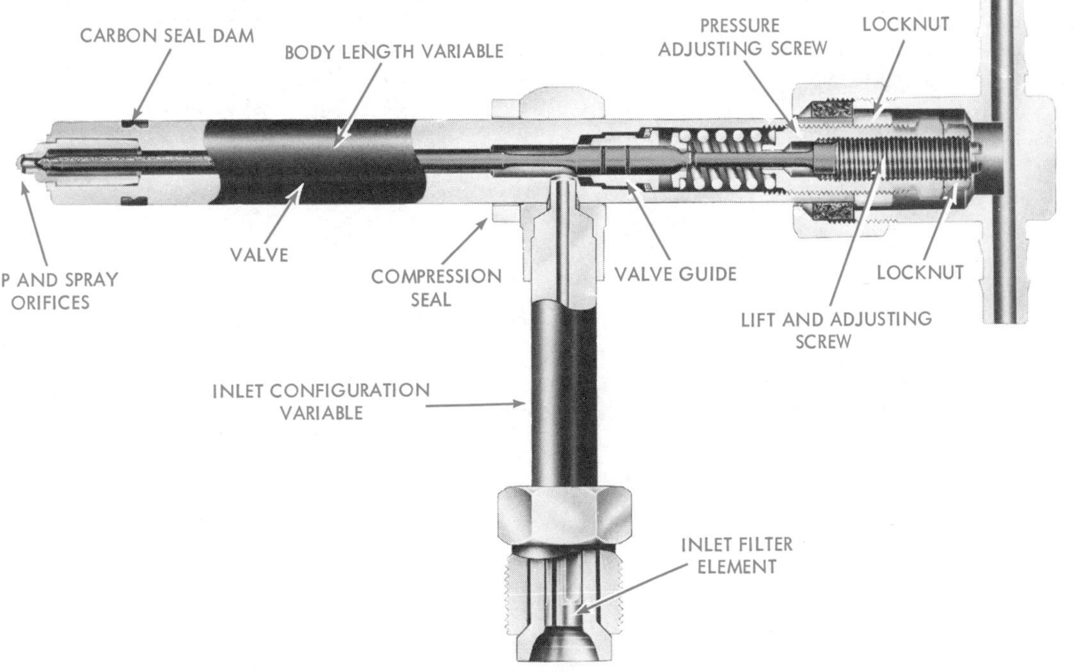

Fig. 14-38. Roosa Master pencil type injector. (Stanadyne/Hartford Division)

Checking On Your Knowledge

The following questions give you the opportunity to check up on yourself. If you have read the chapter carefully, you should be able to answer the questions. If you have any difficulty, read the chapter over once more so that you have the information well in mind before you go on with your reading.

DO YOU KNOW

1. What are the three main classes of fuel injection systems used on modern diesel engines?
2. How does early injection affect the operation of diesel engines?
3. How does late injection affect the operation of diesel engines?
4. What does atomization of diesel fuel have to do with engine operation?
5. What are the three important functions of a multiple plunger pump?
6. How accurate must pump metering of fuel be on a modern diesel engine?
7. How accurate must pump timing be to obtain the best performance of a diesel engine?
8. How is fuel metering controlled on an APE multiple plunger pump?
9. What does *timed for port opening* mean on an APE pump?
10. What are advantages of the distributor type pump over the multiple plunger?
11. How are G.M. unit injectors actuated?
12. What are the differences between a *hole-type* and *pintle-type* nozzle and where are each used?

Burning the Fuel

Chapter

15

In the preceding chapter you learned how the fuel is injected into the engine cylinder. The purpose of this chapter is to tell you how the inside of the cylinder is designed to make the fuel burn properly. Chapter 9 explained the general principles of combustion, but now we will go into the special case of combustion in a diesel engine, inside what is called the combustion chamber.

First, you'll learn what conditions must be met for fuel to burn inside the cylinder of any diesel engine. Then we'll look into the *solid-injection* principle on which most of today's engines work.

The solid-injection principle was first applied to simple open combustion chambers, but the demand for high-speed, small-bore engines for mobile use brought about the development of more complex designs. The most important of these special designs are the turbulence chambers, precombustion chambers, air cells and energy cells.

Diesel Combustion

Because the diesel engine handles the entire job of vaporizing, mixing, and igniting fuel inside the cylinder in an *extremely short time*, conditions essential for good combustion must be provided. What are these conditions, and how are they attained in practice?

Liquid fuel is converted to vapor form before it will start to burn. Compression of the air charge in the cylinder provides heat for vaporizing, but heat alone is not enough if the fuel is to be completely vaporized and burned in the space of a few hundredths of a second, or in a high-speed engine even a few thousandths of a second.

Vaporization can take place only from the surface of the liquid fuel. Consequently, to obtain the extremely rapid vaporization required, the liquid fuel mass must be broken into a large number of small particles, each presenting its surface to the heat. The particles must be caused to spread over as large an area as possible in order to bring each particle into contact with the heated air and to keep the particles from being blocked by gas generated by themselves and by surrounding burning particles.

Combustion Process

But don't get the false picture of the combustion operation as one in which fuel turns to gas, the gas mixes with air, and the whole mixture is brought to ignition temperature. Actually, while conditions like this exist at times, particularly during the ignition-delay period, they are undesirable because from this kind of burning comes *knock* and rough running.

Under proper conditions, the vapor ignites as it forms, and combustion takes place directly from the surface of the fuel droplet. The steps of vaporization, mixing, and ignition still occur, but with extreme rapidity and in thousands of individual, scattered spots. Each fuel particle, dashing along its own path, absorbs heat as it goes and throws off vapor which combines with the nearest air molecules. The process has been described as resembling a shower of miniature meteors, trailing fiery gases in their wake.

Search for Oxygen. As long as the rushing particles meet fresh supplies of oxygen, the burning continues. The motion through the air removes the burned gases and exposes an ever-fresh surface to the oxygen until the particle is completely consumed.

But the oxygen supply in the cylinder is limited and each succeeding particle of fuel finds an atmosphere weaker in oxygen than did its predecessors. Those arriving near the end of the process find considerable difficulty in searching out the oxygen needed for rapid and complete combustion. Thus the process tends to slow up during the later stages, and the less complete combustion reduces engine power and efficiency.

Stages in the Combustion Process. Combustion of fuel in a diesel engine can be divided into four stages which can be identified on the pressure-time diagram in Fig. 15-1.

Stage 1 is called the ignition delay period. During this period some preliminary oxidation takes place, but the burning is not sufficient to cause any appreciable rise in temperature or pressure. This period between the beginning of injection and the beginning of rapid pressure rise is dependent upon several factors: (1) the ignition quality of the fuel, (2) the amount of air within the combustion chamber, its temperature and pressure, (3) the speed and amount of fuel injected, (4) the amount of atomization taking place and (5) the amount of turbulence present. The higher the temperature and pressure in the combustion chamber, the finer the atomization of the fuel particles and the higher the turbulence, the shorter will be the delay period.

Fuel with greater amounts of light molecules will decrease the delay period.

Stage 2 is called the rapid burning period. During this period the fuel already injected and the fuel being injected during this period, will increase rapidly in pressure and temperature. The rate of combustion is dependent upon the degree to which the fuel becomes mixed with air

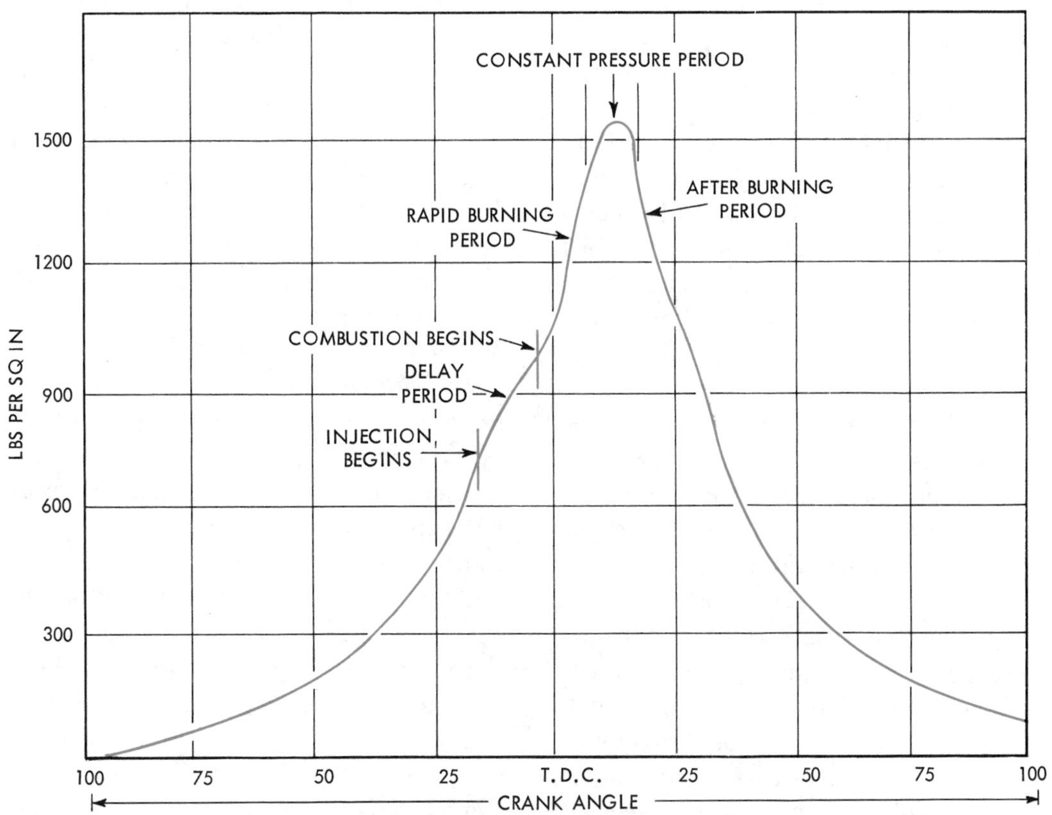

Fig. 15-1. Pressure-time diagram locating the stages of combustion in a compression-ignition (C.I.) engine.

by atomization, vaporization, turbulence, and the molecular make-up of the fuel. Combustion knock in a diesel engine is directly related to the amount of ignition delay and will take place at the end of the second stage (the period of rapid burning).

Stage 3 is the period of constant or controlled pressure rise. Injection usually ends just prior to the end of this period. Theoretically the rate of injection and the rate of burning should be such that the pressure remains quite constant during this stage.

Stage 4 is called the after-burning period. During this stage the piston is rapidly moving downward. The heavier molecules of fuel will finally vaporize, mix with oxygen and burn. Although the volume above the piston is getting greater during this period, useful pressure is still applying torque to the crankshaft.

All burning should be completed before the exhaust valve is opened or it will lead to a smoky exhaust.

Combustion Requirements

Successful combustion inside the cylinders of a diesel engine depends on the prevalence of the following four conditions:

1. Fine atomization

2. High temperature for prompt ignition
3. High relative velocity between fuel and air particles
4. Good mixing of fuel and air particles

Atomization, penetration, and dispersion (spreading) of the fuel depend largely on the injection system, which you studied in the preceding chapter.

Compression ratio, cylinder dimensions, and cooling arrangements determine the temperature conditions.

Mixing depends on the proper relation of the injection pattern, the air-intake system, and the shape of the combustion space formed by the cylinder head, the cylinder walls, and the piston crown.

Temperature. The important factors in bringing the temperature up to the level needed for prompt ignition are compression ratio and engine size. In a small-bore engine, the surface of the combustion space is large relative to its volume, and the cool water-jackets lower considerably the temperature of the compressed charge. To offset this, small-bore engines generally use higher compression ratios to produce higher temperatures. Another important reason for higher compression ratios in small-bore engines is the fact that such engines run at high speeds. This reduces the time available for combustion and requires reducing the ignition delay; higher compression shortens the delay period.

Air Motion. The remaining essential factor in combustion is air motion, which is necessary for good mixing. Most combustion spaces, even the simplest, produce some turbulence, and this is sufficient for engines with large cylinder bores operating at low or medium speeds. On the other hand, small-bore, high-speed engines require highly developed combustion chambers to prepare the fuel charge and produce the conditions needed for complete combustion in the fantastically short time available.

Solid Fuel Injection Principles

Most modern diesel engines use the solid-injection principle, in which the spray is produced solely by the pressure of the oil itself. It is only this pressure energy which is available for *atomization* (fineness of spray) and *penetration* (distance the particles travel). These requirements are contradictory to some extent, and therein lies part of the problem of matching the combustion chamber and the fuel-injection system so as to obtain the best combination. Let's now examine some combustion-chamber designs, starting with the simple open chamber.

Open Combustion Chambers

An open combustion chamber may be defined as one in which, at the time of injection, *all the air is contained in a single space.* It thus includes the simplest possible arrangement, consisting of a flat cylinder head and flat piston crown. The open system is also called *direct injection* because the nozzle sprays right into the cylinder space. The oil must be sprayed into this shallow space in such a way as to meet all the air but without striking too heavily on the piston or cylinder walls.

Although the air and fuel are both in motion in all engines, in the open-chamber types the fuel motion is greater, and on it the nature of the combustion largely depends. Fuel is injected in a number of jets arranged to give the best possible coverage of the space.

To help bring the fuel and air together, modern engines of the open-combustion chamber type generally depart from the flat shape of the piston crown and cylinder head. The top of the piston crown is often made concave (hollow like a saucer) and sometimes is shaped so that the combustion space becomes roughly hemispherical (like half a ball). In such designs, the piston approaches the cylinder head closely, allowing only the clearance needed for mechanical reasons. Such changes from the flat disk-shaped combustion chamber make it easier for the spray pattern to fit the combustion space. Cross-sections of some typical designs appear in Fig. 15-2. Fig. 15-2A is a two-cycle design; the others are four-cycle.

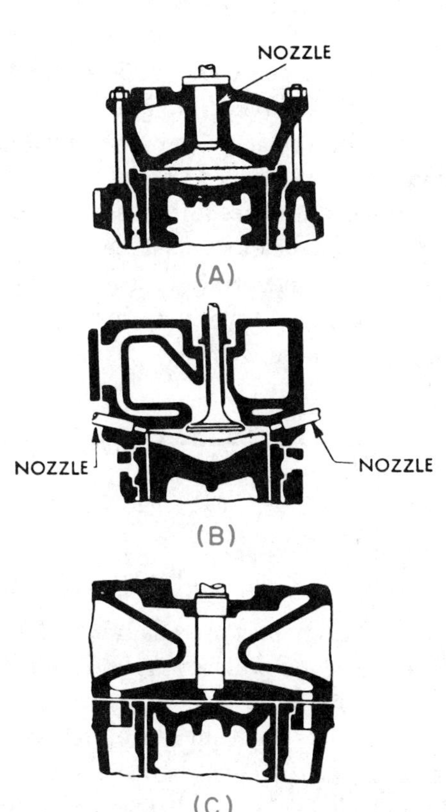

Fig. 15-2. Typical open combustion chambers.

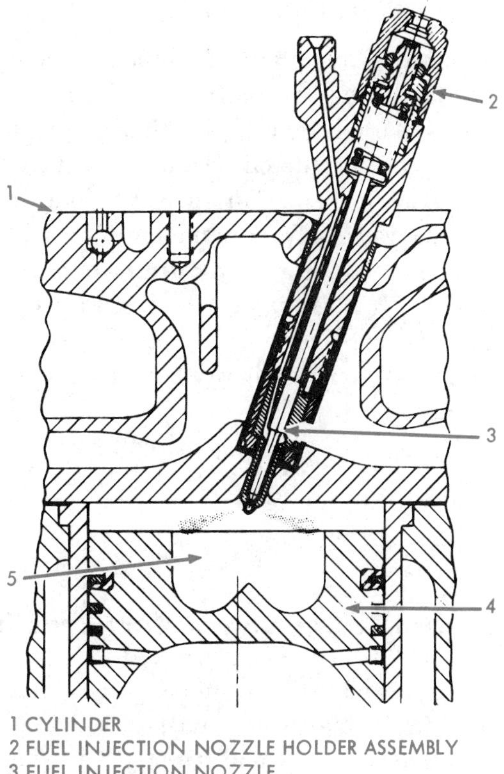

Fig. 15-3. Open combustion chamber with swirl cup. (Allis-Chalmers)

Design Fig. 15-2B uses two spray nozzles arranged oppositely and a concave piston crown. Design Fig. 15-2C has an annular shape combustion space in the piston crown.

The deep cut-out swirl cup on the crown of the piston in Fig. 15-3 is being used widely in 4-stroke cycle and some 2-stroke cycle engines with cylinder bores up to 6″ dia.

The open chamber system has a minimum of heat absorbing surface per unit of combustion space volume, therefore it is easier to obtain high thermal efficiencies with this design.

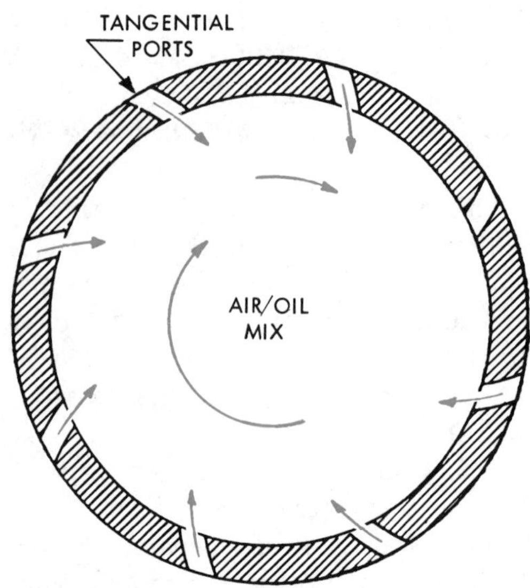

Fig. 15-4. Tangential scavenging ports produce air swirl for two-cycle engine.

Open Chamber Air Motion

Improved mixing can be brought about by producing air currents which cut across the oil sprays, breaking them up and spreading the oil particles in a more turbulent fashion. This may be done in several ways: (1) port mixing, (2) utilizing the motion set up during compression and (3) in two-cycle engines using the swirling action of scavenging ports tangential to the cylinder bore, as in Fig. 15-4.

In general, open combustion chambers find widest use in medium and large-bore engines operating at low and medium speeds. All-around performance, particularly at constant speed, is good. Although the injection pressure must be relatively high in good penetration, the compression ratio can be relatively low and the starting characteristics are good.

Special Design Combustion Chambers

Open combustion chambers are seldom used in engines which must run at high maximum speeds and through a wide speed range, as in mobile applications. The speed change unbalances the correct proportioning of the spray, resulting in incomplete combustion and a smoky exhaust. Such engines use chambers of specialized designs which may be grouped into four main classes: (1) turbulence chambers, (2) precombustion or antechambers, (3) air cells and (4) energy cells. Let's look at each in turn to find their particular qualities.

Turbulence Chambers

In turbulence chamber designs, the fuel is injected into an auxiliary chamber separated from the cylinder by an orifice. This auxiliary chamber, which holds virtually the full cylinder charge at the end of the compression, is shaped to create highly turbulent conditions. Fig. 15-5 shows cross-sections of two typical designs.

Most turbulence chambers approximate a sphere in shape and are located either in the cylinder head, like the Waukesha design, Fig. 15-5A left; or at one side of the block, as in the Hercules design, Fig. 15-5B. The fuel nozzle sprays into the turbulence chamber. In operation, the piston forces the air charge into the turbulence chamber, setting up a rapid rotary motion. The velocity through the throat of the orifice increases as the piston rises, reaching a peak somewhat before top center. Near this point, the spray nozzle injects the fuel into the turbulent air currents and good mixing results as combustion takes place.

An additional feature of the Hercules design, Fig. 15-5 right is that when the piston approaches top center, it begins to cover the air passage between the cylinder and the turbulence chamber, Fig. 15-6. This increases the air velocity in the passage and makes the air flow in the chamber still more turbulent.

Precombustion Chambers

At first glance, you might suppose that the precombustion chamber designs shown in Fig. 15-7 act like the turbulence chambers described in the previous topic. However the basic principle is quite different. In the precombustion chamber system, the auxiliary chamber *does not contain* the full air charge but only a part of it, the remainder being in the cylinder space.

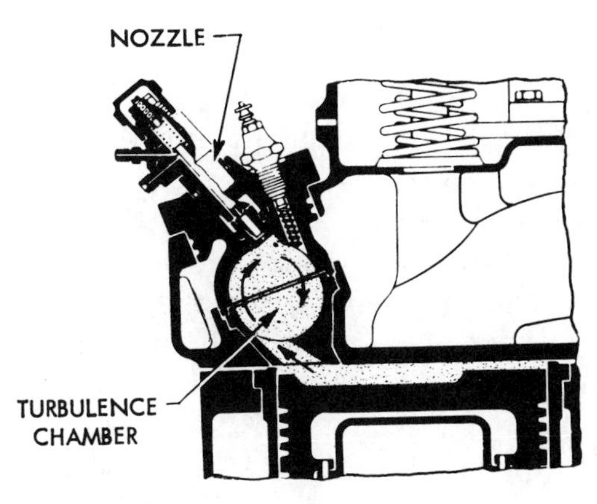

WAUKESHA DESIGN

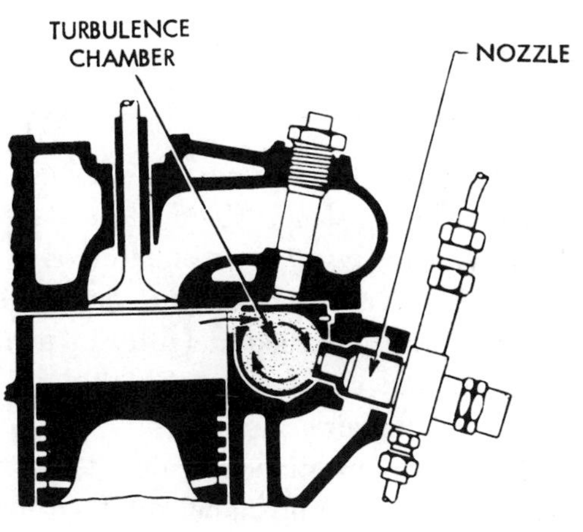

HERCULES DESIGN

Fig. 15-5. Typical turbulence chambers of Waukesha and Hercules designs.

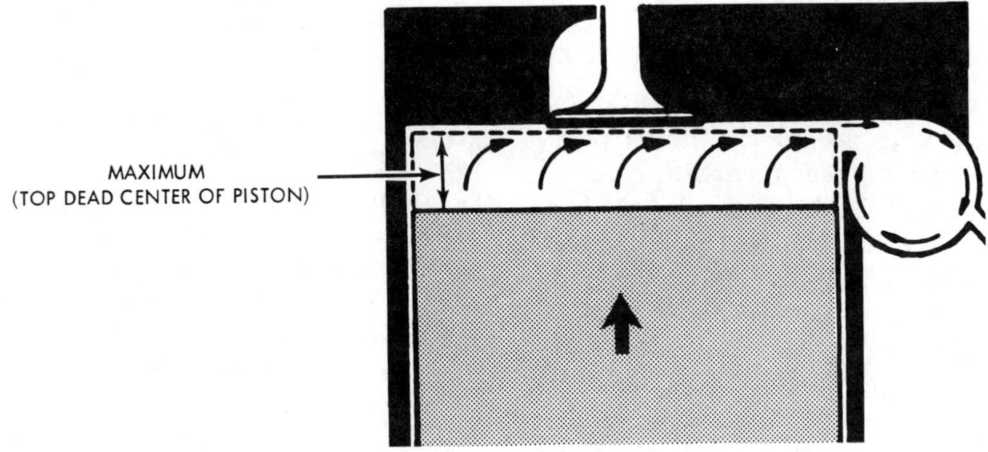

Fig. 15-6. Rising piston forces gas into turbulence chamber.

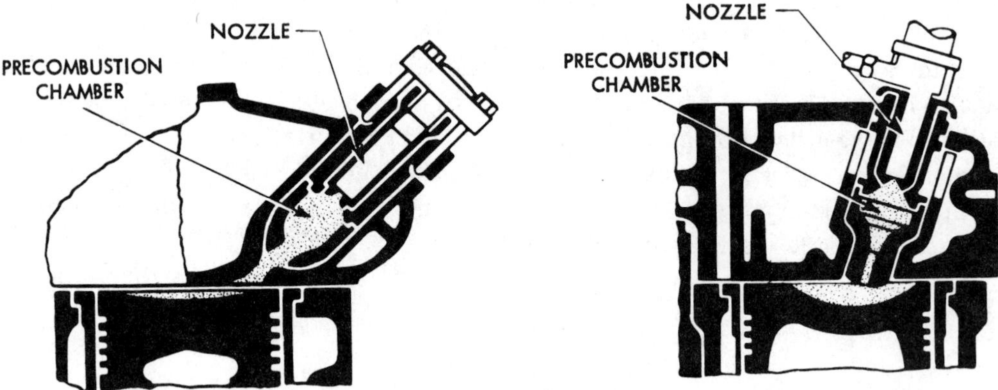

Fig. 15-7. Precombustion chamber types.

The precombustion chamber provides a quiet space in which the fuel starts to burn with an *insufficient* amount of air. This excessively rich mixture explodes, throwing the unburned fuel into the main chamber and mixing it thoroughly with the remaining air.

The precombustion chamber is more or less cylindrical in shape; the fuel nozzle is located at one end of the chamber and directs the oil spray at the throat. The chamber is usually water-cooled and contains about one-third of the air charge after compression.

Engines using precombustion chambers are able to handle successfully a wide range of fuels, but their fuel consumption tends to run slightly higher than with other types.

A problem common to the precombustion type engine is found in its starting characteristics. When a cold engine is being cranked, the larger surface of the precombustion chamber, plus the high velocity through the throat, tends to cool the cylinder charge and causes it to start hard in cold weather.

To remove this problem, most engines of this type have, as standard equipment, some kind of heating device such as a glow coil, manifold heater or coolant heater for use in cold ambient temperatures. These engines will also usually have higher compression ratios, about 18:1 and sometimes as high as 22:1 to improve their starting.

Air Cells

In air-cell systems, the nozzle injects fuel directly into the main combustion chamber, where the burning starts. The air cell is a space provided in the piston or in the cylinder head, Fig. 15-8 A and B, in which a large part of the air is trapped during the compression stroke. This air is at close to maximum pressure when the piston starts its downward stroke and the pressure in the main combustion space starts to fall. The higher pressure in the air cell causes its air to expand and blow out into the main combustion space. By doing so, it creates additional turbulence and insures the complete burning of the fuel charge.

The design of the air cell in the cylinder head is combined with a turbulence chamber, Fig. 15-8.

One disadvantage of the air-cell system is that a portion of the air remans trapped in the cell and takes little or no part in the combustion. Refined designs generally use the air cell in combination with a turbulence or precombustion chamber to obtain better performance than that given by the air cell with an open combustion chamber.

Energy Cells

The *energy cells* used in diesel engines resemble *air* cells, so don't confuse them.

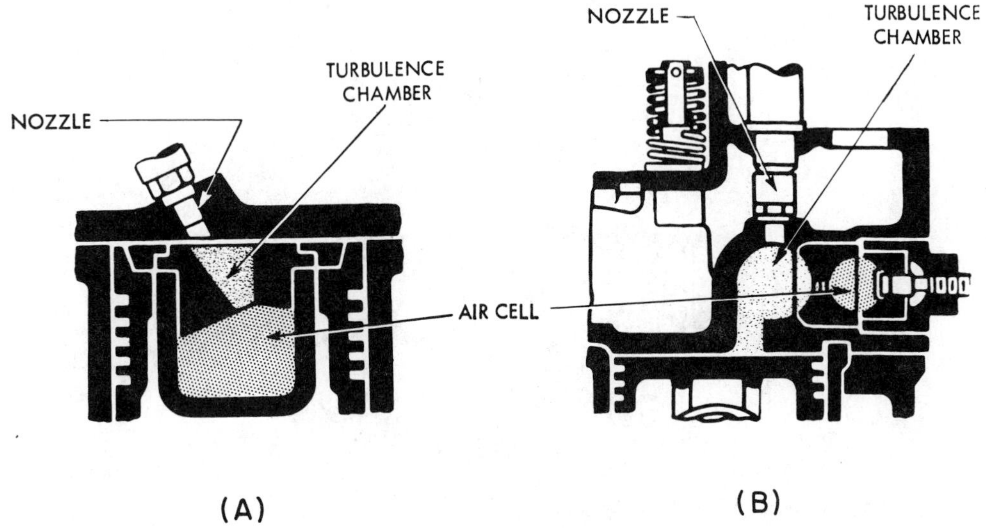

Fig. 15-8. Air cell designs.

The distinction is that fuel is blown *into* the energy cell and burns there, using the air in the cell. It is the *products of combustion* which blow back into the main chamber to create turbulence. You can see that this differs considerably from the air-cell system, in which the cell simply stores and gives up an air charge. The advantage of the energy cell is that the combustion in the cell creates a higher pressure and thus greater turbulence, and leaves no idle air in the cell.

Let's see how energy cell systems work by studying the construction and operation of a typical system, the Lanova design shown in Fig. 15-9. This design employs a combustion chamber consisting of two rounded spaces cast in the cylinder head. The inlet and exhaust valves open into the main combustion chamber. The fuel-injection nozzle lies horizontally, pointing across the narrow section where the lobes join. Opposite the nozzle is the two-part energy cell, which contains less

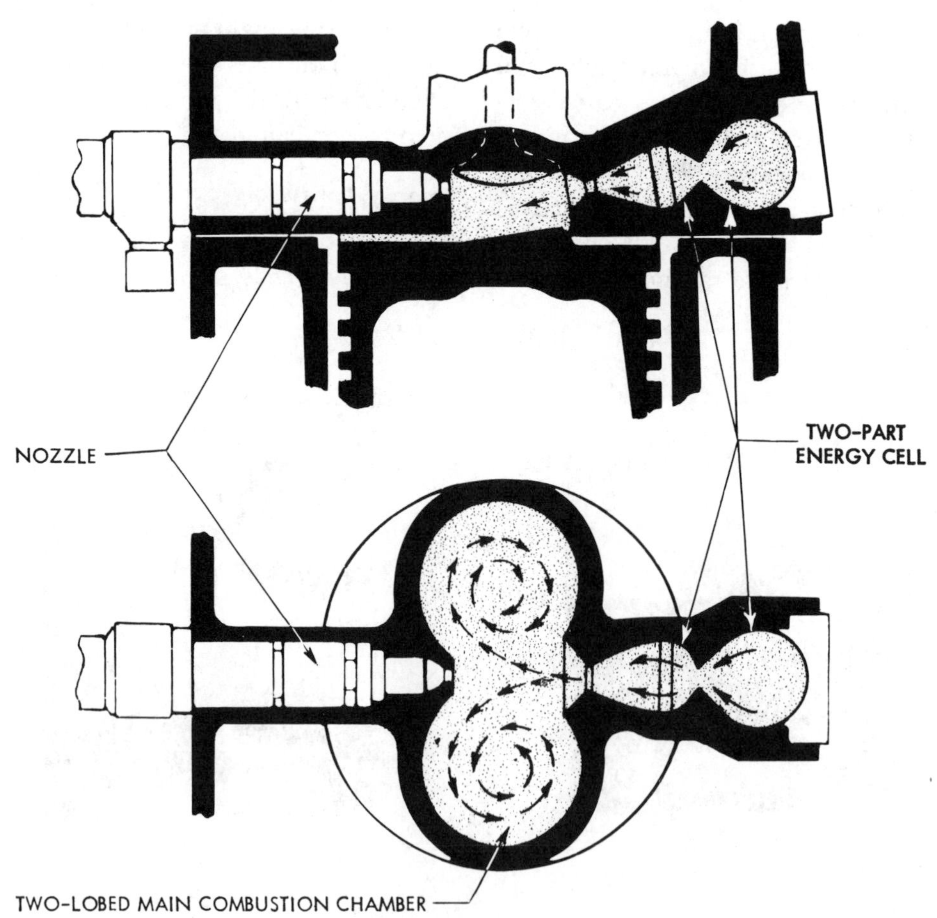

Fig. 15-9. Lanova energy cell.

than 20 percent of the main-chamber volume. This contrasts with the air-cell system, in which the cell contains a large percentage of the total volume.

The action is as follows: During the compression stroke, the piston forces air into the energy cell. Near the end of the stroke, the nozzle sprays oil across the main chamber in the direction of the mouth of the energy cell. While the fuel charge is traveling across the center of the main chamber, between a third and a half of the fuel mixes with the hot air and burns at once. The remainder of the fuel enters the energy cell and starts to burn there, being ignited from the oil already burning in the main chamber.

At this point, the cell pressure rises sharply, causing the products of combustion to backflow at high velocity into the main combustion space. This sets up a rapid swirling movement of fuel and air in each lobe of the main chamber, promoting the final fuel-air mixing and insuring complete combustion. The two restricted openings of the energy cell control the time and rate of expulsion of the turbulence-creating blast from the energy cell into the main combustion space.

Fig. 15-10 shows another design of energy-cell or dual-combination system known as the Ramsey system. The main combustion chamber, shaped roughly to fit the outside of the fuel spray, is cast

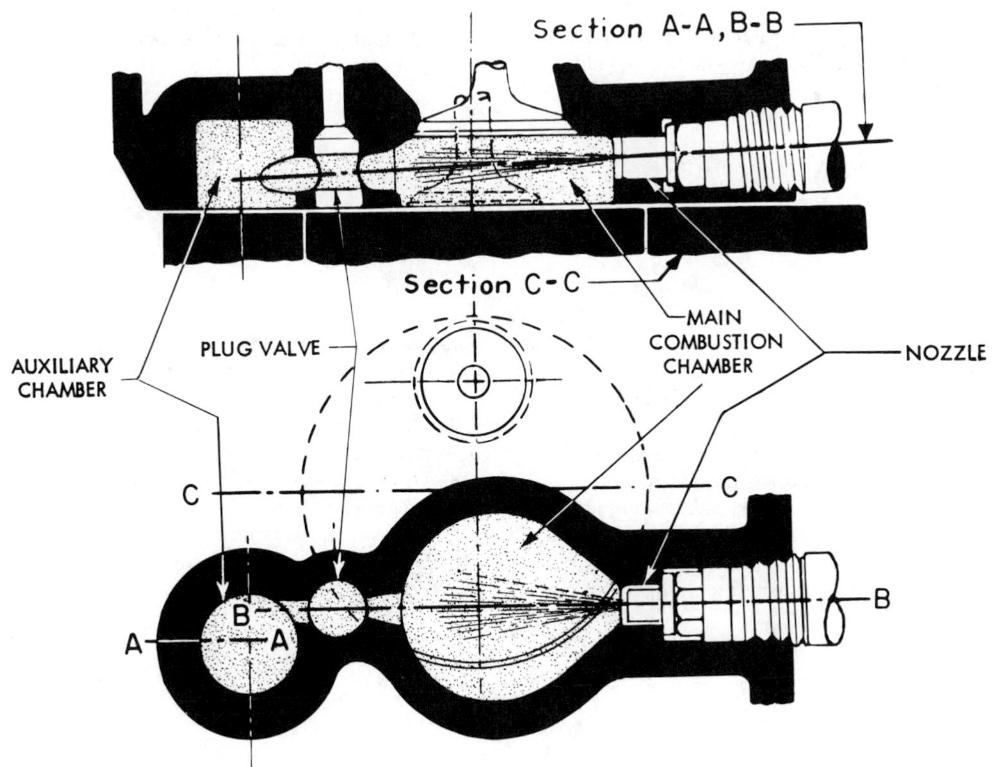

Fig. 15-10. Ramsey energy cell.

into the cylinder head. A passage connects this chamber with an auxiliary chamber which acts as an energy cell. A plug valve in the passage permits the operator to close off the auxiliary chamber in order to increase the compression pressure in the main chamber for easier starting in cold weather. As in the Lavona design, part of the fuel burns in the main chamber and the remainder in the auxiliary chamber. Backflow of the products of combustion from the auxiliary chamber into the main chamber creates turbulence and insures complete combustion.

Energy-cell* combustion systems meet the requirements of high-speed engines and produce high power output without excessively high pressures in the main combustion space.

*These are not to be confused with the use of the term *energy cell* as used to provide direct current of electrons from chemical combustion of gases to produce electricity.

Checking On Your Knowledge

The following questions give you the opportunity to check up on yourself. If you have read the chapter carefully, you should be able to answer the questions. If you have any difficulty, read the chapter over once more so that you have the information well in mind before you go on with your reading.

DO YOU KNOW

1. What is the difference between *atomization* and *vaporization?*
2. Describe the four stages of burning in the diesel combustion process.
3. What are the four combustion requirements for successful burning in a diesel engine?
4. Why is *atomization* and *penetration* contradictory in some respects and make it necessary to match the combustion?
5. Why does the modern diesel engine with an *open* combustion chamber usually have a hollowed out chamber on the top of the piston?
6. What is a swirl chamber?
7. Draw a turbulence chamber and explain how it operates.
8. Draw a precombustion chamber and explain how it operates.
9. Draw an air cell and explain how it operates.
10. Draw an energy-cell and explain how it operates.

Governing

Chapter

16

The purpose of this chapter is to acquaint you with a small but very important part of diesel engines—the governor. We'll give this topic a little extra attention in order to remove the veil of mystery that so often envelops it.

First you'll find out what a governor is—a device that controls the engine's speed or output *automatically*, and does the job far better than hand control. You'll see why some governors are called *constant-speed*, while others are termed *variable-speed, speed-limiting, load-limiting*, and so on.

You will learn the basic principle behind any governor. It is probably much simpler than you suppose.

The governor does its job in two steps: (1) measuring the speed, and (2) changing the engine fuel control. You'll see that a *hydraulic governor* differs from a *mechanical governor* by using hydraulic power to achieve the second step.

In order to understand how different governors act, you'll get to know the meanings of some governing terms, such as *speed regulation, speed-droop, hunting*, and *isochronous*.

Mechanical governors are the simplest. You'll see their construction and will follow their actions when the engine load changes.

Taking up hydraulic governors, we'll start with the simplest possible type to show how a *servo-motor* utilizes hydraulic power. But you'll learn that such an elementary hydraulic governor is characterized by hunting and is therefore not practical.

Practical hydraulic governors avoid hunting by using a speed-droop mechanism. You'll learn how speed-droop is produced and why it prevents hunting.

Commercial hydraulic governors employ either *permanent* or *temporary* speed-droop. You'll see why temporary speed-droop or *compensation* is used in the more refined isochronous governors which keep the engine at exactly the same speed from full load to no load. We'll trace in detail the action of both types of hydraulic governors.

The basic types of mechanical and hydraulic governors are often modified to suit special kinds of service. We'll therefore have a look at some of the important modifications, such as *overspeed governors*, *pressure-regulating governors*, and others.

Quite different from all of these governors is the electric load-sensing governor which responds directly to changes in the electric load imposed on the generator driven by the engine, and adjusts the fuel supply to maintain constant speed.

And finally you'll see how the different kinds of governors are actually applied in service. For example, how the engine govnors in electric power plants cause their engines to share the total load, and what variable-speed governors do on automotive diesel engines.

Governors

An engine governor is commonly a speed-sensitive device that automatically controls or limits the speed of the engine by adjusting the amount of fuel fed to the engine. The usual kind of governor adjusts the rate of fuel supply in such a way as to keep the engine running at a *steady speed* regardless of the amount of load. Such a governor is called a *speed governor*.

Principle of Speed Governors

The basic principle which makes a speed governor work is that any change in load immediately causes a change in speed. The power developed by any internal-combustion engine depends upon the amount of fuel burned in the cylinders (up to the engine's capacity). In other words, if fuel injects at a faster rate (pounds of fuel oil or cubic feet of gas per stroke), the engine develops more power. If the power developed by the engine exceeds the power required (the surplus power goes into acceleration and the engine speeds up. On the other hand, if the load increases so as to exceed the power developed, the engine slows down.

If an engine runs with a *fixed throttle*, which means that the rate of fuel flow is constant, the engine will speed up when the load decreases and will slow down when the load increases. This might even go so far that the engine would run away when the load was taken off or might stall (come to rest) when the load was increased.

To keep the engine running at a steady speed, the flow of fuel must vary in such a way that the power developed is just equal to that needed at that speed.

One way of trying to do this would be for the engine operator to stand by at all times and to adjust the throttle whenever he noticed the speed change, or, if the engine had a tachometer (speed indicator), whenever he saw the needle move. When the speed fell, he would open the throttle farther in order to supply more fuel to the cylinders, and when the speed rose, he would reduce the fuel flow.

But, even with the greatest skill, he

couldn't do this fast enough to maintain a reasonably uniform speed unless the load stayed constant or changed very slowly. The function of the speed governor is to do this kind of job rapidly, accurately, and automatically. First, the governor notes the change in engine speed, then it adjusts the rate of fuel admission to suit.

Flywheels

Flywheels are used to regulate speed. When the power produced by an engine exceeds the power needed, the surplus power causes the engine to speed up. How much it will accelerate depends upon the inertia of the flywheel and the other rotating parts, because the surplus power is converted into kinetic energy. When the load increases beyond the power developed, the rotating parts give out kinetic energy and slow down. In either case the flywheel reduces the amount of speed change.

According to physics the kinetic energy of a rotating part is proportional to its inertia and to the square of its rotating speed. Thus the heavier the flywheel and the larger its diameter, the greater will be its inertia. Therefore, a large flywheel will absorb a given amount of surplus power as kinetic energy with less increase in speed than a small flywheel. If the engine power is deficient, there will be less decrease in speed. Thus a larger flywheel causes an engine to run steadier.

Flywheels serve two purposes. One is to supplement the action of the governor by preventing a change in speed during the time taken by the governor to change the rate of fuel flow and by the cylinders in developing the corresponding power. The other purpose is to smooth out the momentary speed changes that occur as the pistons go through their cycles of compression, expansion, etc. These *cyclic* speed variations happen because when a piston is on its firing stroke, it tends to speed up the engine, while on the compression stroke it tends to slow it down.

Other Kinds of Governors

The *constant-speed governor* described in the foregoing paragraphs is intended to maintain the engine at a single speed from no load to full load. Other kinds of governors do different jobs, among which are the following:

Variable-Speed Governor: to maintain any selected engine speed from idling to top speed.

Speed-Limiting Governor: to control the engine at its minimum speed or at its maximum speed. Governors for automotive engines often control both the minimum and the maximum speed. A governor which holds the engine at its maximum safe speed is also called an *overspeed governor*. Note that a speed-limiting governor does *not* control the speed when the speed is within the designed limit or limits.

Overspeed Trip: to shut down the engine in case it overspeeds. It is a safety device only.

Load-Limiting Governor: to limit the load applied to the engine at any given speed. Its purpose is to prevent overloading the engine at whatever speed it may be running.

Load-Control Governor: to adjust to amount of load applied to the engine to suit the speed at which it is set to run. It is widely used on diesel locomotives to improve fuel economy and reduce engine strains when the locomotive is not loaded.

Pressure-Regulating Governor: used on an engine driving a pump, to maintain a constant inlet or outlet pressure on the pump.

Torque-Control Governor: to control the maximum speed of the output shaft of a torque-converter attached to an engine, independently of the engine speed.

Electric Load-Sensing Governor: to feel directly, changes in electric load and quickly adjust the fuel quantity. It is used on engines in electric power plants.

Electric Speed-Governor: to control the engine speed by measuring the frequency (cycles per second) of the electric current produced by an engine driven governor.

Speed Governors

In governing engine speed, a governor must first *measure the speed.* All governors, from the simplest to the most elaborate types, include an accurate speed-measuring device. After measuring speed, a governor must *transfer the indication* of the speed-measuring device (when a change in speed occurs) into a movement of the governor terminal or output shaft. This shaft connects to the control rod of the fuel-injection system, thus regulating the amount of fuel injected into the engine cylinders.

Measuring the Speed

This first step of governor action is performed in practically all diesel-engine and gas-engine governors by a *centrifugal ball-head.* In the usual design, shown in Fig. 16-1, a pair of flyweights located on opposite sides of a shaft are rotated by the engine through gears. The rotating weights produce a centrifugal force; this force is opposed and balanced by a spring, generally called a speeder spring. Fig. 16-2 shows the balanced position at normal speed.

Note that the ball-arms are vertical. If the speed increases, Fig. 16-3A, the centrifugal force of the flyweights also increases and the weights move out from the axis of rotation. The outward movement of the weights lifts the ball-arm toes, thus increasing the opposing spring force. Since the spring is stiff compared to the centrifugal force, a balance is reached at a point where the spring force equals the centrifugal force at some new position of the flyweights a little farther out.

The reverse action takes place if the speed falls, Fig. 16-3B. The centrifugal force of the flyweights decreases and the speeder spring pushes them inward until a new balance is reached. Thus, for any given speed, the flyweights will take a definite position a certain distance from the axis of rotation.

Operating the Throttle

The second step of governor action is moving the fuel-control mechanism of the engine. If the fuel-injection system (Chapter 14) is the jerk-pump type, the fuel-control mechanism changes the de-

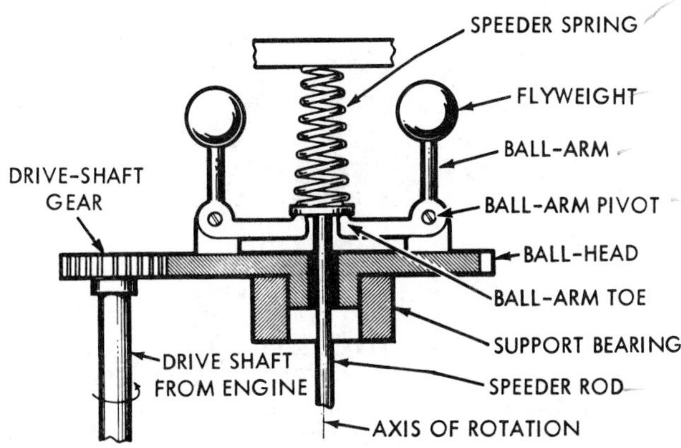

Fig. 16-1. Centrifugal ball-head. (Woodward Governor Co.)

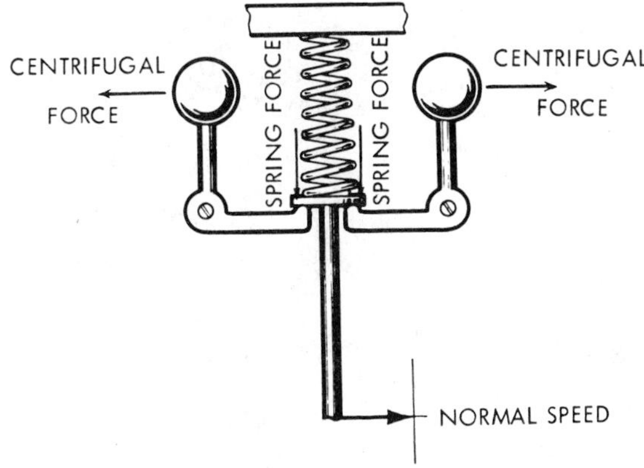

Fig. 16-2. Ball-arms are vertical when ball-head forces are at normal speed. (Woodward Governor Co.)

livery of the pump. If it is the common-rail type, the control mechanism changes the fuel flow to the injectors. We will use the term *throttle* to mean the fuel-control mechanism of the engine, whatever it is.

Mechanical and Hydraulic Governors Differ

Governors fall into two types, depending on the kind of force used to move the engine throttle. If the force comes *directly* from the speed-measuring device, it is a *mechanical governor;* if the force comes *indirectly*, it is a *hydraulic* or *relay governor*. In mechanical governors, the movement of the speeder rod itself actuates the linkage to the throttle. In hydraulic governors, the speeder rod actuates a small valve controlling a fluid under pressure; this fluid pushes on a piston or servomotor to actuate the throttle.

Speed Adjustment

In most governors, the spring force which resists the centrifugal force of the

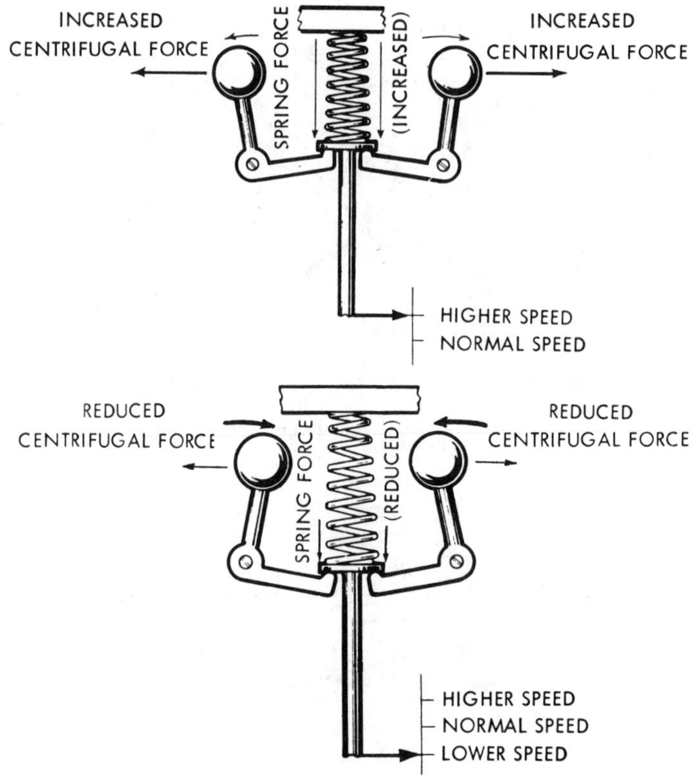

Fig. 16-3. Ball-arms move out with the ball-head at higher speeds, and in when ball-head drops to lower speeds. (Woodward Governor Co.)

flyweights can be adjusted by the engine operator in order to adjust the control speed of the engine.

Fig. 16-4 shows construction of a speed adjuster. The upper end of the speeder spring fits into a speeder plug which may be adjusted up or down by means of a knob or lever on the outside of the governor. On governors used in electric power plants, the speed adjusting screw is often operated by a small reversible electric motor controlled from the switchboard.

The reason why a change in the speeder-spring force changes the control speed of the engine is simple. You saw

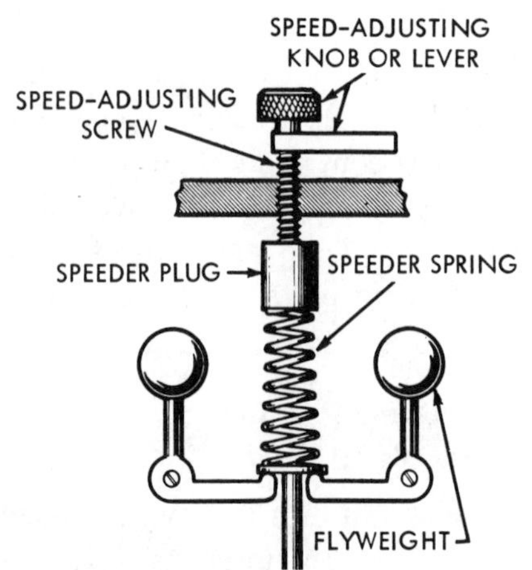

Fig. 16-4. Speed-adjusting screw and mechanism. (Woodward Governor Co.)

that the forces in a centrifugal ball-head balance each other at some certain speed. Now, if we change the spring force, we destroy this balance. To reach a new balance, the speed must change.

Suppose we reduce the spring force by turning the speed adjusting screw (Fig. 16-4) so as to raise the speeder plug, and thus reduce the compression of the spring. For the reduced spring force to balance the centrifugal force of the flyweights in the same vertical position as before, less centrifugal force is needed and therefore less engine speed. In other words, reducing the spring force causes the engine to run at a lower speed when carrying the same load as before.

Similarly increasing the spring force causes a higher engine speed for the same load, because more centrifugal force is needed to balance the spring.

Some governors use an independent spring for adjusting speed. Any spring which, in one way or another, resists the flyweight force will affect the engine speed at which the governor forces balance. The independent spring is usually located outside the governor head for convenience, and is generally softer than the speeder spring for fine adjustment.

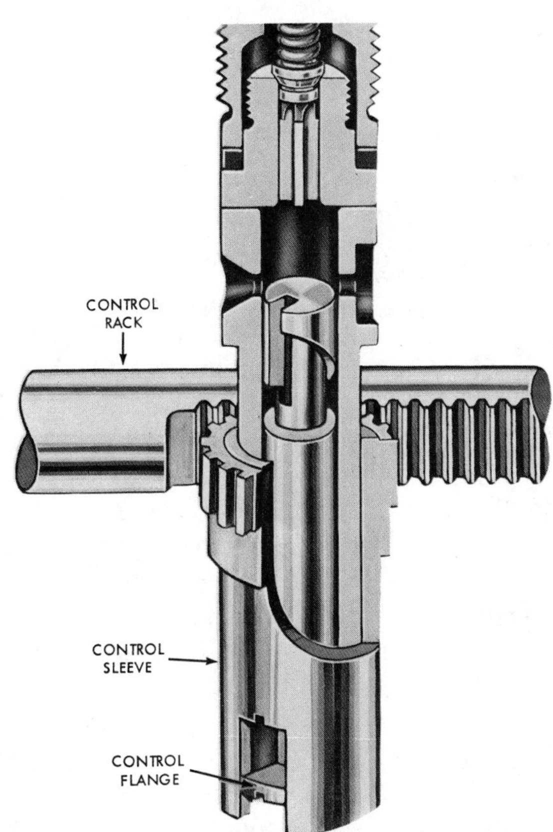

Fig. 16-5. Plunger rotation. (American Bosch AMBAC Industries, Inc.)

Fig. 16-5 shows the geared *rack and pinion* in an actual governor designed for controlling the amount of fuel by rotating the control sleeve and plunger when the control rack is moved to the left or right by governor action.

Definition of Terms

In order to understand why different types of governors are needed for different kinds of jobs, you will need to know the meanings of several terms that are used in describing the characteristics of governor action. The principal characteristics which determine the degree of governor control of the engine may be briefly defined as follows:

Speed regulation (technically, steady-state speed regulation) of an engine is the change in its sustained or final speed, when the load is changed from full rated load to zero load or vice versa, without adjusting the governor (such as by changing the speeder-spring tension). This change in speed is expressed as a percentage of rated full-load speed, as follows:

$$\text{Speed regulation, in percent} = \frac{(\text{no-load speed} - \text{full-load speed})}{\text{rated full-load speed}} \times 100$$

Note that speed regulation refers to a change in *steady* speed. This means that the governor and the engine are given enough time to reach a stable position and speed for the given load.

For partial changes in load, the change in steady speed is proportional to the speed regulation. To illustrate, suppose an engine runs at 600 rpm at full load and that, when all the load is removed, the engine speed increases to a steady speed of 624 rpm. Since the speed increase between full load and no load is 24 rpm, the speed regulation is:

$$\frac{24}{600} \times 100 = 4 \text{ percent}$$

Knowing the speed regulation, it is easy to compute the change in speed that would occur for any partial change in load. Assume that the load changes by only one-quarter of rated load, as would be the case if the load changed from full load to three-quarter load, or from half-load to one-quarter load. The resulting change in speed would be a quarter of the change from full load to no load, that is a quarter of 24 rpm, or 6 rpm. We can put this in tabular form as shown in Table 16-1.

Speed regulation is important for many reasons, one of which is that it determines how two or more engines driving the same load will share any change in load. The speed regulation of an engine is directly related to the speed-droop of the governor.

Speed-droop of a governor is the change in governor rotating speed which causes the governor's output shaft (fuel-control rod) to move from its full-open throttle position to its full-closed throttle position

TABLE 16-1 SPEED CHANGES AS LOAD CHANGES

LOAD	SPEED, RPM	SPEED CHANGE PERCENT
FULL RATED LOAD	600	0
3/4 RATED LOAD	606	1
1/2 RATED LOAD	612	2
1/4 RATED LOAD	618	3
NO LOAD	624	4

or vice versa. Every speed governor is a device which responds to changes in speed. Its speed-droop shows how big a speed change is needed to cause the governor's control shaft to travel through its full working range.

Speed-droop of a governor differs from the speed regulation of an engine in that the droop may be either *permanent* or *temporary*. A governor must have speed-droop to prevent false motions or overcorrection.

If the speed-droop is *permanent*, the governor's output shaft comes to rest in a different position for each speed. This correspondingly affects the engine's speed regulation; the engine's final speed is different for each amount of loading. On the other hand, if the speed-droop is *temporary*, the governor's output shaft always comes to rest at the *same speed*, and so the engine's final (steady) speed remains constant, regardless of load. Since this is often advantageous, many hydraulic governors are designed to use temporary rather than permanent speed-droop.

Isochronous governor (same time or identical speed governor) is one which regulates the engine speed so that the speed regulation is zero percent; the engine's steady speed is timed to be exactly the *same* at any load from no load to full load. The speed-droop of such a governor is temporary.

Momentary speed changes, also termed *transient speed changes*, are the short-lived speed changes which occur immediately after a sudden change in load. Such speed changes usually exceed the final speed change corresponding to the speed-droop. Momentary speed changes are expressed as the percent increase or decrease in speed, referring to the speed at the instant of the load change.

Hunting is the repeated and sometimes rhythmic variation of speed due to *overcontrol* by the governor. Hunting is a continuous problem, that is, the uneven surges keep on repeating themselves. However, such periodic speed variations may occur through no fault of the governor. If the governor is responsible, the hunting will cease if the engine throttle is blocked in a fixed position. Then, if the surging of oscillations continue, the governor is not the cause.

Stability in a governor is its ability to maintain speed with either constant or varying loads without hunting.

Dead band, sometimes called *sensitivity*, is the change in speed required before the governor will make a corrective movement of the throttle. This hesitation results from the lag in governor action caused by friction and lost motion in the governor mechanism.

Promptness is the speed of action of the governor, which depends upon its power relative to the work it must do. The greater the power, the shorter the time required to overcome the resistances.

Work capacity denotes the power of the governor as shown by the amount of work it can do at its output shaft. The work capacity, in inch-pounds, equals average force, in pounds, exerted by the governor at its terminal (final) level multiplied by distance, in inches, through which the terminal lever moves in its full travel. Thus, a governor whose terminal lever exerts a force of 4 pounds over a distance of 3 inches has a work capacity of 12 in-lb.

The *corrective action* of every speed governor is initiated by a change in engine speed. It is called *corrective action* because it *corrects* the engine speed. The smaller the *dead band*, the sooner the governor will commence the corrective action after the speed begins to change. The greater the *promptness* of the governor the faster it will shift the engine throttle to its new position. *Dead band* determines the time lag before the governor starts to act, while *promptness* determines how rapidly it completes its corrective action.

Both dead band and promptness are therefore important factors in governor performance. Some hydraulic governors combine small dead band with high work capacity (promptness) to such a degree that they will respond to a change in speed of less than $\frac{1}{100}$ of 1 percent, and can shift the fuel-control mechanism from full-load position to no-load position, or vice versa, in less than a quarter of a second.

Engine speed deviation is any change in speed from normal. The amount of speed deviation for a given change in load depends on the characteristics of the engine as well as those of the governor. Briefly, the conditions that determine speed deviation are the following:

1. The time required to correct the rate of fuel injection to correspond with the new load. This depends on the governor, as previously explained.

2. The inertia of the flywheel and other rotating parts. This depends on the construction of the engine and the apparatus which it drives.

3. The time required for the engine's power output to respond to the change in rate of fuel injection. This depends on the number of engine cylinders, the speed, and the type of fuel-injection system. For example, a two-cylinder, 200 rpm engine takes longer to respond than an eight-cylinder, 1200 rpm unit.

Ideally, if it were possible (1) to detect instantly the amount of load change, (2) to correct the rate of fuel injection instantly and accurately, and (3) to obtain immediate power response from the engine cylinders, the speed would never change, no matter how rapid or how great were the load changes. That is impossible because of practical limitations.

Some delay always occurs between the moment the load changes and the moment when the engine responds to the correct rate of fuel injection. The flywheel inertia, by absorbing or supplying energy, reduces the speed change during the delay period, but the longer the delay, the more will be the speed change.

Therefore, precise speed control requires (1) a fast acting, accurate governor and (2) sufficient flywheel inertia to maintain the speed within required limits until the engine has had time to respond to the new fuel setting.

Mechanical Governors

You learned that when the ball-head (speed-measuring device) of a governor speeds up, the centrifugal force of the flyweights becomes greater than the opposing spring force, and vice versa. Mechanical governors utilize these forces directly to operate the fuel-changing mechanism. The fuel-changing mechanism of a diesel or dual-fuel engine is shown schematically in Fig. 16-6.

Power Device. This is represented by the speeder rod transmitting power directly from the flyweights.

Linkage. Connects the power device and the fuel-control valve.

Fuel Control Valve. Instead of the elaborate designs actually used in engines, the schematic figures show a simple gate valve to make the fuel opening plainly visible.

Now let's look at the combined action of the speed-measuring device and the fuel-changing mechanism in such an elementary mechanical governor when a load increase occurs and when a load decrease occurs.

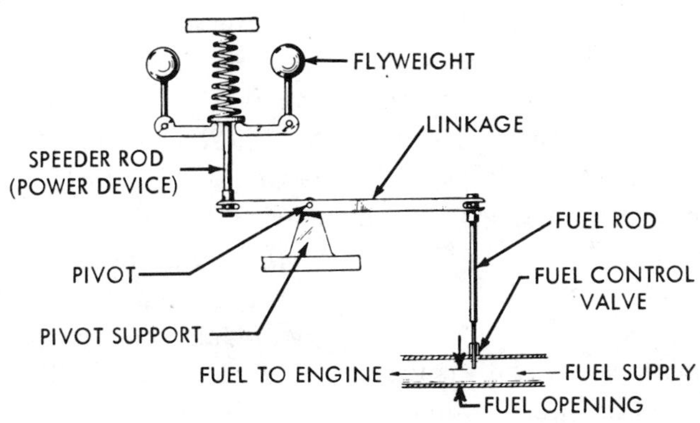

Fig. 16-6. Schematic arrangement of fuel-changing mechanism. (Woodward Governor Co.)

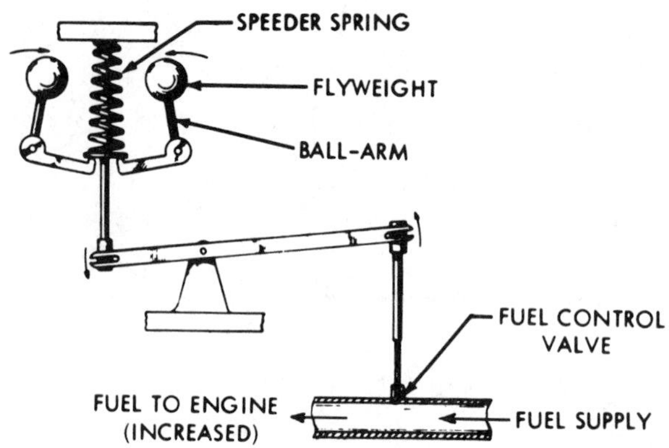

Fig. 16-7. Effect of load increase on mechanical governor. (Woodward Governor Co.)

Load Increase (Fig. 16-7):

1. Load is applied to the engine and its speed decreases.
2. Decreased engine speed results in decreased ball-arm speed.
3. Decreased ball-arm speed results in reduced centrifugal force of the flyweights, allowing the speeder spring to force the flyweights in and push the speeder rod down.
4. As the speeder rod moves down, the fuel valve opens to increase fuel.
5. The fuel increase supplies additional power to the engine to carry the increased load.
6. Engine speed picks up, but does not reach original speed because, if it did, the throttle would not be open wide enough to supply the increased load. The throttle opening can only be widened by the flyweights moving *in* toward their axis of rotation by reason of slower speed.

Load Decrease (Fig. 16-8):

1. Load is dropped from the engine and its speed increases.
2. Increased engine speed, with corresponding higher ball-arm speed, results in increased centrifugal force acting against the speeder spring to raise the speeder rod.
3. As the speeder rod moves up, the fuel valve closes to reduce the fuel, which reduces the engine output to match the reduced load.
4. The engine speed drops, but not to the former load speed because it is mechanically impossible to maintain reduced throttle with the flyweights back in original position. The smaller fuel opening necessary for the reduced load requires the ball-arms to be *out* from the vertical by reason of higher speed than before.

Note in the foregoing governor actions that the final, steady speed of the engine was less after the load was picked up, and that the final speed was higher after the load was dropped off. In the load-increase example, governor action was insufficient to restore engine speed *up* to the former (unloaded) level while the load was being carried. Likewise, in the load-decrease example, governor action failed to restore engine speed *down* to the former (loaded) level while the load continued removed. In other words, at different loads the engine runs at different speeds, but at any particular load the fuel valve always

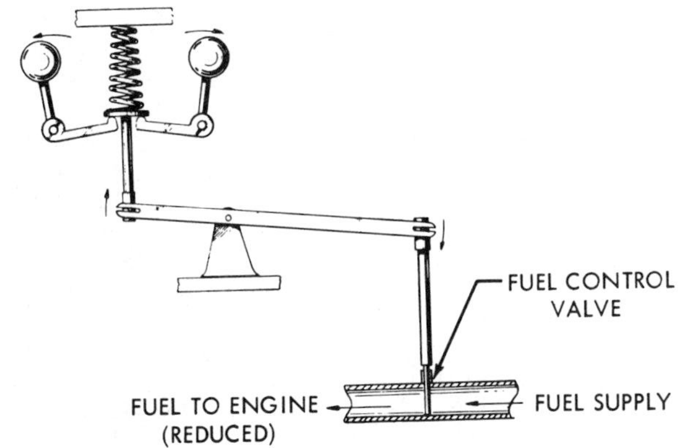

Fig. 16-8. Effect of load decrease on mechanical governor. (Woodward Governor Co.)

comes to rest in the same position, so that the engine runs at a certain speed which corresponds to the load.

This inability to completely restore original steady speeds after changes in load is called permanent speed-droop. It is an inherent characteristic of all mechanical governors because the power of the flyweights moves the throttle *directly* by mechanical means. For many types of service, a reasonable amount of permanent speed-droop is quite satisfactory.

Advantages

Favorable as well as unfavorable characteristics are to be found in mechanical governors. Advantages are:

1. They are inexpensive.
2. They are satisfactory where it is not necessary to maintain exactly the same speed, regardless of load.
3. They are extremely simple, with few parts.

Disadvantages

Along with favorable characteristics, mechanical governors have shortcomings that limit their application:

1. They have large dead bands, since the speed measuring device must also furnish the force to move the engine fuel control.
2. Their power is relatively small unless they are excessively large.
3. They have an unavoidable speed-droop, and therefore cannot truly provide constant speed where this is needed.

Typical Mechanical Governors

The Pickering mechanical governor, Fig. 16-9, is used on small high-speed engines. Note that it employs a single outside spring which acts both as speeder spring and speed adjuster spring. The free use of antifriction (ball) bearings improves the governor's sensitivity.

Fuel injection pumps of small engines often include a built-in governor for the sake of compactness. Depending on the class of service, such governors may be of the constant-speed, variable speed, or torque-control types.

Fig. 16-10 shows a cutaway of an actual Cummins standard automotive type gov-

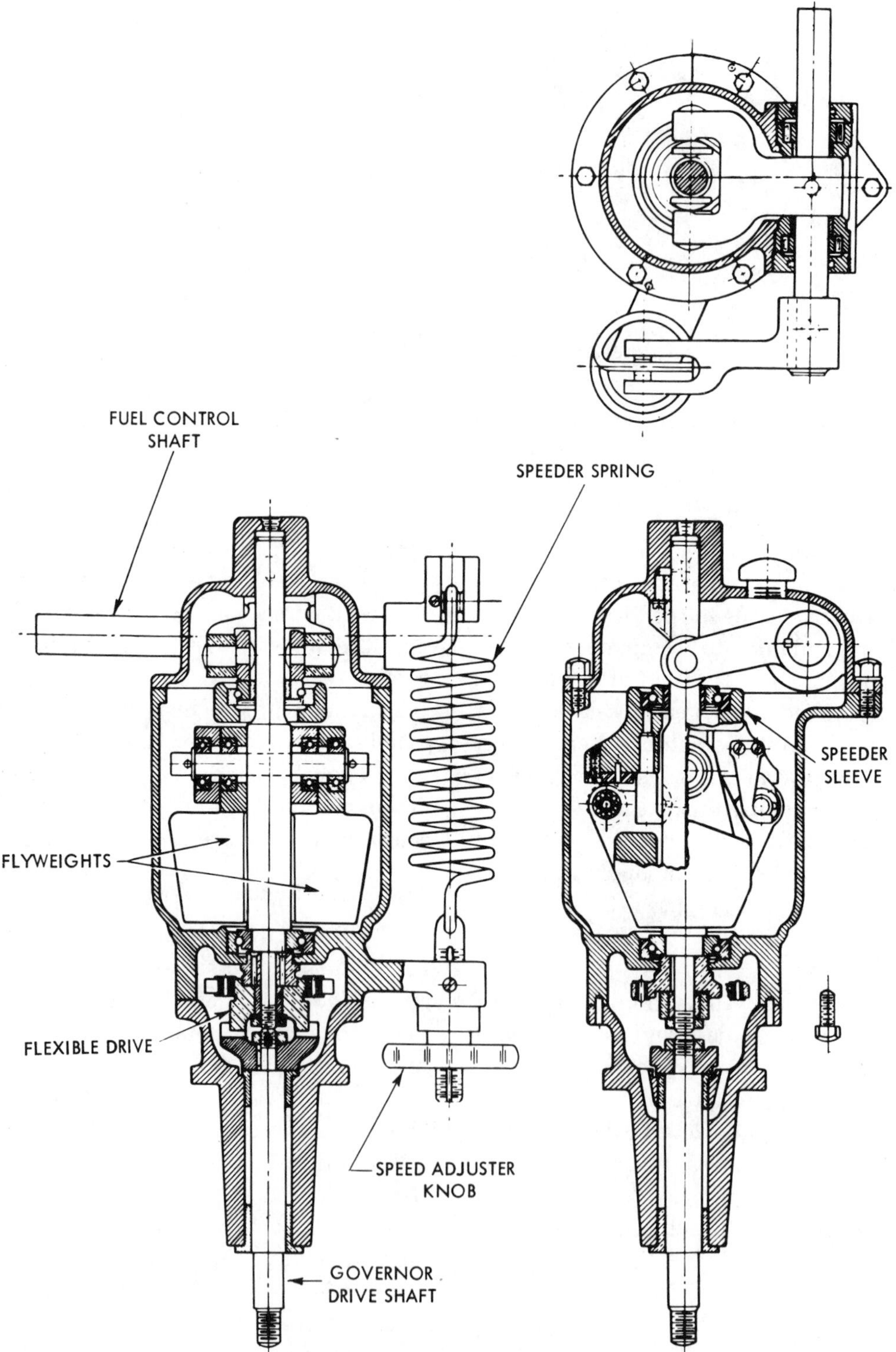

Fig. 16-9. Pickering mechanical governor.

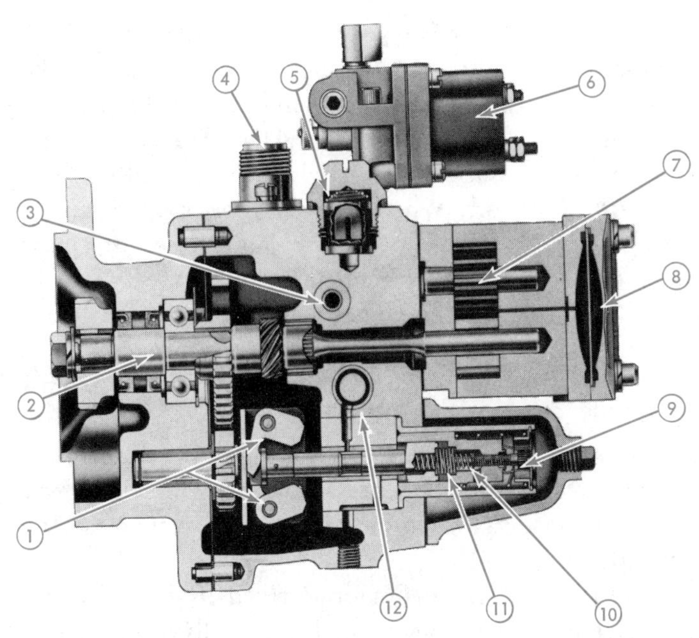

Fig. 16-10. Cummins automotive type governor. (Cummins Engine Co.)

ernor, built into the fuel injection system. This is a speed-limiting governor which controls the minimum or idle speed, and the maximum or high speed, each speed having a corresponding spring to oppose the force of the governor flyweights.

When idling, the flyweight force is balanced by the resisting force of the idling springs, and the engine runs at a constant minimum speed. When the truck driver steps on the throttle to increase the fuel flow, the engine runs at an intermediate speed determined by the throttle position. If the speed should rise to the set maximum, the resisting force of the high-speed spring will balance the governor flyweight force, and the engine will run at a constant maximum speed.

Hydraulic Governor Principles

Although hydraulic governors have more parts and are usually more expensive than mechanical governors, they find use in many applications because: (1) they are more sensitive, (2) they have greater power to move the fuel-control mechanism of the engine, and (3) they can be made isochronous (timed for identical speed for all loads).

In hydraulic governors, the power

which moves the engine throttle does *not* come from the speed-measuring device. Instead, it comes from a hydraulic power piston, or *servo-motor.* This is a piston which is acted upon by a fluid under pressure, generally oil under the pressure of a pump. By using appropriate piston size and oil pressure, the power of the governor at its output shaft (work capacity) can be made sufficient to operate promptly the fuel-changing mechanism of the largest engines.

The speed-measuring device, through its speeder rod, is attached to a small cylindrical valve, called a *pilot valve.* The pilot valve slides up and down in a bushing which contains ports that control the oil flow to and from the servo-motor. The force needed to slide the pilot valve is exceedingly small; consequently, a small ball-head is able to control a large amount of power at the servo-motor.

Whether the governor is mechanical or hydraulic, the engine should be equipped with a separate overspeed trip lever. This prevents a runaway should the governor become inoperative.

Elementary Hydraulic Governor

Fig. 16-11 shows, schematically, the principle of an elementary hydraulic governor. Note that the land (raised portion) of the pilot valve is the same width as the port. When the governor is operating at control speed, the land closes the port, and there is no flow of oil.

If the governor speed falls, due to increase of engine load, the flyweights move in and the pilot valve moves down. This opens the port to the power piston and connects it to a supply of oil under pressure. This pressure oil acts on the power piston, forcing it upward to increase the fuel.

If the governor speed rises, due to decrease of engine load, the flyweights move out and the pilot valve moves up. This opens the port from the power piston to the drain into the sump. The spring above the power piston forces the power piston down, thus decreasing the fuel.

Note that the port stays closed at only one certain speed, and that the throttle can be in any position, from full load to no load, at this speed. Ideally the engine

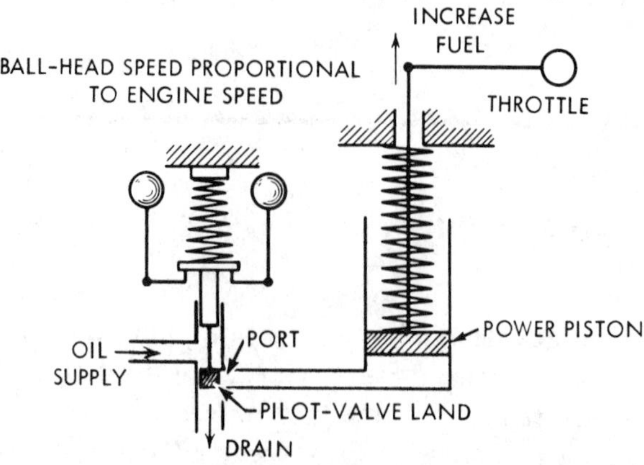

Fig. 16-11. Elementary principle of hydraulic governor. (Woodward Governor Co.)

should run at exactly the same speed at any load and the governor would be termed *isochronous*.

Defect of Elementary Hydraulic Governor

Unfortunately, the simple hydraulic governor just described has a serious defect which prevents its practical use. It is inherently unstable, that is, it keeps moving continually, making unnecessary corrective actions. In other words, *it hunts*.

The cause of *hunting* is the unavoidable time lag between the moment the governor acts and the moment the engine responds. The engine cannot *instantaneously* come back to the speed called for by the governor. Therefore, if the engine speed is below the governor control speed, the pilot valve is moving the power piston to increase the fuel. By the time the speed has increased to the control setting so that the pilot valve is centered and the power piston has stopped, the fuel has already been increased too much and the engine continues to speed up. The *overspeed* then opens the pilot valve the other way, so as to decrease the fuel. But by the time the engine speed falls to the right value, the fuel control has again traveled too far, the engine *underspeeds*, and the whole cycle repeats itself, again and again. Therefore, some means of stabilizing the governor must be added to make it satisfactory. The usual way is to provide speed-droop, as explained in the next section.

Hydraulic Governor with Permanent Speed-Droop

In order to obtain stability, most hydraulic governors employ *speed-droop*. Speed-droop gives stability because the engine throttle can take only one position for any one speed. Therefore, when a load change causes a speed change, the resulting governor action ceases at the particular point that gives the amount of fuel needed for the new load. In this way speed-droop prevents unnecessary governor movements and over-correction (hunting).

Note that to prevent hunting, the *speed-droop must be enough* to take care of the unavoidable delay while the engine itself is responding to the governor action. If the speed-droop is insufficient, some hunting will still occur while the engine is returning to its steady speed after the first momentary speed change.

The hydraulic governor about to be described employs *permanent* speed-droop, and therefore is not isochronous. Later you'll learn about isochronous hydraulic governors; these employ *temporary* speed-droop to give stability.

Permanent speed-droop can be obtained in several ways, one of which is to connect a lever between the power piston and the speeder spring so that, as fuel is increased, the speed setting is decreased. Reducing the speeder-spring force reduces the engine speed, and increasing the spring force increases the engine speed.

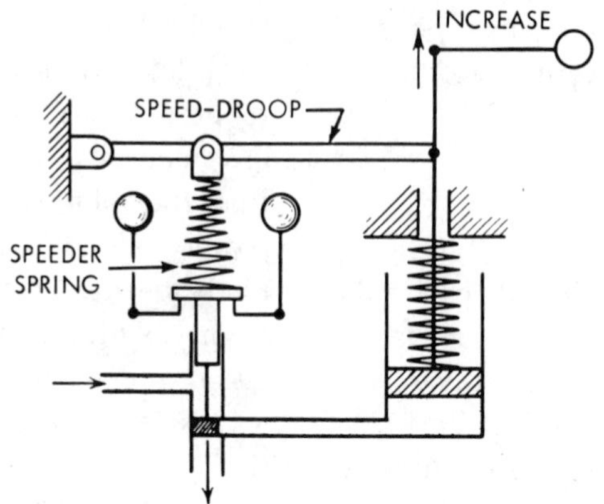

Fig. 16-12. Principle of speed droop lever. (Woodward Governor Co.)

The speed-droop lever, Fig. 16-12, does this.

Let's trace the action of a hydraulic governor with permanent speed-droop, as shown in Fig. 16-13. A simple pilot-valve plunger, attached to the end of the speeder rod, slides in a bushing with drilled control ports. Oil lines from these ports connect to both sides of a power piston on the fuel control rod. The valve ports are just closed when the ball-arms are in vertical position and the engine is running at the desired speed and load. It works as follows.

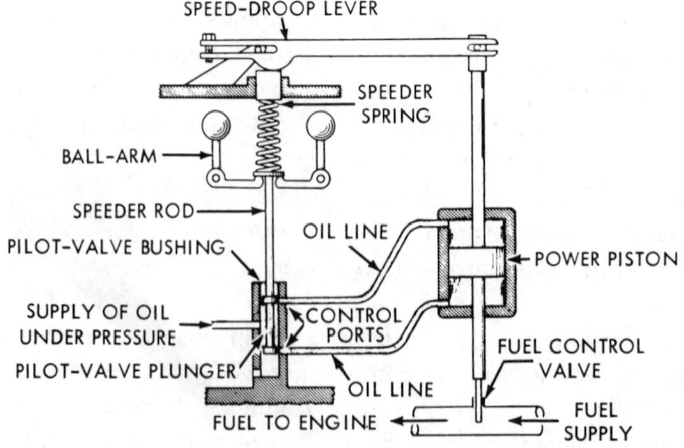

Fig. 16-13. Arrangement of hydraulic governor with speed-droop lever. (Woodward Governor Co.)

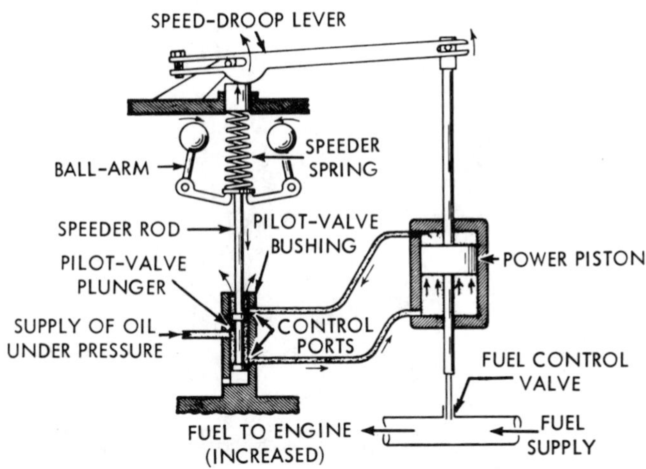

Fig. 16-14. Effect of load increase on hydraulic governor with speed-droop lever. (Woodward Governor Co.)

Load Increase (Fig. 16-14):

1. Load is applied to the engine and its speed decreases.

2. As the engine speed decreases, the ball-arms move in, lowering the pilot-valve plunger.

3. Lowering of the pilot-valve plunger opens the ports. Oil under pressure flows through the lower port to the underside of the power piston to force it up and increase the fuel. The oil on top of the piston flows out of the upper port and escapes from the top of the bushing to a sump.

4. As the power piston moves up, it pushes the speed-droop lever up, reducing the speeder-spring force.

5. The reduced speeder-spring force allows the ball-arms to move out, thus raising the pilot-valve plunger and slowing the further upward movement of the power piston.

6. When the ball-arms reach vertical position, the control ports will be closed and the power piston will stop moving up.

7. Since the speeder-spring force is reduced as fuel or load is increased, a balance is reached with less flyweight force, that is, with lower engine speed.

8. The reduction of engine steady speed caused by load increase is the speed-droop.

Note that the speed-droop lever prevents over-corrections by stopping the corrective movement of the power piston *before* the engine has returned to its previous speed.

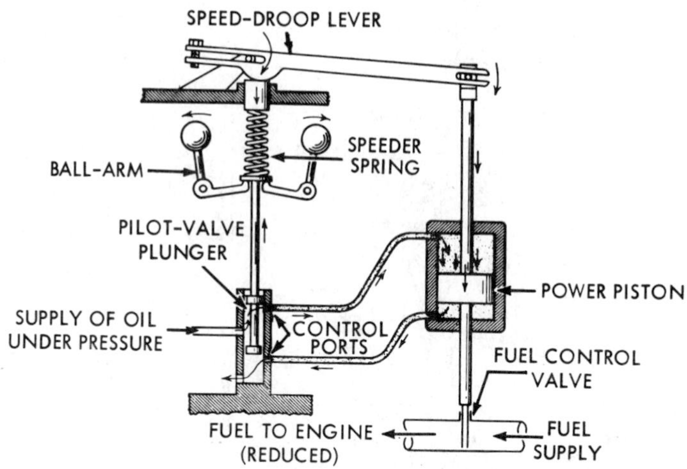

Fig. 16-15. Effect of load decrease on hydraulic governor with speed-droop lever. (Woodward Governor Co.)

Load Decrease (Fig. 16-15):

1. Load is dropped off the engine and its speed increases.

2. As the engine speed increases, the ball-arms move out, raising the pilot-valve plunger.

3. Raising the pilot-valve plunger opens the control ports. Oil under pressure flows through the upper port to the upper side of the power piston and forces it down to reduce fuel. The oil under the piston escapes through the lower port to the sump.

4. As the power piston moves down, it pulls the speed-droop lever down, increasing the speeder-spring force.

5. The increased speeder-spring force pushes the ball-arms in, thus lowering the pilot-valve plunger and slowing the further downward motion of the power piston.

6. When the ball-arms reach vertical, the control ports will be closed and the power piston will stop moving.

7. Since the speeder-spring force is increased as load is reduced, it requires more weight force (higher engine speed) to balance the spring force.

8. The increase of engine steady speed caused by load decrease is the speed-droop, the effect of which is to prevent hunting by stopping the corrective movement of the power piston *before* the engine has returned to its previous speed.

Advantages

A number of favorable characteristics are to be found in hydraulic governors having permanent speed-droop. Advantages are:

1. They are relatively inexpensive.

2. They are accurate and sensitive, giving good speed control.

3. They are simple and have few parts, as far as hydraulic governors go.

4. They are more powerful than mechanical governors of similar dimensions.

Disadvantages

Offsetting their favorable characteristics, hydraulic governors having permanent speed-droop exhibit several shortcomings that detract from their use value. Disadvantages are:

1. They are not isochronous (do not provide the same speed for all loads).

2. Their speed-droop is not conveniently adjustable. (Adjustment is made inside such governors.)

Fig. 16-16, which is a cross-section of the Woodward SG Governor, shows you the internal construction of an actual small hydraulic governor that employs permanent speed-droop in order to obtain

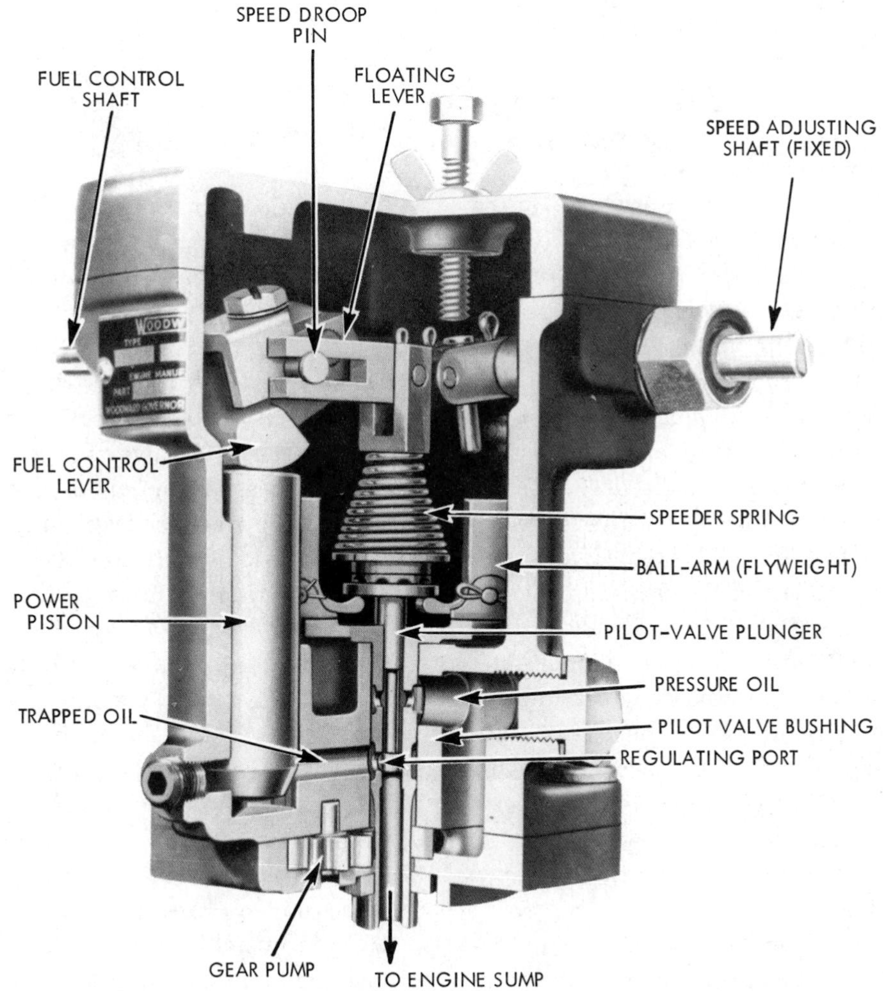

Fig. 16-16. Woodward SG governor with permanent speed-droop. (Woodward Governor Co.)

stability. The various elements are named to correspond with Figs. 16-13, 16-14, and 16-15.

Note, that the actual governor has a *single-acting* power piston. The oil pressure works on only *one* end of the piston, the bottom end. Thus oil pressure moves the piston *up*. The spring force in the engine linkage acts downwards against the power piston to move it *down* when the pilot valve plunger opens the regulating port and relieves the oil pressure.

Fig. 16-16 also shows the gear pump which puts 120 psi pressure on the oil. The ball-head is rotated by the governor drive shaft through the pilot valve bushing. The amount of speed-droop may be changed by shifting the position of the speed-droop pin in the slotted floating lever. The speed is adjustable by turning the speed-adjusting shaft; this changes the compression of the speeder spring.

Isochronous Hydraulic Governor

The hydraulic governor described in the previous article cannot maintain its engine at constant speed regardless of load because it employs *permanent* speed-droop to prevent hunting. It is not isochronous.

One kind of *isochronous hydraulic governor* is able to maintain exactly constant speed without hunting. It does this by employing speed-droop to give stability while the fuel is being corrected, and then gradually removing the droop as the engine responds to the fuel change and returns to its original speed. Thus, the speed-droop for an isochronous hydraulic governor is *temporary*. The use of temporary speed-droop to prevent over-correction of the fuel supply is called *compensation,* and it requires two actions:

1. Droop application, as the fuel supply is changed.

2. Droop removal, as the engine responds to the fuel change and returns to the original speed.

These actions can be accomplished by a combination of two hydraulic pistons and a needle valve. The droop action is applied by means of two pistons which are connected by an oil passage. The droop is removed by permitting the escape of oil from the connecting passage through the needle valve, which permits the oil to leak to a sump. Centering springs return the droop-applying piston back to its original position.

The parts required in a governor, Fig. 16-17, to apply and remove the temporary speed-droop (compensation) are:

1. A transmitting piston (*actuating piston*), to transfer the motion of the fuel-changing mechanism to a responding piston.

2. A spring-loaded responding piston (*receiving piston*), which acts on parts of the governor to cause speed-droop.

3. An adjustable needle valve in the connecting oil passage of the two pistons, which allows oil to leak off to the sump.

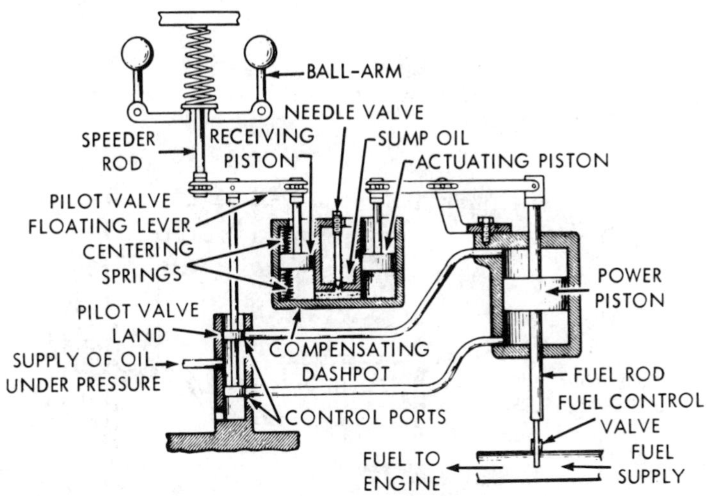

Fig. 16-17. Arrangement of isochronous hydraulic governor. (Woodward Governor Co.)

Let's again trace in principle the action of a governor—this time, an isochronous (temporary speed-droop) hydraulic governor—as the engine load changes.

Load Steady (Fig. 16-17):

1. The engine is running at normal speed under steady load.

2. The ball-arms are vertical and the pilot-valve floating lever is horizontal.

3. The control ports in the pilot-valve bushing are covered by the lands on the pilot-valve plunger.

4. The receiving compensating piston is in normal position.

5. The power piston and fuel rod are stationary. (The position shown corresponds to approximately one-half fuel.)

Load Increase (Fig. 16-18):

1. Load is applied to the engine and its speed decreases.

2. The ball-arms move in, lowering the pilot-valve plunger to admit oil pressure under the power piston.

3. Oil pressure moves the power piston up, Fig. 16-19, which pushes the actuating piston down.

4. The actuating piston displaces oil to the receiving piston and forces the receiving piston up, compressing the upper spring and lifting the pilot-valve plunger, which closes the control ports to stop further movement of the power piston.

This action is very rapid, and the small opening of the needle valve prevents appreciable leakage. Therefore, the oil displaced by the actuating piston causes a corresponding movement of the receiving piston.

5. The power piston has now moved up, increasing the fuel supply to bring the engine speed back to normal.

6. As the engine responds to the fuel change, Fig. 16-20, its speed gradually returns to the original speed, and the ball-arms gradually return to the vertical position. At the same time, the upper spring

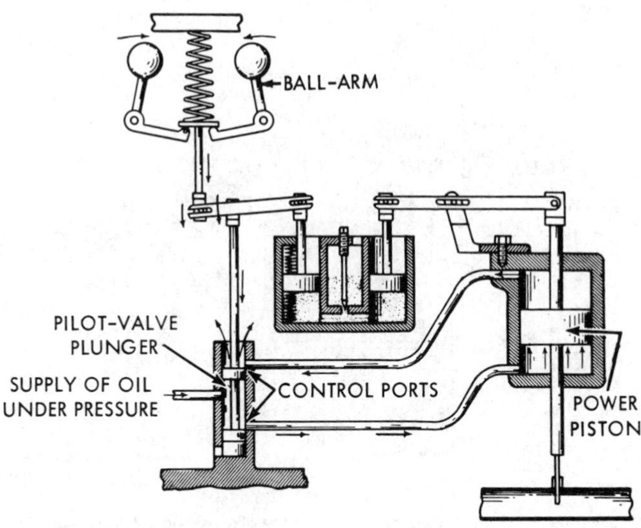

Fig. 16-18. First effect of load increase on isochronous hydraulic governor. Ball-arms move in and lower the pilot valve plunger. (Woodward Governor Co.)

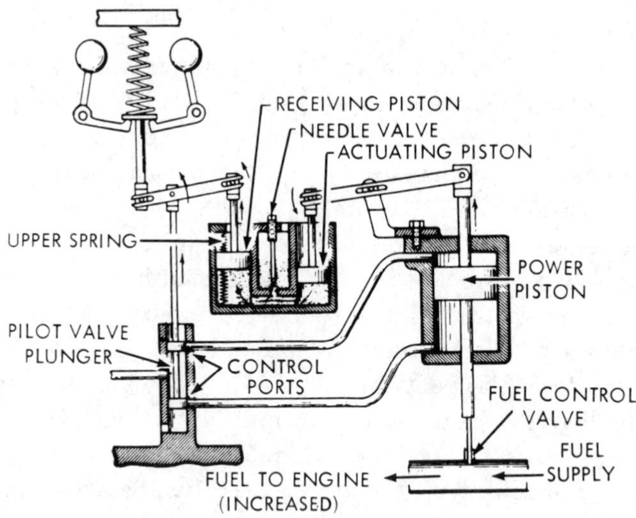

Fig. 16-19. Second effect of load increase on isochronous hydraulic governor. Power piston moves up and re-centers pilot valve plunger. (Woodward Governor Co.)

starts pushing the receiving piston down to its normal position. The floating lever then tilts about the pilot-valve pivot pin.

7. The rate at which the receiving piston moves down is determined by the adjustable opening of the needle valve. If the needle valve opening is correct, the rate of return of the receiving piston will match exactly the rate of return of the ball-arms to vertical.

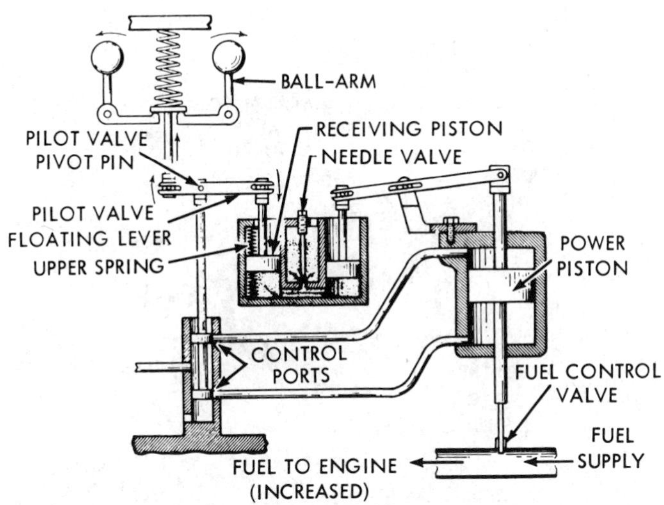

Fig. 16-20. Final effect of load increase on isochronous hydraulic governor. As engine speed rises to normal, leakage through needle valve restores pilot valve floating lever to normal horizontal position. (Woodward Governor Co.)

8. At completion of the cycle, the engine will be running at its original speed, the ball-arms will be vertical, the floating lever will again be horizontal, the control ports will be closed, the receiving piston will be back to its original position, and the power piston will be at a new position supplying fuel for increased load.

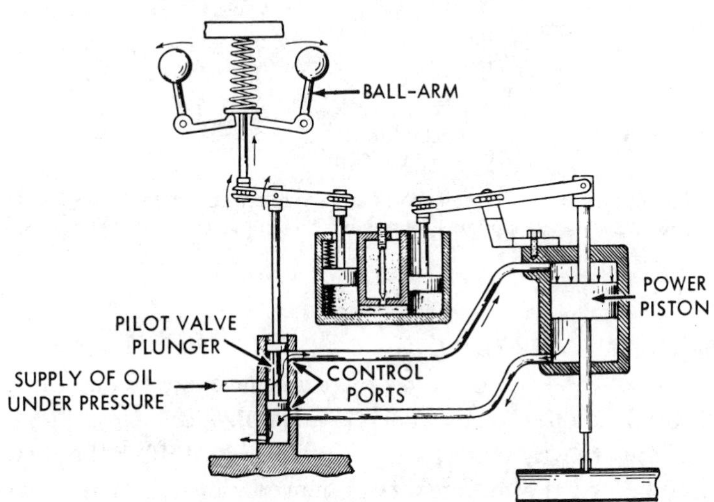

Fig. 16-21. First effect of load decrease on isochronous hydraulic governor. Ball-arms move out and lift the pilot valve plunger. (Woodward Governor Co.)

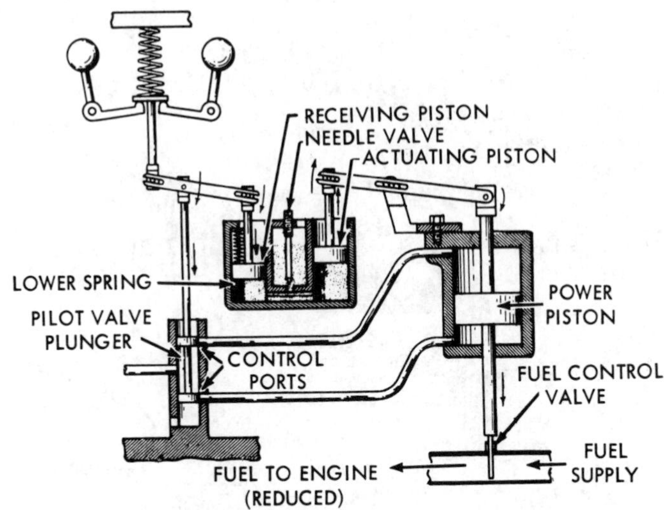

Fig. 16-22. Second effect of load decrease on isochronous hydraulic governor. Power piston moves down and re-centers pilot valve plunger. (Woodward Governor Co.)

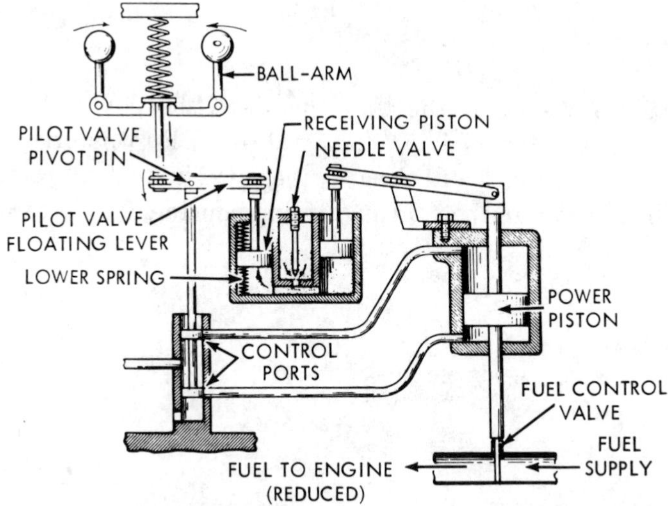

Fig. 16-23. Final effect of load decrease on isochronous hydraulic governor. As engine speed falls to normal, leakage through the needle restores pilot valve floating lever to normal position. (Woodward Governor Co.)

Load Decrease (Figs. 16-21 16-22, 16-23):

When load is dropped off, the governor goes through exactly the same steps as for load increase, but all the movements are in the opposite direction. You can follow this by studying Figs. 16-21, 16-22, and 16-23. In Fig. 16-21, the ball-arms have moved out and lifted the pilot-valve plunger.

In Fig. 16-22, the power piston has moved down to reduce the fuel supply and lift the actuating piston. This has caused the receiving piston to move down to lower the pilot-valve plunger and close the control ports.

In Fig. 16-23, oil leakage through the needle valve has permitted the receiving piston to return gradually to its original position in the same time which the engine took to return to its original speed and the ball-arms to become vertical again.

In the governor just described, the temporary speed-droop (compensation) is applied by changing the effective length of the connection between the speeder rod and the pilot valve. Changing the length of this connection causes speed-droop for the same reason that changing the compression of the speeder spring causes speed-droop. Each method of causing speed-droop changes the engine speed at which the pilot valve seals the control ports.

Strictly speaking, only isochronous governors are true *constant-speed* governors. All mechanical governors and some hydraulic governors have permanent speed-droop and are non-isochronous, because the engine steady speed is slightly different at different loads. However, the speed change from full load to no load (speed regulation) is usually only 3 to 8 percent. Since such governors hold the engine speed substantially constant, they too are classed as constant-speed governors.

Governor Modifications

The basic types of governors previously described are often modified or fitted with auxiliary devices to meet special requirements. Let's look at some of them.

Isochronous Governors with Permanent Speed-Droop

Many electric power plants contain several units generating alternating current. Each unit must take a share of the load while the power system holds a constant frequency (cycles/sec). This requires the use of isochronous (temporary speed-droop) governors fitted with a mechanism to provide adjustable permanent speed-droop. Such governors employ both temporary and permanent speed-droop mechanisms. The temporary speed-droop is always effective, in order to prevent hunting. The permanent speed-droop mechanism can be adjusted, as required, from no droop (isochronous) to about 5 percent droop in steady speed.

Fig. 16-24 shows how adjustable permanent speed-droop is added to the isochronous hydraulic governor described earlier. The speed-droop lever introduces permanent speed-droop by changing the force of the speeder spring. It works in exactly the same manner as the hydraulic speed-droop lever.

The speed-droop is adjusted by shift-

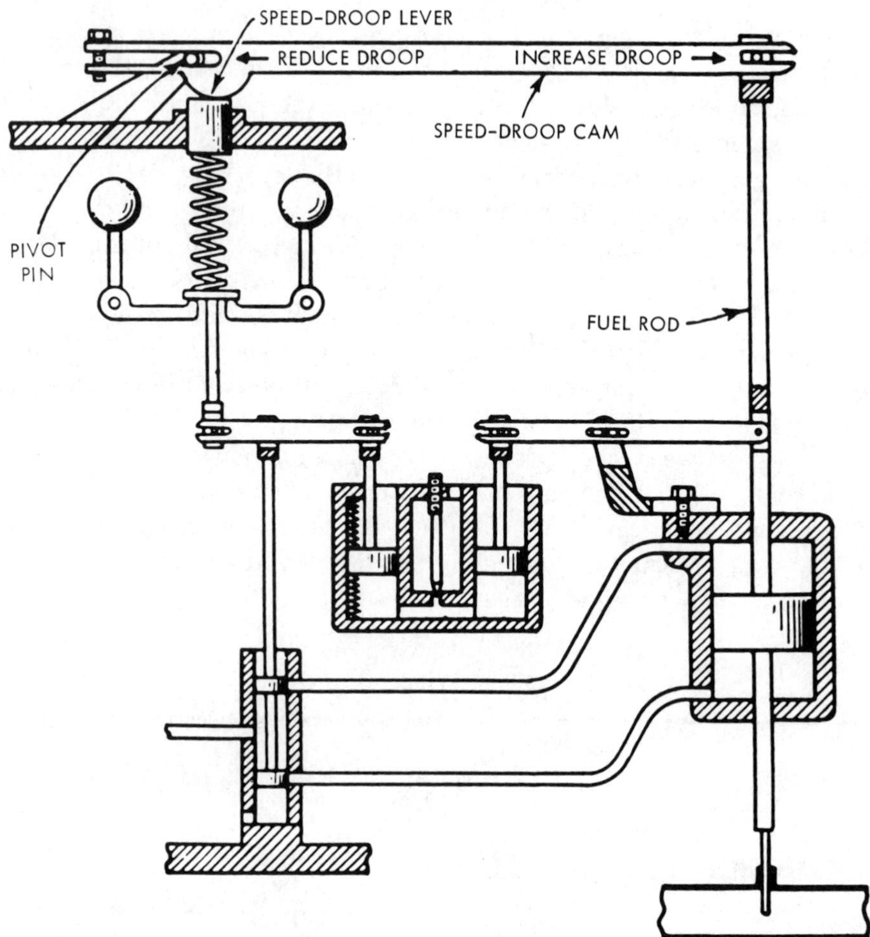

Fig. 16-24. Speed-droop lever provides adjustable permanent speed-droop in isochronous hydraulic governor. (Woodward Governor Co.)

ing the speed-droop lever endwise to change its leverage. If the lever is shifted to the right, a given movement of the fuel rod causes a greater movement of the speed-droop cam, thus a greater change in spring force and a greater speed-droop. If the speed-droop lever is shifted to the left, the cam has less effect on the spring and speed-droop is reduced. If the lever is shifted all the way to the left, the nose of the cam comes directly under the pivot pin, whereupon the cam has no effect on the speeder spring, the permanent speed-droop is eliminated and the governor becomes isochronous.

An actual isochronous hydraulic governor with adjustable permanent speed-droop is the Woodward UG8, shown in simplified cross-section in Fig. 16-25 and schematically in Fig. 16-26. A speed-droop

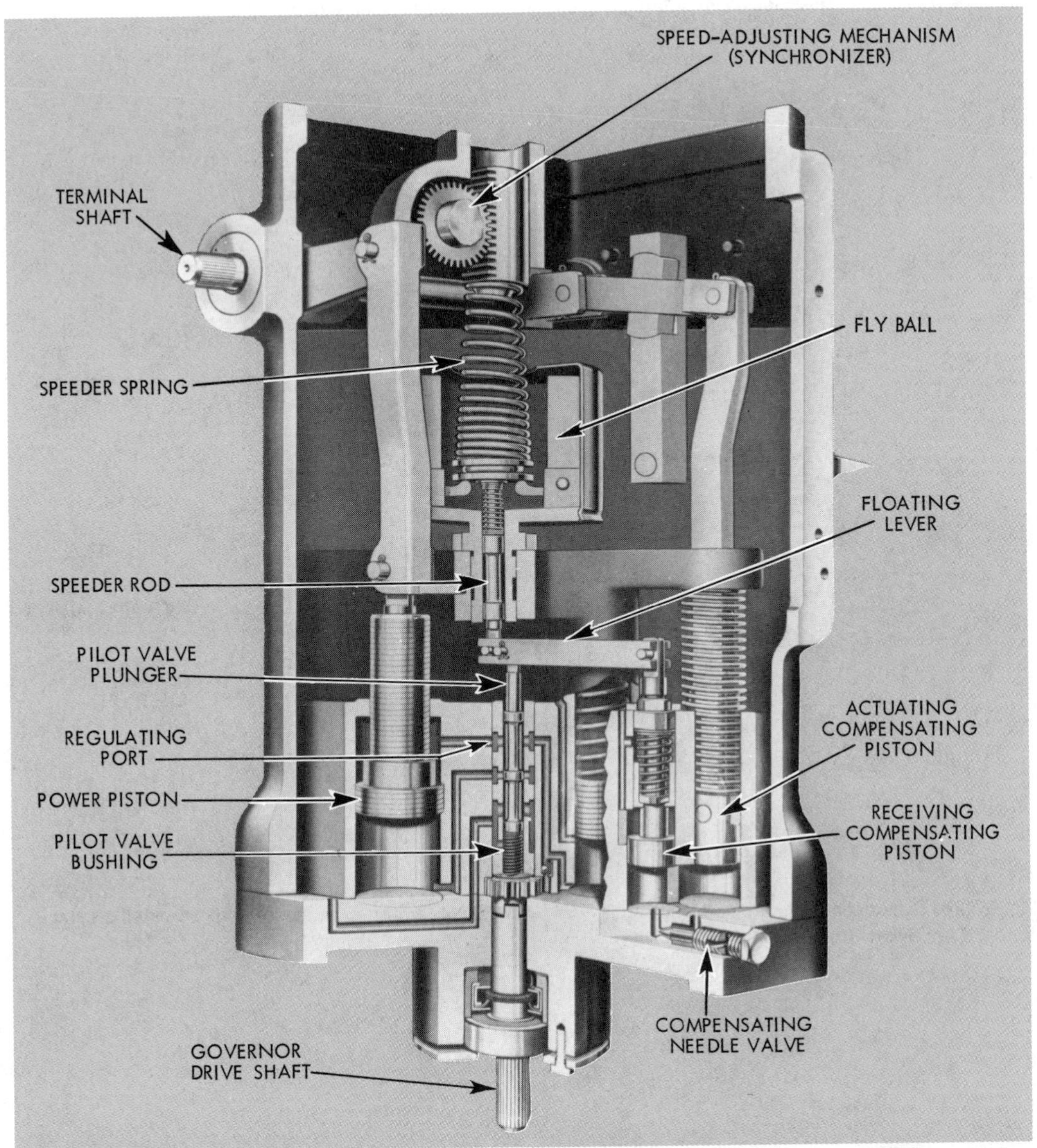

Fig. 16-25. Cross-section of Woodward UG8 isochronous hydraulic governor with adjustable speed-droop. (Woodward Governor Co.)

lever connected to the governor terminal shaft varies the compression of the speeder spring when the terminal shaft rotates in response to a load change. The effective length of this lever can be adjusted by an external knob, so that the amount of permanent speed-droop may be adjusted while the engine is running.

The floating lever changes the position of the pilot valve relative to the speeder

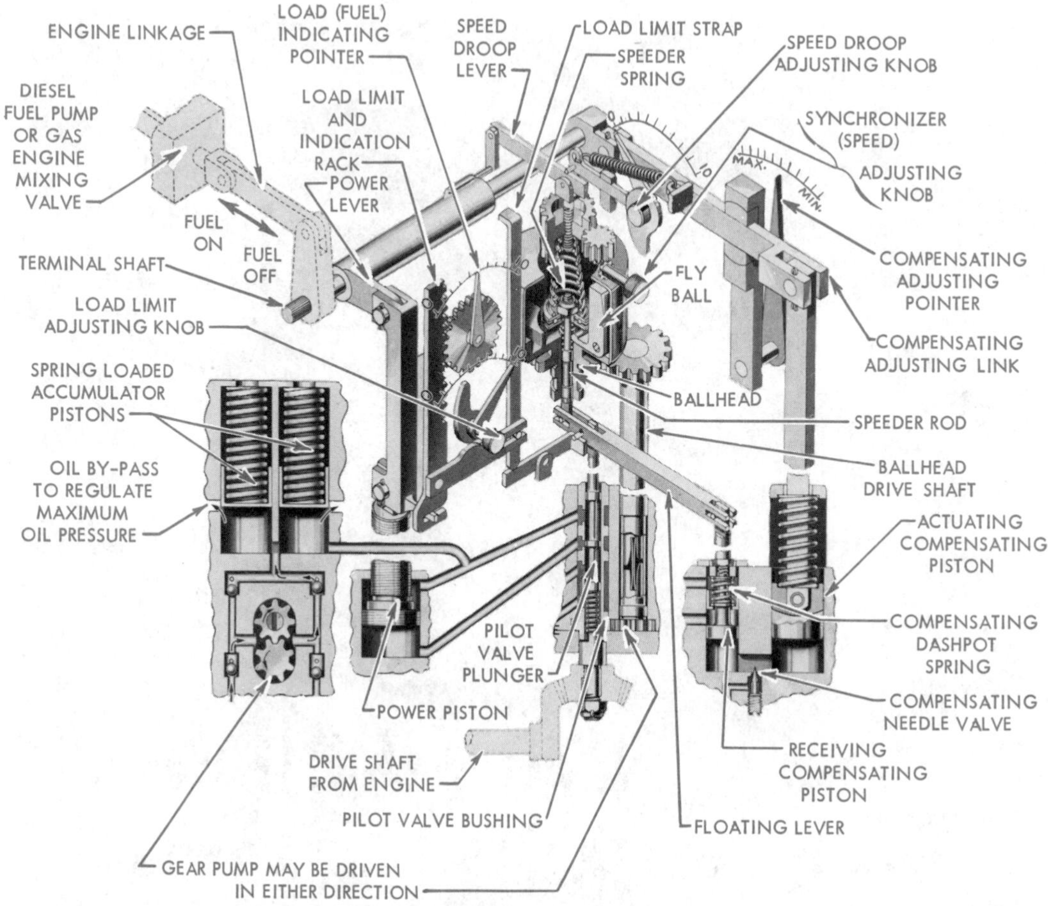

Fig. 16-26. Schematic diagram of Woodward UG8 isochronous hydraulic governor with adjustable speed-droop. (Woodward Governor Co.)

rod, and thus introduces the temporary speed-droop which makes the governor isochronous (if the permanent speed-droop is adjusted to zero). The compensating pistons and needle valve gradually remove the temporary speed-droop in the manner previously described.

Like the Woodward SG governor, the UG8 is also single-acting, but instead of using a spring to return the power piston it uses *continuous* oil pressure. The effective area of the piston's upper surface is half that of the lower surface; thus an equal amount of force is available for movement in each direction.

The more recent Woodward PG, Fig. 16-27, is a versatile type of isochronous hydraulic governor which lends itself to many modifications. It employs a compensating system which is entirely differ-

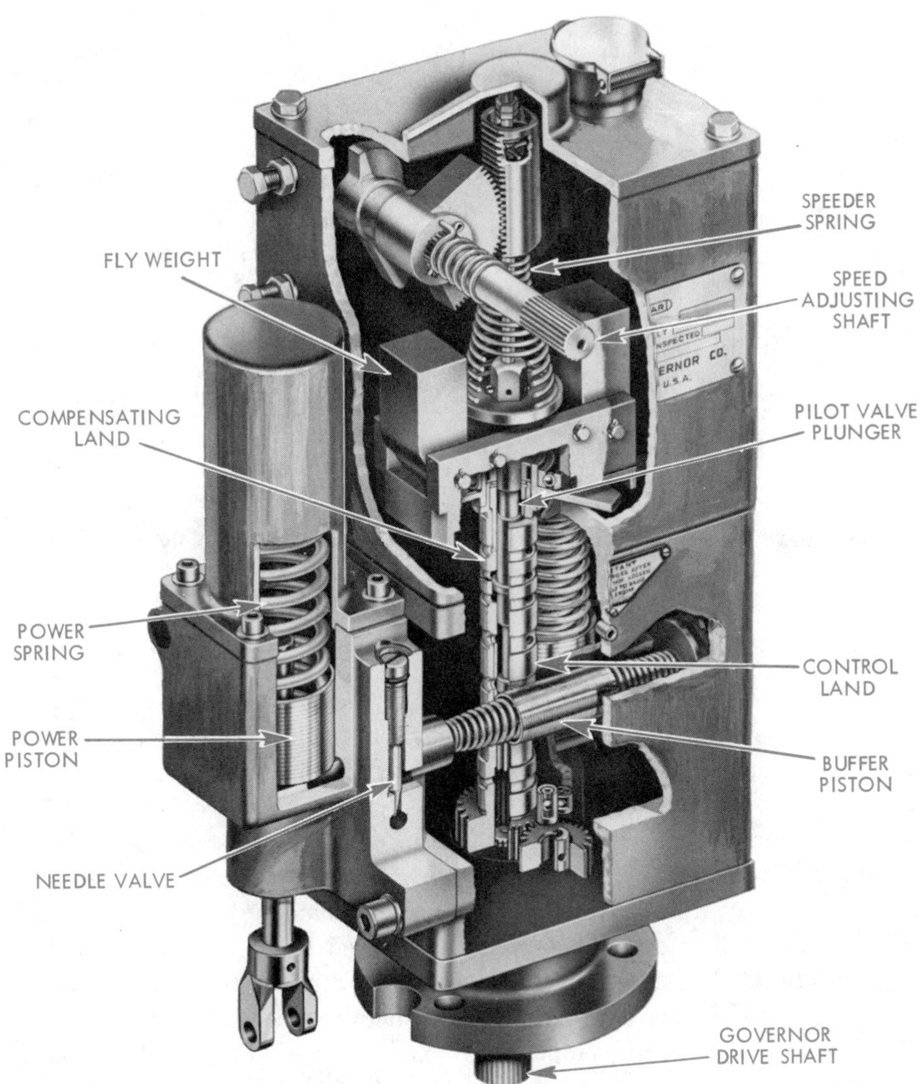

Fig. 16-27. Woodward PG isochronous hydraulic governor, pressure compensation model. (Woodward Governor Co.)

ent than the UG8 governor previously described. Instead of transmitting the movement of the power piston back to the pilot valve through a system of levers and compensating pistons, the PG does the job directly by oil pressure. This is called *pressure compensation* and is shown schematically in Fig. 16-28.

The compensating system consists of a buffer piston, buffer springs, and needle

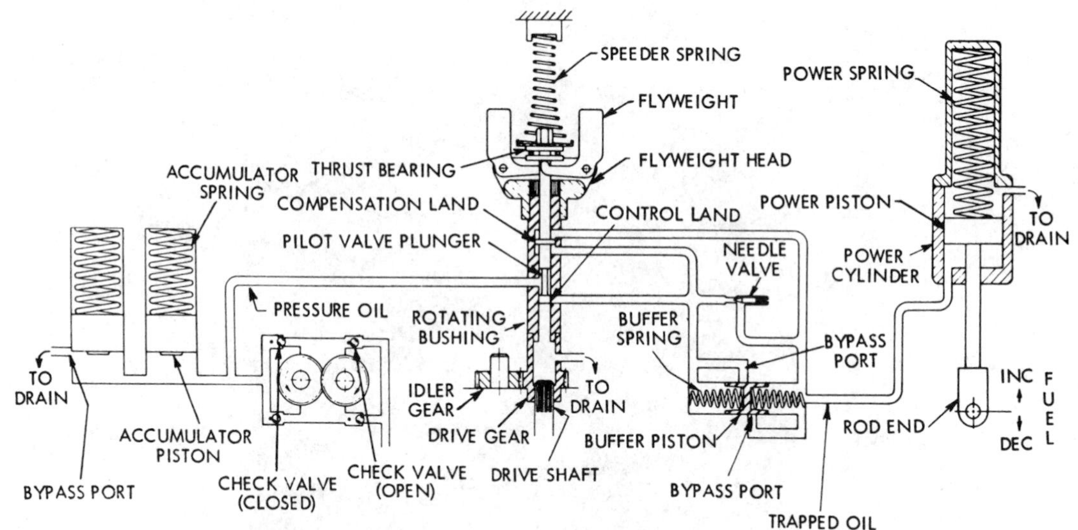

Fig. 16-28. Schematic of pressure compensation system. (Woodward Governor Co.)

valve. Briefly, here's how it works. On an increase in load, engine speed decreases, the governor flyweights move in and the speeder spring pushes down the pilot valve plunger. The control land on the plunger opens the control port, permitting pressure oil to flow into the buffer cylinder, pushing the buffer piston to the right. This in turn, forces the trapped oil on the right to lift the power piston and increase the fuel.

Since the movement of the buffer piston compressed the right buffer spring and released the left spring, the oil pressure on the left side of the piston exceeds that on the right side. These two differing pressures are transmitted to the lower and upper sides of the compensation land on the pilot valve plunger. The pressure difference produces an upward force which raises the plunger back to its centered position and closes the control port even though the engine speed is still below normal.

By the time the control port is closed, the power piston has been lifted to the position where the engine is receiving the increased amount of fuel required for the increased load; the engine speed therefore continues to increase to normal. While the speed is increasing, the gradual passage of oil through the needle valve equalizes the oil pressure on each side of the buffer piston, and above and below the compensation land on the pilot valve plunger. If the needle valve has been properly set, these oil pressures become equal at the moment the engine speed again becomes normal.

At the completion of the cycle, the engine is running at its normal speed, the buffer piston is in central position, the pilot valve control port is in closed position, and the power piston is up at a new position supplying increased fuel for the increased load.

The pressure compensating system of the PG governor increases its sensitivity and also makes it possible to add many auxiliary gadgets to adapt it for use on diesel locomotives, pipeline pumping units and other highly specialized applications.

By merely adding adjusting screws to load the buffer springs, this governor can be made to ignore small, quick changes in engine speed such as might be caused by misfiring of a spark-ignited two-cycle gas engine running at low-load or by the intermittent load of a directly connected reciprocating pump.

Variable-Speed Governors

A governor which can control the engine speed at several values between minimum speed and rated speed is called a *variable-speed governor.* It resembles a constant-speed governor in its ability to hold the engine at steady speed. It differs in that the operator can adjust the governor to maintain any one of several desired steady speeds.

Speed adjustment is obtained in most variable-speed governors by changing the spring force that resists the centrifugal force of the flyweights. You will remember how an adjustment of the spring force is used to make minor changes in the control speed of a constant-speed governor. The speed-adjusters of constant-speed governors work through a range of only a few percentage points change in speed. But the same principle is applied in many variable-speed governors to permit control speeds as low as one-sixth normal.

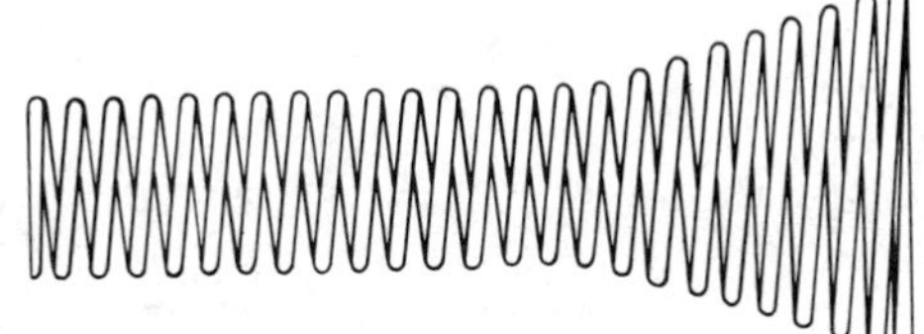

Fig. 16-29. Trumpet-shaped speeder spring.

To work well through a great range, a special trumpet-shaped or conical spring, Fig. 16-29, is needed. An ordinary cylindrical spring does not give as good control at reduced speed as at full speed. The reason is that, at reduced speed, a cylindrical spring which is stiff enough (large force per inch of compression) to match the flyweight force at high speed, is too stiff to match the greatly reduced flyweight force at reduced engine speed. The trumpet-shaped spring is so wound that at light loads (low speeds) all the turns are active and the spring is quite soft, while at high loads (high speeds) the larger turns touch each other and go out of action, leaving the smaller turns to provide a stiffer spring.

Two-Speed Mechanical Governors

A governor that controls engine speed at idling speed and at high speed by using two different springs is called a *two-speed mechanical governor.* One is a stiff spring suitable for high speed; the other is a soft spring, which is better for low speed because it provides greater sensitivity and less speed-droop. The springs may be arranged to act either singly or in combination.

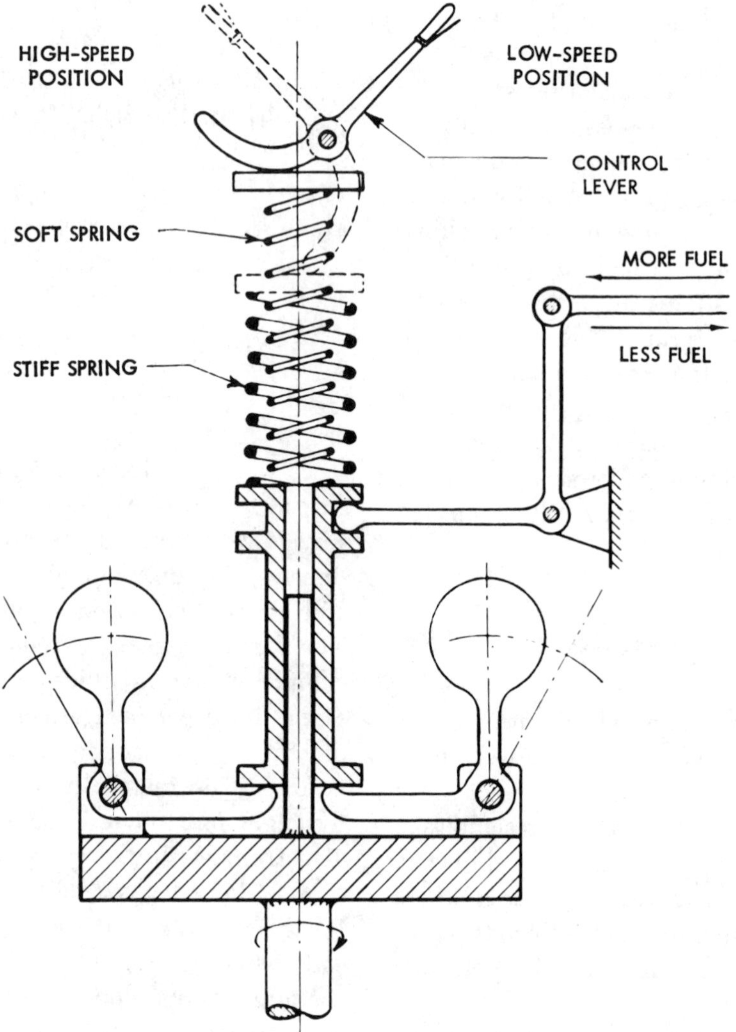

Fig. 16-30. Two-speed mechanical governor.

An example of such a governor is shown, in elementary form, in Fig. 16-30. For low-speed operation, the control lever is put in the position shown, in which only the soft inner spring is acting. For high-speed operation, the operator pushes the lever to the left, thus putting both springs into action.

Another two-speed mechanical governor used in automotive service, was described previously (Fig. 16-10). It controls the engine at idling speed and at maximum speed. The idling speed control causes the engine to run at a low *constant* speed which it would not do if there was merely an idling stop on the fuel control; this is discussed further later in this section. The maximum speed control prevents engine damage due to excessive speed.

Overspeed Governors and Trips

Overspeed governors are employed as safety devices to protect engines from damage due to overspeeding from any cause. When an engine is equipped with a regular speed governor, the overspeed governor will function only in the event that the regular governor fails to operate.

If the overspeed control merely slows the engine down, but allows it to continue to run at a safe operating speed, it is termed an *overspeed governor.* If it brings the engine to a full stop, however, it is called an *overspeed trip.*

Overspeed governors and trips of all types use the principle of resisting the centrifugal force of a flyweight by means of a spring. The spring is loaded to a force which will overbalance the centrifugal force of the flyweight until the engine speed rises above a desired maximum. When this speed is reached, the centrifugal force overcomes the spring force and puts into action the controls which slow down or stop the engine.

Overspeed governors generally act on the fuel supply to the engine cylinders, like regular governors. So do many overspeed trips. However, it is possible for an engine to overspeed on account of some kind of fuel entering the cylinders independently of the fuel injection system. For example, oil or gas might accidentally enter the engine's air intake. Some overspeed trips guard against such contingencies and shut down the engine by holding open the exhaust valves or by closing off the air supply.

Torque-Converter Governors

By means of a torque-converter, an engine's power (torque multiplied by speed) is converted to a different combination of torque and speed to suit the requirements of the load. Under certain conditions the resisting torque may be quite low. It is then possible for the torque-converter to reach an excessive and unsafe speed. Here's where the *torque-converter governor* comes in—it reduces the engine fuel, thus cutting the engine speed and thereby preventing the converter's output shaft from exceeding pre-determined maximum speed.

Fundamentally, the governor consists of two speed-governors. The first is driven by the engine as usual, the second by the output shaft of the torque-converter. The second governor leaves the fuel control to the first while the torque-converter speed is below maximum. But if for some reason the torque-converter reaches maximum speed, its governor reduces the engine fuel and keeps the converter speed from rising further.

Pressure-Regulating Governors

Engines which drive pumps in services where the pumping pressure should be kept constant are enabled to deliver power as desired through the use of *pressure-regulating governors.* Pumping units on oil pipe lines and gas pipe lines are examples of installations for such service.

An ordinary variable-speed governor can be converted into a pressure-regulating governor by adding a device which *varies the speed setting* of the governor in accordance with the pumping pressure. The pumping pressure acts on a piston or diaphragm balanced by a spring and connected to the speed-adjuster of the governor. If the pressure drops below the desired value, the resulting movement of the diaphragm moves the speed adjuster to increase the speed setting. (In some designs, the diaphragm operates a pilot valve which controls a servo-motor connected to the speed adjuster, in order to obtain more power and greater sensitivity.) As a result of the increased speed setting, the engine runs faster and the pumping pressure increases to its original value. On a rise in pumping pressure, the engine speed setting is reduced, causing the pressure to return to normal.

The Woodward PG-PL governor works on the same principle, but is arranged for remote control by an air-pressure signal. The air-pressure signal may be given either by hand or by an automatic controller worked by pipeline pressure, rate of flow, or some other quantity which is to be regulated. In the governor itself, the air-pressure signal raises or lowers a diaphragm connected to a pilot valve plunger. The pilot valve feeds oil to or from a piston which changes the compression of the governor speeder spring.

Load-Limiting Governors

Too much fuel can seriously damage an engine, due to lugging; this causes engine overheating and extreme pressure on the bearings. A governor, whether isochronous or not, responds only to speed changes. Therefore, if an engine slows down because of excessive load, the governor increases the fuel supply. To prevent excessive fuel in such cases, most engines are fitted with a fixed collar or stop on the fuel control shaft to limit the maximum amount of fuel supply, even if the governor calls for more fuel. This is called a *fuel-limiting stop.*

Fuel-limiting stops, however, have one fault. If a very heavy load is put on an engine so that the fuel supply reaches the limit, the engine speed will drop below the governor control speed. The engine will then operate at the throttle setting fixed by the stop as if it had no governor. A second effect is that, when running at reduced speed due to overload, the engine cannot develop as much power as at full speed. Thus, just when maximum power is demanded, developed power drops off.

Load-limiting governors prevent a loss of engine speed due to heavy load by limiting the *load* to the safe capacity of the engine. They cannot always be used because it is not always feasible to limit the load, but an important application of load-limiting governors is on diesel-electric locomotives. Here the load-limit device is an attachment to the governor which, when the governor terminal shaft reaches the maximum safe fuel position, operates an electric control which reduces the output of the electric generator driven by the engine. The electric control is a rheostat which reduces the generator excitation. This reduces the load on the engine enough to permit the engine to run at full speed and to develop its full safe power.

Load-limiting governors serve another important purpose where the locomotive carries a number of engines, some in better mechanical condition than others. By limiting the fuel consumed by each engine, the governors permit the locomotive to utilize the full power capability of each without danger of overloading.

Load-Control Governors

When the above scheme is carried a step further in connection with the variable-speed governors used on diesel-elec-

TABLE 16-2 PERFORMANCE OF A TYPICAL DIESEL-ELECTRIC LOCOMOTIVE USING LOAD-CONTROL GOVERNORS

OPERATING FEATURES	THROTTLE POSITION							
	1	2	3	4	5	6	7	8
ENGINE SPEED, RPM	275	350	425	500	575	650	725	800
PERCENT OF FULL SPEED	34	44	53	63	72	81	91	100
HORSEPOWER SETTING OF LOAD-CONTROL GOVERNER	40	80	210	400	600	800	1040	1280
PERCENT OF FULL-SPEED HORSEPOWER	3	6	16	31	47	62	81	100

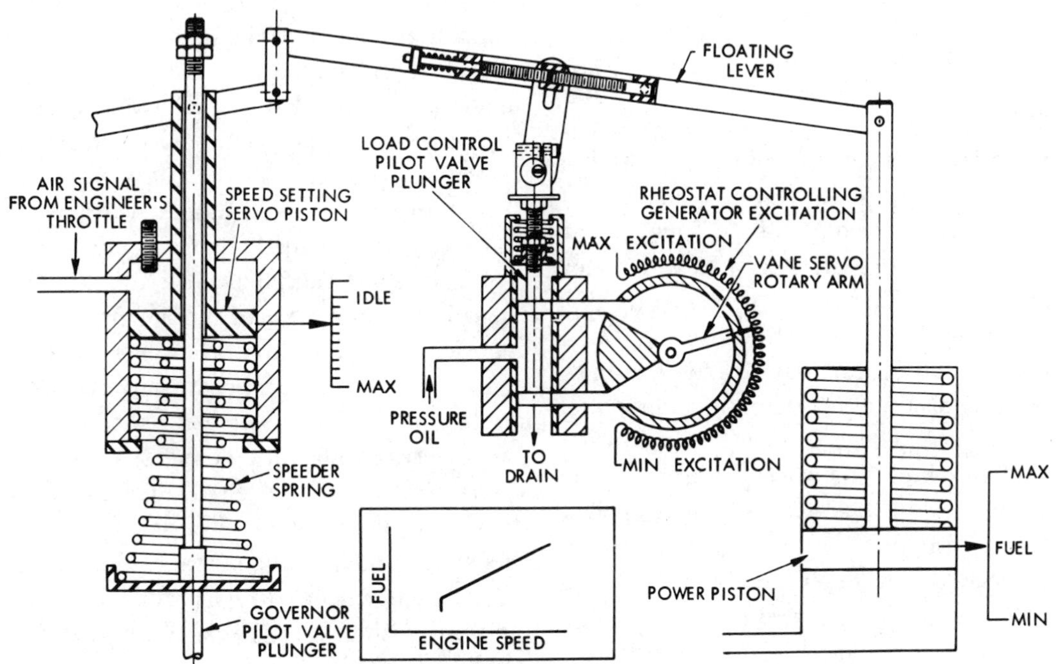

Fig. 16-31. Schematic of Woodward PG load control mechanism. (Woodward Governor Co.)

tric locomotives, we have what are called *load-control governors.* For each particular speed setting, the governor not only maintains the engine speed called for, but also adjusts the rheostat controlling the generator excitation in such manner that the amount of load applied to the engine is the amount predetermined to be suitable for the particular speed. Thus, when partial power is required, the diesel engine not only runs at reduced speed, but also injects less fuel per stroke. This greatly reduces the engine strain and maintenance cost.

Table 16-2 shows the engine speed and horsepower at the eight throttle positions of a typical diesel-electric locomotive using load-control governors.

Fig. 16-31 is a simplified schematic of the load control mechanism as fitted to the previously described Woodward PG governor (Figs. 16-27 and 16-28). An air signal from the engineer's throttle positions the speed-setting servo piston to load the speeder spring at the desired speed setting. If the actual speed differs from the desired speed, the governor flyweights move the governor pilot valve plunger to reposition the power piston to deliver more or less fuel and thus raise or lower the engine's speed.

Up to here the action has been only like that of a variable speed governor. But now the load control mechanism begins to work. Note that for each *speed* setting there is but one *fuel* setting at which the load-control pilot valve plunger is centered. The engine will run at this speed setting only if the electrical load on the generator equals the engine's power output at the corresponding fuel setting.

Here's how this governor adjusts the electrical load to equal the engine's power output. Suppose that the electrical load exceeds the engine's power; the engine will start to slow down and the flyweights will cause the power piston to rise and deliver more fuel. The rising of the power piston lifts the floating lever, which lifts the load-control pilot valve plunger. This causes pressure oil to be forced into the upper chamber of the rotary servo motor, pushing its arm down to reduce the excitation of the generator and thus decrease the electrical load. The reduced load causes the engine's speed to rise again. The governor flyweights now speed up, lowering the power piston to deliver less fuel. The reduction of the field excitation current and the engine fuel quickly balance where the engine speed has returned to the set speed and the power piston to its original position. This is the position at which the load control pilot valve plunger is again centered. Now the engine runs at the desired speed, with the electrical load matched to the fuel quantity preset for that speed.

Electric Load-Sensing Governors

These are quite different from load-control governors, which change the load to suit each engine speed. Load-sensing governors sense changes in load and adjust the fuel supply to maintain the *same* speed at the new load. They respond faster to load changes than speed governors because the load change itself works the governor mechanism *directly*. In a speed governor the load change must first cause a speed change (however small) before the governor mechanism responds.

The load-sensing element consists of an electrical network connected to the generator which the engine drives. The network measures the electrical load and delivers a small proportional current to a solenoid within the governor. The armature within the solenoid coil moves the fuel-control mechanism.

Load-sensing governors always include a speed-sensing element in addition to the load-sensing element; otherwise the engine speed would gradually float away from steady speed.

Electric Speed-Governors

One form of electric governor is the load-sensing governor just described. Another form responds to engine speed as measured by the *frequency* of the electric current produced by the engine-driven generator. An electrical network measures this frequency; when it departs from normal the network delivers an electric signal to a solenoid which operates the governor's fuel control mechanism. The frequency measuring network takes the place of the flyweights and speeder spring in non-electric governors.

How Governors Are Used

Constant-speed governors are used on engines which must maintain steady speed while supplying a varying load. Mechanical governors all have speed-droop and therefore cannot maintain exactly the same engine speed at different loads. However, the speed regulation with such governors is generally fairly close (from 3 to 8 percent). This is close enough for many kinds of service, such as driving a line-shaft in a machine-shop or driving an electric generator to supply loads which permit some variations in voltage and frequency (cps).

Speed Adjusters for Electric Generation

Governors for engines driving electric generators are always fitted with accurate *speed adjusters,* which are needed:

1. To adjust the engine speed to conform with the generator's rated speed.

2. To synchronize one alternating-current generator with another to cause the generators to produce exactly the same cycles of voltage before they are switched together or connected in parallel.

3. To regulate the electrical power output of each engine when operating in parallel. When engines are coupled together, the speed adjuster is really a *load adjuster,* because it can adjust the load on the engine without substantially changing its speed. The reason is that the electrical forces of alternating-current generators compel *all* the coupled engines to keep in step and to change their speeds in unison.

Turning the speed adjuster of one engine toward *higher speed* increases the spring force resisting the flyweight centrifugal force, pushes down the speeder rod and moves the fuel linkage toward more fuel. If the engine were running alone, this would increase its speed. But since the speed of this engine is electrically interlocked with the speeds of the coupled engines, its own speed increases only a trifle. Nevertheless, the engine *tries* to push ahead. The increased fuel and power advances the generator's *phase angle,* which in turn increases its electrical output.

Hydraulic governors are coming into more and more general use for electric generating units in order to obtain more accurate frequency control, closer voltage regulation, and better load sharing. The last depends upon the speed droops of the governors.

Governor speed-droop determines how a common load divides among two or more engines when the load changes. The common load, whether connected mechanically or electrically, compels the engines to run always at the same relative speeds. Suppose, for example, two engines having full-load speeds of 800 and 1200 rpm are running at full load, driving alternating-current generators connected in parallel, and supplying current at a frequency of 60 cps. If the load falls off and the engine speeds increase enough to raise the generator frequency 5 percent to 63 cps, *both* engines will have speeded up 5 percent. Thus, the first engine will speed up to 840 rpm (800 + 5 percent of 800) and the second engine will speed up to 1260 rpm (1200 + 5 percent of 1200).

Now let's suppose that both engines have the *same* speed regulation because both governors have the same speed-droop, 10 percent from full load to no load. In that case, both governors, running at 5 percent overspeed, will set their fuel controls for half-load $\left(\frac{5\%}{10\%} = \frac{1}{2}\right)$ and each engine will continue to carry its proper share of the common load.

But, suppose the governors have *different* speed-droops, so that the first engine has a speed regulation of 5 percent and the second engine 10 percent. If a load decrease speeds up both engines 5 percent, as before, the first engine's governor will cut off *all* its fuel and it will deliver no power, while the second engine will develop half-power. So you see that the governors of engines connected together must have equal speed-droops if each engine is to take automatically its correct share of the common load when the load changes.

Adjustable Speed-Droop

Ordinary mechanical governors have fixed and fairly large speed-droops. Consequently, if two or more engines have different makes of mechanical governors, the speed-droops will probably differ considerably. Therefore, if the engines are coupled to a common load, they will not take their proper shares of the load when it varies.

In the hydraulic governor with speed-droop the droop is adjustable. This permits adjusting the governors on all the engines for equal droop so as to obtain correct *load sharing*. Although the speed-droop type of hydraulic governor, under favorable conditions, is stable with a speed-droop as little as 1 percent, such a low droop is practicable only if the engine is running *singly*.

If coupled with other engines, good load sharing generally requires a speed-droop of 2 or 3 percent. In any event, some speed-droop must always be present, and if this type of governor is used to generate alternating current, the frequency will vary with the load. Consequently, electric clocks on the system will not keep accurate time unless the watch engineer (or some automatic device) makes frequent adjustments on the speed adjusters of all the engines.

Isochronous hydraulic governors have no speed-droop and will work well on engines running singly. However, they are unsuited to engines running in parallel (except in the special case next described), because the absence of speed-droop prevents proper load sharing, as you learned above.

Isochronous hydraulic governors with speed-droop attachments find wide use. These governors not only employ temporary speed-droop to give them stability, but also have adjustable permanent speed-droop to control load sharing. Since the permanent speed-droop can be made quite small (because it is not needed for stability), engines so equipped can be made to regulate quite closely and still share the load correctly.

Furthermore, such governors permit *parallel operation at constant frequency.* The governor of one engine is set for zero speed-droop (isochronous) and all the other governors are given equal speed-droop of sufficient amount to insure correct load sharing. All changes in load are then automatically taken by the engine with the isochronous governor, which maintains constant speed. This compels the other engines to maintain their same speeds, and consequently the electrical frequency of the plant remains unchanged. Care must be taken, however, that the engine with the isochronous governor has sufficient *capacity* to take *all* the load changes.

Electric Load-Sensing Governors

These governors find their greatest use in certain electric power plants where constant frequency must be maintained despite rapid and large changes in load. The load-sensing governor also has a unique advantage over speed-governors in the way it can divide the load among several generating units operating in parallel. The governors on all the engines can be set for the same isochronous speed. Also, the load-sensing elements can be so adjusted that the engines will always share the load among them in proportion to their capabilities, throughout the entire load range.

Variable-Speed Governors

In some kinds of service the engine operator must be able to change the engine's speed while it is running. Examples of such service are engine-driven *pumps and compressors,* subject to variable demand and therefore requiring variable output. Usually the most efficient way to vary the output of such a machine is to vary its speed.

A variable-speed governor not only permits changing the engine speed, but also keeps the engine under governor control at whatever speed the operator selects. For instance, a variable-speed governor for an engine normally rated at 800 rpm might be adjustable to hold the speed steady at any point between 200 and 800 rpm. If the engine drives a pump, the pump output at the low speed will be one-fourth that at the high speed.

Diesel locomotive engines use variable-speed governors to which load-control devices are often attached.

Automotive Diesel Governors

Three types of governors find use on trucks, buses, and taxis: two-speed, variable-speed, and torque converter. Two-speed governors perform two necessary functions: (1) limitation of maximum engine speed and (2) control of idling speed. Variable-speed governors go further; they not only limit maximum speed and control idling speed, but also maintain any desired constant road speed between these limits.

Maximum speed limit is required on an automotive diesel because a diesel can be more easily overspeeded than a gasoline engine. When the driver opens the throttle wide on a gasoline engine, and the speed increases, the pressure drop through the carburetor increases rapidly. This reduces the engine's torque and so retards the acceleration. On the other hand, the diesel engine, on wide-open throttle, continues to inject the same amount of fuel per stroke as the speed increases, and therefore may quickly reach a damaging speed.

Idling speed controls are necessary with diesels. While a gasoline engine will idle at approximately constant speed against a fixed fuel stop, most diesels will not. The reason is that the gasoline engine is naturally self-regulating, because the amount of fuel mixture entering the cylinders depends on *time* as well as on throttle position. At a fixed throttle position, if the engine begins to slow down, more fuel enters per stroke and the engine speeds up again. But if it speeds up above the desired idling speed, less mixture enters per stroke and the engine slows down.

This regulating action does not exist on a diesel engine. If we place an idling stop on the diesel pump-control shaft and set it for a low no-load speed, the fuel delivered per stroke will decrease slightly (because of pump leakage, etc.) if the speed decreases. This may cause the engine to stall. To prevent stalling, we would have to set the idling speed higher, much higher proportionately, than in a gasoline engine. But a higher idling speed would be objectionable. It will not only waste fuel but will also spoil the braking effect of the engine while drifting. *Drifting* is what happens when the driver has taken his foot off the control pedal while slowing down or while descending a hill.

But by putting the idling speed under the control of a governor, we can get steady idling at a lower speed because the governor automatically supplies more or less fuel to keep the engine running at the desired constant speed. The lower idling speed reduces fuel consumption and improves the braking effect. Also, since no fixed fuel stop is used, the driver can *completely cut off the fuel* when the vehicle is drifting above the idling speed, thus further increases the braking effect and reduce fuel consumption when descending a hill.

Variable-speed governors for automotive diesels govern the engine speed over the complete speed range of about 6 to 1. The driver's foot pedal connects to the governor speed-adjuster springs, and at each position of the foot pedal the engine runs at a certain corresponding speed. When the driver wants to change speed, the governor alters the fuel flow *gradually* while the engine speed is changing. This results in better combustion, because it feeds the proper amount of fuel for each speed.

Torque-converter governors used on diesel-driven equipment fitted with torque-converters prevent the output shaft of the converter from running away if something unusual happens. For example, this might occur on a diesel-driven hoist if the hoisting cable broke while the engine was working hard at lifting the load.

Checking On Your Knowledge

The following questions give you the opportunity to check up on yourself. If you have read the chapter carefully, you should be able to answer the questions. If you have any difficulty, read the chapter over once more so that you have the information well in mind before you go on with your reading.

DO YOU KNOW

1. What is the purpose and principle of a speed governor?
2. What is the relationship between a flywheel and a governor and how do they work together?
3. Identify and explain the purpose of nine other types of governors.
4. How does a speed governor measure the speed and operate the throttle?
5. How do mechanical and hydraulic governors differ?
6. Define the terms *speed regulation, speed droop, isochronous, hunting, dead band* and *work capacity.*
7. What are the advantages and disadvantages of mechanical governors?
8. What is the principle upon which hydraulic governors operate?
9. What are the advantages and disadvantages of a permanent speed droop hydraulic governor?
10. How does a pressure-regulating governor operate?
11. Explain the use of speed adjusters for electric generating units.
12. Explain the three automotive diesel governors, their purpose, and principle of operation.

Chapter

17

Diesel Engine Testing, Instrumentation and Design

In Chapter 8 some of the basic terms and principles of engineering related to diesel engines were introduced. A study of the measurable characteristics of gases—pressure, volume, and temperature, and their relationship during the combustion process in the engine were included in Chapter 9. The properties of diesel fuels were investigated in Chapter 10. The purpose of Chapter 11 was to make clear to you how the amount of power which an engine produces and the amount of fuel it burns are related. All are involved in engine testing and design.

Another purpose of this chapter is to assist you in understanding what testing instruments are used in evaluating engine design, how the instruments are used, and what are the results of such testing. Examples of testing problems are included.

Engine Testing—Torque, Horsepower, and Loading Devices

Since Dr. Rudolph Diesel first proposed the compression-ignition engine in 1892, many changes in design and materials have taken place. Although the diesel engine produced today is one of the most durable and efficient portable power plants available, progress concerned with improvement and the development of a better engine goes on continually.

Some of the principal reasons for a continuous research and development program are: (1) a diesel engine is still far from a perfect engine (35 percent thermal efficiency), (2) the weight per horsepower is quite high, (3) the diesel engine affects our environment (noise and harmful emissions), (4) tests must be evaluated to determine engine operating information (for example: at what speed should the engine be run to obtain the best efficiency and engine life), (5) after failures in the field, testing of engines and parts must lead to re-design (field service representatives search for reasons for failures and report their findings to the research and development division), and (6) comparison with competition leads to other testing.

Engine Dynamometer Test Cells

The test consoles in Fig. 17-1 have two sets of instrumentation and controls. Two different engines are set up in their individual test cells as seen through the protective glass. The test consoles contain the dynamometer controls, precision instrumentation for measuring torque, speed, turbocharger rpm, temperatures, and manifold pressures. They have, as well, the more conventional gages for indicating such details as oil pressure, fuel temperature and water pressure to the dynamometer.

Fig. 17-2 shows an engine installed in a *test cell.* This engine is connected with twin intake and twin exhaust systems and has a heat exchanger in the front of it for waterjacket temperature control. The heat exchanger permits testing over a wide range of jacket-water temperatures.

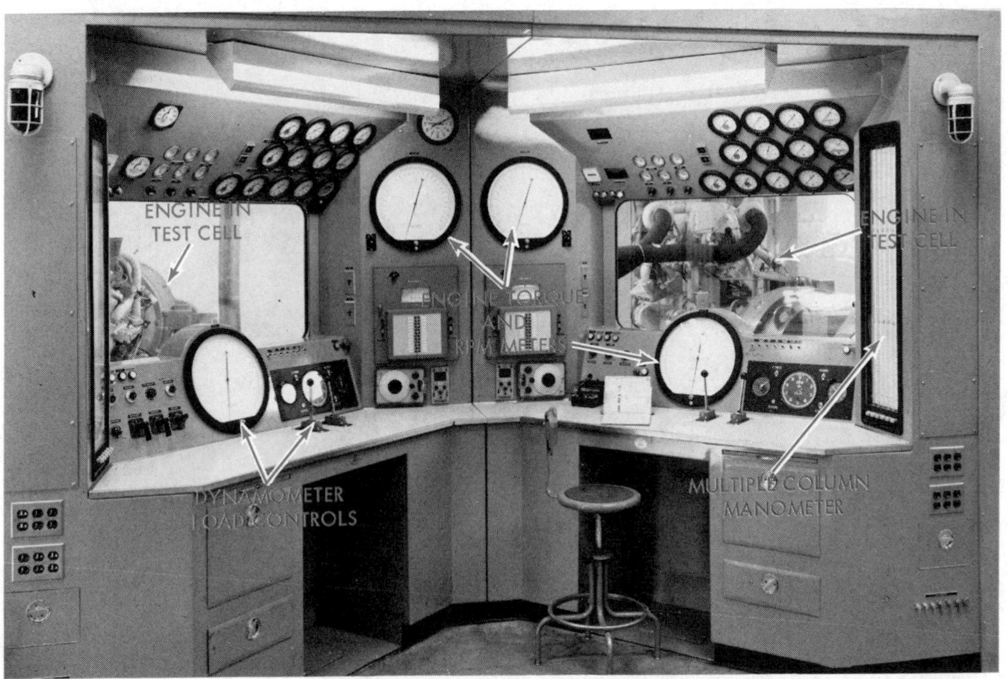

Fig. 17-1. Twin test consoles for controlling and obtaining data from the two diesel engines observed through safety windows. (Allis-Chalmers, Engine Div.)

Fig. 17-2. V-type diesel engine set-up in one of the dynamometer test cells behind the test console of Fig. 17-1. (Allis-Chalmers, Engine Div.)

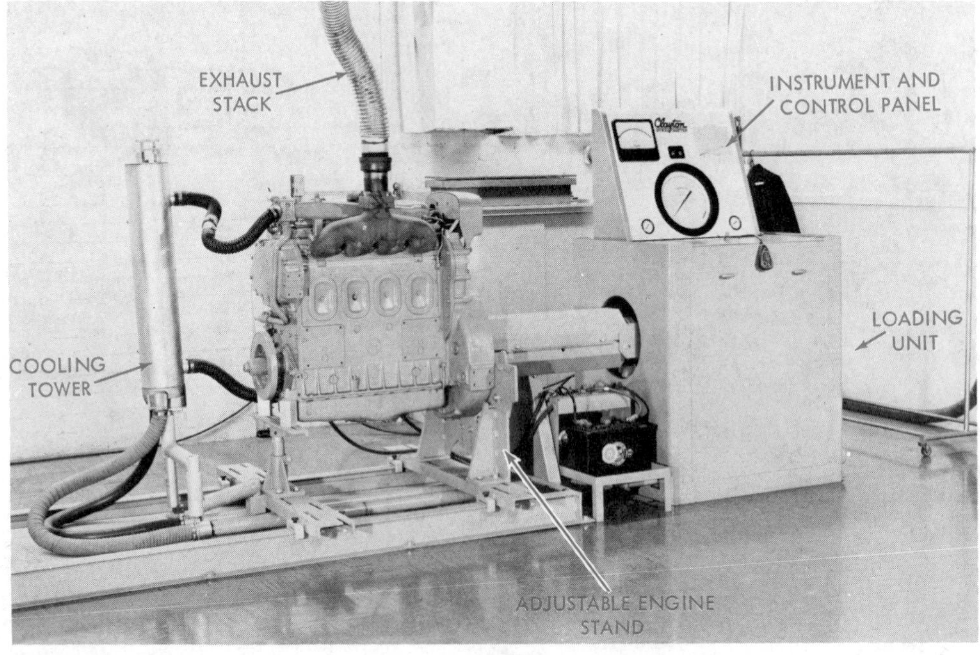

Fig. 17-3. Side view of hydraulic type engine dynamometer with engine installed, ready to run. (Clayton Manufacturing Co.)

Fig. 17-3 shows a side view of a Clayton hydraulic-type engine dynamometer which is used for research as well as for running-in new and rebuilt diesel engines. This model is made in 200, 300, 500 and 700 horsepower absorption capacities.

The dynamometer in Fig. 17-3 has a fast-change adjustable engine stand and a cooling tower (heat-exchanger) for accurate control of the engine water temperature.

Measurement of Engine Torque and Horsepower

To measure engine torque and horsepower a dynamometer is required.

There are several kinds of dynamometers. These may be divided into (1) chassis type, which measures rear wheel horsepower or horsepower available to push the vehicle down the road; (2) engine dynamometers which measure only the horsepower at the flywheel end of the engine, and (3) pressure take-off (PTO) dynamometers which are used mainly on farm tractors and measure the horsepower at the power take-off shaft of a tractor.

Another way of dividing types of dynamometers is by the method by which power is absorbed. The two major groups here are the hydraulic type, which usually uses a water-absorption unit, and the electric dynamometer, which can operate as a generator to absorb the power output of the engine or as an electric motor which can drive the engine at various speeds to measure friction losses.

The *Chassis Dynamometer* is used both in service testing, and in research and development. Most chassis dynamometers use an hydraulic type absorption unit.

The absorption unit is quite similar to a centrifugal pump as shown in Fig. 17-4. The rotor contains vanes or blades which throw water outward toward the rim of the rotor. The rotor is connected to the drive roller of the dynamometer.

Unlike the centrifugal pump, however, the absorption unit's outside housing (or stator) which encloses the rotor also has vanes or blades. The vanes of the stator oppose the flow of rapidly moving water and, since the stator revolves on its own bearings and is completely separated from the rotor, the shearing of the water between the two parts will cause the stator to turn. It cannot turn much, however, because a torque arm is connected to the stator which will apply pressure to a torque bridge. The bridge records the pressure electrically, which is then reflected on the horsepower gage in units of rear-wheel horsepower.

A schematic diagram of a typical chassis dynamometer is illustrated in Fig. 17-5. Other instrumentation is visible in the schematic, such as a speed meter for measuring rear wheel rpm. It is operated from the idle roll which drive an electric tachometer generator.

Other parts are necessary, for when power is absorbed, heat is generated. To dissipate the heat the water in the power absorption unit is circulated through a heat exchanger which is controlled at a maximum temperature by two solenoid operated valves. Double sets of rollers or bogies are available for diesel trucks with tandem drive axles. Fig. 17-6 shows a single axle tractor being tested on a dynamometer equipped with bogie rolls and an inertia flywheel.

The inertia flywheel simulates actual on-the-road driving conditions and is nec-

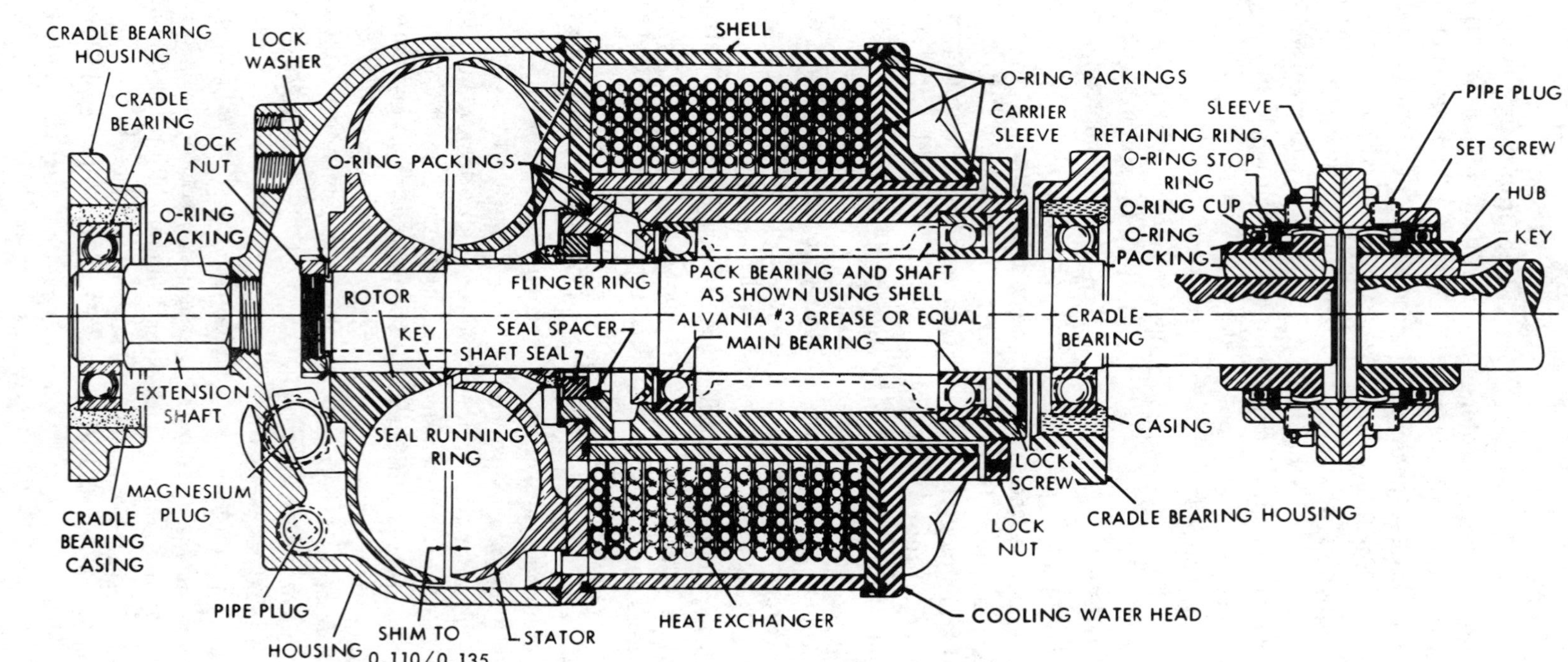

Fig. 17-4. Cutaway view of absorption unit for hydraulic type chassis dynamometer. (Clayton Manufacturing Co.)

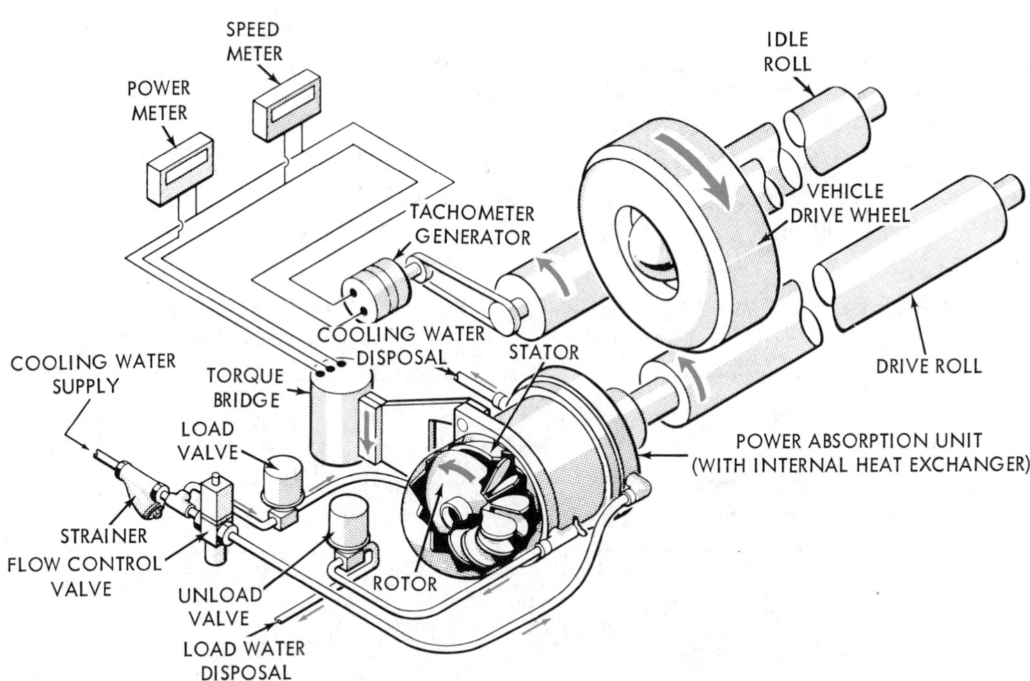

Fig. 17-5. Schematic diagram of typical hydraulic type chassis dynamometer. (Clayton Manufacturing Co.)

Fig. 17-6. Chassis dynamometer testing a diesel powered truck. (Clayton Manufacturing Co.)

essary for checking shift points on automatic transmission vehicles.

There are numerous tests which can be made by utilizing the chassis or engine dynamometer. Tests may be made at variable or constant speeds.

The variable speed tests may be run at full-load to check maximum horsepower and fuel consumption at various speeds, or it may be run at part-load for various reasons. Constant speed tests are concerned mainly with finding fuel consumption at different loads and for evaluation of the effectiveness of the fuel injection system.

Fig. 17-7 illustrates torque and horsepower curves for a normally aspirated engine. Notice the torque drops off sharply from 1400 rpm and horsepower has not peaked at a governed speed of 2800 rpm.

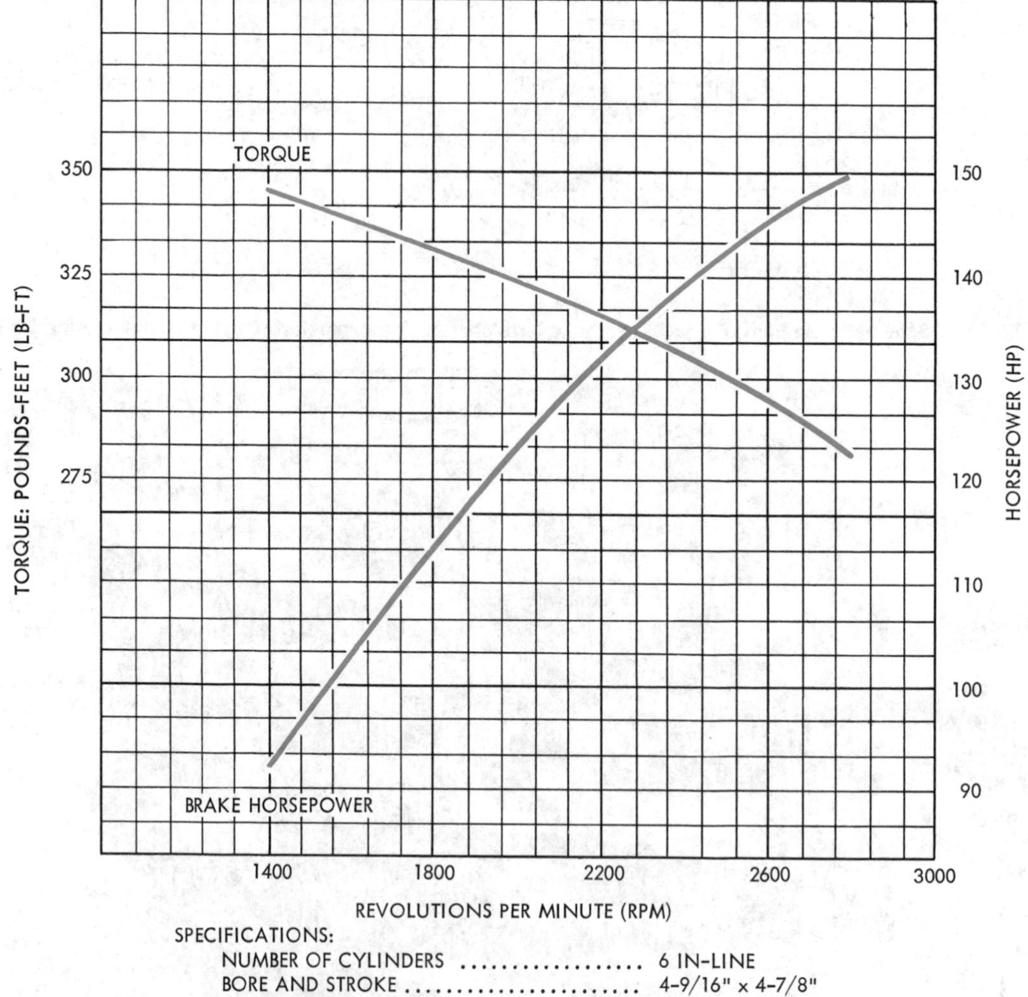

SPECIFICATIONS:

NUMBER OF CYLINDERS	6 IN-LINE
BORE AND STROKE	4-9/16" x 4-7/8"
DISPLACEMENT (CU IN)	478
MAX GOVERNED RPM	2800
BASIC ENGINE HP AT GOVERNED RPM	150
BASIC ENGINE TORQUE (FT-LB)	345 @ 1400

Fig. 17-7. Torque and horsepower curves for a normally aspirated engine.

Fig. 17-8 shows the same engine as Fig. 17-7 with the exception that it is supercharged. Notice the torque peaks at about 1800 rpm and the horsepower is reaching its peak at 2800 rpm.

Compare the torque and horsepower values of this engine as a normally aspirated version and as a supercharged version.

From the various kinds of torque and horsepower tests, the manufacturer will usually rate the engine at *gross*, *intermittent* and *continuous duty* ratings.

These curves are identified to protect

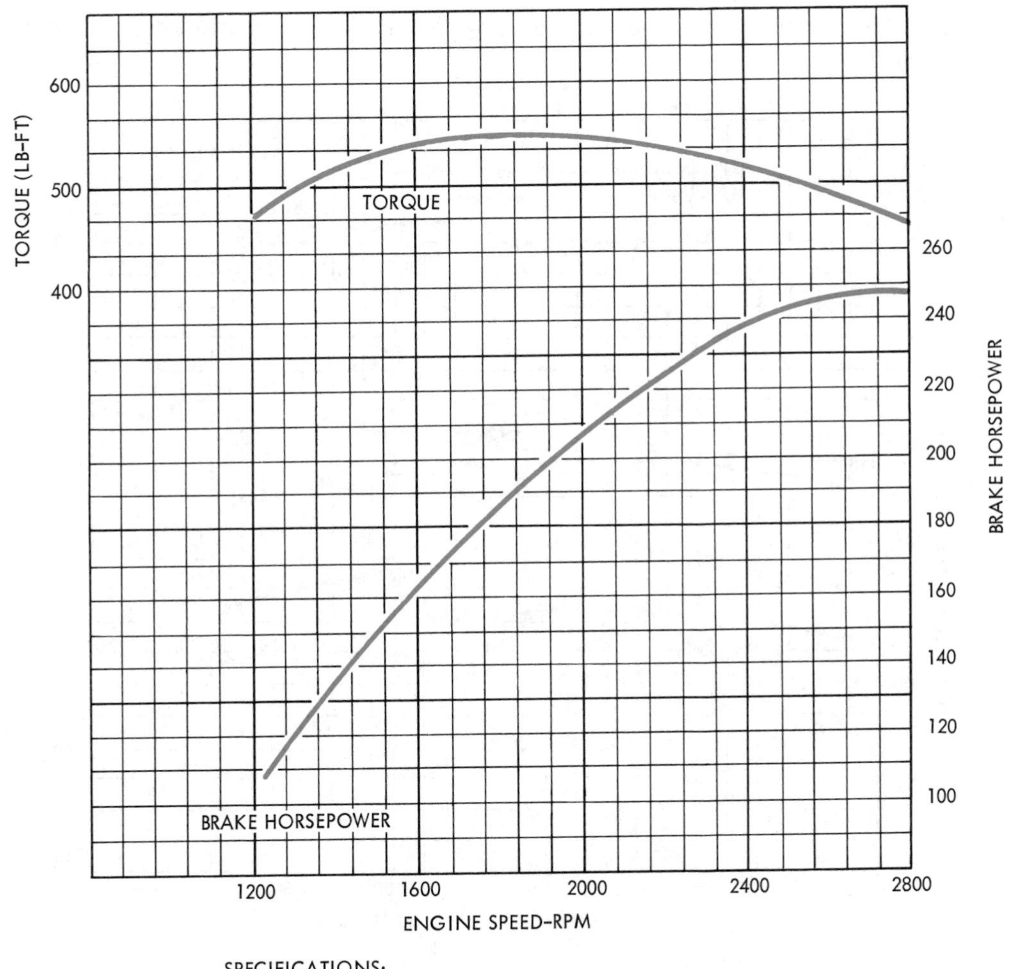

SPECIFICATIONS:

NUMBER OF CYLINDERS .... 6 IN-LINE
BORE AND STROKE .... 4-9/16" x 4-7/8"
DISPLACEMENT (CU IN) 478
MAX GOVERNED RPM 2800
BASIC ENGINE HP AT GOVERNED RPM... 250
BASIC ENGINE TORQUE (FT-LB) .... 550 @ 2000

Fig. 17-8. Torque and horsepower curves for same engine as in Fig. 17-7, but supercharged.

the engine from being run at maximum and shortening the expected life of the engine. Fig. 17-9 illustrates the three curves. Generally the continuous duty curve will be 60 to 70 percent of the gross curve.

Measuring Heat Rejection to the Cooling System

One example of a test which utilizes the engine being run at certain loads and speeds on the dynamometer is the measurement of heat rejection to the cooling water. Fig. 17-10 illustrates an approximate layout of the heat exchanger, the thermocouples, and the flow meters. The exception is that there is no expansion tank—just a vent pipe in the engine cooling system at the water outlet of the engine, and the heat exchanger is insulated. The heat exchanger is selected to produce an approximate temperature rise

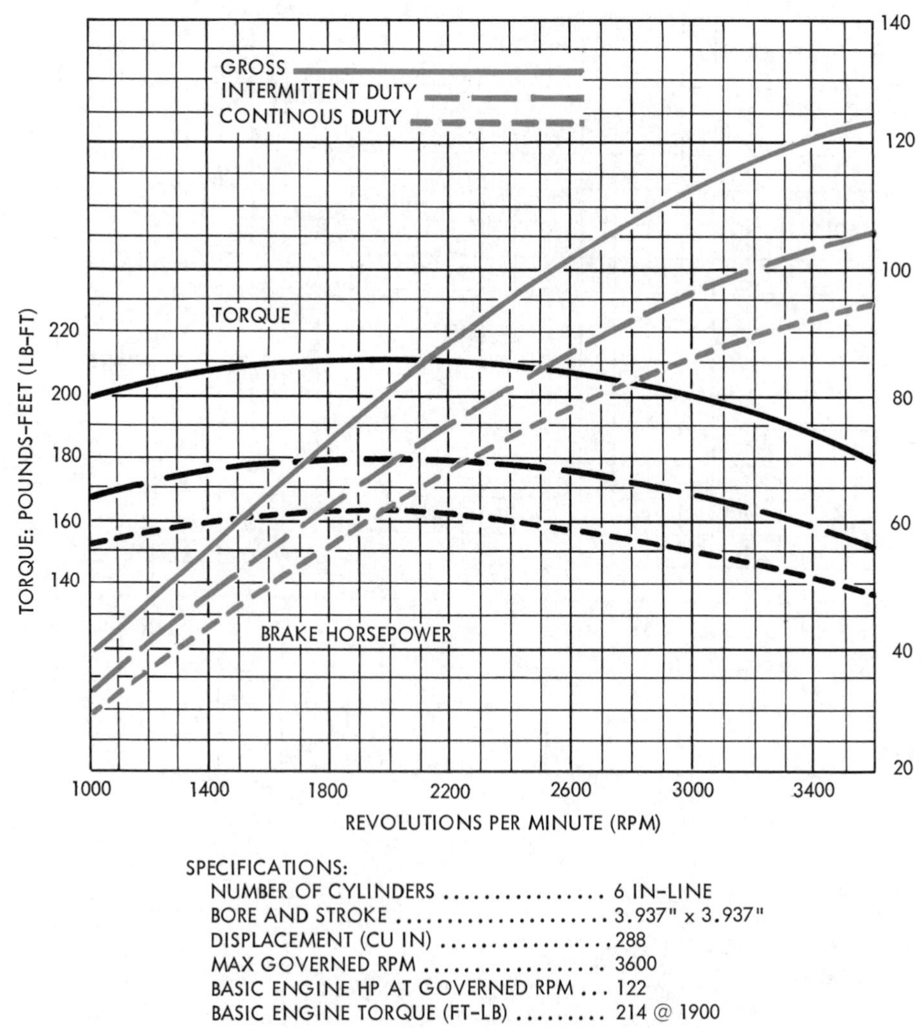

Fig. 17-9. Curves of gross, intermittent, and continuous duty horsepower ratings.

of 50°F on the raw water, based on an engine heat rejection of 27 to 37 Btu/bhp-min and an engine outlet temperature of 190 to 200°F.

The procedure includes: (1) connecting the heat exchanger to the engine and cooling tower to raw water supply; and (2) installing the instrumentation at points shown in Fig. 17-10. The heat exchanger and water pipe are insulated, usually with an asbestos covering. (3) The engine thermostat must be locked open and the by-pass line plugged (if one is used). (4) Run the engine at the load and speed at which the heat rejection is to be measured.

Adjust the raw water flow so the engine coolant outlet flow stabilizes at 190 to 200°F. (5) Take the flowmeter and inlet and outlet temperature readings for both the engine coolant and the cooling tower circuits. (6) Calculate the heat in-

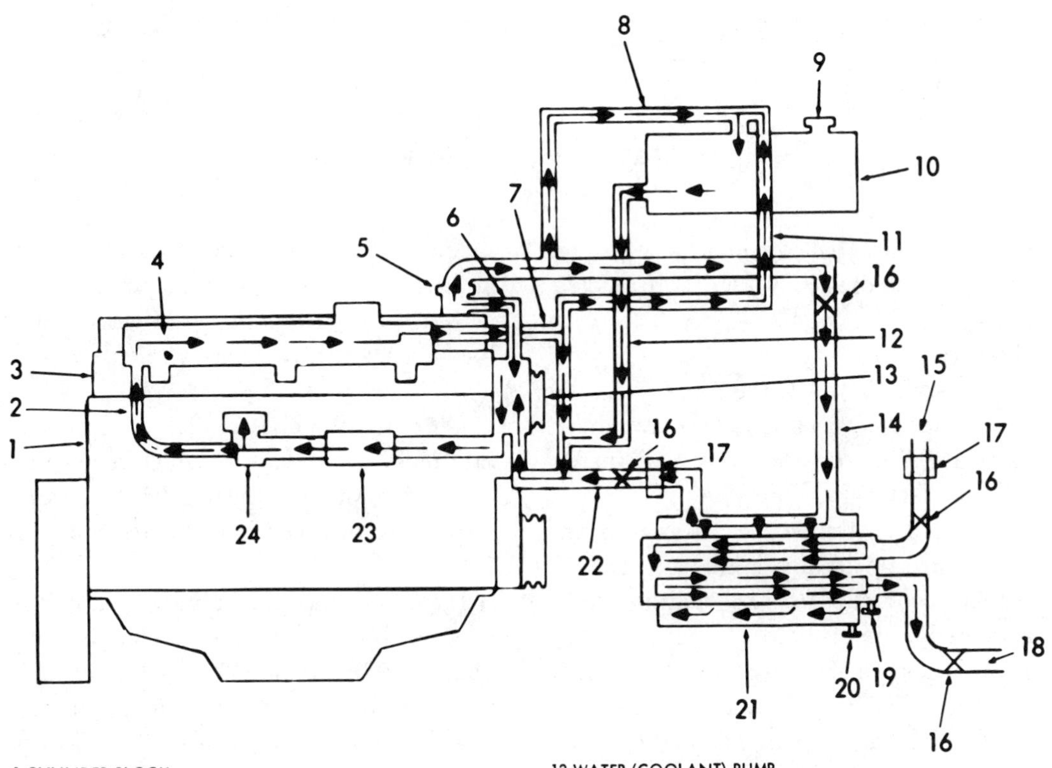

1 CYLINDER BLOCK
2 PIPE (COOLANT INLET TO EXHAUST MANIFOLD)
3 CYLINDER HEAD
4 WATER COOLED EXHAUST MANIFOLD MANIFOLE
5 THERMOSTAT HOUSING
6 BY-PASS HOSE
7 PIPE (COOLANT OUTLET TO PUMP INLET)
8 VENT LINE (EXHAUST MANIFOLD TO EXPANSION TANK)
9 FILLER CAP (VENTED)
10 EXPANSION TANK
11 VENT LINE (COOLANT OUTLET PIPE TO EXPANSION TANK)
12 PIPE (COOLANT RETURN TO PUMP INLET)
13 WATER (COOLANT) PUMP
14 PIPE (THERMOSTAT HOUSING TO HEAT EXCHANGER INLET)
15 WATER FROM COOLING TOWER
16 THERMOCOUPLES
17 FLOWMETER
18 WATER DISCHARGE -- TO COOLING TOWER
19 WATER DRAIN COCK
20 ENGINE COOLANT DRAIN COCK
21 HEAT EXCHANGER
22 PIPE (HEAT EXCHANGER TO ENGINE INLET)
23 OIL COOLER
24 COOLANT TO BLOCK INLET ELBOW

Fig. 17-10. Heat exchanger and instrumentation for measuring heat rejection to cooling water. (Allis-Chalmers, Engine Div.)

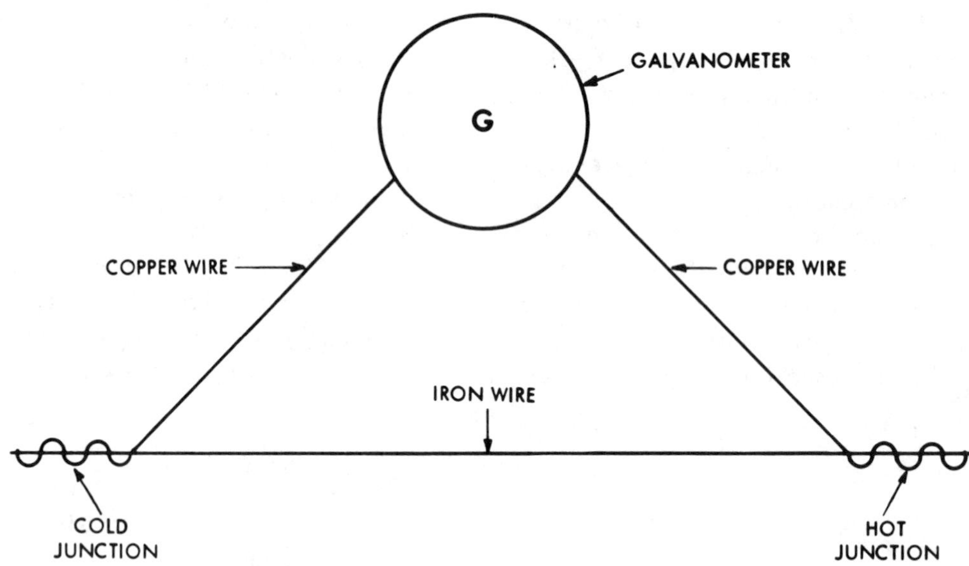

Fig. 17-11. Thermocouple circuit principle.

put and heat output of the heat exchanger. Theoretically, the two calculations should be equal.

Thermocouples

Although thermocouples are explained in a general way earlier, a thorough investigation of this important test instrument is necessary.

If two such unlike metals as iron and copper are physically joined together in a thermocouple, and two thermocouples arranged in an electrical circuit as in Fig. 17-11, they can be used to indicate temperature. If heat is applied to the right hand joint while the other remains cold, a small electrical voltage will be generated. This can be read on an accurate galvanometer which is calibrated in degrees temperature. The greater the temperature difference between the hot and cold junctions, the greater the voltage generated.

The unlike metals used for one common thermocouple is constantin and iron. *Constantin* is an alloy of copper and nickel and has a temperature range of approximately 100°F to 2000°F.

A constantin and copper thermocouple has a temperature range of about 0°F to 600°F.

Chromel which is an alloy of chromium and nickel, and *Alumel* which is an alloy of aluminum and nickel, have a broad temperature range from about 100°F to 2600°F.

Platinum and platinum-rhodium have a very high temperature range, from about 300°F to 3000°F.

A typical thermocouple installation is one described previously in this chapter, to measure heat rejection in an engine cooling system. In this test Constantin and copper would probably be used. The hot junction in this test would be inserted into the water and the cold junction would be in the galvometer. Connecting wires from the thermocouple to the instrument are usually insulated with an asbestos covering.

Pressure Measurement

A knowledge of the cylinder pressures under operating conditions is a necessary prerequisite for determining the mean effective pressure (mep) and *indicated horsepower* (ih) of diesel engines. A practical engineer or technician can use a mechanical indicator for checking combustion conditions of a diesel engine. Operating factors affecting firing efficiency, such as point of firing, ignition lag of the fuel, rate of pressure rise after ignition occurs (burning rate), and compression and maximum firing pressures can be evaluated.

The mechanical indicator for high-speed engines in Fig. 17-12 has a Chronomatic drum that holds an attached band of paper upon which a complete change of pressure on a time base is recorded (pressure-time diagram) for a full engine

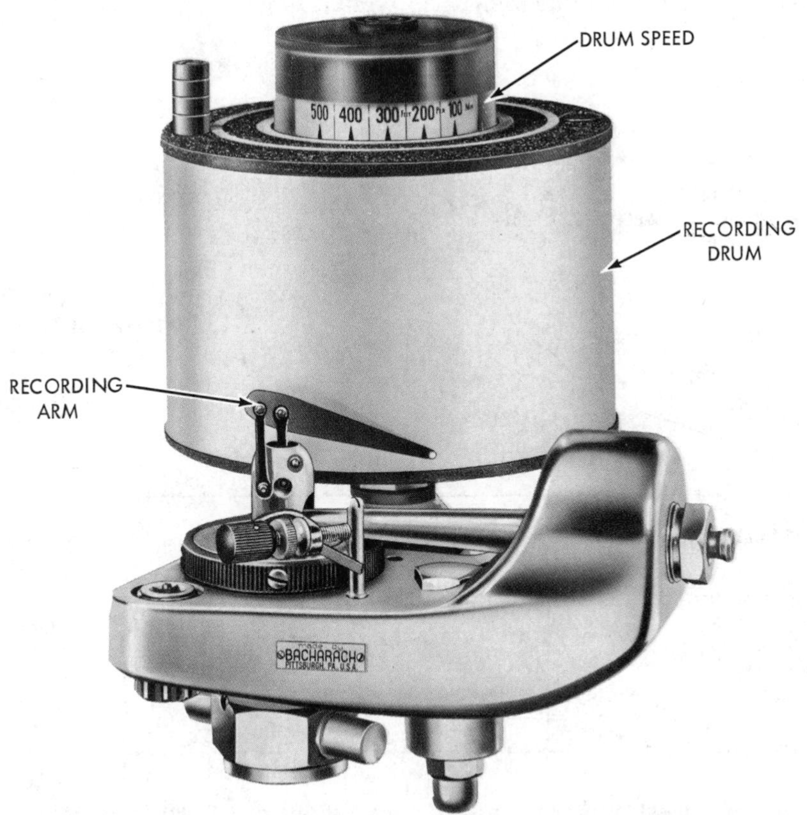

Fig. 17-12. Chronomatic engine-pressure indicator. (Bacharach Instrument Co.)

cycle. The scale of the speed indicator on top of the drum is from 100 to 500 feet per minute (fpm). This allows the drum speed at which the diagram is taken to be selected so the diagram may be spread out sufficiently to permit convenient study of certain pressure events.

As is evident from the diagram in Fig. 17-13, a drum speed of 400 fpm gives a clear picture of the combustion characteristic of a diesel operating at 1000 rpm. Referring to Fig. 17-14, it can be seen that for a complete pressure-time diagram of a four-cycle diesel running at 400 rpm, a drum speed of 200 fpm is best suited.

To obtain top dead center (TDC) marking, cut the fuel off on the cylinder being tested after the firing diagram has been taken. This allows a compression diagram, Fig. 17-15, to be drawn at the same drum speed and engine speed as the firing diagram.

A line (X-distance) is drawn parallel to the atmospheric line across this compression diagram. The top dead center (TDC) position is the mid-point on this line between the intersections with compression and expansion lines on Fig. 17-15. This TDC line may be transferred to the firing diagram, Fig. 17-16, by drawing a single parallel line at the identical pressure as the line in Fig. 17-15. Referring to Figs. 17-15 and 17-16 will show that this horizontal line is located a distance *X* above the atmospheric line. Measure off distance *Y* on the firing diagram to get TDC.

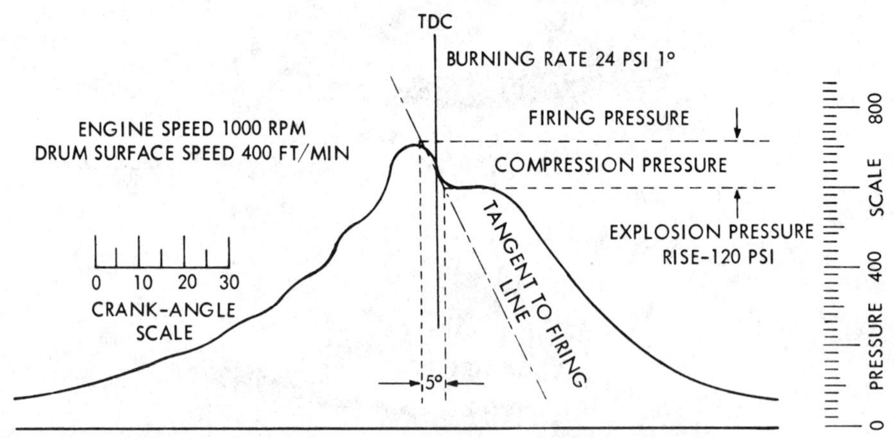

Fig. 17-13. Firing diagram at high drum speed to magnify combustion phase. (Bacharach Instrument Co.)

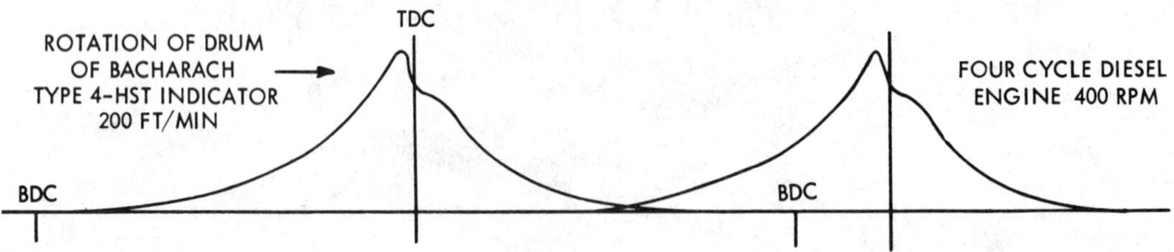

Fig. 17-14. Four-cycle diesel engine compression diagram at 400 rpm. Drum speed 200 fpm. (Bacharach Instrument Co.)

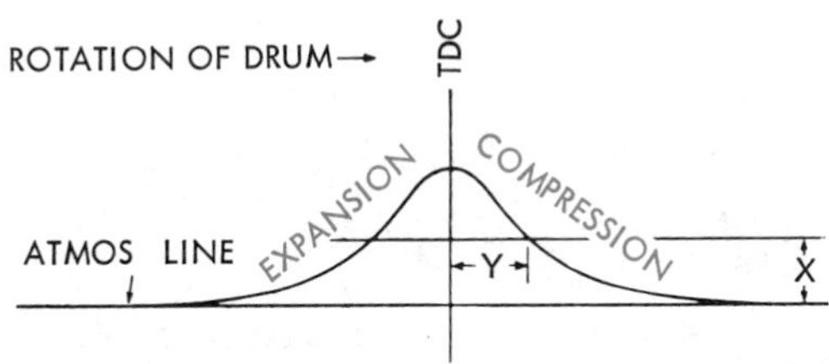

Fig. 17-15. Compression diagram peripheral speed of chronomatic drum 150 fpm. Engine speed, 300 rpm. Indicator scale 600 psi. (Bacharach Instrument Co.)

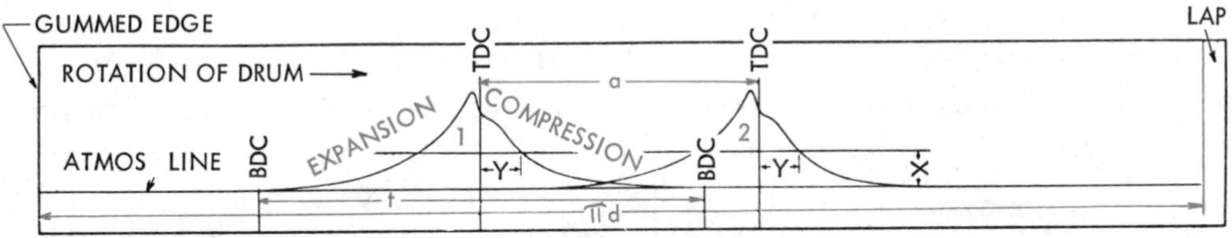

Fig. 17-16. Firing diagram for four-cycle diesel engine. Peripheral speed of chronomatic drum 150 fpm. Engine speed 300 rpm. Conrod crank angle 4:1. Indicator spring scale 600 psi. MEP 75.5 psi. (Bacharach Instrument Co.)

Bottom dead center (BDC) positions may be established by measuring half the distance (t) to either side of top dead center. Referring to Fig. 17-16, a four-cycle engine, the distance *t* may be calculated by the equation: $t = \frac{\pi d - a}{2}$ wherein *a* is the distance between similar top dead center points of two successive engine cycles. For two-cycle engines the distance *t* is equal to distance *a*.

Determining Mean Effective Pressures

There are three ways of finding MEP from the pressure-time diagram. The first two methods would involve constructing a pressure-volume diagram from the pressure-time diagram and then determining mean-effective pressure either through calculating the enclosed area (calculus problem), or by using a *planimeter* to determine the area.

The method of determining MEP directly from the pressure-time diagram area. (cards, like those illustrated in Figs. 17-17 and 17-18) for the following pressure measurements are usually available with the test equipment.

The method of determining MEP directly from the pressure-time diagram is based on the fact that, if the rotational speed of the engine to that of the indicator drum is kept at a constant ratio, distances measured along the crank circle from top or bottom dead center bear the same ratio to distances measured from top or bottom dead center on the pressure-time diagram.

Practical application of this direct method of determining MEP requires that the pressure-time diagram is taken at a drum surface speed in feet per minute numerically equal to one-half the engine speed in revolutions per minute. With this relationship between drum

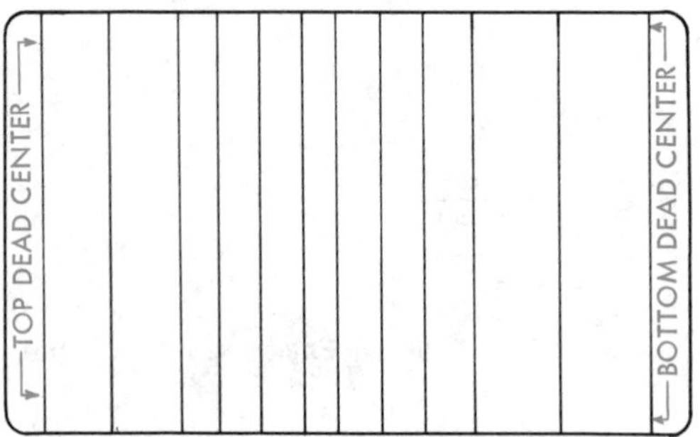

Fig. 17-17. Base scale for Conrod ratio 4:1. (Bacharach Instrument Co.)

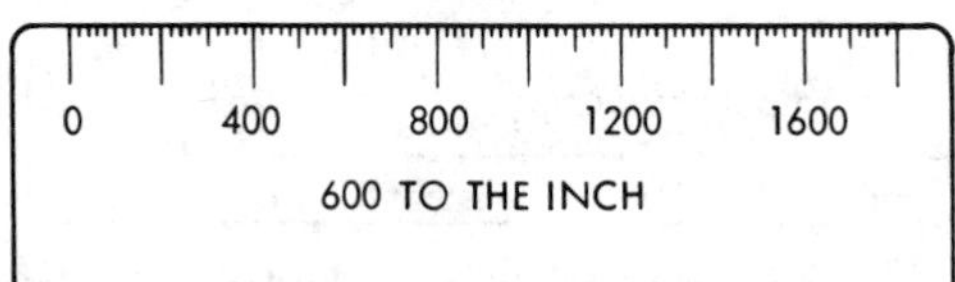

Fig. 17-18. Pressure scale for indicator spring 600 psi. (Bacharach Instrument Co.)

speed and engine speed, the compression-expansion humps of two successive engine cycles will follow each other as shown in Fig. 17-16, and the distance *t* between two successive bottom dead center positions will be approximately six inches.

With bottom dead center positions (BDC) marked with perpendicular lines extending from the atmospheric line to the edge of the card (see Fig. 17-16), place the base scale (Fig. 17-17) underneath the indicator card. Position it so that the TDC and BDC lines of the base scale meet the extended TDC and BDC lines of one of the two compression-expansion humps on the diagrams. In order to accomplish this, it will be necessary to place the base scale on an angle to the indicator card.

When in position, the points of the base scale should be transferred to the indicator card. Perpendicular lines should then be drawn which cut the expansion and compression lines. Then, measure each ordinate with the exception of BDC and TDC ordinates with the pressure scale, Fig. 17-18. The sum of the compression ordinates is subtracted from the sum of the expansion ordinates and one decimal place is pointed off in the result. The figure obtained is the *mean effective pressure*.

It must be understood that this example will only work for a scale for a connecting rod crank ratio of 4:1. Instructions can be obtained from the company for construction of a scale for other Conrad-crank ratios.

Piezoelectric Pressure Indicators

Another type of cylinder pressure indicator utilizes a principle called *piezoelectric*. Electrical properties of many piezoelectric materials, such as quartz, change when the material is subjected to physical force or stress. When a quartz transducer (pressure indicator) is subjected to pressure in an engine cylinder or a fuel injection line, a very small voltage will be generated which increases as the pressure increases. If this small voltage is converted to a high-characteristic voltage by an amplifier and fed into an oscilloscope, a pattern on the scope can represent a pressure-time diagram of an engine.

Fig. 17-19 illustrates in the lower left corner a piezoelectric pressure gage with two adaptors. The top of the figure shows the gage installed in the adaptor. The maximum pressure range of this gage is 10,000 psi.

Fig. 17-20 shows a typical piezoelectric pressure indicator installation showing the gage, the low-noise extension, the

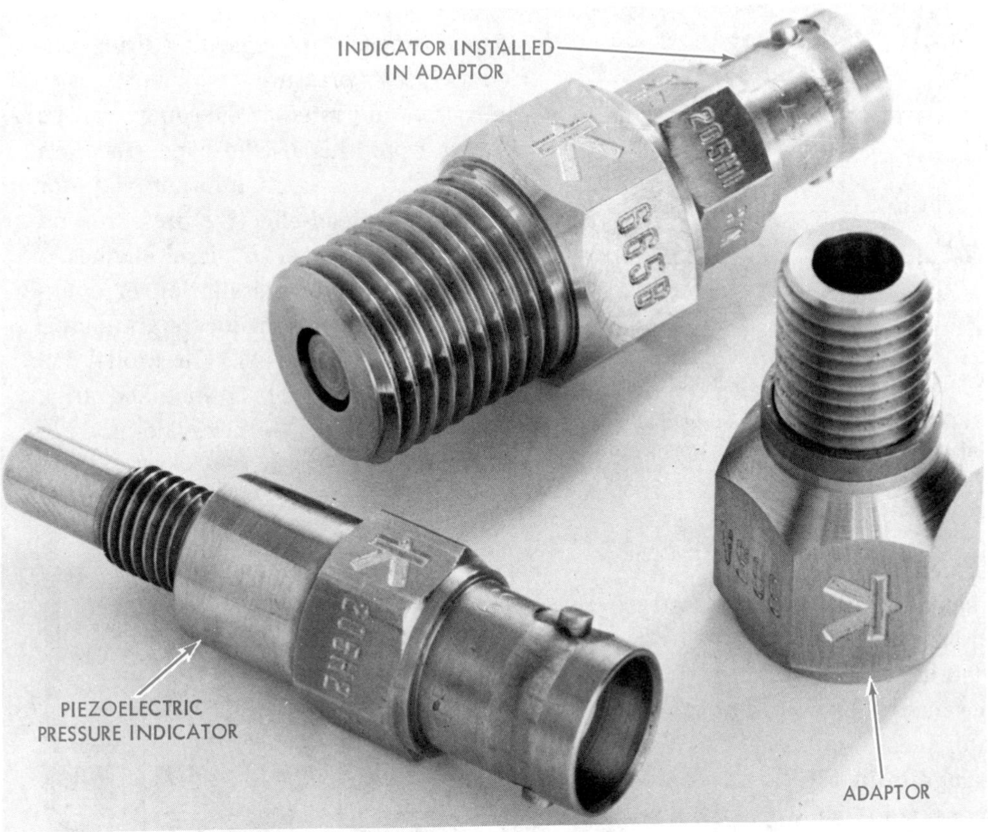

INDICATOR INSTALLED IN ADAPTOR
PIEZOELECTRIC PRESSURE INDICATOR
ADAPTOR

Fig. 17-19. Piezoelectric pressure gage and two adaptors. Pressure range to 10,000 psi. (Kistler Instrument Co.)

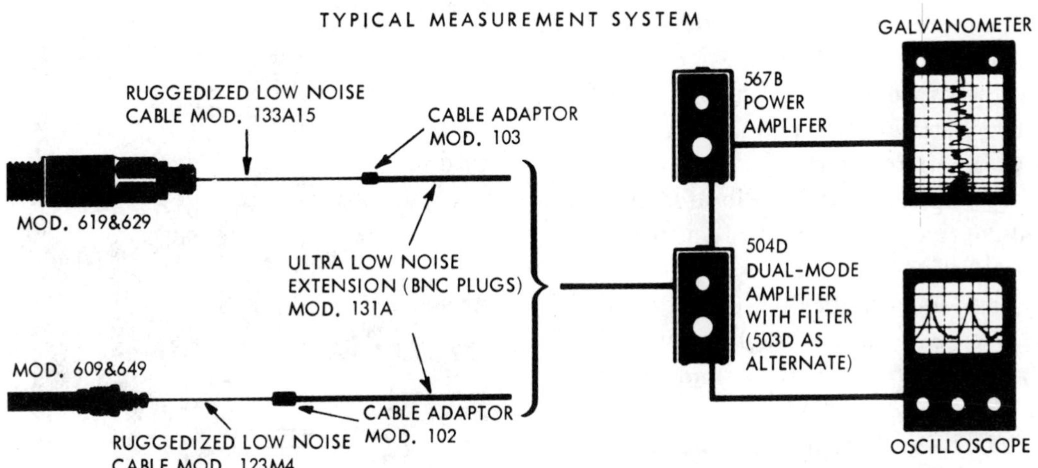

Fig. 17-20. Typical piezoelectric pressure gage installation with oscilloscope or galvanometer. (Kistler Instrument Co.)

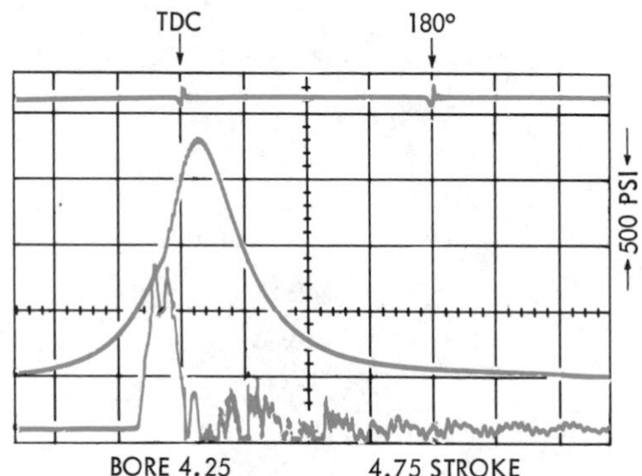

Fig. 17-21. Diesel cylinder pressure-time diagram and fuel-injection pressure through piezoelectric pick-up to oscilloscope.

amplifier and the oscilloscope.

The signal must be calibrated so a certain horizontal distance will be a certain amount of time and a certain vertical distance will represent a certain amount of pressure.

The diagram in Fig. 17-21 is from an oscilloscope trace showing the cylinder pressure-time diagram of a diesel engine. The lower trace in the picture shows the fuel injection pressure build-up and its dissipation. The straight line trace with the two slight oscillations at the top of the picture, Fig. 17-21, identifies TDC of the piston and BDC. The two oscillations in the trace are 180° apart; therefore each vertical grid-line equals 45° of crankshaft rotation.

The oscilloscope is calibrated prior to taking the pressure measurement so that each horizontal grid-line in this picture equals 500 psi. Peak firing pressure in this engine appears slightly past TDC (approximately 8° past) and peaks at approximately 1760 psi.

The fuel injection pressure corresponds to the piston position and peaks at approximately 1300 psi.

Operating factors regarding firing efficiency, rate of pressure rise, burning rate, fuel pump and injector efficiency, can be studied from this oscilloscope trace. In addition, it serves as another way for determining mean-effective pressure and indicated horsepower of diesel engines.

Two methods of installation of cylinder pressure-type transducers are shown in Figs. 17-22 and 17-23. The Model 619 shown in Fig. 17-24 is installed in a special cylinder head which is designed to receive the transducer.

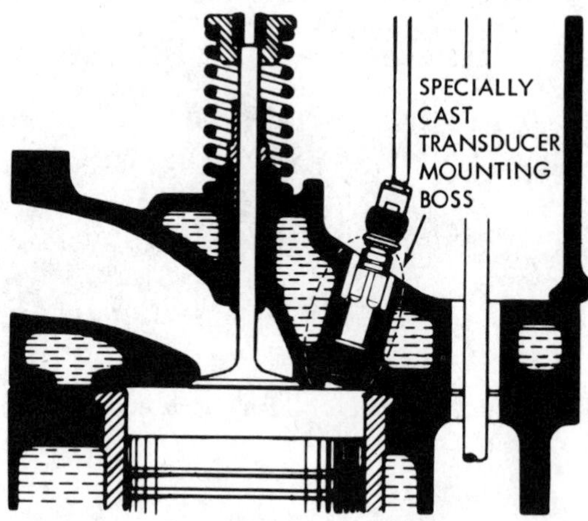

Fig. 17-22. Installment through cylinder head. (Kistler Instrument Co.)

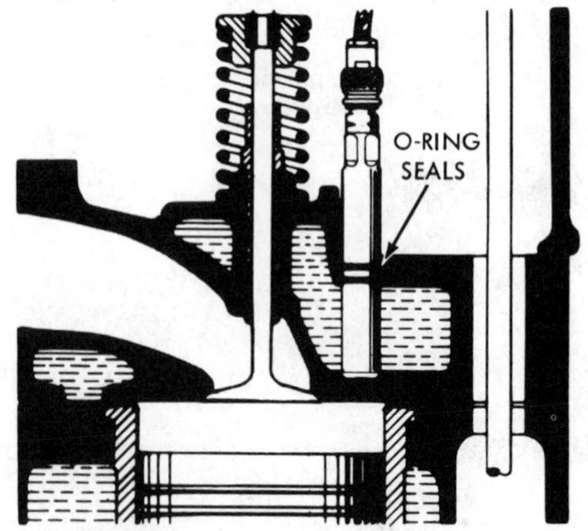

Fig. 7-23. Installation through water jacket in normal production engine. (Kistler Instrument Co.)

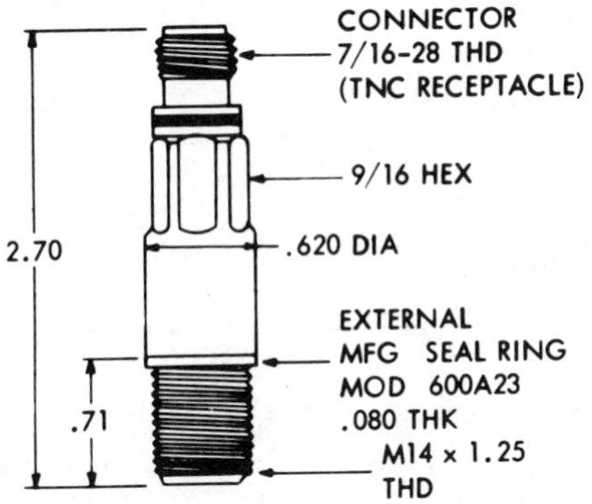

Fig. 17-24. Pressure pick-up of Model 619. (Kistler Instrument Co.)

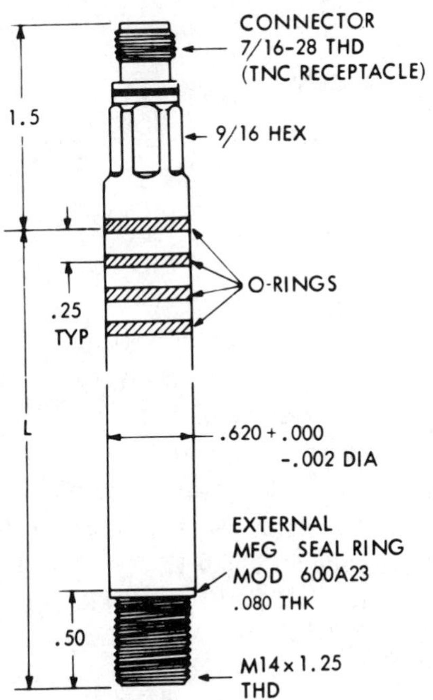

Fig. 17-25. Pressure pick-up of Model 629. (Kistler Instrument Co.)

Model 629, shown in Fig. 17-25, may be installed in a normal production engine. This transducer is installed through the water jacket of the combustion chamber as shown in Fig. 17-23. The lower end of the transducer is screwed into a threaded hole into the combustion chamber and the upper end contains O-rings to seal against coolant leakage from the water jacket.

Since extreme heat can affect the performance of the transducer, this particular installation allows it to be cooled by the coolant water in the cylinder head.

Some transducer models have water cooling jackets incorporated in them, whereas others have fins for cooling by air.

Measuring Exhaust Back-Pressure

The total restriction of the entire exhaust system of a normally aspirated engine at the exhaust outlet with the engine running at full speed, and under full load conditions, must not exceed 3 inches mercury (3″ Hg) back-pressure.

The total restriction of the entire exhaust system of a turbo-charged engine must not exceed 1″ Hg back-pressure. Back-pressure in exhaust systems is usually checked with a mercury U-tube manometer as shown in Fig. 17-26.

It should be noted in Fig. 17-26 that the manometer scale is graduated in inches both above and below the zero mark, and each inch is divided into tenths. When the manometer is installed for use, sufficient mercury must be put into the U-tube manometer so that height in both columns align with the zero mark on the scale.

Since this is a differential pressure manometer, add the height of the liquid in

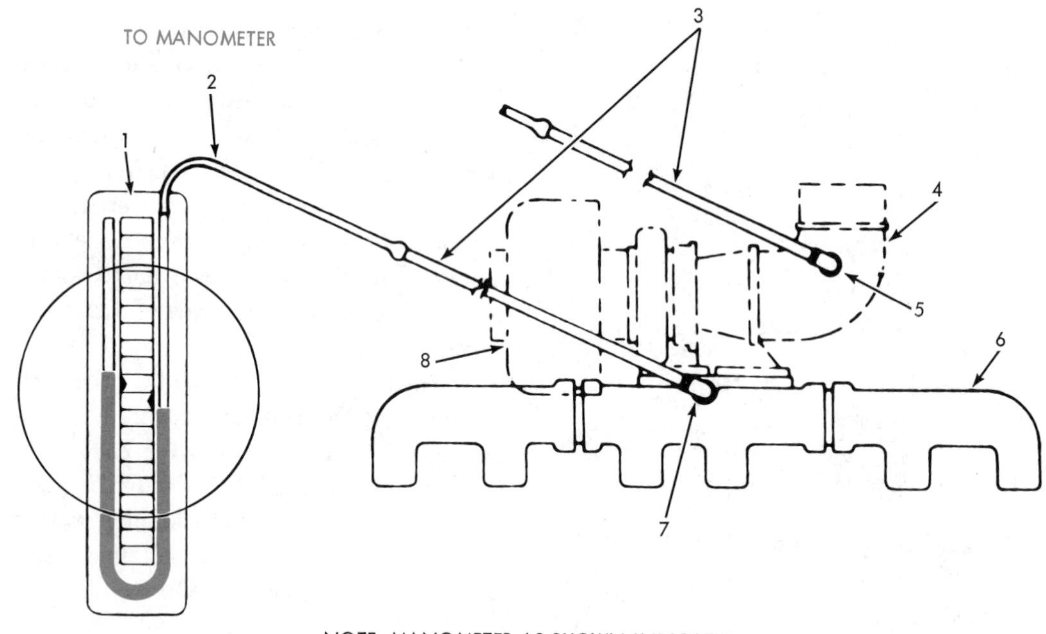

Fig. 17-26. Checking exhaust back-pressure. (Allis-Chalmers Mfg. Co.)

both columns to obtain the final figure. For example, in Fig. 17-26, if the liquid is 1.5 inches high in the left column (the one open at atmospheric pressure) and 1.5 inches low in the right column (the one attached to the exhaust system) it means the manometer indicates 3 inches of back-pressure *above* atmospheric pressure.

It should be noted that the U-tube manometer is a differential pressure gage. Therefore, if mercury is 1.5 inches high in the right hand column (Fig. 17-26) and 1.5 inches low in the left column, the manometer would be indicating 3 inches of vacuum (3″ Hg) as the pressure level of the exhaust system.

To convert the mercury reading in inches to actual pressure in pounds per square inch (psi) each 1 inch of mercury is equivalent to 0.491 psi, this 3″ Hg is equal to 3 × 0.491, or 1.473 psi.

Measuring Crankcase Pressure

The manometer is a very versatile and useful instrument in the area of engine testing. For example, some industries use the manometer to test how well the piston rings are sealing out *blow-by* of combustion gases around the cylinders.

A new or rebuilt engine is installed on an engine dynamometer and run at gradually increasing loads and speeds. The crankcase ventilation system is sealed so no pressure can escape and a water-type U-tube manometer is connected to the crankcase and the crankcase pressure is read as the engine is run on the dynamometer for a prescribed time.

The water filled manometer is used instead of a mercury filled manometer because it is much more sensitive to very small variations of pressure above or below atmospheric pressure.

Where the mercury manometer indicates 1 psi for every 2.0359″ Hg, or (1″ Hg equals 0.491 psi) the water manometer indicates 1 psi for every 27.6798 inches of water (″H_2O) 1″ H_2O equals 0.0361 psi.

Other Engine Tests and Calculations

Although the dynamometer may be involved in many kinds of testing done on diesel engines, it will not always be the principal testing device used in research and development experiments on engines. *Intercooler efficiency* is an example of a test which utilizes the dynamometer as an accessory instrument with the main objective being to determine the efficiency of a turbocharger with intercooler as compared to the turbocharger without an intercooler.

Intercooler Efficiency

Although this test would be broadened to include other objectives and can develop information under actual conditions, the purpose of its use at this time is to illustrate a sample test, using a single objective where equipment, procedure, and data can be analyzed.

After following the Test Procedure, sample results will be indicated in Table 17-1. This table shows the difference between the efficiency of the engine with and without intercooler.

The equipment used for this test includes (1) a dynamometer, (2) a multiple column mercury manometer, (3) a millivolt pyrometer (°F), (4) an air flow nozzle, (5) a water flow meter (5.0-50 gal/min range), and (6) thermocouple instruments. Fig. 17-27 shows a schematic of the instrumentation of the engine.

The testing procedure follows:

DESCRIPTION OF METHOD

1. Engine is connected to the dynamometer
2. Pressure and temperature points on the engine are established
3. Test points are then connected to the manometer and the millivolt pyrometer by means of tubes and thermocouples respectively. Thermometers are also installed

TABLE 17-1 A COMPARISON OF TEST RESULTS ON A SIX-CYLINDER TURBOCHARGED ENGINE WITH INTERCOOLER AND WITHOUT INTERCOOLER.

	With Intercooler 170 bhp at 2200 rpm	Without Intercooler 170 bhp at 2200 rpm	With Intercooler 187 bhp at 2200 rpm
BRAKEHORSEPOWER	170.5	170.2	187.5
FUEL CONSUMPTION (lb/bhp-hr)	.395	.415	.400
TEMPERATURES, °F			
Air Cleaner Outlet	90	93	96
Intake Manifold (before intercooler, avg)	(296)	(316)	(330.3)
(Front-center-rear)	(298-299-291)		(332-334-325)
Intake Runners			
(Intercooler, avg)	(224)		(224)
(Front-center-rear)	(227-220-227)		(250-238-245)
TEMPERATURE	71-79-64		82-96-80
Intercooler Water In	197		207
Intercooler Water Out	200		209
Intercooler Water Flow (gal/min)	16.5		17
Radiator Top Tank	193		204
Radiator Bottom Tank	182		192
Exhaust Manifold	1245	1342	1352
Pressures ("Hg)			
Intake Manifold (before Intercooler)	28.3	33.00	33.2
Intake Manifold (after Intercooler)	28.2		32.9
Exhaust Manifold	26.3	26.3	29.8
Intercooler Efficiency (%)	72.7		70.2

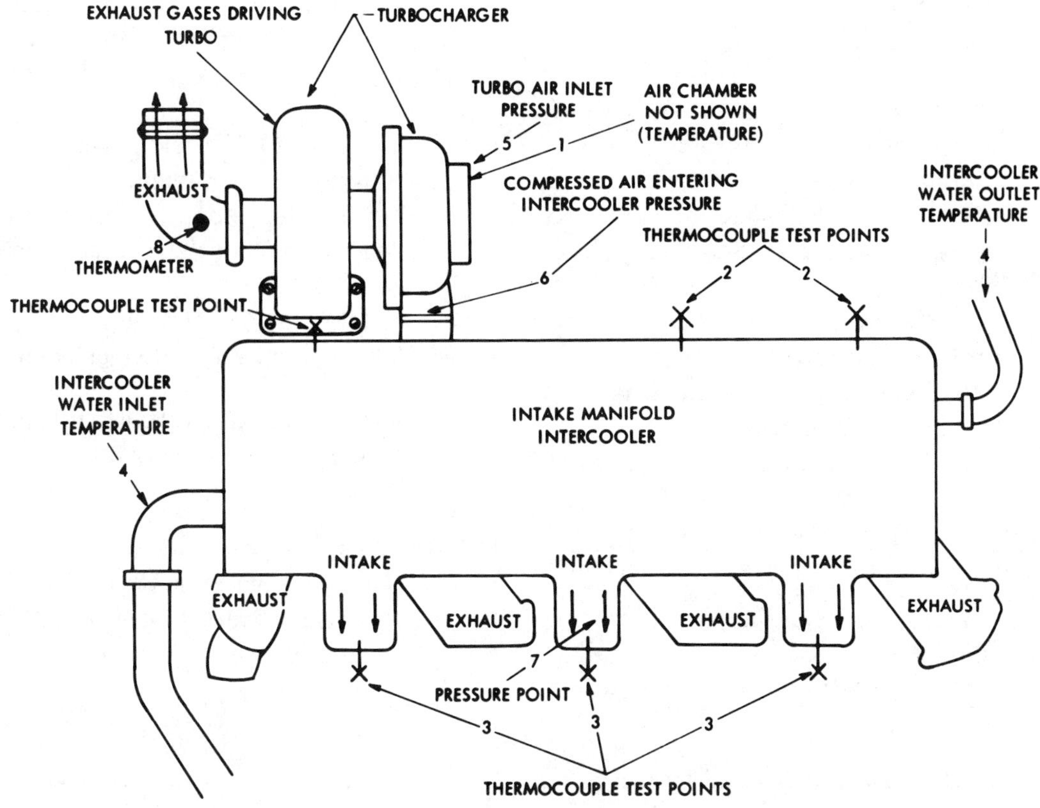

Fig. 17-27. Instrumentation for checking the efficiency of diesel engine with and without intercooler.

4. Referring to Fig. 17-27, temperatures are taken at the following locations:
   a. Air intake (around the air cleaner area)
   b. Intercooler air inlet (manifold cover at 3 locations)
   c. Intercooler air outlet (bottom of 3 *intake* runners)
   d. Exhaust manifold outlet
   e. Engine crankcase oil
   f. Engine coolant (*in* and *out* of radiator)
   g. Intercooler coolant (*in and out*)
5. Pressures are taken at the following locations:
   a. Turbo air inlet
   b. Intake manifold, air *in*—ahead of the intercooler
   c. Intake manifold, air *out*—below intercooler core in the center runner
   d. Exhaust manifold
   e. Air flow

6. Inlet air temperatures should be regulated to 90 to 95°F range for 2200 rpm full load runs

7. Radiator top tank temperatures are held above 190°F to assure full open thermostat position. A 7 lb pressure cap should be used to assure full out-

put of water pump at these higher temperatures

8. A flow meter should not be used for checking water flows during the test runs. This will cause added restriction in the normal system. Pressures should be established on previous runs. (Flows can be later determined by use of a flow meter in 5.0 to 50 gal/min range with setting to the pressures previously established.)
   a. Water flows obtained with the above flow meter are:
      1.0″ Hg pressure.... 7.36 gal/min
      4.5″ Hg pressure....15.95 gal/min
      7.2″ Hg pressure....20.61 gal/min
      8.5″ Hg pressure....23.06 gal/min
9. After finding the pressure of the air flow, the volume in cubic feet per minute can be determined by means of a graph, which is provided by the maker of the air flow nozzle. The air flow at 2200 rpm in this example was 395 cu ft/min
10. Test runs should be made with this diesel turbocharged engine:
    a. Fuel flow set to obtain 170 bhp at 2200 rpm without the intercooler installed.
    b. Fuel flow set to obtain 170 bhp at 2200 rpm with the intercooler installed
    c. Fuel flow set to obtain 188 bhp at 2200 rpm with the intercooler installed

NOTE: This test identifiies one speed and two loads only. Under actual conditions, data should be gathered from a number of speeds and loads.

SAMPLE CALCULATIONS OF THE TORQUE AND THE BRAKE MEAN EFFECTIVE PRESSURE

Torque was calculated by means of the following formula:

$$T = \frac{5252\ (\text{brake horsepower})}{\text{rpm}}$$

Data used:

Bhp = 170.2

rpm = 2200

Therefore:

$$T = \frac{5252\ (170.2)}{2200} = 406\ \text{lb-ft}$$

Bmep was found by using the following formula:

$$\text{Bmep} = \frac{1{,}008{,}000\ \text{Bhp}}{D^2\ LMN}$$

Data used:

Bhp = 170.2

D = 4.25″

L = 4.75″

M = 6.0

N = 2200

Therefore:

$$\text{Bmep} = \frac{(1{,}008{,}000)\ (170.2)}{(4.25)^2\ (4.75)\ (6)\ (2200)} = 152\ \text{lb/in}^2$$

Torque, Bmep, and Heat Transfer

Another calculation that can be made from the same data is *torque* and *brake mean effective pressure.* These calculations follow.

After that a sample calculation of *heat transfer* between water coolant and turbocharged air is indicated. The data used below was obtained at 2200 rpm with the fuel flow set to obtain 170 horsepower.

Airflow at 2200 rpm was checked by a flow meter and was found to be 395 cu ft/min.

SAMPLE CALCULATION OF HEAT TRANSFER BETWEEN TURBOCHARGED AIR AND WATER COOLANT

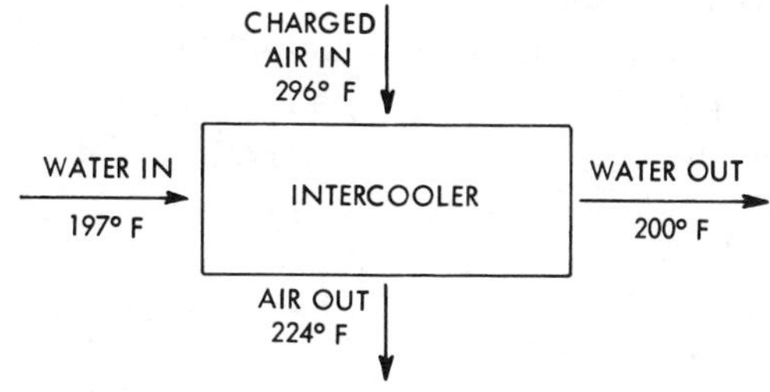

Other data used:

Specific weight of air at 300°F .... 0.052 lb/cu ft
Specific heat of air at standard conditions 0.240 btu/lb-°F
Specific weight of water at 200°F .... 60.1 lb/cu ft
Specific heat of water at standard conditions 1.0 btu/lb-°F
Amount of cu ft/gal of water 0.1337

Air flow = (395 cu ft/min) (0.052 lb/cu ft) (72°F temp change) (0.240 btu/lb-°F) = 360 btu/min heat transfer

Water flow = (16.5 gal/min) (60.1 lb/cu ft) (0.1337 cu ft/gal) (3°F temp change) (1.0 btu/lb-°F) = 398 btu/min heat transfer

$$\frac{\text{Actual turbocharged air temp drop}}{\text{Theoretical maximum temp drop}}$$

(The theoretical maximum temperature drop would be the difference between the entering turbocharged air temperature and the entering water coolant temperature)

From the above formula:

$$\frac{296\text{-}224}{296\text{-}197} = \frac{72}{99} = 72.7\% \text{ efficiency}$$

Discussion and conclusion of the test results will include further explanation for clarity. For example, the sample calculation of the heat transfer between the turbocharged air and the water coolant may be explained in respect to accuracy. According to the Law of the Conservation of Energy, the heat transfer of the air (360 btu/min) and the water (398 btu/min) should be the same. That is, the energy given up by the turbocharged air should have been wholly absorbed by the water. This law applies to perfect or ideal situations, where there is not any loss of energy to the atmosphere by way of radiation and conduction. It is apparent by the above values that there was some loss of energy to the surrounding air through radiation and conduction.

Also, after analyzing the data, additional conclusions may be gleaned from the test project. For example, in this test project it may be pointed out that the exhaust temperature (1352°F) and intake manifold boost pressure (33.3″Hg) at 187 Bhp output with the intercooler installed were approximately the same as those when operating at 170 bhp without the intercooler.

From this it can be seen that the power output of the engine has been increased without any sacrifice in the exhaust temperature. This should promote a longer engine life cycle.

The fuel consumption comparison indicates that fuel intake is less with intercooler than without it.

If this test were run at different speeds and loads further useful information may be found.

Another test involving the loading of a diesel engine has the objective of de-

Fig. 17-28. Diesel-electric set installed in test cell. (Allis-Chalmers)

termining the expected capabilities of a diesel-electric set. This includes a production test of the diesel engine and a test of the electric generator.

Since the electric generator can be used as a loading device for the engine and the wattage output can be observed, and converted to horsepower, there is no need of a conventional dynamometer in this case. Fig. 17-28 shows the diesel-electric set installed in a test call for evaluation of engine and generator capabilities.

Diesel Engine Design

Although the diesel engine change in design is extremely slow when compared to those of many other products—TV, for example, there have been remarkable improvements over the past twenty years in the diesel engine. Some of the reasons for a continuous research and development program carried on by all diesel engine

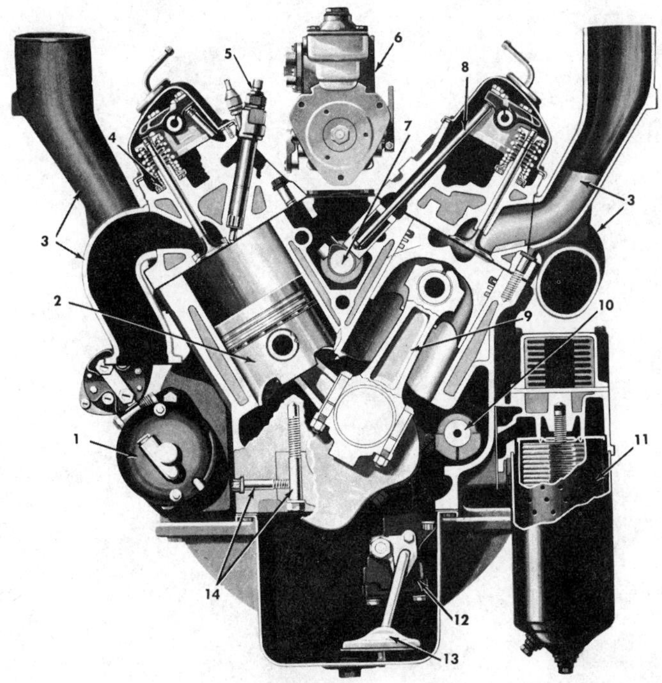

1 STARTER
2 PISTON
3 MANIFOLDS
4 VALVE ROTATOR
5 FUEL INJECTION NOZZLE ASSEMBLY
6 FUEL INJECTION PUMP ASSEMBLY
7 CAMSHAFT
8 PUSH ROD
9 CONNECTING ROD ASSEMBLY
10 BALANCE SHAFT AND WEIGHT ASSEMBLY
11 OIL FILTER ASSEMBLY
12 ENGINE OIL PUMP ASSEMBLY
13 OIL INLET SCREEN AND TUBE ASSEMBLY
14 MAIN BEARING CAP BOLTS

Fig. 17-29. Sectional view of Toro-Flow short-stroke diesel engine. (General Motors Corp.)

TABLE 17-2 TORO-FLOW ENGINE DATA

ENGINE MODEL	D351	D478	DH478	D637	DH637
Type and No. of cylinders	60° V-6	60° V-6	60° V-6	60° V-8	60°V-8
Stroke Cycle	4	4	4	4	4
Displacement (Cu In)	351.2	477.7	477.7	637	637
Bore and Stroke (In)	4.56x3.58	5.125 x3.86	5.125x3.86	5.125 x 3.86	5.125x3.86
Compression Ratio	17.5:1	17.5:1	17.5:1	17.5:1	17.5:1
Firing Order	1-6-5-4-3-2	1-6-5-4-3-2	1-6-5-4-3-2	1-8-4-3-6-5-7-2	1-8-4-3-6-5-7-2
Cylinder Nos. Front to Rear					
Left Bank	1-3-5	1-3-5	1-3-5	1-3-5-7	1-3-5-7
Right Bank	2-4-6	2-4-6	2-4-6	2-4-6-8	2-4-6-8
Gross Brake Horsepower at rpm	130 at 3200	150 at 3200	170 at 3200	195 at 2600	220 at 2800
Governor Full Load Setting rpm	3200	3200	3200	2600	2800
Approx No Load Setting rpm	3450	3450	3450	3000	3050
Idle Speed Setting rpm	625-650	625-650	625-650	625-650	625-650
Fuel Grade Required	Refer to Current GMC Diesel Fuel Oil Service Bulletin				

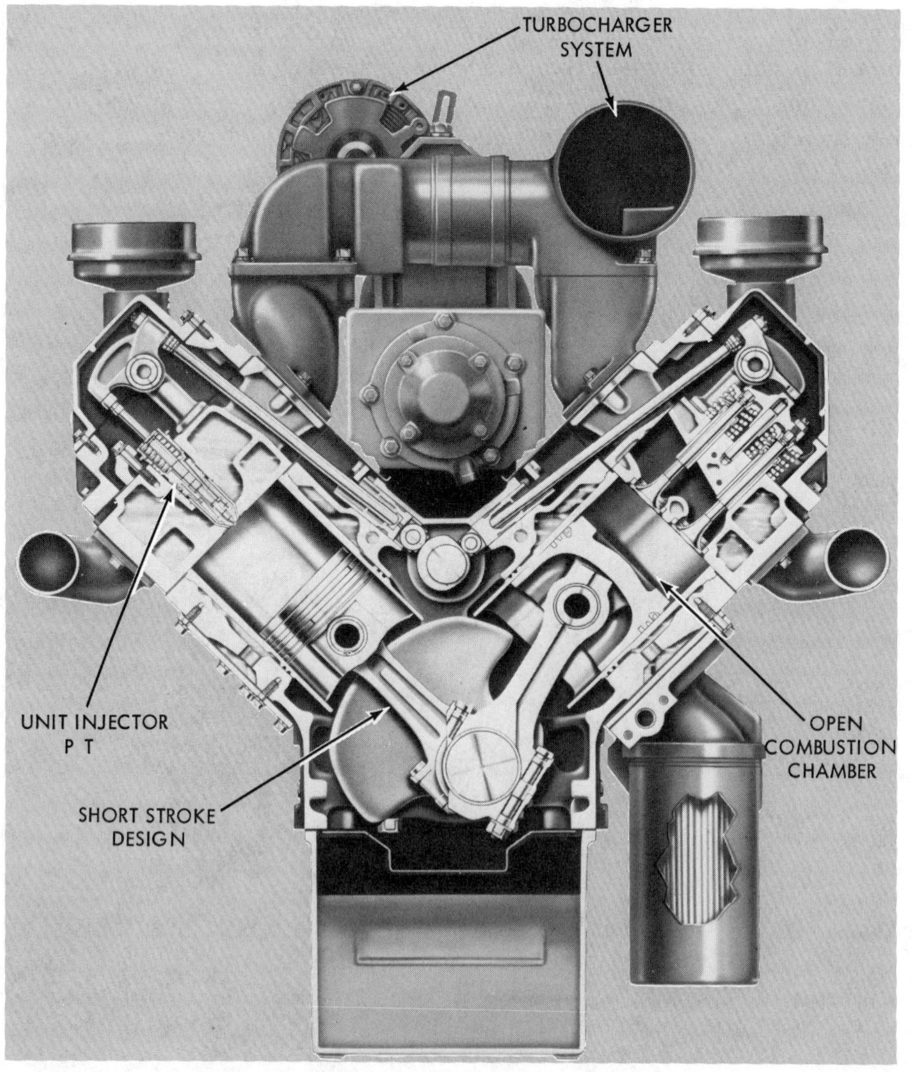

Fig. 17-30. V-type 903-C high-powered turbocharged high-speed short-stroke diesel. (Cummins Engine Co.)

manufacturers were stated at the beginning of the chapter.

Reduction of Weight per Horsepower

The turbocharger has made significant improvement in increasing the horsepower while keeping the weight and size of the engine approximately the same. The addition of the intercooler to the turbocharger has helped to make further gains.

Another approach to improving the weight-to-horsepower ratio was indicated recently in regard to a completely new

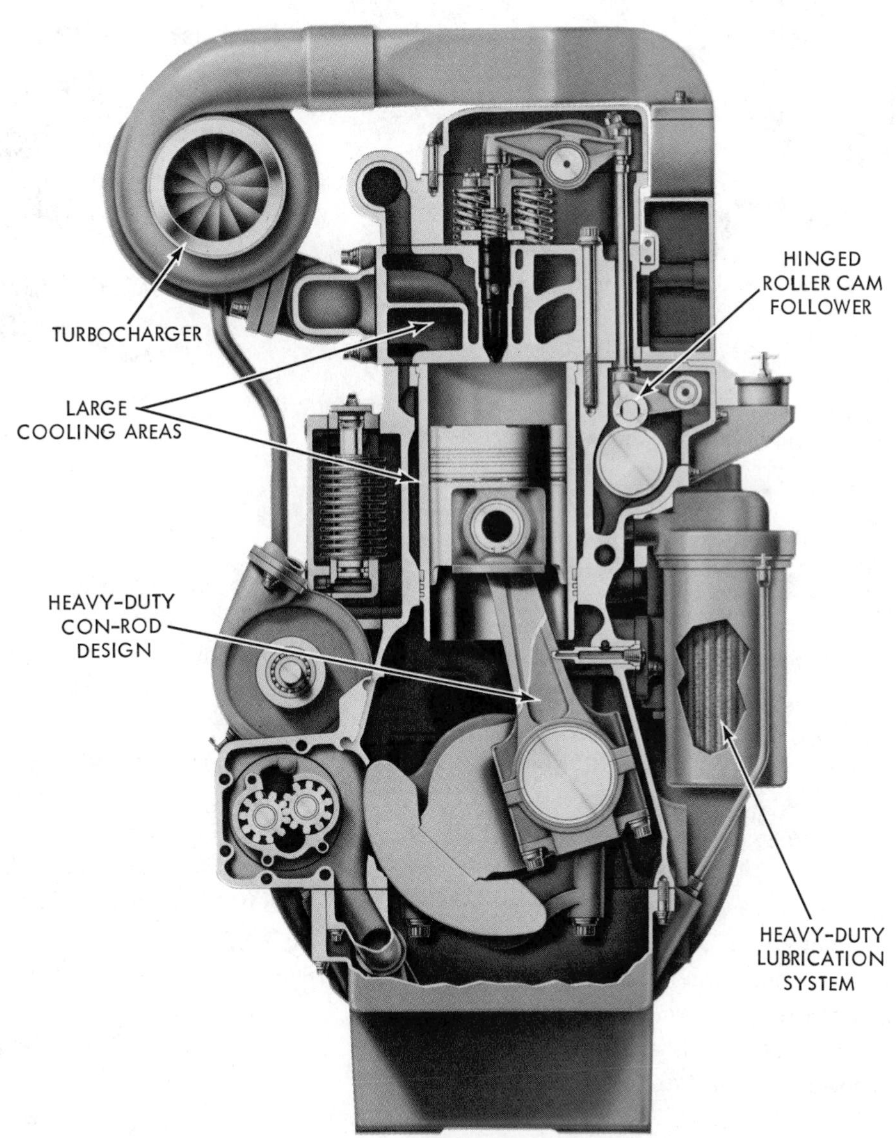

Fig. 17-31. Six-cylinder in-line turbocharged diesel truck engine. (Cummins Engine Co.)

Fig. 17-32. Eight-cylinder V-type Series 100 American Bosch pump. (American Bosch AMBAC Industries, Inc.)

General Motors diesel engine which is presently being field tested. The engine has 75 percent fewer moving parts than the conventional diesel. It is more quiet-running and in a comparison of the conventional engine in the 325 horsepower class, which weighs 4500 to 5000 pounds, the experimental engine weighs approximately 1700 pounds. G.M. officials claim the emission qualities on the engine are low enough to pass the strict 1975 California air pollution standards.

The movement toward the short-stroke design for diesel truck engines has followed the earlier trend in automotive gasoline engines. The G.M. Toro-Flow diesel in Fig. 17-29 is an example of this trend. The engine is produced in three 60° V-6 models, and two V-8 models, all with a 17.5:1 compression ratio. All engines are four-stroke cycle and four of the five engines use the same (over-square) piston size (5.125″) and the same stroke (3.86″).

Table 17-2 shows comparative specifications for all models.

Other major diesel truck engine manufacturers have made similar design changes in engines. Fig. 17-30 shows a V-type 903-C cutaway of a Cummins diesel. This is a turbocharged high-speed diesel designed specifically for the greater power needs of on-the-highway diesel tractors.

The in-line six-cylinder turbocharged

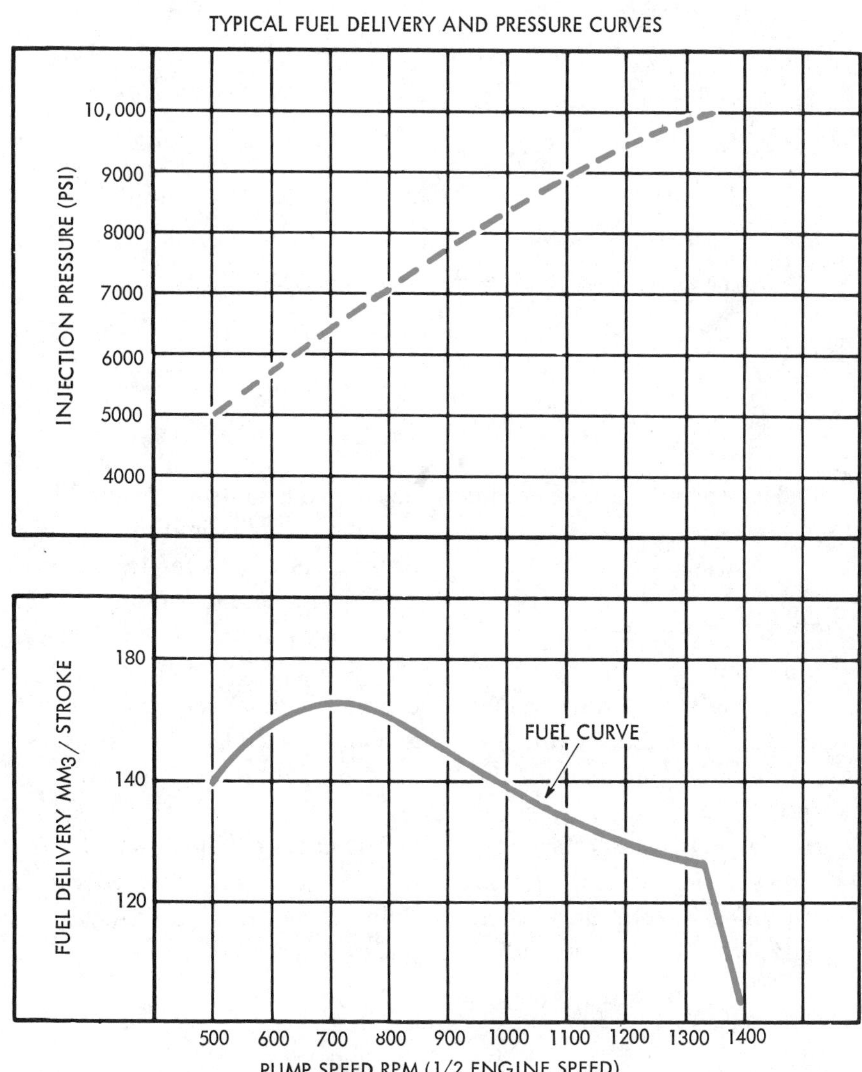

Fig. 17-33. Typical fuel delivery and pressure curves charted on American Bosch APE-BB injection pump. (American Bosch AMBAC Industries.)

TABLE 17-3 BASIC APE 8-V PUMP DATA

No. cylinders............8	Fuel.........No. 2 diesel
Plunger Dia........9-13 mm	Lubrication......Engine oil
Fuel Delivery mm^3 / Stroke .. to 350	Mounting Flange ... Special
Engine Piston Displacement (in^3)	Drive.......1/2 Engine Speed
Per cylinder...... to 245	Timing Advance...Optional
Total..........to 1960	Governor....Variable speed
Max Pump RPM 1300	Overall length w/gov 19-7/8"
Max Injection Pressure 10,000 (PSI)	Weight (average).....59 lb

Cummins diesel has the more conventional long stroke design and is used on many on-the-highway trucks, Fig. 17-31.

Design of Parts and Accessories

Fuel injection equipment has also been improved. One example of this is the new 100 series American Bosch which is designed for application on present and future medium horsepower high-speed diesel engines. Fig. 17-32 shows the eight-cylinder V-type version of this pump. Available accessories include an electric shut-off, single-lever control including mechanical shut-off, manifold, pressure-actuated smoke-limiting device, and an excess fuel device for improved starting.

The smoke limiter, of aneroid (non-liquid pressure-measuring) design, operates from engine manifold pressure and controls fuel delivery during rapid engine acceleration and under various load conditions to meet exhaust emission standards.

The chart in Fig. 17-33 shows the fuel delivery and pressure curves and the Table 17-3 contains basic data on this pump.

Checking On Your Knowledge

The following questions give you the opportunity to check up on yourself. If you have read the chapter carefully, you should be able to answer the questions. If you have any difficulty, read the chapter over once more so that you have the information well in mind before you go on with your reading.

DO YOU KNOW

1. Identify six reasons for continuous research and development programs concerned with improvement of diesel engines.
2. Identify the different kinds of dynamometers and the purposes of each.
3. Explain how a water-type dynamometer operates.
4. What is the purpose of an inertia flywheel on a chassis dynamometer?
5. Explain the principle of a thermocouple.
6. Explain two ways to determine mean effective pressure of a diesel engine.
7. What are the potential advantages of a piezoelectric pressure indicator over a mechanical indicator?
8. Identify the causes of excessive exhaust back-pressures.
9. Why are diesel engines rated at a lower horsepower output than maximum and what is that approximate rating?
10. State the purpose of an intercooler.

Chapter

18 High-Compression Gas-Burning Engines

The purpose of this chapter is to acquaint you with certain kinds of internal-combustion engines which resemble ordinary oil-burning diesel engines because they all use high compression, but which differ because they can burn *gaseous* fuels. These types of engines were developed in order to produce low-cost power from cheap fuel gases.

You will learn about three distinct types of such engines: (1) *gas-diesels,* (2) *dual-fuel engines,* and (3) *high-compression spark-ignited gas engines.*

The two latter types of engines work on the principle of compressing a *lean mixture* of gas and air. To understand this principle, you'll learn about *ignition temperature, composition limits, pre-ignition,* and *detonation.*

You'll see how the dual-fuel engine uses a spray of *pilot oil* to ignite the mixture of gas and air, while the high-compression gas engine employs an *electric spark* for that purpose. You'll also learn the other special features which distinguish the gas-burning engines from ordinary diesels.

Selecting the right type of engine for a particular job depends upon the nature of the gas supply.

High Compression Increases Burning Efficiency

Pipe lines now convey natural gas throughout the continent and make this ideal fuel widely available for use in internal-combustion engines. (Chapter 10 described natural gas and other gaseous fuels.)

Gas engines of the older type run on nearly perfect mixtures of gas and air; that is, the ratio of fuel to air is close to the theoretical combining ratio, as in gasoline engines.

These *perfect-mixture* engines must employ low compression when using ordinary gaseous fuels. If we try to use too much compression, the higher pressure and the accompanying higher temperature cause the extremely inflammable mixture to start burning of its own accord. This *self-ignition* results in severe knocking, loss of power, and possible damage.

Consequently, when gas engines of the older type run on natural gas, their compression ratio is limited to about 5 to 1, or about 120 psi compression pressure. This low compression ratio results in low thermal efficiency, which means high fuel consumption.

Knowing that higher compression reduces fuel consumption and operating cost, engineers have long sought ways to burn gaseous fuels at higher compression. It is only recently that they have discovered how to burn gaseous fuels at a compression volume ratio as high as 14 to 1, like that of a diesel engine, and thus achieve diesel engine efficiencies which exceed those of low-compression gas engines by about 40 percent.

Three ways have been found to burn gas at higher compression. These three ways, together with the engines adapted to accomplish the goal, are here mentioned in brief, preliminary to fuller descriptions later in the chapter.

Gas-diesel engines do *not* admit the gas with the intake air and then compress the gas-air mixture, as other engines do. Instead, they compress air alone. At the top end of the compression stroke, they inject the gas at high pressure into the cylinder just at the moment it is to fire. Thus, the gas cannot fire ahead of time. They also inject with the gas a small amount of fuel oil, called *pilot oil,* to assist the ignition and burning of the gas so that it burns smoothly and as fast as it is injected.

Dual-fuel engines admit the gas at the same time as the air and then compress the gas-air mixture at diesel compression ratio. At the top end of compression, they inject fuel oil which the high temperature of the gas-air mixture ignites to fire the mixture. Despite the high compression ratio, self-ignition is prevented in dual-fuel engines by employing very *lean mixtures,* in which the proportion of fuel to air is far less than that in the nearly perfect mixtures used in ordinary gas engines. Dual-fuel engines can run on gas and oil together, in various proportions; hence the name.

Dual-fuel engines are sometimes called, rather confusingly, **gas-diesel,** on the theory that when they run on gas they are gas engines and when they are set to run solely on oil they are diesel engines. The true gas-diesel engine is the earlier type described above; it does not compress the gas in the engine cylinder.

High-compression spark-ignited gas engines, like dual-fuel engines, compress a mixture of gas and air to high pressure, preventing self-ignition by using a *lean mixture.* However, they do not use any fuel oil at all, not even for ignition purposes. Instead, they ignite the gas-air mixture by electric spark.

Gas-Diesel Engines

The gas-diesel engine is so named because it works on the true diesel principle of (1) compressing air only, (2) injecting the fuel at the top end of the compression stroke, and (3) depending solely on the high temperature of the compressed air to ignite the pilot oil and thus fire the gas fuel. It was the first type of high-compression engine to burn gas fuel but the later types (dual-fuel and spark-ignited) have superseded it because they are simpler and cheaper.

The main commercial development of the gas-diesel is the Nordberg large two-cycle engine. The general design and the compression pressure are the same as those of the Nordberg two-cycle oil-diesel engine. Its special features are: (1) a three-stage gas compressor which raises gas to the injection pressure, (2) an individual pump for each cylinder to pump the pilot oil, (3) a fuel-injection valve to admit gas and pilot oil simultaneously to each cylinder, and (4) a hydraulic mechanism to open the fuel-injection valve at the proper time and to vary the valve lift under governor control so as to admit the quantity of gas needed to carry the engine load.

Fig. 18-1 shows the arrangement. The gas compressor is automatically regulated to supply fuel gas at approximately 1100 psi. The gas flows through a cooler and a storage bottle to the fuel-injection valves in the engine cylinder heads. A pilot-oil pump for each cylinder delivers a small, fixed amount of fuel oil to each injection valve.

When the fuel valve opens, the pilot oil is blown into the cylinder by the compressed gas, where it meets air which has been compressed to a pressure of about 480 psi and to a temperature of about 1000°F. This temperature causes the pilot oil to ignite; the resulting flame ignites the gas and also stabilizes its combustion.

An actuator pump, driven from the camshaft, connects through tubing to an actuator on the cylinder head, which times the opening of the injection valve and also controls the amount of valve lift. The governor varies the amount of hydraulic fluid delivered at each stroke of the actuator pump, thus controlling the lift of the fuel-injection valve and the amount of gas injected.

The amount of pilot oil needed to ignite the gaseous mixture and stabilize its combustion is about 5 percent of the total fuel at full load, as measured by heating value (Btu). A typical example is as follows:

The fuel efficiency of this gas-diesel engine, in terms of the heating values of the gas and pilot oil consumed per unit of power output, is approximately the same as when running as an oil-diesel. However, the engine costs more than an oil-diesel because of the three-stage gas com-

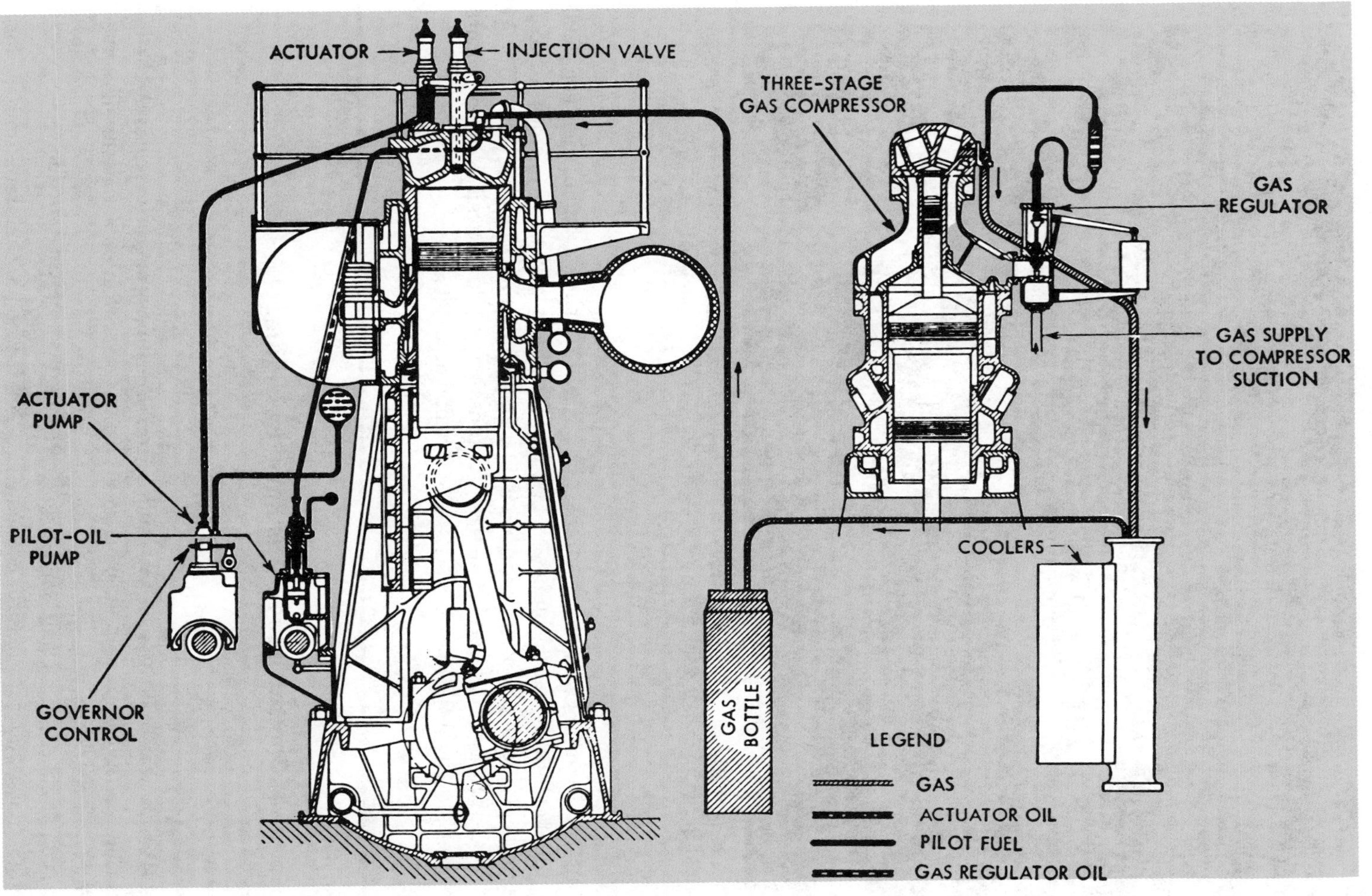

Fig. 18-1. Cross-section of gas-diesel engine. Engine construction shown at left; automatically regulated fuel-gas compressor shown at right. Schematic of piping indicates flow of gas and pilot oil, and also the hydraulic circuit to actuate and control the fuel-injection valve. (*Power* Magazine)

Fig. 18-2. Nine of eighteen Nordberg two-cycle, 21.5" gas-burning diesel engines with a total of 64,500 Bhp, at an aluminum metals plant of Reynolds Metals Co. (Nordberg Mfg. Co.)

pressor, the hydraulic actuating mechanism, and the additional oil pumps; it is therefore suitable only for large sizes. It takes several hours' time to change over a gas-diesel to an oil-diesel.

Fig. 18-2 shows nine of eighteen 3600-hp Nordberg gas-diesel engines installed in an aluminum smelting plant in Arkansas where local availability of natural gas favored their use.

Ignition and Combustion in Gas Burning Engines

The reason why mixtures of gas and air can be highly compressed in modern dual-fuel and spark-ignited engines is because such engines run on *lean mixtures.* Although much research is still needed on the ignition and combustion of lean mixtures under engine conditions, the following simplified explanation will serve.

Speed of Chemical Reaction

You learned in Chapter 9 that the form of combustion known as burning is a chemical process in which fuel combines chemically with the oxygen of the air and produces heat. The speed at which a chemical combining process takes place (reaction speed) increases rapidly when the *temperature* of the combining substances is increased. Thus, at room temperature, the reaction speed of natural gas and air is practically zero and there is no combustion. At higher temperatures, the chemical combining process goes faster. Consequently, heat is produced more rapidly, causing the temperature of the mixture to rise. The rise in temperature speeds up the reaction further, and finally the reaction may become so rapid that the heat produced makes the gases hot enough to give off light and thus become flame.

Reaction speed also increases as the *pressure* of the mixture increases. This is natural, because at higher pressure the mixture becomes denser and, consequently, the reacting molecules are squeezed closer together.

Ignition Temperature

If a theoretical or *perfect* mixture of gas and air is heated gradually in a container of some sort, a certain temperature will be reached at which the reaction speed will become so rapid that some of the mixture will *self-ignite*. It will burst into flame and cause the entire mixture to burn rapidly. This is called the *ignition temperature* of the mixture.

In an engine which compresses a gas-air mixture, both the pressure and temperature of the mixture increase during the compression stroke. Since higher pressure and higher temperature each speed up the chemical reaction, the ignition temperature of a highly compressed perfect mixture is lower than that of the same mixture at lower pressure.

Ignition Limits of Various Compositions

The same thing is true of mixtures whose composition is somewhat richer or leaner (more gas or less gas, respectively) than a perfect mixture. However, the further a mixture departs from the perfect composition, the longer it must be kept at the ignition temperature before it starts to burn. This is called *ignition delay*. Finally a limit is reached on either side of the perfect mixture beyond which the mixture becomes either too rich or too lean to self-ignite. These limits we call the *explosive limits*, and mixtures within this range we call *explosive mixtures*.

Although mixtures just outside the explosive range will not *self-ignite*, they will burn if ignited by an *outside* source of high temperature, such as an electric spark or a flame. The further they depart from the explosive limit, however, the slower they burn. Finally they become so rich or so lean that they will not support combustion at all; these are the *limits of flammability*. The limits become wider as the mixture temperature increases. Mixtures within this range are called *flammable mixtures*. Fig. 18-3 will help to make this all clear.

Natural gases vary in composition. The lower, or lean, explosive limit (that is, the leanest mixture which will self-ignite) runs about 4.6 to 5.2 percent gas; the upper explosive limit (richest mixture) runs about 13.9 to 14.9 percent gas. In high-compression gas-burning engines, we are concerned only with the lean limit.

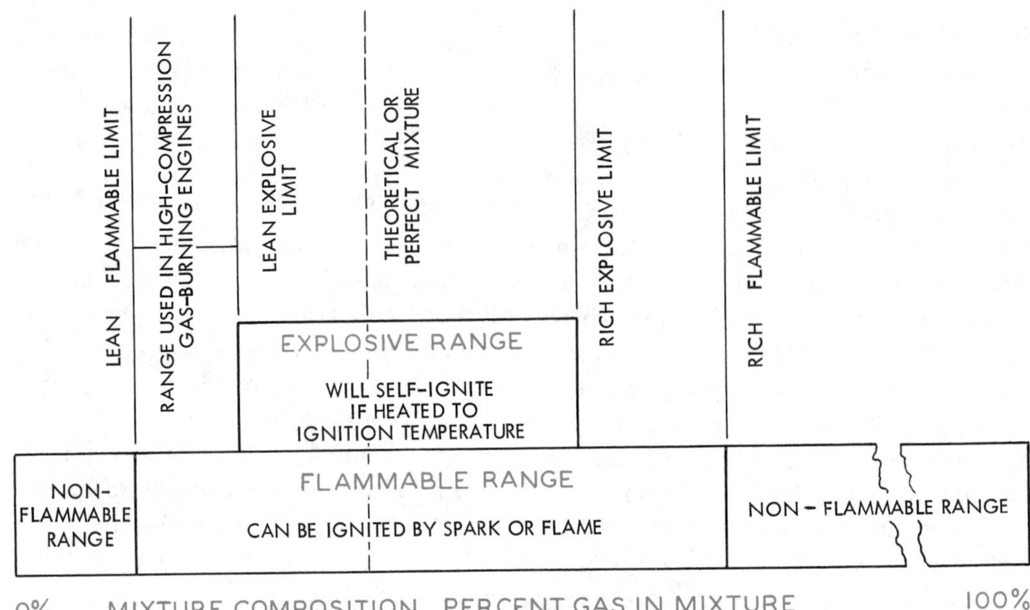

Fig. 18-3. Fuel composition and limits of ignition diagram.

Self-Ignition in Engines

Under certain conditions, some of the unburned mixture in an engine cylinder may be raised to so high a temperature and pressure that it fires of itself. It self-ignites before it is ignited. Self-ignition may appear in two forms — *pre-ignition* and *detonation.*

Pre-ignition. The form of self-ignition called *pre-ignition* may take place in a dual-fuel engine *before* the pilot oil is injected, or in a spark-ignited engine *before* the spark occurs. (*Pre* means *before.*) During the compression stroke, both the pressure and the temperature of the mixture are raised. That portion of the mixture which is in contact with a hot spot, such as a carbon deposit, valve, piston top, or spark plug electrode, becomes hotter than the rest of the mixture and may reach ignition temperature. It then self-ignites and its flame causes the rest of the mixture to burn while the piston is still rising. Pre-ignition is harmful because it greatly reduces engine output, creates excessive pressures and forces, and causes an annoying knock. In modern engine designs, careful attention is given to avoiding hot spots and thus preventing pre-ignition.

Detonation. The form of self-ignition called *detonation* may occur with a lower compression pressure than that which would cause pre-ignition, and it is therefore more likely to happen. Detonation is self-ignition in which a distant portion of the mixture that has not yet been reached by the igniting flame fires suddenly of itself because of the influence of the high pressure and temperature of the other portion of the mixture which has already burned. Note that the *time* when detonation occurs is *after* the spark occurs in a spark-ignited engine, and *after* the pilot oil commences burning in a dual-fuel engine.

The characteristic feature of detonation is a shock wave which travels within the cylinder at a speed higher than that of sound (supersonic) as a result of particularly rapid combustion. The ensuing sudden and excessive rise in pressure produces a sharp metallic *knock* or *ping*. Not only is the sound of detonation objectionable; it also increases the engine pressures and heat losses, reduces the engine's power and efficiency, and may damage it.

You have probably encountered detonation occasionally when operating an automobile. It occurs in a gasoline engine when the operating conditions are unfavorable. For example, when low-octane gasoline is used or when the spark is advanced too early detonation will occur.

The process of detonation is not fully understood, but the usual explanation is that after part of the mixture has ignited, the resulting combustion pressure supercompresses the still-unburned mixture. The supercompression has two effects: (1) it raises some of the unburned mixture (often called *end-gas*) to its self-ignition temperature, causing it to fire explosively before the normal combustion flame reaches it; and (2) the heat produced by the supercompression decomposes part of the unburned mixture into other highly combustible gaseous compounds which then burn violently.

Time delay is also involved in the detonation process. When the end-gas is exposed to a given temperature and pressure produced by the supercompression, it self-ignites only after a definite (though short) period of time has elapsed. The higher the temperature or pressure, the shorter is this time delay.

Principles

Now let's see what these facts lead to in the several kinds of gas-burning engines.

Ordinary *low-compression gas engines* draw into their cylinders and compress a mixture which is nearly perfect and is therefore well within the explosive range. A perfect mixture of typical natural gas and air contains about 9.4 percent gas; an air-fuel ratio of about 9.64 to 1:

$$\frac{\text{Air}}{\text{Fuel}} = \frac{100.0 - 9.4}{9.4} = \frac{9.64}{1}$$

The mixture gets hotter when it is compressed, but if the compression ratio is low enough, the mixture does not heat up to the ignition temperature during the compression stroke. Consequently, the mixture does not burn until it is ignited by an electric spark at the end of the compression stroke. The high temperature of the spark causes the adjacent gas to start burning, whereupon the combustion spreads quickly through the whole mixture.

But if we try to use high compression like that of a diesel, the hottest part of the mixture reaches its ignition temperature before the end of the compression stroke, and pre-ignition takes place. Even if we keep the compression somewhat below that of a diesel so as to avoid pre-ignition, detonation occurs after the spark has ignited the part of the mixture nearest the spark plug. The resulting ill effects make the use of high compression impractical in an engine which employs a mixture of perfect or nearly perfect composition.

These facts led to the recent development of engines using extremely lean gas-air mixtures.

Diesel engines have always had to use extremely lean mixtures. Here's the reason. A diesel injects its oil fuel at the end of the compression stroke into air heated by compression above the ignition temperature of the oil. In an extremely short time, the oil particles must vaporize, find the oxygen and combine. This can happen only if there is a large excess of air, which means a large air-fuel ratio and an extremely lean mixture. At partial loads, the mixture becomes still leaner because less fuel is injected into the same amount of air.

Now suppose we change the diesel engine into a *dual-fuel engine* by admitting gas along with the air, *before* compression, while continuing to inject some oil at the end of compression. Natural gas has about the same heating value per pound as fuel oil; consequently, if the engine produces the same power as before, it will use about the same weight of gas as it did oil, and the air-fuel ratio will remain about the same for the gas mixture as for the oil mixture. Since the air-fuel ratio of a diesel engine at full load is about 21 to 1, the corresponding gaseous mixture will be extremely lean — about 4½ percent gas:

$$\frac{100\%}{21 + 1} = 4\tfrac{1}{2} \text{ percent}$$

This percentage is just below the lean explosive limit.

After this mixture has been taken into the cylinder, compression raises its temperature to about 1000°F, which would cause it to self-ignite *if* the mixture ratio were within the explosive range. But since it is outside this range, it does not ignite of its own accord. At the proper time near the end of the compression stroke, oil is injected. When the oil droplets hit the heated gas-air mixture, the oil begins to burn, just as it does when it is injected into heated air in a diesel engine. The high temperature of the burning oil droplets ignites the gaseous mixture at numerous points, and the added heat causes the whole of the mixture to burn rapidly.

The least amount of oil which will satisfactorily trigger the combustion of the gaseous mixture is about 4 percent of full-load fuel in dual-fuel engines. The percentage of pilot oil is always expressed on a heat-value basis.

Dual-fuel engines operate in like manner when a larger proportion of oil fuel is injected. In fact, the proportions of oil and gas may be changed instantly on a running engine from, say, 5 percent pilot oil and 95 percent gas all the way to 100 percent oil and no gas. This ability to use the two fuels in various proportions is quite useful in applications where sufficient gas is not always available.

Spark-ignited high-compression gas engines also apply the foregoing principle of using lean mixtures just outside the explosive range in order to avoid pre-ignition and detonation. But instead of using a spray of pilot oil to ignite the compressed gaseous air mixture, they use an electric spark. However, an electric spark is unable to ignite the gas-air mixture as effectively as the pilot oil; this fact leads to differences in design and operation.

Dual-Fuel Engines

Like the gas-diesel engine the dual-fuel engine was also developed from the diesel engine, but much later. It appeared during World War II, when low-cost natural gas was becoming widely available at the same time that the price of diesel fuel oil was rising rapidly. Dual-fuel engines met this cost problem by permitting the efficient use of gas fuel instead of oil in engines that were only slightly modified from ordinary diesels.

Dual-fuel engines compress a gas-air mixture to diesel pressures. Self-ignition (pre-ignition and detonation) is prevented by employing an extremely lean mixture, below the lean explosive limit, as you learned in the previous section.

Power Rating

The amount of load which a dual-fuel engine can carry is limited by detonation. The reason is simple. As the load is increased, more fuel must be introduced to develop the increased power. Thus, if the load is increased on a dual-fuel engine (when it is running with the maximum proportion of gas and the minimum proportion of pilot oil), the governor causes more gas to be admitted. Since the amount of air in the cylinder remains constant, the ratio of gas to air increases. At some point of increased load, the gas-air mixture ratio becomes rich enough to enter the explosive range, whereupon detonation sets in.

The power rating of a dual-fuel engine is therefore limited by its tendency to detonate. Generally, the commercial power rating of a dual-fuel engine is the same or slightly more than that of the same size engine running as a diesel on oil fuel only.

Dual-Fuel Engine Features

The special features that distinguish dual-fuel engines from ordinary diesel engines (sometimes called oil-diesels to avoid confusion) are: (1) means for admitting the gas fuel, (2) means for reducing the air flow at partial loads, (3) modifications of the oil-injection system, (4) governor modifications, and (5) gas valves and proportioning schemes.

Admitting the Gas Fuel

Unsupercharged four-cycle dual-fuel engines generally admit the gas fuel into the intake air before the air reaches the cylinder itself. This may be done in two ways: (1) the gas can enter the main air stream to all the cylinders at the entrance to the air-intake manifold, or (2) the gas can enter the individual air passage leading to each cylinder. Most builders prefer not to admit gas to the main air stream because of the difficulty of getting even distribution to all the cylinders.

Fig. 18-4 shows the simple method used by Cooper-Bessemer and others for in-

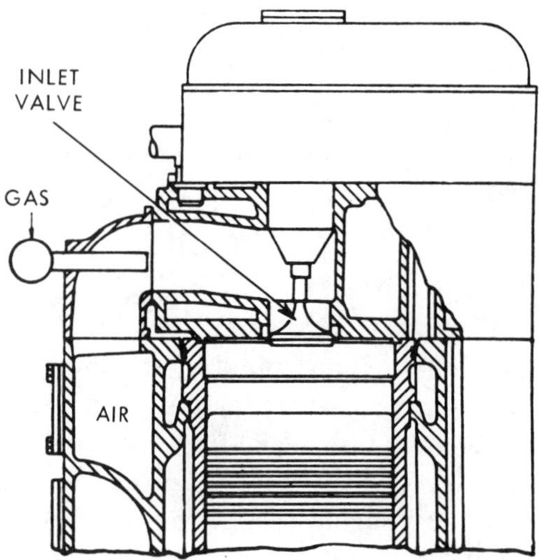

Fig. 18-4. Nozzle in air-intake elbow admits fuel gas in four-cycle dual-fuel engine. (*Power Magazine*)

troducing the gas into the air-intake passage leading to each cylinder. The gas flows from a header to a slotted or perforated tube inserted in the elbow which connects the air manifold to the cylinder head. Although the gas flows into the intake passage continuously, even when the inlet valve in the cylinder head is closed, the inrush of air when the inlet valve opens seems to tear apart the gas accumulation and produce a uniform combustible mixture.

In another design the gas is likewise admitted into each intake elbow, but instead of being allowed to enter freely, the gas is restrained by an automatic plate-type valve, shown at *M* in Fig. 18-5. The plates are held closed by light springs and are opened by the suction created by the flow of air into the cylinder when the main inlet valve opens. Consequently, gas enters the elbow only when there is an air flow to carry the gas into the cylinder.

In two-cycle engines, on the other hand, the gas should not join the scavenging air *before* the air reaches the cylinder. As you know, a two-cycle engine uses air flow to clear the cylinder of the burned gases; consequently, part of the scavenging air passes out with the exhaust. Therefore, if the gas charge were mixed with the entering scavenging air, part of the gas would go to waste. For this reason, two-cycle engines admit the gas *directly to the cylinder* through a *timed* valve which does not open until the exhaust ports or valves have closed.

Fig. 18-6 shows the Fairbanks-Morse design. The upper camshaft opens the gas-injection valve as soon as the rising piston has covered the exhaust ports. Further piston motion then compresses the mixture of gas and air.

Two-cycle engines require higher gas-injection pressures than four-cycle engines because the gas must be admitted quickly on the up-stroke, after the exhaust ports or valves have closed and before the compression pressure builds up too high. Furthermore, it is more difficult to mix the gas and air thoroughly in a two-cycle engine, not only because of the shorter time available, but also because the gas does not inject into a rapidly flowing air stream as in a four-cycle engine. A higher gas-injection pressure helps the mixing. Good mixing is essential because if part of the mixture is richer than the average, it may cause detonation.

Gas admission in four-cycle supercharged engines also must be timed because all supercharged engines use large valve overlap. In such engines, the inlet valve may open as much as 160 degrees

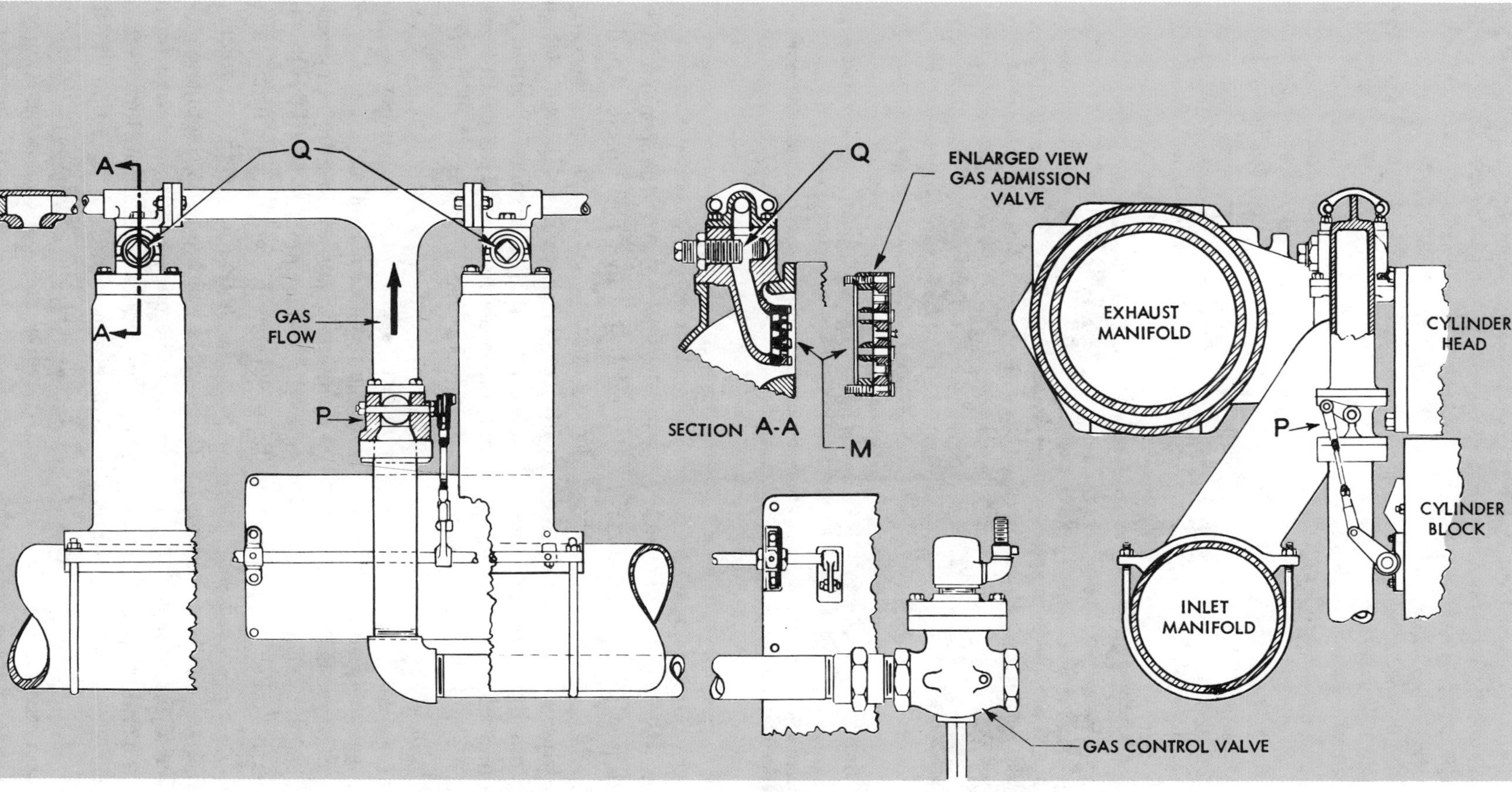

Fig. 18-5. Gas-admission valve and gas control on four-cycle non-supercharged dual-fuel engine. (*Diesel Power*, publication)

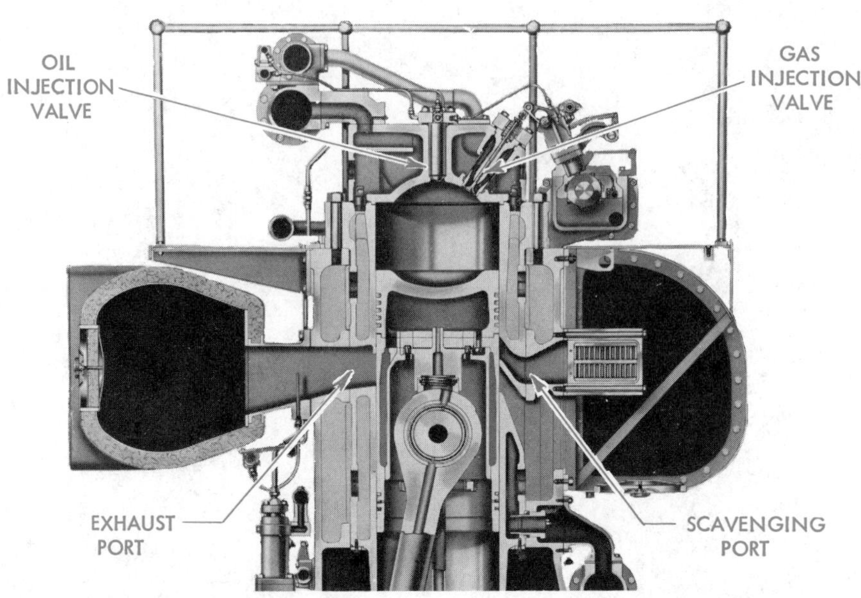

Fig. 18-6. Gas-injection valve in cylinder head of two-cycle dual-fuel engine. (Colt Industries, Fairbanks Morse Motor and Generator Operations)

before the exhaust valve closes. Consequently, if fuel gas joined the supercharged air *before* the exhaust valve closed, the air which blew through the exhaust valve during the scavenging process would carry unburned gas with it. By admitting the gas only *after* the exhaust valve has closed, all of the gas stays in the cylinder and is used.

Timing of the gas admission is accomplished in several ways. One method is to use a gas-injection valve directly in the cylinder head, similar to the two-cycle design in Fig. 18-6. However, this introduces another valve into the already crowded four-cycle cylinder head.

To avoid this, several other designs are used in four-cycle supercharged engines. The method shown in Fig. 18-7 uses a cam-actuated gas-admission valve to feed the gas into the supercharged air stream while the inlet valve is still open but after the exhaust valve has closed. Another timing method combines the gas valve with the intake valve. In the design shown in Fig. 18-8, a collar mounted on the intake valve stem controls the gas. The intake cam is so shaped that it at first opens the valve to admit air only, and later opens the valve farther to permit gas to enter as well as air.

A unique way of locating the gas-injection valve directly in the cylinder head without introducing another valve is to make double use of the customary check valve which admits compressed air to start the engine. Such a design for a dual-fuel engine is shown in Fig. 18-9.

The engine is started like an oil-diesel, using oil fuel only, and the air-starting check valve receives compressed air on each power stroke. After the engine has

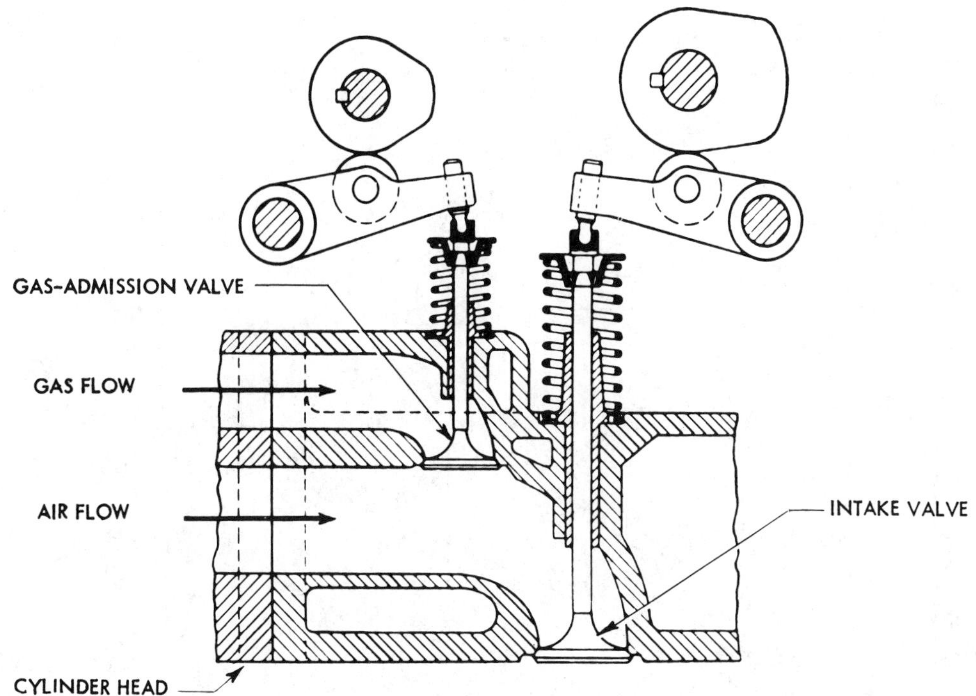

Fig. 18-7. Timed gas-injection valve feeds gas into intake airstream of four-cycle supercharged engine. (*Power Magazine*)

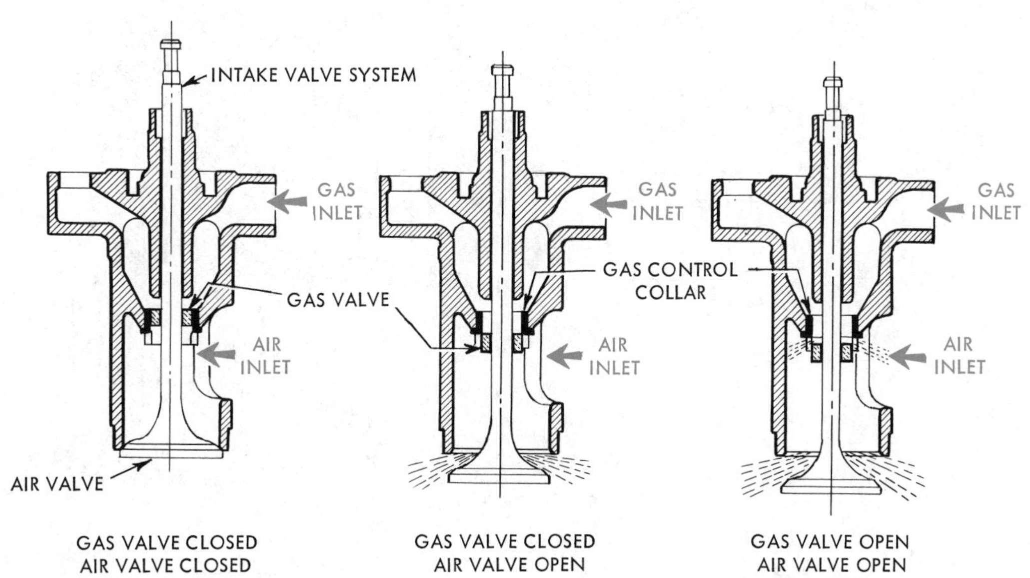

Fig. 18-8. Gas-timing valve mounted on stem of air-intake valve of four-cycle supercharged engine. (Worthington Corp.)

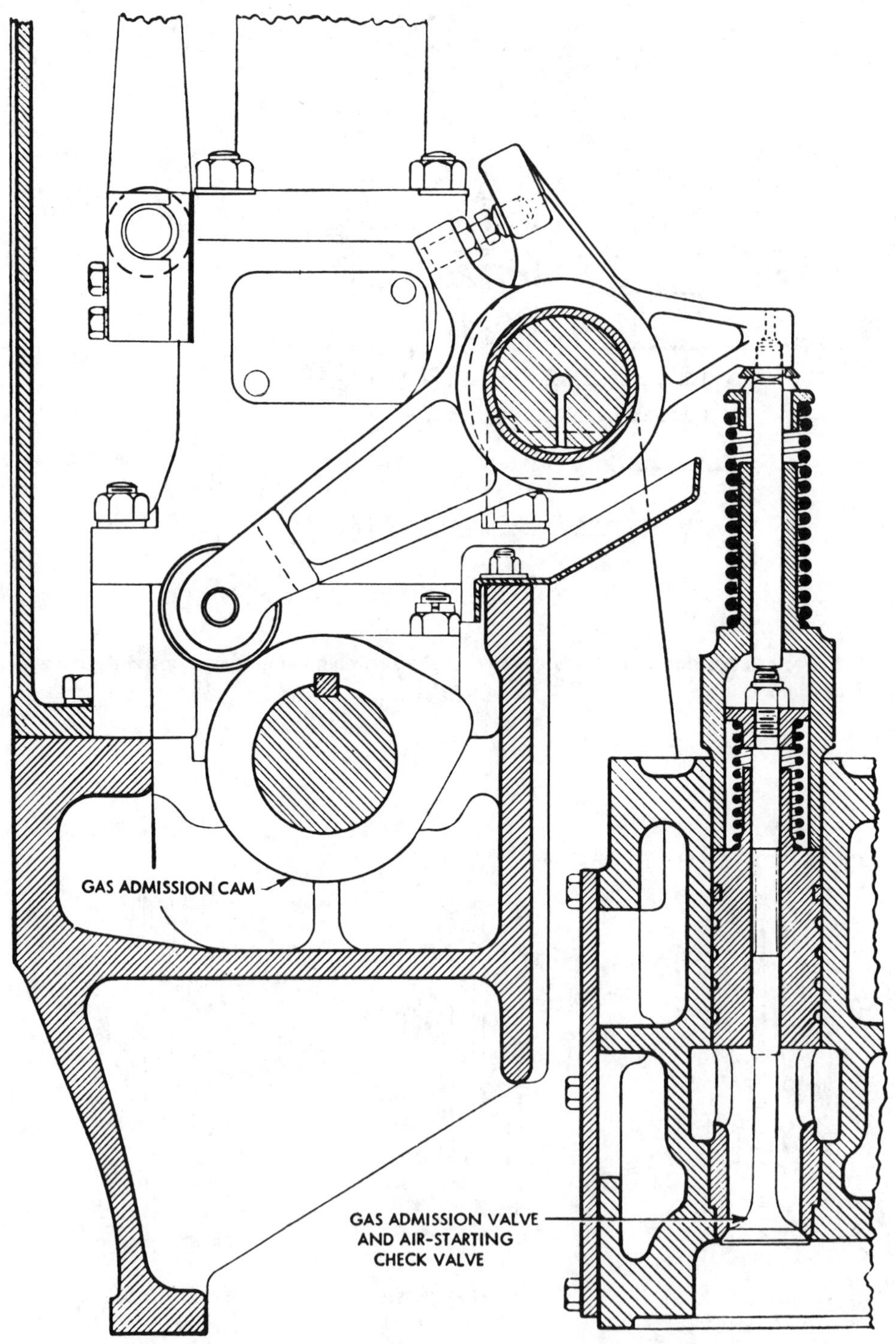

Fig. 18-9. Gas-timing valve feeds gas directly into cylinder of supercharged engine. The gas-timing valve uses air-starting check valve already in cylinder head.

been started and the starting air has been cut off, it is switched to dual-fuel operation. This turns on the gas and puts the gas-admission cam into action. This cam is so timed that it admits the gas during the period between the closing of the exhaust valve and the subsequent closing of the air-intake valve.

Reducing the Air Flow at Partial Load

Early dual-fuel engines showed excellent fuel economy when running at full load, but at partial loads the fuel consumption was extremely high. The cause was found to be that at partial loads the excessively lean mixture ignited and burned so slowly that the combustion was not completed and a large part of the gas passed out unburned. The excessively lean mixture at partial loads resulted from the fact that, as in an ordinary diesel engine, the quantity of air admitted remained the same at all loads; consequently, an air-fuel ratio of say 20 to 1 at full load rose to about 40 to 1 at half-load.

Modern engines, therefore, use devices to reduce the air flow when the load falls off. By decreasing the air quantity at partial loads they maintain a good flammable mixture at all loads.

There are several ways of reducing the air flow. The simplest method, used on naturally-aspirated (non-supercharged) four-cycle engines, is to throttle the incoming air by means of a damper (butterfly valve) in the intake air passage. Two-cycle engines reduce the air flow by decreasing the output of the scavenging blower or by opening a by-pass valve on the blower. On turbocharged engines some of the exhaust gas is by-passed around the gas turbine, causing the turbocharger to run slower and thus deliver less air.

The control of air quantity to suit the load is generally *automatic*. One way of doing this is to operate the air control from the fuel control shaft of the engine governor. Another way makes use of the fact that when the load falls off, the temperature of the exhaust gases drops (because less fuel is being burned). A temperature-responsive element in the exhaust manifold moves a pilot valve connected to a servo-motor which works the air control and maintains a satisfactory air-gas ratio.

The foregoing throws light on the importance of the air-gas ratio in high-compression gas-burning engines. The mixture ratio must be lean enough to prevent pre-ignition or detonation, but not so lean as to prevent complete combustion. As you learned earlier in this chapter, the theoretical or perfect mixture ratio of typical natural gas and air is about 9.64 to 1. This ratio is much too rich for a high-compression engine.

Permissible ratios vary with engine cylinder size, inlet air temperature, gas composition and other factors. For example, the air-gas ratio for a particular set of conditions in a dual-fuel engine might be adjusted for a rich limit of 15.5 to 1 at full load to avoid detonation, and a lean limit of 20 to 1 at no load to avoid quenching the combustion.

In dual-fuel engines, yet another limit controls the amount of air-throttling which can be employed for the purpose of enriching the mixture at partial loads. This limit is reached if the throttling so reduces the temperature after compression that the pilot oil fails to ignite.

Modifications of the Oil-Injection System

Earlier models of dual-fuel engines employed the same oil-injection pumps and nozzles as the corresponding oil-diesels. However, experience showed that full-size pumps and nozzles worked irregularly when the pump delivery was cut down to the tiny quantities of oil actually needed for pilot ignition. To obtain *uniform injection* of pilot oil, the pumps had to be adjusted to deliver more than the minimum amount required for ignition purposes. This increased operating costs because oil fuel was far more expensive than gas in most localities where dual-fuel engines were used.

This difficulty led to various modifications of the oil-injection system. One simple solution was to substitute injector nozzle tips of smaller capacity, capable of pulling only one-half of the oil-diesel rating. In one make of engine, such smaller nozzle tips reduce the pilot-oil consumption to 4 percent of the total full-load fuel, compared to 6 percent with full-size tips. However, this arrangement is not feasible for engines which may be required to shift without notice to full-oil operation.

A more flexible arrangement uses two sets of oil-pump plungers of different sizes, the smaller size delivering pilot oil only, and the larger size having sufficient capacity for full-power operation on all-oil. Some designs employ two independent pumps for this purpose. Other designs such as Fig. 18-10, use a duplex-type pump which has two pumps of different sizes built in one casing. The small-diameter pilot plunger rides on top of the main plunger and always delivers its full-stroke capacity. The pilot-oil pump discharges into the oil-injection line through its own small delivery valve.

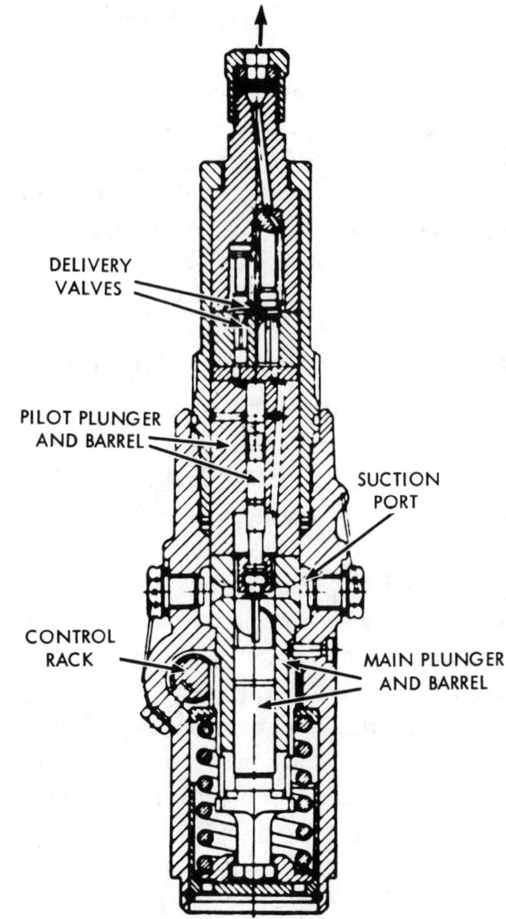

Fig. 18-10. Duplex-type oil-injection pump with pilot plunger and main plunger.

When the engine operates entirely on gaseous fuel (except for the pilot oil), the rack on the main plunger is at zero position and the main plunger therefore pumps no oil. When operating on all-oil, or with a combination of gas and oil in excess of the pilot-oil requirement, the governor actuates the main plunger rack to the desired setting. The total oil fuel delivered will then be the constant fuel-stroke capacity of the pilot plunger plus the additional oil delivered by the main plunger to meet the load demand.

Pilot-Oil Control. Some oil pumps, like the one just described, deliver the same amount of pilot oil at all engine loads (when running on gas fuel). Other designs put the pilot-oil delivery under governor control. At light loads, more pilot oil is injected in order to thoroughly burn the cooler mixture. At heavy loads, less pilot oil is delivered because the mixture is hot and more combustible. Such control is intended to reduce the consumption of pilot oil to the minimum required to assure satisfactory ignition at all loads.

Another control which has been used increases the oil delivery momentarily when the load changes suddenly or when the engine is shifted from one fuel to the other. Its purpose is to stabilize the combustion while conditions are changing.

Governor Modifications

Governors of dual-fuel engines must be able to control the delivery of either gas or oil or a combination of both. In installations where sufficient gas to carry the engine load is always assured (or if sufficient advance notice of impending gas shortage is always given), the governor linkage may be arranged for manual transfer from one fuel to the other.

For running on gas, the governor connects to a throttle valve in the gas supply; for running on oil, the governor connects to the usual diesel oil-injection regulating system. The engine operator shifts a fuel-selection lever which moves a pivot in the governor linkage and thus transfers the governor's control action from one fuel to the other. When he shifts to gas, the oil-injection pump automatically continues to feed a fixed amount of pilot oil.

This scheme of manual fuel selection may be extended to permit the operator to adjust the governor linkage to feed *both* gas and oil in any desired proportions. If the amount of available gas fuel shows signs of diminishing below the selected quantity, the operator can adjust the fuel-selection lever to provide the additional amount of oil fuel required to meet the load demand.

In many plants, however, the uncertain supply of gas makes manual fuel proportioning impractical. Automatic proportioning is therefore used. One method of accomplishing this is to use the first half of the governor stroke for controlling gas, and the second half for controlling oil. During the *first* part of the governor travel (starting at the no-load position), a slip arrangement holds the oil-pump racks in pilot-oil position, and the governor regulates the position of the *gas-throttle valve* to supply enough gas to carry the load. However, if no gas is available, or if the gas supply is insufficient for the engine loading, the governor travels further in its search for fuel. It then gets into the *second* part of its range, where it starts to move the *oil-pump racks* and thus feeds more oil fuel. In this part of the range, the gas throttle remains wide open. The engine then burns all the gas that is available together with sufficient oil, under governor control, to develop the required power.

Safety Precautions

Gas-burning engines, like diesel engines, require protection from damage caused by low lubricating oil pressure, high water jacket temperature or overspeed. If such a protective device shuts down a gas-burning engine it is important, for safety's sake, that the gas-admission valve be closed at once to prevent gas accumulating in or around the engine. Furthermore, dual-fuel engines should be stopped quickly if the fuel-oil supply should fail, because there would be no pilot oil injected to ignite the gas. For these reasons, most gas-burning engines are fitted with *automatic* devices to shut off the gas supply whenever the engine stops or if the fuel-oil supply (to a dual-fuel engine) fails.

Gas Valves and Proportioning Arrangements

There are two common systems for regulating the engine's total intake of gas and dividing it among the cylinders. One method uses a single butterfly valve in the entrance to the main gas manifold to throttle the gas flow to suit the engine load, together with individual proportioning valves or nozzles in the gas-intake lines to each cylinder. These proportioning valves or nozzles are adjusted to fixed settings to divide the gas flow equally

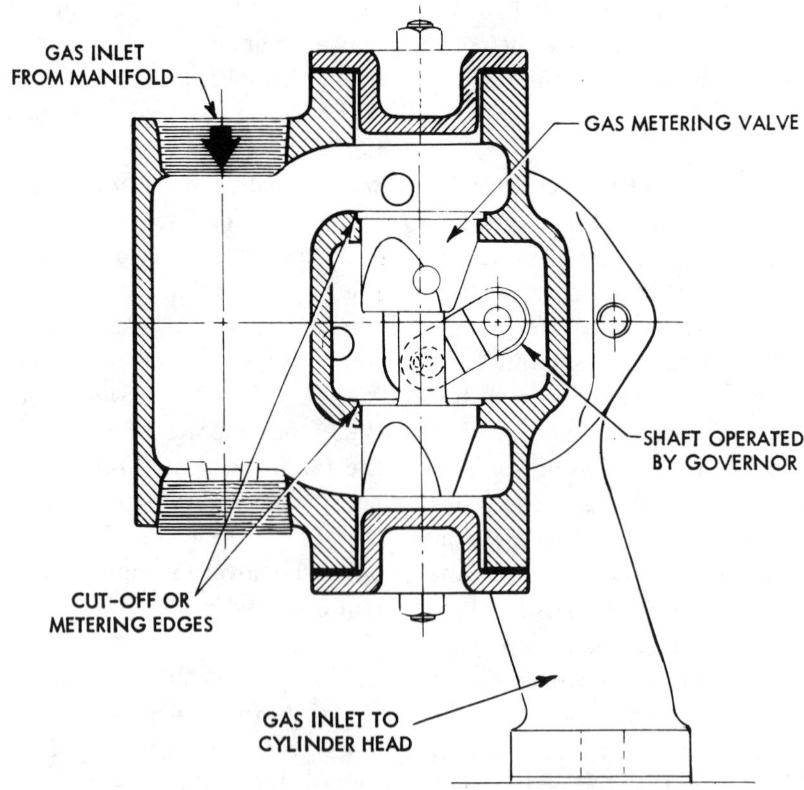

Fig. 18-11. Individual gas-metering valve for each cylinder. (Worthington Corp).

among all cylinders. (Fig. 18-5 shows such an arrangement. *P* is the governor-controlled butterfly valve in the entrance to the gas manifold; *Q* is the individual proportioning valve for each cylinder.)

The other method employs individual gas-metering valves for each cylinder, all of the metering valves being controlled by the governor through a common shaft. The Worthington design of Fig. 18-11 illustrates this type of gas-metering valve. It comprises two tapered plugs, which permit the valve to move freely and to meter accurately. Each valve is individually adjusted to provide equal gas flow to all cylinders, while the governor controls the *total* amount of gas by moving all the valves simultaneously.

High-Compression Spark-Ignited Gas Engines

The modern high-compression gas engine, using spark ignition, developed surprisingly, not from the long-known low-compression gas engine, but from the much newer dual-fuel engine. Engineers learned from the dual-fuel engine that a very lean mixture of gas and air could be highly compressed without pre-igniting or detonating. The dual-fuel engine ignited this gaseous mixture at the right moment by injecting a small amount of pilot oil. However, in localities where cheap gas was always available, the cost of the pilot oil greatly increased the operating expense.

Why not ignite the compressed gas-air mixture with a spark instead of with pilot oil? Tests quickly showed that the kind of electric ignition system used on low-compression gas engines would not do. Therefore, the electric ignition system was strengthened to provide (1) the higher voltage required to cause a spark to jump across the denser mixture caused by the higher compression, and (2) the increased heat energy needed to ignite a lean mixture as compared with a perfect mixture.

Even the strongest spark cannot fire a lean gas-air mixture as effectively as a burning spray of pilot oil. A spark produces heat energy in limited amount and at only one point, from which the flame must spread to the rest of the mixture. A spray of pilot oil not only supplies far more heat energy to make the gaseous mixture ignite and help it burn, but also provides numerous igniting points throughout the mixture so that the flame fronts have a shorter distance to travel. As a result of the poorer ignition process, spark-ignited high-compression engines generally require somewhat richer mixtures than dual-fuel engines and show a greater tendency to detonate.

Factors Promoting Detonation

The major factors which promote detonation in high-compression spark-ignited gas engines are:

1. Engine overloaded
2. Cylinders unequally loaded
3. Compression ratio too high
4. Compression pressure too high (in supercharged engine)
5. Mixture too rich
6. Mixture not uniform
7. Spark timed too early
8. Combustion chamber too elongated
9. Turbulence insufficient
10. Entering air temperature too high
11. Jacket water temperature too high
12. Fuel gas has low ignition temperature (high hydrogen content)

Many engine builders, to cope with variations in gas composition and changing operating conditions, use lower compression ratios on their high-compression spark-ignited engines than on their corresponding dual-fuel engines. Permissible compression ratios are less in ordinary two-cycle engines than in four-cycle, because in most two-cycle engines the combustion chamber is not fully scavenged, nor is there as much time to cool it.

Small engine cylinders can use higher compression ratios than large cylinders of the same design. In the present stage of development, compression ratios for large two-cycle engines range from 7.5 to 9.5 (based on trapped volume, the volume in the cylinder after the ports have closed).

For four-cycle engines the compression ratios commonly range from 10.5 to 12.5. The Fairbanks, Morse two-cycle, *opposed-piston,* spark-ignited gas engine uses a ratio of 12.5. Compression *pressure* depends not only on the compression ratio but also on the pressure in the cylinder when compression begins. In highly supercharged four-cycle engines the compression pressure may reach 600 psi.

Present-day lean-mixture spark-ignited engines, although much more efficient than low-compression gas engines, generally consume slightly more fuel than their dual-fuel counterparts. The reasons are: (1) some of the gas does not burn fully, and (2) if the compression is made somewhat less, the fuel efficiency necessarily becomes lower.

Special Features

As you have just learned, spark-ignited engines tend to use lower compression than dual-fuel engines. Some other features follow:

Electric ignition system of high voltage (25,000 to 30,000 volts) is required. Shielded spark plugs take the place of the fuel-oil nozzles in the cylinder heads. Spark plugs often contain resistors to reduce burning of electrodes. Spark plugs must be well cooled to prevent pre-ignition. (See Chapter 19 for details.)

Gas admission is accomplished in the same way as in dual-fuel engines.

Air flow is automatically reduced for operation at partial loads, because if the mixture is too lean it will not ignite and burn properly. Air-flow control is more important on spark-ignited engines than on dual-fuel engines, since an electric spark is not as good an ignitor as a spray of burning pilot oil. The mixture ratio must be held within closer limits.

Some engine builders consider accurate air control so essential that they use individual air-throttling valves on each cylinder. Many engines today carry the automatic air control to such a point that they *proportion* the air flow to the gas flow throughout the entire range, and thus maintain a nearly constant air-gas ratio at *all* loads.

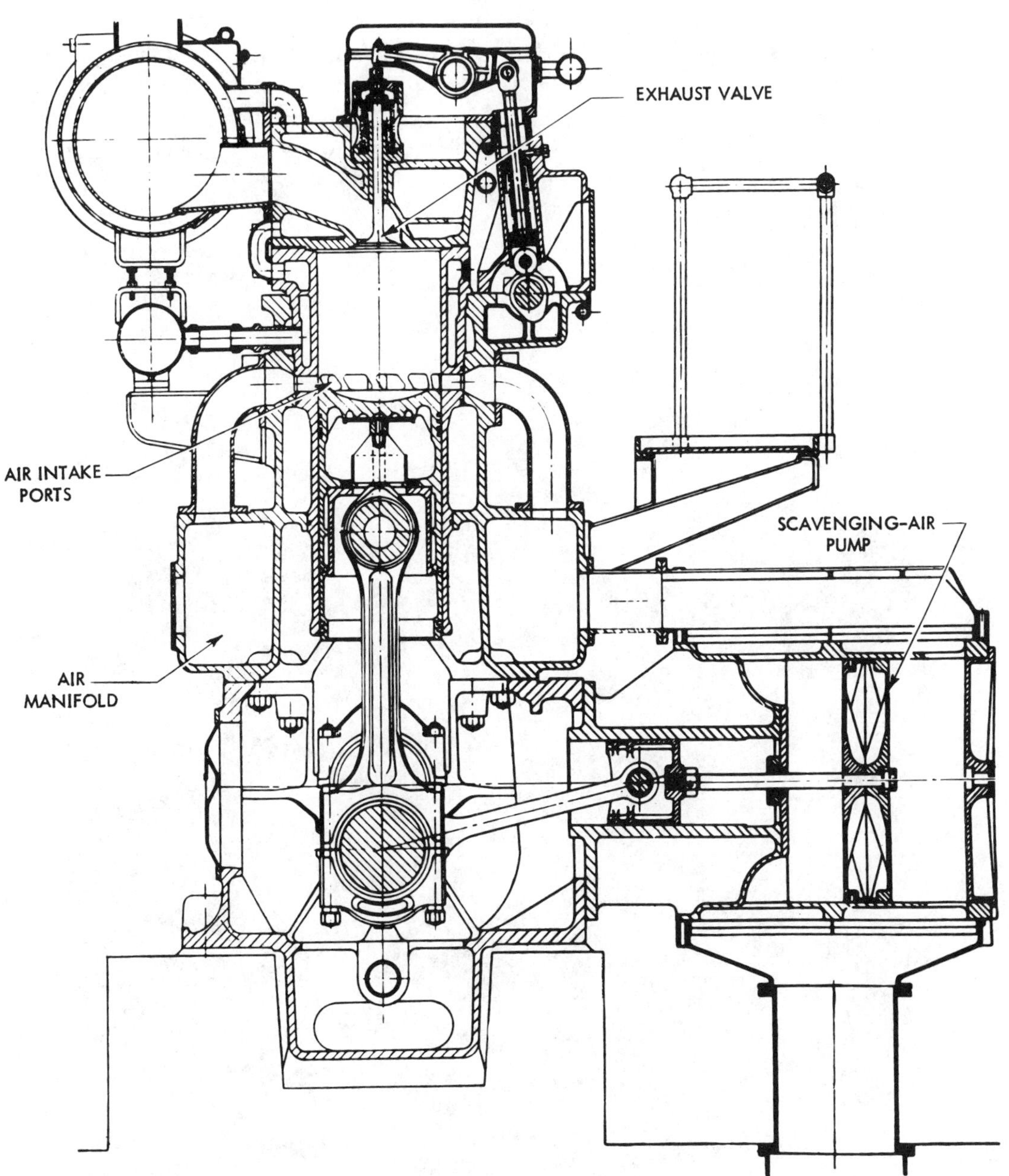

Fig. 18-12. Cross-section of Worthington two-cycle spark-ignited high-compression gas engine. (Worthington Corp.)

Gas proportioning is accomplished in the same ways as in dual-fuel engines. Still another method used on some spark-ignited gas engines is a single mixing valve which meters *both* the gas and air so as to maintain a substantially constant air-gas ratio at all loads. This valve is operated by the engine governor and adjusts the flow of gas (and air) to suit the load.

Intake air cooling is used on practically all supercharged engines and on many that are not supercharged because, all other things being equal, the lower the temperature of the air entering the engine cylinder, the more power it can develop without detonation. Another benefit is that if the intake air is cooled enough it is possible to use a higher compression ratio and earlier spark timing and thus improve fuel efficiency. The air may be cooled by passing it around water-

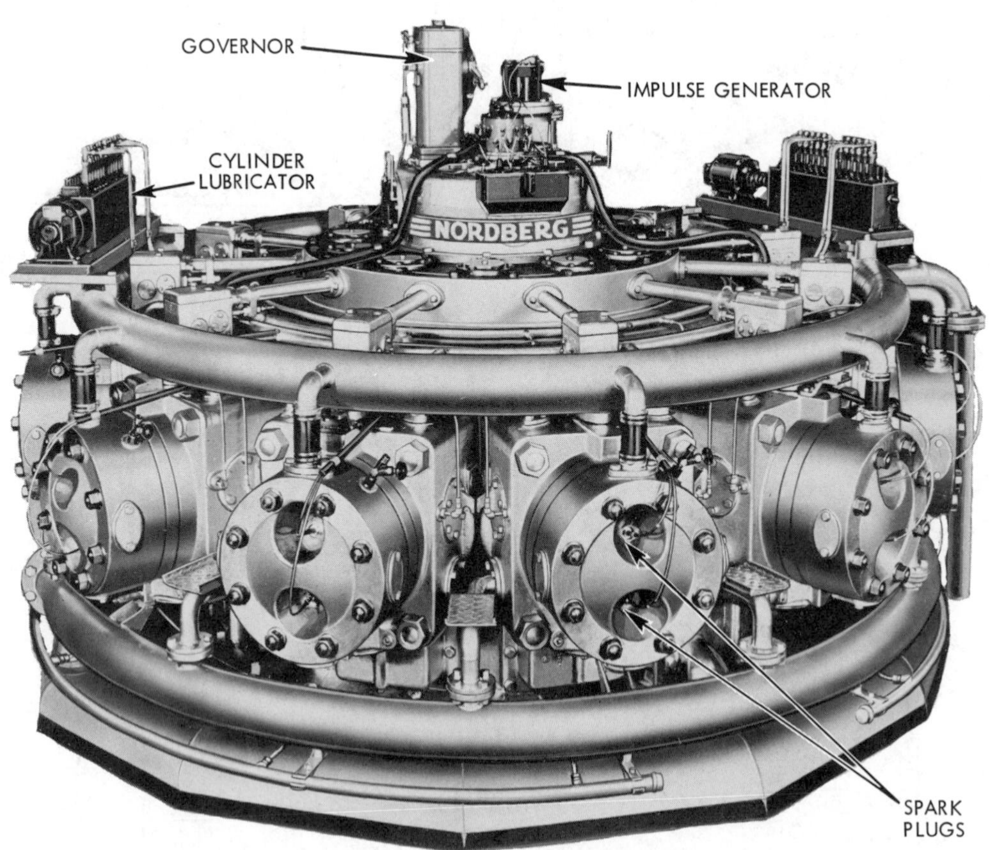

Fig. 18-13. Nordberg twelve-cylinder radial two-cycle spark-ignited high-compression gas engine. (Nordberg Mfg. Co.)

cooled pipes or through a water spray. It can be further cooled by expanding it (after a supercharger has compressed it) in an *air turbine*. This device, called a turbocooler, is described more fully later in this chapter, in connection with the Cooper-Bessemer KSV engine.

Typical Examples

Fig. 18-12 is a cross-section of a 1600 bhp, 320 rpm Worthington, two-cycle engine, with 8 cylinders, 16″ dia × 16″ stroke. It is of the uniflow type; the intake air enters the cylinder through ports and the exhaust gas leaves through valves in the cylinder head. Gas enters through a cam-actuated valve (not shown) in the cylinder head.

Fig. 18-13 shows the 2125 bhp, 400 rpm, Nordberg, two-cycle radial engine, with 12 cylinders, 14″ bore × 16″ stroke. Ignition current is supplied to two spark plugs in each cylinder by a special impulse generator which has no cams,

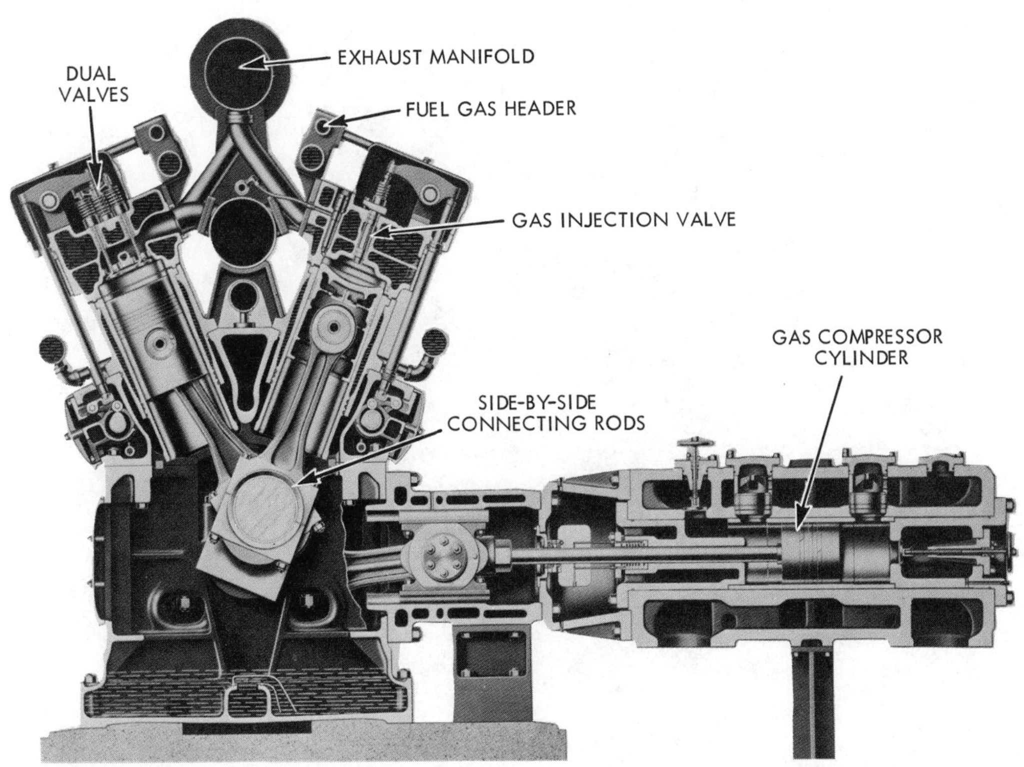

Fig. 18-14. Cross-section of Ingersoll-Rand KVT four-cycle spark-ignited high-compression gas engine and gas compressor. (Ingersoll-Rand Co.)

breaker points, brushes, or collector rings. The cylinders receive scavenging air and discharge exhaust gases through opposite rows of ports. Fuel gas enters the cylinder through a cam-actuated gas valve.

This 12-cylinder engine uses a different design of restraining linkage for the master bearing than the 11-cylinder radial engine described earlier. Two directly opposite connecting rods are rigidly con-

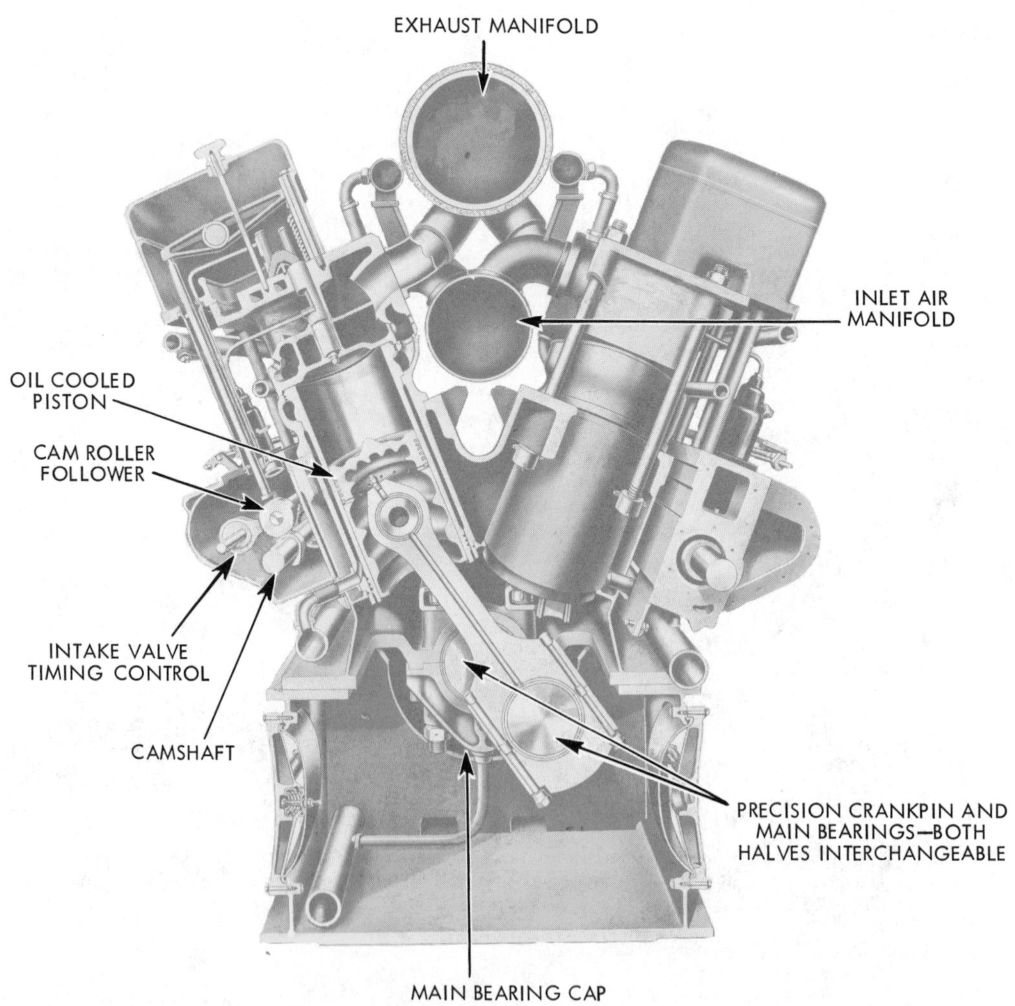

Fig. 18-15. Cross-section of Nordberg Supairthermal four-cycle high-compression gas engine. (Nordberg Mfg. Co.)

nected to extending knuckle pins, each of which carries a small restraining crank. A large restraining link connects the two cranks. This linkage permits the main crankpin with its master bearing to revolve freely, while keeping the master bearing in its proper position.

Going from two-cycle engines to four-cycle, Fig. 18-14 shows the Ingersoll-Rand KVT 16″ bore × 22″ stroke, V-type, high-compression gas engine with built-in gas compressor for use on gas pipelines. It has dual inlet and exhaust valves and dual spark plugs. Either of two supercharging systems are available: the *pulse system* or the *constant-pressure system.*

The pulse system uses a divided exhaust manifold and a compression ratio of about 8.5; it gives greater power capacity, but at a slight sacrifice in fuel efficiency. The constant-pressure system (with a single exhaust manifold as shown in Fig. 18-14) uses a higher supercharge pressure and closes the inlet valve early; the resulting high expansion ratio improves the fuel economy at a slight sacrifice of power capacity. The compression pressures of both systems are the same, 450 psi. The 16-cylinder engine, running at 330 rpm, is rated either 4500 bhp with a fuel rate of 6900 Btu/bhp-hr on the pulse system, or 4000 bhp with a fuel rate of 6300 Btu/bhp-hr on the constant-pressure system.

Many engine builders offer high-compression engines of the same cylinder size and general construction in all three types, namely, diesel, dual-fuel and spark-ignited. Following are two examples of such engines.

Fig. 18-15 is a cross-section of the Nordberg Supairthermal four-cycle turbocharged engine, 13½″ bore × 16½″ stroke, V-type. The crankshaft is underslung, the removable main-bearing caps are *below* the shaft. A jet of lubricating oil from the drilled connecting rod cools the piston crown. The Supairthermal engine was the first commercial American engine to utilize the principle of *variable* timing of the inlet valve.

At *heavy loads* an automatic device shifts the inlet valve linkage to cause the inlet valve to close earlier in the suction stroke; this reduces the charge volume, and prevents excessive compression and firing pressures without affecting the long expansion ratio needed for good efficiency. At *light loads*, the inlet valve is closed later in the stroke so as to maintain the full compression pressure; the resulting higher compression *ratio* improves the fuel economy and also facilitates starting. When used as a dual-fuel engine, the inlet valve timing is automatically adjusted by another device which lowers the compression pressure by 60 to 80 psi (to avoid detonation) when the engine is shifted to run on gas instead of oil.

Figs. 18-16 and 18-17 show a Cooper-Bessemer KSV four-cycle pulse-turbocharged engine, 13½″ bore × 16½″ stroke, V-type. The 16-cylinder size is rated at 4900 bhp at 514 rpm, and employs 8 exhaust manifolds to obtain uniform scavenging. The spark plugs (two per cylinder) are submerged in oil to keep them cool and prevent pre-ignition.

A unique feature of this engine is the use of a turbocooler, consisting of an air turbine driving an air blower. Its function is to feed the engine with the coolest possible air charge to increase the load it can carry without detonation. Here's how it works, Fig. 18-18. First, an exhaust-

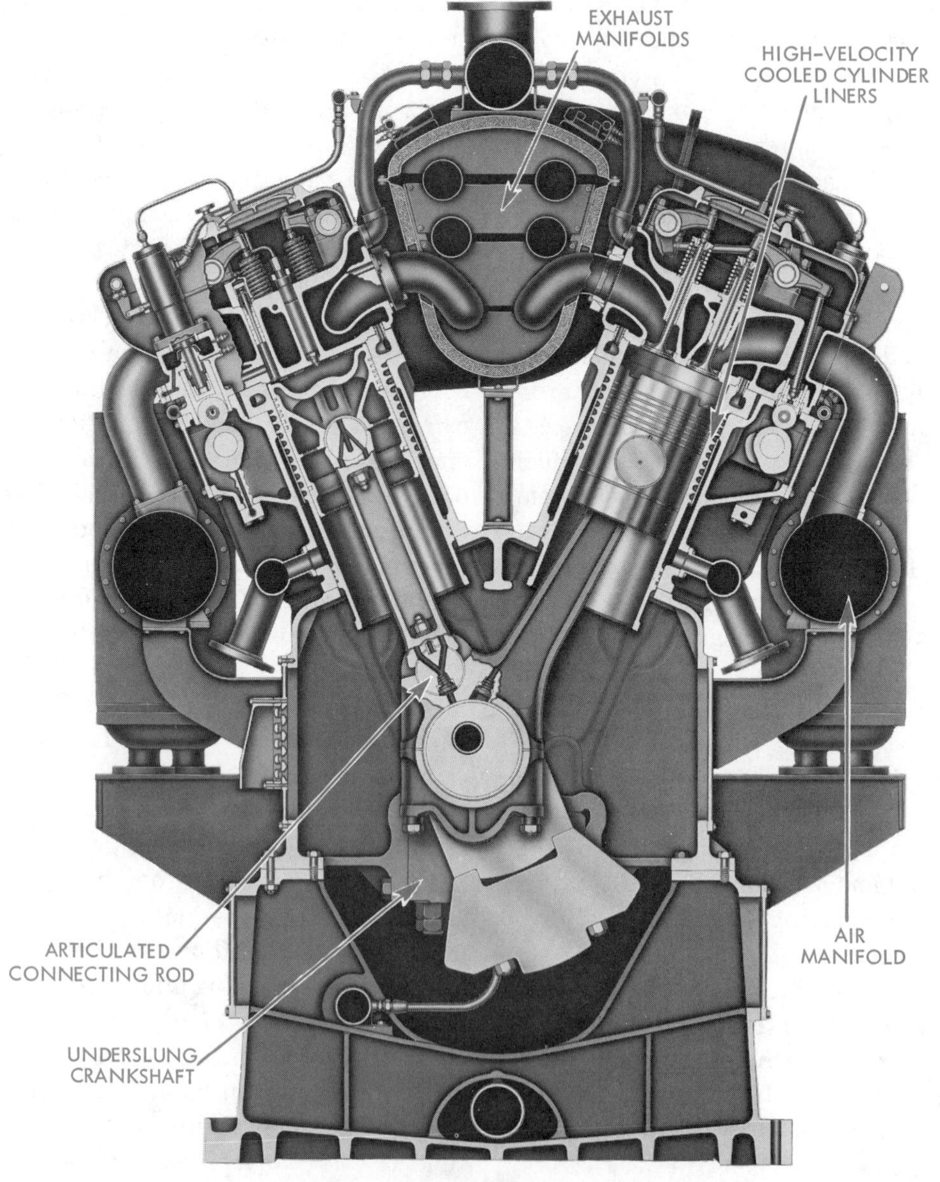

Fig. 18-16. Transverse cross-section of Cooper-Bessemer KSV four-cycle high-compression engine. (Cooper-Bessemer Div., Cooper Industries, Inc.)

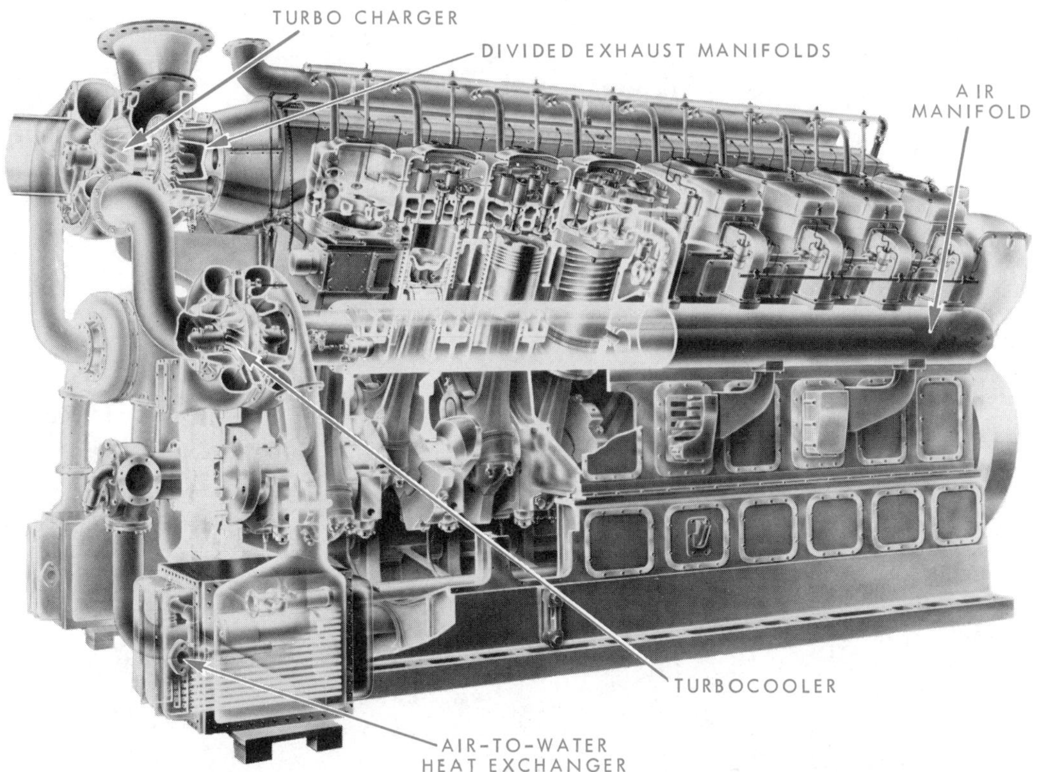

Fig. 18-17. Longitudinal cross-section of Cooper-Bessemer KSV four-cycle high-compression engine. (Cooper-Bessemer Div., Cooper Industries, Inc.)

driven turbocharger boosts the air pressure to about 2.7 times atmospheric pressure. Next, the blower wheel of the turbocooler compresses the air further. Following the double compression the air passes through an air-to-water heat exchanger to remove most of its acquired heat. After being cooled by the water, the air enters the expander wheel of the turbocooler; here it drops in pressure to produce the power to drive the connected blower wheel already mentioned. The expansion cools the air further, below the temperature to which the available water can cool it. In one test the turbocooler alone lowered the temperature of the air entering the cylinders from 132°F to 106°F, for a substantial increase in volume of air.

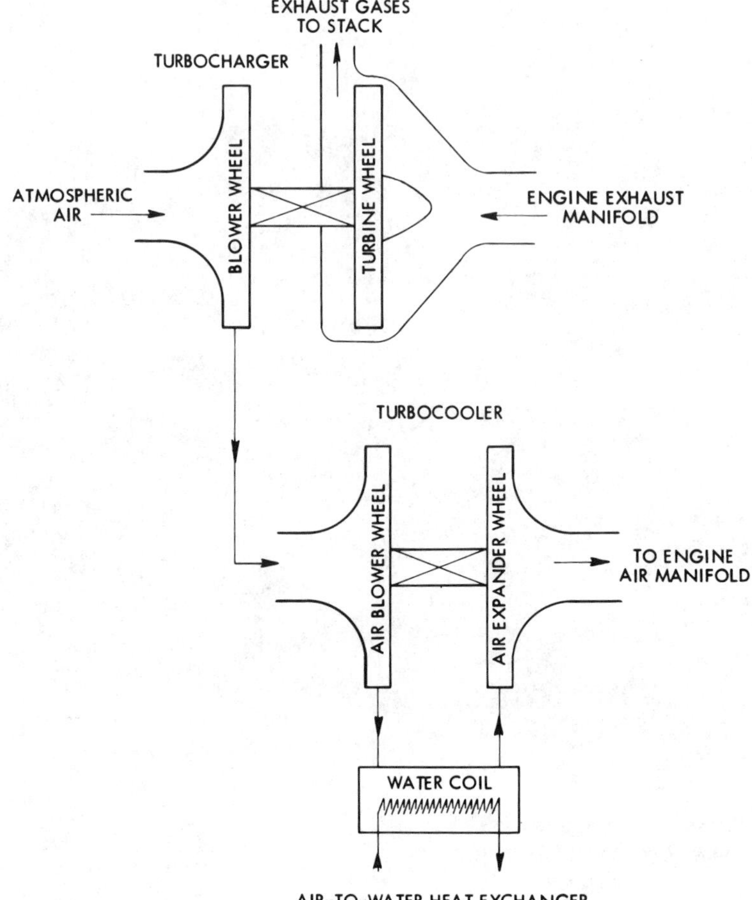

Fig. 18-18. Schematic of air turbocooled system.

Uses of High-Compression Gas-Burning Engines

You have seen how the ordinary diesel engine, which burns oil fuel efficiently, is now supplemented by related types of high-compression engines which efficiently burn gas, either alone or in combination with oil.

How these gas-burning engines are applied depends upon (1) the relative prices of oil and gas, (2) whether the gas is continuously available, and (3) the amount of time available to change over from gas to oil.

The gas-diesel engine, which injects the gas at the time of firing runs on about 95 percent gas fuel. If gas becomes unavailable or too costly, the engine can be changed over to an ordinary oil-diesel by exchanging certain parts. The change-over takes several hours' time.

The dual-fuel engine, on the other hand, can instantaneously switch from about 95 percent gas and 5 percent oil to no gas and 100 percent oil, or to any desired ratio of gas and oil in between. It is therefore adaptable to all sorts of situations. But if oil is expensive and gas is cheap, the cost of even the small amount of pilot oil increases the total fuel cost by a large percentage.

The spark-ignited engine uses no pilot oil; it runs on gas only. Obviously, it can be used only where the gas supply is adequate and uninterrupted.

Cheap gas is always available in the gas-compression stations on gas pipelines. Here *spark-ignited* high-compression engines find an ideal application.

Aluminum-reduction plants well illustrate the application of *gas-diesel* and *spark-ignited* engines. These plants convert aluminum oxide (alumina) to metallic aluminum by means of an electrochemical process which requires huge amounts of electric power. This power must be cheap because it takes about 10 kw-hr to produce a pound of aluminum.

The first engine-powered aluminum plant was built in Arkansas, where cheap natural gas was available. Its electric power plant had 78,000 kw capacity, of which half was in gas-diesel engines and half in low-compression perfect-mixture gas engines. However, neither type of engine showed the lowest possible power cost. The gas-diesels, although very efficient because of their high compression, required pilot oil, which was relatively expensive. The low-compression gas engines, on the other hand, though consuming no pilot oil, used more gas because of their lower efficiency.

Consequently, the second engine-powered aluminum plant, installed in Texas, utilized high-compression, lean mixture, spark-ignited gas engines. These engines, of the two-cycle type with a compression pressure of about 260 psi, were not only highly efficient (using about 8000 Btu/net kw-hr), but also consumed no oil fuel at all. To reduce the space occupied and the cost of this huge plant, the engines employed a unique radial design, with 11 cylinders grouped around a single crank on a vertical shaft.

Fig. 18-19 shows one of the three engine rooms, each containing forty engines producing 1000 kw net output each. Later, two more engine rooms were built, each containing thirty-seven 12-cylinder radial engines, producing 1100 kw net output each. The resulting total capacity, exceeding 200,000 kw, made this the world's largest internal-combustion engine power plant up to that time. It consumed 50,000,000 cu ft of gas daily.

The *dual-fuel engine*, with its ability to burn gas and oil in varying proportions, suits several important applications, such as the following:

1. In localities where plenty of cheap natural gas is available much of the time, *but not always*, dual-fuel engines are a logical answer to power needs in the absence of any cheaper source. Gas supply companies generally face much greater demands from their customers in cold weather, and sometimes cannot furnish all the gas that is wanted. Since the user

Fig. 18-19. Forty 1000 Kw high-compression spark-ignited radial gas engines in one engine room. (Nordberg Mfg. Co.)

of dual-fuel engines can readily shift to oil at such times, he can often obtain a lower gas price than that offered to general consumers who require gas at the time of the gas company's peak load.

2. Sewage-disposal plants employ dual-fuel engines to great advantage. Modern sewage-treatment plants use the activated sludge process in which air and bacteria digest the sewage, and convert it into harmless material like humus. The digesting process produces a gas having a heat value of about 650 Btu/cu ft, high-heat value. Although this gas contains considerable hydrogen sulfide (H_2S), which is corrosive, it burns readily in a

gas or dual-fuel engine. Such engines can provide much of the power needed to drive pumps, blowers, etc. Also, the heat in their jacket water and exhaust gas can heat the buildings and the sludge digesters.

However, the supply of sewage gas is not uniform, and at certain times the engines need more fuel than that available in the form of sewage gas. Dual-fuel engines fit this situation well and have come into wide use. They burn sewage gas alone (except for pilot oil) when it is available; at other times they burn all the sewage gas on hand and automatically make up the shortage with oil fuel.

Checking On Your Knowledge

The following questions give you the opportunity to check up on yourself. If you have read the chapter carefully, you should be able to answer the questions. If you have any difficulty, read the chapter over once more so that you have the information well in mind before you go on with your reading.

DO YOU KNOW

1. Explain the different principles of operation which enable the three types of internal combustion engines to burn gaseous type fuels.
2. What causes ignition limits in high-compression gas burning engines?
3. Explain the difference between pre-ignition and detonation in gas burning engines.
4. What is a dual-fuel engine?
5. What are 10 major factors which promote detonation in high-compression spark-ignited engines?
6. Explain the principle and function of an air turbocooling system.
7. Where are high-compression gas engines used and why are they used instead of conventional diesel engines?

Chapter

19 Auxiliary Systems

To run an internal-combustion engine, you must supply it with fuel and air, remove the spent gases, lubricate its cylinders and bearings, and circulate cooling water through its jackets. Also, you need some way of putting the engine into motion in order to start the firing of the fuel. If it has spark plugs they must be fired by high-voltage electricity, accurately timed. In addition, many engines are provided with protective devices which act when something goes wrong, and either give an alarm signal or shut the machine down. Sometimes engines are fully automated, they are started, loaded, and stopped automatically.

The foregoing services are provided by auxiliary equipment assembled into systems. The best-built engine will not work satisfactorily unless the various auxiliary systems do their jobs too.

The purpose of this important chapter is to explain what each auxiliary system must do, how the various systems work, and what the engine operator must do to keep them working right.

In this chapter you will learn about the lubricating system, the cooling system, the fuel-supply system, the air-intake system, the exhaust system, the starting system, the electric ignition system, the alarm and shutdown system and the automatic starting and load-control system.

Lubricating System—Large Engines

Chapter 7 explained the general principles of lubricating systems. For small engines, all or most of the lubricating system is built in or attached to the engine itself. For larger engines, the amount of space occupied by the accessory equipment such as tanks, coolers, and pumps is so large that these parts are mounted separately for the sake of convenience and accessibility.

In *large engines,* a separate mechanical *forced-feed* system is usually installed to oil the cylinder walls. Also, in many large engines, the pressure-circulating system carries oil to the piston crown for cooling purposes. Large engines, furthermore, require oil coolers to reduce the oil's temperature before it is recirculated.

In the typical design, Fig. 19-1, a gear pump takes its suction from the sump and forces the oil through a strainer and oil cooler to an engine header which supplies all the bearings and other working parts. A passage through the wristpin and piston supplies oil to cool the piston crown, and a telescopic pipe returns the oil to the crankcase. A mechanical forced-feed lubricator lubricates the cylinders.

The piping arrangement for a very complete lubrication system is shown in Fig. 19-2.

Lubricating Oil Storage

All engine lubricating systems provide for oil storage. The storage gives the oil a chance to settle before it is recirculated and also supplies make-up for the oil lost during engine operation. The reservoir

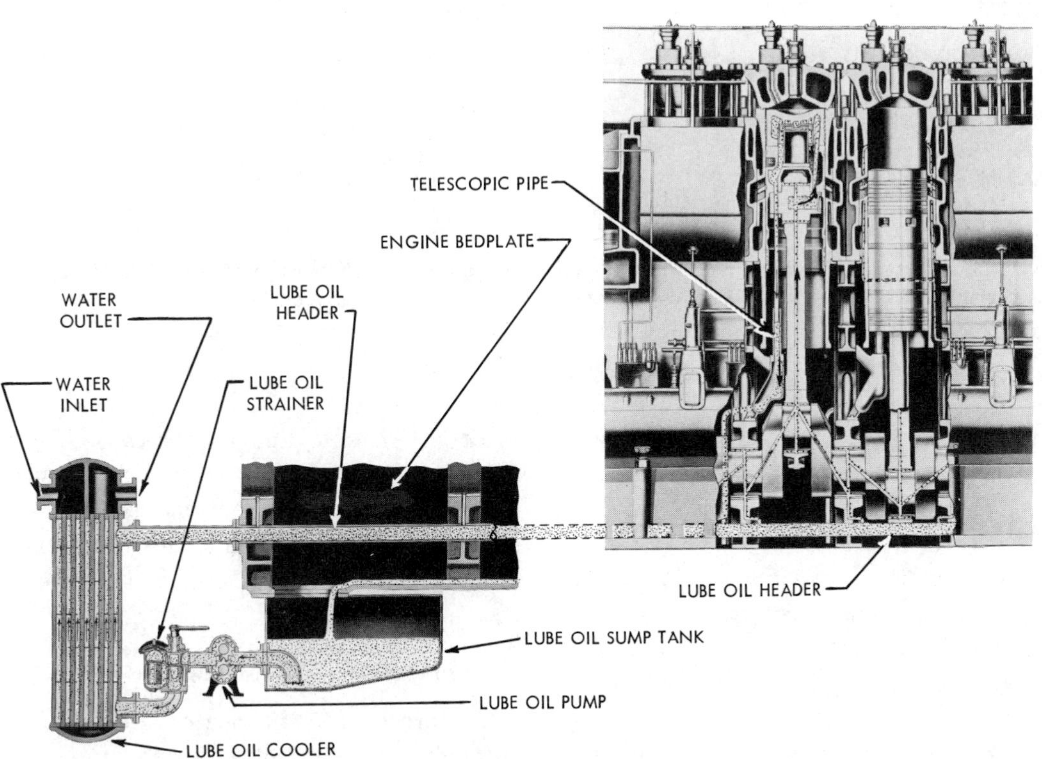

Fig. 19-1. Pressure lubrication system of large engine with oil-jacketed pistons. (Nordberg Mfg. Co.)

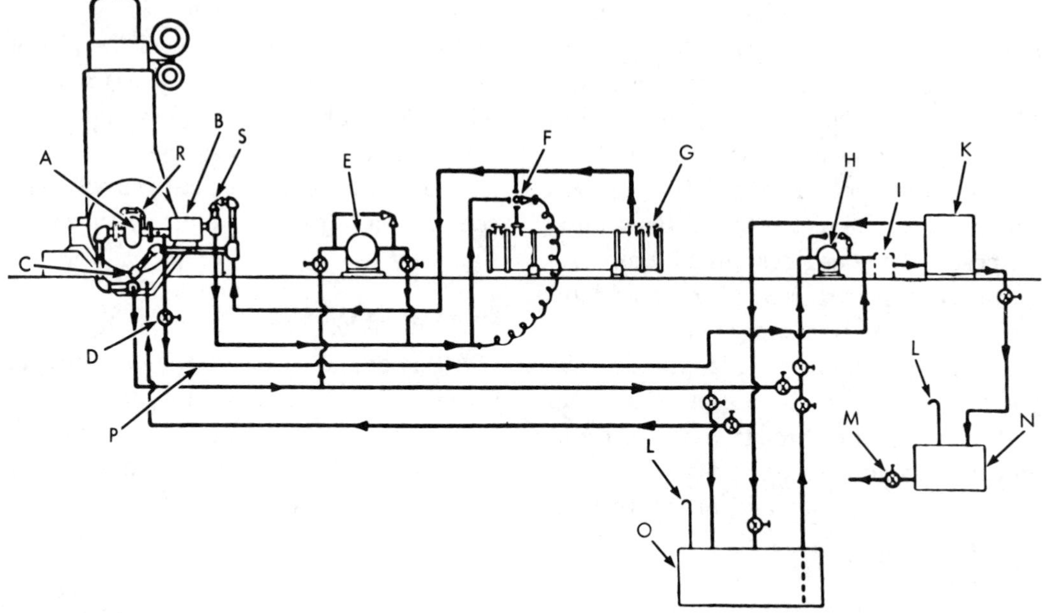

A LUBRICATING-OIL PUMP
B LUBRICATING-OIL STRAINER
C PRESSURE HEADER
D VALVE TO CONTROL FLOW TO LUBRICATING-OIL FILTER
E BEFORE-AND AFTER-PUMP (IF REQUIRED)
F AUTOMATIC TEMPERATURE REGULATOR (IF REQUIRED)
G LUBRICATING-OIL COOLER
H AUXILIARY LUBRICATING-OIL TRANSFER PUMP (IF REQUIRED)
I HEATER (IF REQUIRED)
K LUBRICATING-OIL FILTER
L OPEN VENT
M DRAIN
N SLUGE TANK (IF REQUIRED)
O DRAIN TANK (IF REQUIRED)
P LINE USED WHEN NO AUXILIARY LUBRICATING-OIL TRANSFER PUMP IS INSTALLED
R, S RELIEF VALVES

Fig. 19-2. Piping arrangement of lubricating oil system. For pump protection set relief valve R at 75 psi. For cooler by-pass (cold oil), set relief valve S at 15 psi. (Diesel Manufacturers' Assn.)

to which the oil drains from the bearings and cylinders is called a *sump.*

Wet and Dry Sumps

Many engines have what is known as a *wet* sump; that is, the oil is stored in the crankcase itself or in a sump which is part of the crankcase. The vigorous motion of the crankshaft and connecting rods creates air currents which agitate the oil's surface and tend to produce crankcase "fog." To avoid excessive lubrication of the cylinder walls caused by this fog, some engines use a *dry* sump. In this arrangement, a storage tank outside the engine is used, and the oil drains out of the crankcase as soon as it reaches it.

Mechanical Forced-Feed Lubricators

A forced-feed lubricator forces oil through tubing directly to one or more entrance points, or feeds, in each cylinder wall. The number of feeds per engine cylinder depends on the cylinder's diameter.

The lubricator consists of a group of small reciprocating pumps mounted in a single casing, which serves as an oil reservoir. Each pump connects by tubing to a cylinder feed. The pumps supply a measured quantity of oil at each stroke, and are driven from the engine by an eccentric, or by a crank and ratchet (notched wheel), or by gearing. The amount of oil fed by each pump cylinder is usually separately adjustable. In most cases, a sight feed keeps the oil flow visible. Fig. 19-3 shows a typical forced-feed lubricator.

Often, oil which is particularly suited for lubricating the cylinder is used in the forced-feed lubricator. Of the oil that is fed to the cylinders, part stays on the walls, part is burned by exposure to hot gas, and part drains away. In some cases, the excess oil, with its impurities, is caught and drained off either to waste or to a separate tank for purification. In most cases, however, the excess oil drains off the cylinder walls and falls into the bearing oil in the crankcase.

Oil Storage Tanks

In large stationary and marine installations oil-storage tanks are used to hold

Fig. 19-3. Typical forced-feed lubricator for lubricating diesel cylinders. (McCord Lubricator Co.)

the oil drained from the engine crankcase and to store new or purified oil. This permits draining the crankcase of used oil and refilling it promptly with fresh oil. Also it allows the used oil to be purified at a convenient time. Tanks and piping are made of black iron or steel rather than brass, copper, or galvanized metal, because copper and zinc (found in brass and galvanizing) increase the tendency of oil to oxidize. Take care to prevent water and dirt entering the oil during storage. It's surprising how clean new oil may become badly contaminated in this way.

Oil Coolers

Because the ability to transfer heat is lost rapidly when their cooling surfaces become dirty, oil coolers should be ample in size and arranged for easy cleaning. Coolers are generally of the shell-and-tube type, in which the oil passes through tubes enclosed in a casing; water circulates in the casing around the tubes. The used oil tends to deposit sludge in the tubes, and impure water may deposit scale or dirt on the outside of the tubes.

When a cold engine is started, the oil should be heated to its normal working temperature as soon as possible. In many engines, particularly small ones, *jacket* water is used in the oil cooler and a thermostatic control causes the water to heat rapidly as soon as the engine starts. Consequently, the hot jacket water transfers heat to the cold oil, so that the latter warms up quickly.

The 531 cubic inch diesel tractor engine in Fig. 19-4 utilizes this approach for cold engine operation and to transfer the heat from the oil to the cooling system when the engine oil is hot.

Notice the lubricating oil filter is also an integral part of the engine.

If the lubricating oil is cooled independently of the jacket water (as often occurs with larger engines), a by-pass should always be provided around the oil cooler. With the by-pass valve open, the oil cooler does not remove engine heat from the oil; therefore the oil warms up more quickly.

Lubricating Oil Treatment

Various methods are employed to remove some or most of the contamination that occurs in service. This is called oil treatment. It may be carried out in any one of three ways, or in a combination of them. These ways are known as: (1) the batch treatment, (2) the continuous total treatment, and (3) the continuous by-pass treatment.

In *batch treatment*, the entire oil charge in the engine system is withdrawn at one time and is treated. The treated oil is then stored for later replacement in the engine.

Continuous total treatment is practical only where the rate of oil circulation is not rapid; otherwise the equipment required to treat the entire charge as it circulates becomes too large and expensive.

Continuous by-pass treatment is the most common method. Here a portion of the flow (say 10 to 20 percent) is by-passed to the treating equipment. With this arrangement, the condition of a portion of the oil charge is being restored continuously. This keeps the total amount of impurities down to a safe level and avoids the need for large treating facilities.

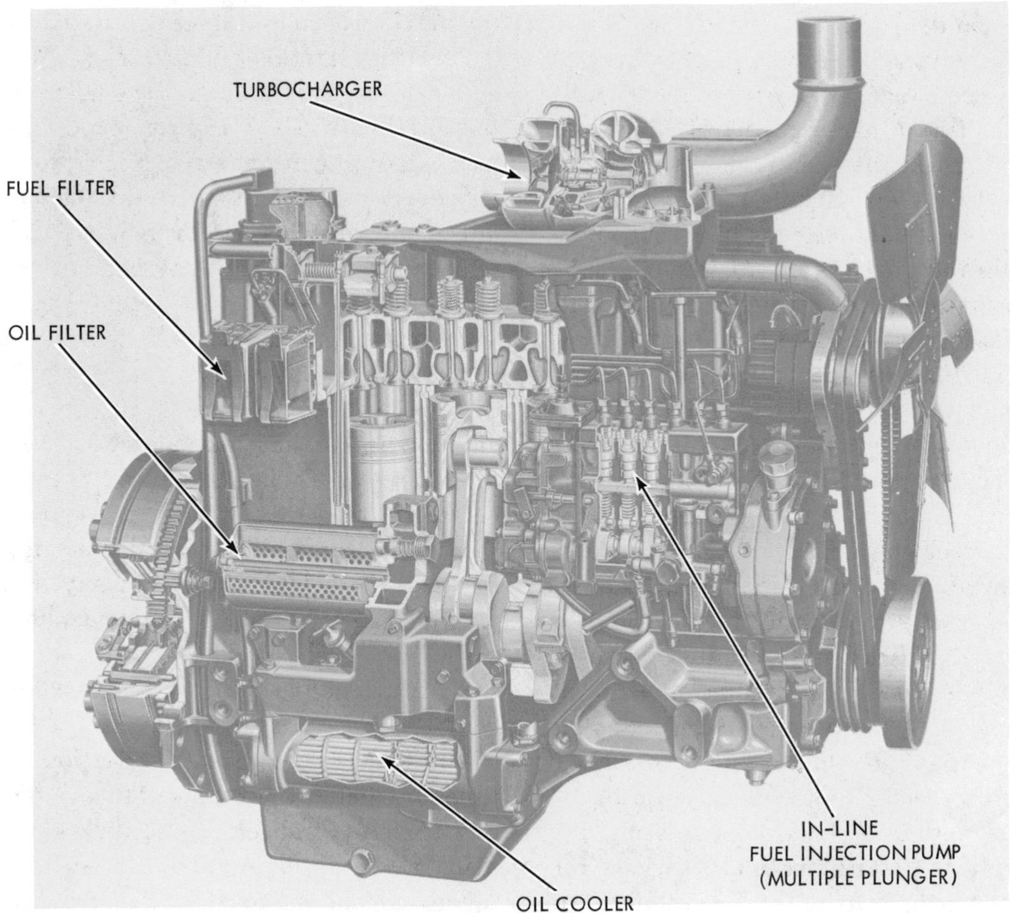

Fig. 19-4. Lubricating oil cooler integrated with cooling system on small diesel engine. (Deere & Company)

Oil Treating Equipment

Many types of treating equipment are available. The subject would fill a book by itself, so let's briefly look at the more important types and get an idea as to how they work.

Settling is the simplest method of removing impurities, but it does not remove those that are dissolved in the oil. If the oil is withdrawn from the engine, placed in a tank where it is heated, and then left undisturbed, all the carbon, sludge, dirt and water will settle to the bottom of the tank, leaving clean oil above it. It is a slow process and requires a large stationary tank.

Centrifuging is similar in principle to settling, in that it depends on the fact that the contaminants it removes are heavier than the oil itself. A centrifuge rotates the oil at high speed in a bowl. The centrifugal force created throws the heavy contaminants outward toward the circumference of the bowl, where they are removed. Since the centrifugal force is several thousand times that of gravity, the centrifuge separates the impurities much faster than a gravity settling tank. Heating the oil reduces its viscosity and improves the cleaning-up process. Some of the dissolved acid can be scrubbed out by introducing water with the oil.

Filtering is widely used, and in many different forms. However, filters may be divided into two general classes—*ab*sorbent and *ad*sorbent.

In absorbent filters the oil is passed through materials such as cotton or glass fiber, cellulose, yarn or waste, or between closely stacked disks; the foreign particles are caught and held or absorbed. Absorbent filters primarily remove the suspended particles in the oil. In some cases, they also tend to *kill* the acids dissolved in the oil by means of *chemical action* between the acid and the filter material.

Adsorbent filters work on a quite different principle, known as molecular adsorption. Certain materials, such as fuller's earth, bauxite, and bentonite, are able to draw out the oil and collect on their surfaces most of the contaminants, including the dissolved ones. The contaminant molecules adhere to the molecules of the filtering material. Since this action takes place only on the surface of the filtering material, the latter must be in porous or powder form. Such filters effectively remove the acids in the oil, but many of them, unfortunately, also remove the useful detergent additives found in heavy duty oils.

Re-refiners, sometimes called chemical reclaimers, are oil-purifying units which employ high-temperature distillation in addition to the adsorption principle described above. By heating the oil to temperatures of several hundred degrees, they drive off not only the water (steam) but also the volatile portion of any fuel oil which may be mixed with the lubricating oil. The successful handling of such equipment requires considerable care on the part of the operator.

Mechanical (absorbent) filtration is usually employed in a continuous by-pass or shunt arrangement. Settling treatment is always done in batches. Centrifuging and adsorbent filtration may be handled by batch or by-pass methods. Various treating methods may be combined to obtain best results.

Cooling System

When fuel burns in the cylinders of a diesel engine, only about one-third of the fuel's heat energy changes into mechanical energy and then leaves the engine in the form of brake horsepower. The rest of the heat shows up in (1) hot exhaust gases, (2) frictional heat of the rubbing surfaces, and (3) heating of the metal walls which form the combustion chamber, the cylinder head, cylinder, and piston. The cooling system's job is to remove the unwanted heat from these parts, so as to prevent the following problems:

1. Overheating and resulting breakdown of the lubricating oil film which separates the engine rubbing surfaces.

2. Overheating and resulting loss of strength of the metal itself.

3. Excessive stresses in or between the engine parts due to unequal temperatures.

Cooling is often provided also for the exhaust manifold to prevent its reaching an uncomfortable temperature.

Although cylinder heads and cylinders are sometimes cooled by a strong draft of air, as in a few very small engines and in some airplane-type engines, these parts are generally provided with jackets through which cooling water is circulated. Pistons transfer their heat to the cylinder walls and to the lubricating oil. Many engines use oil coolers to remove the heat in the lubricating oil.

Table 19-1 illustrates what becomes of the heat developed in a typical diesel engine, in terms of Btu per brake horsepower-hour (bhp-hr). Such a table is called a *Heat Balance*.

TABLE 19-1 HEAT INPUT AND DISPOSAL IN A DIESEL ENGINE
(THREE-QUARTERS TO FULL LOAD)

THERMAL TRANSFER	BTU/BHP-HR	PERCENT
HEAT INPUT	7367*	100.0
HEAT DISPOSAL:		
USEFUL WORK (SHAFT OUTPUT)	2544	34.6
TO COOLING WATER AND LUBRICATING OIL	2194	29.8
TO EXHAUST GASES	2259	30.6
LOST BY RADIATION, ETC	370	5.0
TOTAL HEAT DISPOSAL	7367	100.0

* FUEL CONSUMPTION: 0.39 LB/BHP-HR; HEAT VALUE OF FUEL 18,890 BTU/LB

TABLE 19-2 APPROXIMATE HEAT TRANSFERED TO COOLING SYSTEM
(BTU/BHP-HR, AT FULL LOAD)

COOLANT	FOUR-CYCLE ENGINES			
	NONSUPERCHARGED WATER-JACKETED EXHAUST MANIFOLD		TURBOCHARGED OIL-COOLED PISTONS	
	DRY PISTONS	OIL-COOLED PISTONS	DRY MANIFOLD	COOLED MANIFOLD
JACKET WATER	2200-2600	2000-2500	1450-1750	1800-2200
LUBRICATING OIL	175- 350	300- 600	300- 500	300- 500
TOTAL	2375-2950	2300-3100	1750-2250	2100-2700

COOLANT	TWO-CYCLE ENGINES		
	LOOP SCAVENGING OIL-COOLED PISTONS	UNIFLOW SCAVENGING OPPOSED PISTON	OIL-COOLED PISTONS VALVE-IN-HEAD
JACKET WATER	1300-1900	1200-1600	1700-2100
LUBRICATING OIL	500- 700	900-1100	400- 750
TOTAL	1800-2600	2100-2700	2100-2850

SOURCE: DIESEL ENGINE MANUFACTURES' ASSOCIATION

The heat balance of an engine and the amount of heat absorbed by the cooling water vary with the type of engine and the design of cylinders, exhaust manifold, pistons, lubricating-oil system, and any other equipment which may be cooled directly or indirectly by the circulating water. Table 19-2 shows the approximate amounts of heat absorbed by the cooling systems of various types of engines.

EXAMPLE 1. Find the amount of heat which must be absorbed by the cooling system of a 300 hp engine when running at full load. The fuel consumption is 0.38 lb/bhp-hr, the heat value of the fuel being 19,200 Btu/lb. The heat input lost to the cooling system is 30 percent.

Heat input = 300 × 19,200 = 2,188,800 Btu/hr

Heat lost to cooling system = 0.30 × 2,188,800 = 656,640 Btu/hr

Expressed in Btu/bhp-hr, this is:

Heat lost to cooling system =

$$\frac{656{,}640 \text{ Btu/hr}}{300 \text{ bhp}} = 2189 \text{ Btu/bhp-hr}$$

An adequate cooling system, properly controlled, is essential to satisfactory engine performance. Defective cooling systems and careless control have caused failures of many engines.

Good cooling depends on (1) how much the water's temperature rises in its passage through the engine, (2) what its temperature is when it leaves, and (3) how pure the water is.

Temperature Rise. Let's look at temperature rise first. This ties directly to water flow, and both are important. There's just so much heat to remove. If we have a high flow rate, we absorb the heat in a lot of water, and its temperature goes up only a little. If we cut down the flow rate, less water must remove the same amount of heat. So the temperature rise is thereby greater.

EXAMPLE 2. Find the gallons of water it is necessary to pump per minute to cool a diesel engine putting out 160 bhp and losing 2600 Btu/bhp-hr to its cooling system, if (1) the water temperature rises 15°F in the engine, (2) the water temperature rises 45°F. Proceeding, we set forth:

Heat lost to cooling system = 160 × 2600 = 416,000 Btu/hr

$$\text{Heat lost to cooling system} = \frac{416{,}000}{60} = 6933 \text{ Btu/min}$$

(a) Since 1 lb of water absorbs 1 Btu for each 1 degree F temperature rise, the amount of water required to absorb 6933 Btu per minute with a 15°F rise is:

$$\text{Amount of water} = \frac{6933}{15} = 462 \text{ lb/min}$$

Then, as there are 8.33 lb of water in a gallon, the gallons of water required per minute are:

$$\text{Gals of water per min} = \frac{462}{8.33} = 55.5 \text{ gpm}$$

(b) The pounds of water required to absorb 6933 Btu per minute with a 45°F rise in temperature would be:

$$\text{Amount of water} = \frac{6933}{45} = 154 \text{ lb/min}$$

Therefore, the gallons of water required per minute would be:

$$\text{Gals of water per min} = \frac{154}{8.33} = 18.5 \text{ gpm}$$

Note that case (a), with temperature rise one-third that of case (b), requires three times the rate of flow.

What difference does temperature rise make? In answer, it can be said that the greater the temperature difference between the entering and leaving water, the greater the difference between the temperatures at the bottom and the top of the cylinders. This means more distortion from temperature expansion and greater stresses in the metal. If we keep the temperature rise down by boosting the rate of flow, we gain another benefit because the more rapid flow through the water jackets reduces the risk of vapor pockets and local hot spots.

Temperature is easier to measure than rate of flow. Since the two are related, we usually gage cooling effectiveness in terms of temperature rise. Good practice limits temperature rise to 10 to 15°F at full load.

Leaving Temperature. If we avoid heat distortion and local hot spots, we can safely use a higher off-engine water temperature. In other words, it's better to run an engine with small temperature rise and high leaving temperature than with large temperature rise and low leaving temperature. A high temperature in the water jackets improves engine efficiency, reduces corrosion, and improves lubrication. Also, if the water is cooled outside the engine and recirculated, a high off-engine temperature gives a greater temperature difference to work with in the cooling equipment. This reduces its size and cost.

Water Purity

We might have high flow rate and small temperature rise and still run into trouble—if the water is impure. Most natural waters carry scale-forming impurities. At temperatures met in engine jackets, the dissolved minerals separate out in solid form and coat the metal surfaces. Sea water will form salt deposits if the water outlet temperature exceeds approximately 120°F, which is too cool for efficient operation.

Scale deposits cut the rate of heat flow from the combustion space into the water jackets. This causes the metal temperatures to run higher and may allow local over-heating, with cracks and failures as a possible result. The now commonly used closed cooling system largely avoids such problems. We'll look at this in a moment.

Weather Conditions

Another factor that must be taken into account in many plants is weather. In cold climates, water may freeze and lubricating oil may thicken, particularly when the engine is at rest. In such cases, auxiliary heating systems are used or draining is performed to prevent damage.

Open Cooling Systems

In an open cooling system, the water leaving the engine jackets is either not returned at all or is exposed to the air before being recirculated, as shown by diagram in Fig. 19-5. In Fig. 19-5A the water goes to waste or to some process in which it is used and not returned. In Fig. 19-5B the warm water is cooled in a cooling tower or spray pond by exposing it to the air and letting a small part of the water evaporate. The rest of the water loses heat to make the vapor; after having been cooled in this way, it is pumped back to the engine for re-use. Note that the cooling-tower or spray-pond arrangement, although sometimes called a *closed circuit* because the water

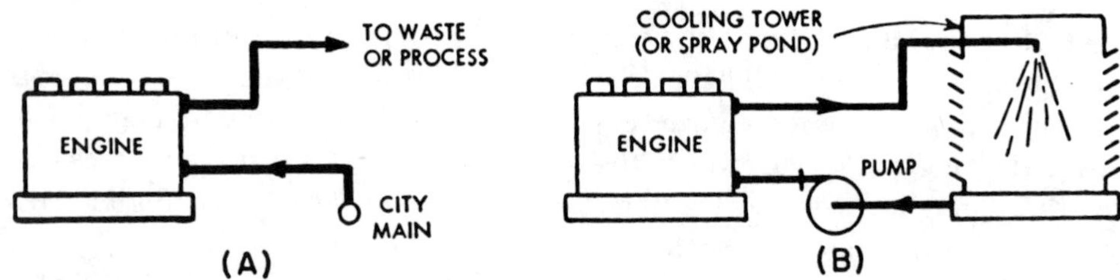

Fig. 19-5. Elementary diagrams of open cooling systems. Continuous addition of fresh water in A, and need for make-up in B, bring danger of scale and deposits.

is recirculated, is an *open system* because the water is open to the air.

Water impurities, and the resulting danger of scale and sediment deposits in the engine jackets, constitute a problem in open systems. If the water is not returned to the engine, the new water continually brings in fresh foreign matter. If the water is recirculated after some of it has been evaporated in a cooling tower or spray pond, the unavoidable loss (½ to 1 gal/bhp-hr) requires additional make-up water, which brings in more impurities. Since the vapor doesn't carry away the impurities, they concentrate continuously and may eventually deposit more scale than if the water had been run to waste. Open systems may be safely used, however, where the water is exceptionally pure and the supply is ample.

Closed Cooling Systems. The jacket water is recirculated through closed heat exchangers in closed cooling systems. (A heat exchanger is a device that transfers heat from one fluid to another, as from water or oil to water or air.) Thus, in closed systems, the same water remains in the system indefinitely and is recooled without exposure to air. If it is pure at the start, it stays pure. The heat exchanger may be *water-to-water* (shell-and-tube type) or *water-to-air* (radiators and evaporative coolers). Fig. 19-6 shows some elementary, closed cooling systems.

With water-to-water heat exchangers, the secondary water (often called 'raw' water) may pass through only once if the supply is ample and there is no need to conserve it. See Fig. 19-6A. However, the loss of raw water may be cut 90 to 95 percent by recirculating it after it is cooled by evaporation in a tower or spray pond as in Fig. 19-6B. Scale deposits are not as serious in heat exchangers as in engine jackets, and heat exchangers can be easily cleaned. The raw water should pass through the *tubes* of the heat exchanger because the inside of a tube can be cleaned more easily than the outside.

Fig. 19-6C shows the radiator type of closed system, in which air blown by a fan cools the water within the tubes of the radiator. The water is not exposed to the air and does not evaporate.

An evaporative cooler, Fig. 19-6D is in a sense a combination of a water-to-water exchanger and a cooling tower. Jacket water flows through the coils while raw water is sprayed over them. Air blown over the coils by a fan evaporates some of the raw water on the coil surface, thus cooling the jacket water within. The heat absorbed in evaporating 1 gallon of wa-

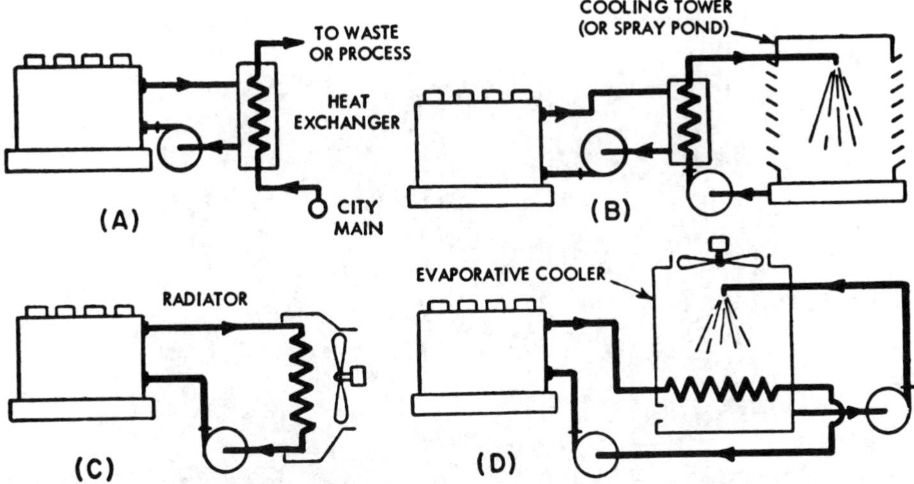

Fig. 19-6. Elementary diagrams of closed cooling systems. A is with heat exchanger. B is with heat exchanger and cooling tower. C is with radiator. D is with evaporative cooler. Note that in all cases the jacket water recirculates without exposure to air.

ter is sufficient, approximately, to cool 100 gallons of water 10°F.

Whether it is a large or small diesel engine, two critical areas in cooling are the injector and the valves. Fig. 19-7 is a cross-sectional view illustrating water cooling of the fuel injector and the valves on a series 71 General Motors diesel.

To insure efficient cooling on this engine, each fuel injector is seated in a thin walled copper tube which extends through the water space in the cylinder head as shown in Fig. 19-7. The lower end of the copper tube is pressed into the cylinder head and spun over to prevent water leaks. The upper end is flanged and sealed with a Neoprene seal.

In addition to being surrounded by cooling water, cooling of injector tubes, valve seats, and valve guides is further assured by the use of water nozzles in the cylinder head. Fig. 19-7 shows the nozzle positioned in such a manner that cool water entering the cylinder head is directed against the sections which are subject to the greatest heat.

Some trucks use thermostatically controlled shutters in front of the radiator to assist in warming up the engine and holding it at a precise operating temperature.

Another way of controlling engine operating temperature is by using a thermostatically-controlled hydraulic-driven fan that operates when engine cooling is needed. Fig. 19-8 shows a sectional view of a hydraulic fan used on some trucks and motor coaches.

The fan drive consists of a driving and a driven torus member much like the fluid coupling in an automatic transmission. Speed of the fan is variable, depending upon operating temperature. When the engine is cool, little or no power is transmitted to the driven torus and the fan idles. When the engine is hot, full power

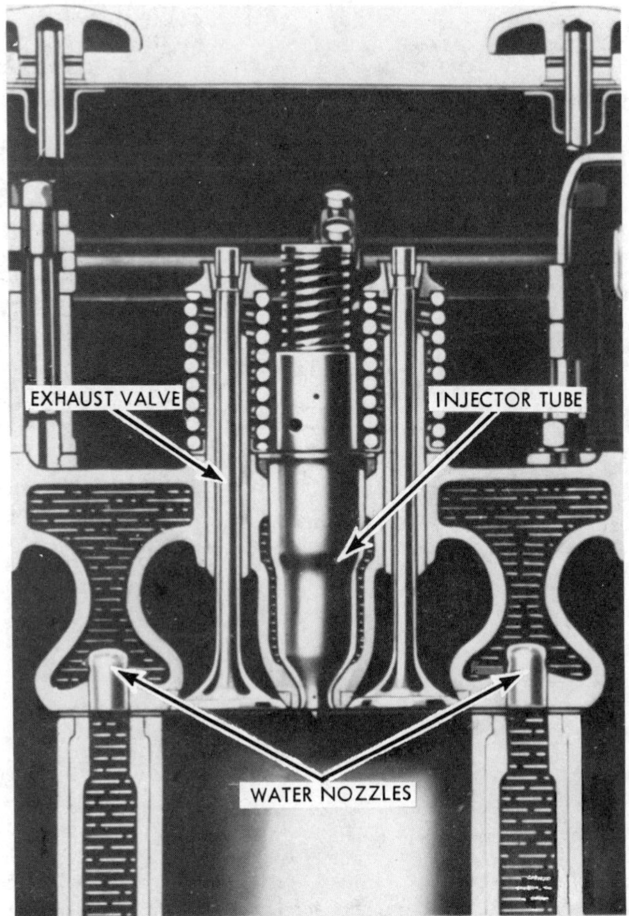

Fig. 19-7. Water cooling injector and valves in a cylinder head. (General Motors Corp.)

is transmitted to the driven torus.

A thermostatic valve is installed in the water return between the radiator and the water pump. The amount of oil retained in the torus housing is controlled by the valve, which is actuated by water temperature.

Under cold operation, the valve is open, permitting oil to return to the crankcase. This does not leave enough oil to tie the two torus members together. As the engine coolant reaches operating temperature and valve starts to close, retaining more oil in the torus housing, power is gradually transmitted to the fan.

Steam Systems

High jacket temperatures are used in steam (sometimes called *vapor phase*) systems. These jacket temperatures are high enough to cause part of the jacket water to change into low-pressure steam when it reaches an overhead tank called a *flash tank*, Fig. 19-9. The heat needed to make the steam is taken from the re-

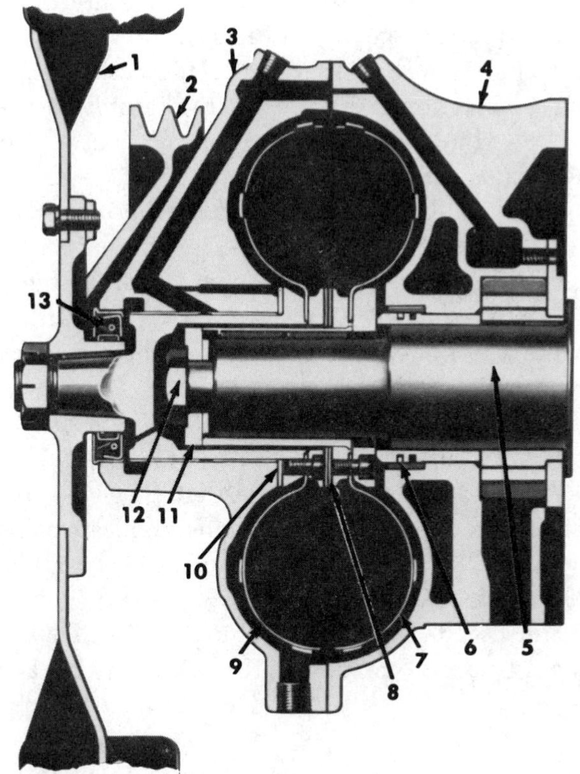

1 FAN BLADE ASSEMBLY
2 FAN PULLEY (REMOTE BLADE TYPE)
3 DRIVEN TORUS HOUSING
4 DRIVING TORUS HOUSING (CRANKSHAFT FRONT COVER)
5 CRANKSHAFT
6 FLUID SEAL RING
7 DRIVING TORUS MEMBER
8 THRUST WASHERS
9 DRIVEN TORUS MEMBER
10 FRONT THRUST WASHER
11 WASHER
12 BOLT
13 OIL SEAL ASSEMBLY

Fig. 19-8. Sectional view of a hydraulic fan used on a truck engine. (General Motors Corp.)

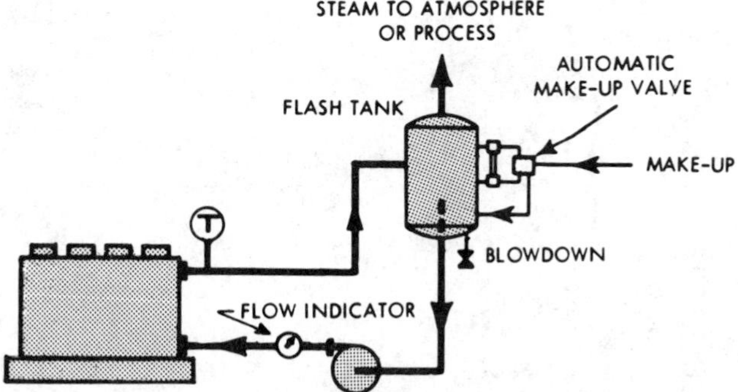

Fig. 19-9. High temperature cooling system: steam discharged (T means thermometer). Jacket heat is converted to steam in flash tank. Continuous make-up is needed.

maining water, which is thereby cooled and then recirculated through the engine. A rapid flow rate (about 30 gal/hr/bhp) prevents steam pockets forming in the engine jackets. The temperature in the engine jackets remains almost constant, regardless of the load, without mechanical control. As the load increases, more steam is produced.

In the simple steam system of Fig. 19-9, the steam leaves the system permanently. Therefore, continuous make-up water is needed to the extent of about ½ gal/bhp-hr. If the make-up water is not pure, scale and sediment will deposit first in the flash tank and then in the engine jackets.

In Fig. 19-10 the loss of steam is prevented by using a condenser to turn the steam into water again, so that practically no make-up is required.

By-Pass Systems

Still other cooling arrangements use *by-passes* to regulate the off-engine water temperature by controlling the temperature of the water *entering* the jackets. This method is far better than changing the rate of flow through the jackets.

To see why, let's first look at the simple control system of Fig. 19-11 which uses no by-pass. The control valve for keeping the off-engine temperature uniform, despite engine load variations, works by changing the flow rate. This has several disadvantages, among which are:

1. The temperature of the incoming water is fixed and is usually lower than desired. Say it is 80°F, and suppose the engine builder recommends an off-engine temperature of 170°F. The temperature rise in the engine will then be 90°F (170 − 80). This rise is much greater than the 10 to 15°F which is desirable to avoid heat distortion.

2. The small flow rate corresponding to the large temperature rise may permit vapor pockets and hot spots to form.

3. Also, at low engine loads, when the

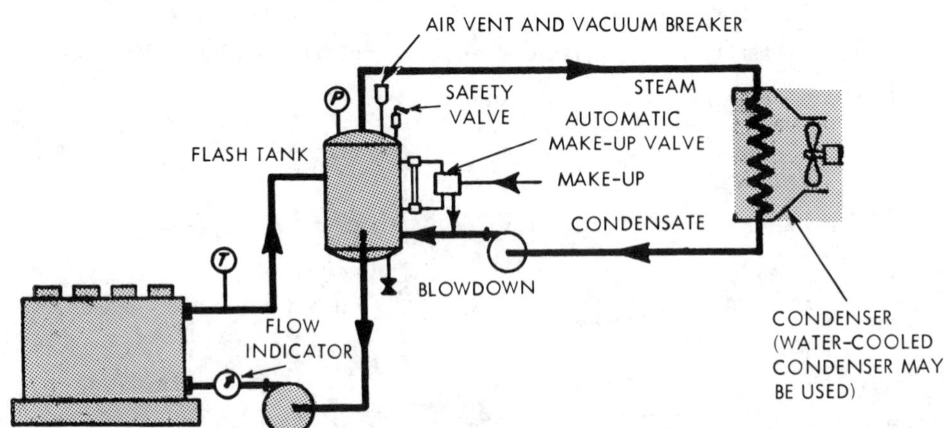

Fig. 19-10. High temperature cooling: steam condensed and returned (letter P means pressure gage). No make-up is needed.

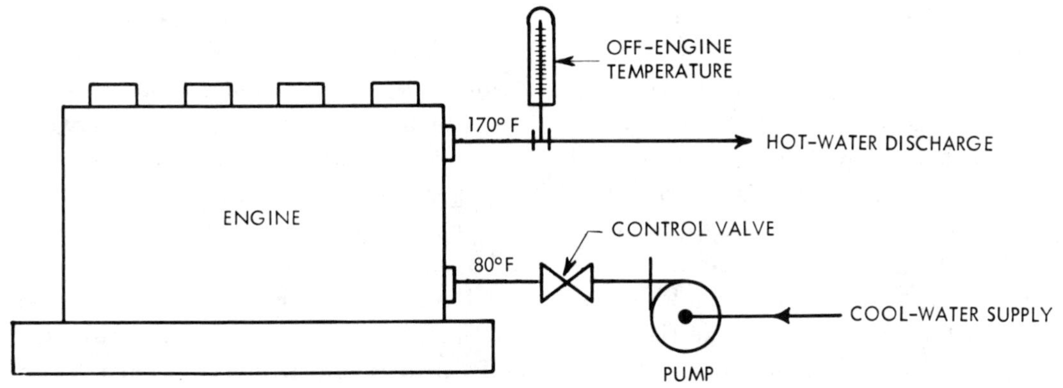

Fig. 19-11. Temperature control by changing flow rate.

flow rate is further reduced, the hazard of vapor pockets and hot spots becomes still greater.

A by-pass system avoids these faults. As shown in Fig. 19-12, part of the hot water off the engine is returned (by-passed) to the inlet piping, where it blends with the incoming cool water to deliver warm water to the engine. If the off-engine temperature is too low, opening the by-pass valve will raise it. Thus, the engine inlet temperature is independent of the water source and can be made high enough to produce only a small temperature rise through the jackets. Furthermore, the flow rate is always high, even at light load, and a high off-engine temperature can be used safely, with resulting benefit to engine efficiency.

Choice of Cooling System

The main factors which determine the best type of cooling system for a particular installation are: (1) cost and avail-

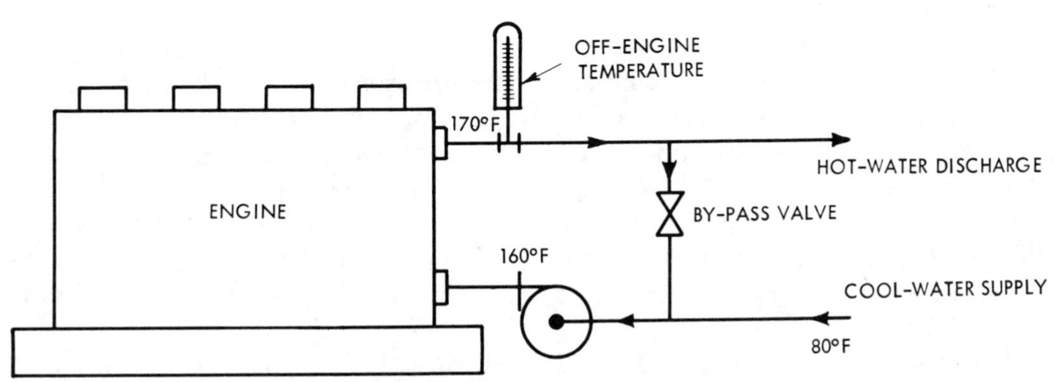

Fig. 19-12. Temperature control using by-pass.

ability of water, (2) water purity, (3) climate, (4) space, and (5) need to regain heat. They affect the choice in the following manner.

Cost and Availability of Water. If there's plenty of *pure* water available at low cost, there's no need to recool. A simple open system is sufficient. Never-the less, a by-pass control should be used to obtain a small temperature rise and a large flow rate. If you have use for the heat, you can use hot water direct from the engine jackets.

Water Purity. If the water is *impure*, use a closed system to keep make-up water out of the engine jackets. If raw water is cheap and plentiful, use a heat exchanger and let the raw water go to waste after it has cooled the jacket water. When the raw water must be conserved, use a heat exchanger and recool the raw water in a cooling tower, spray pond, or evaporative cooler.

Where water is very scarce (whether pure or not), recool the jacket water in a water-to-air radiator. Little or no make-up is required. Radiator systems are universally used in automotive and railroad service.

Climatic Conditions. The size and cost of cooling towers, evaporative coolers, and radiators are affected by climatic conditions. Evaporative cooling depends on the *humidity* of the surrounding air. Moist air can take up less vapor. Hence, more air must be circulated to remove the same amount of heat. *High air temperature* also cuts the capacity because the smaller temperature difference between the water and air causes less direct heat transfer.

The capacity of water-to-air radiators is not affected by humidity because the water is not exposed to the air. But it is directly affected by the air temperature because the cooling is done solely by the difference in temperature between the water and the air. For the same temperature difference, radiators take more fan power than evaporative coolers.

Space Considerations. Cooling towers and evaporative coolers must be located in open spaces where the moist air given off can dissipate freely. It is difficult to use them in enclosed areas, like basements, because vapor clouds form unless ducts are used to carry the air outside.

Radiators can be used more freely since they don't produce vapor, and the warm air they throw off is generally not hot enough to be objectionable. But radiators often cost more than cooling towers or evaporative coolers and take more fan power.

Heat Recovery. Where heat can be used in the form of warm air, as for room heating, a water-to-air radiator system is simple, inexpensive, and efficient. Dampers direct the air flow as needed, into the room in winter, outdoors in summer.

If you want heat in the form of hot water, always use heat exchangers unless the water is unusually pure. Otherwise, the heavy make-up will cause jacket deposits.

Steam systems, which produce low-pressure steam directly from the jacket water, serve where heat is needed in the form of steam rather than in the form of warm air or hot water. Also, steam systems, with their high and relatively uniform jacket temperatures, reduce scale formation and cylinder wear and corrosion, particularly with fuels containing hydrogen sulfide. The high temperatures also tend to reduce fuel consumption.

Fuel-Supply System

The purpose of the fuel-supply system in a diesel plant is to store an adequate supply of fuel oil, clean it, preheat it (if necessary), and deliver it to the fuel-injection pumps of the engine or engines.

A typical fuel-supply system consists of storage tanks; pumps to transfer oil from the delivery point to the storage tanks, and from these tanks to the engine; and strainers, filters, and centrifuges needed to insure clean oil. Instruments and controls also are required to measure the quantity of fuel transferred, to check the oil level in the tanks, and to control the transfer of fuel.

Fig. 19-13 shows the basic parts of a fuel-storage and handling system for a stationary diesel plant. It includes an underground horizontal storage tank and a day tank in the engine room; from the latter, the fuel flows to the engine by gravity. The day tank assures several hours' supply of clean oil ready for use. Using an individual day tank for each engine makes it easier to measure the oil consumed by each unit.

Day tanks are often dispensed with on small engines, particularly in automotive use. The engine is fitted with a built-in transfer pump which lifts oil from the storage tank and feeds it directly to the engine fuel-injection pumps. The transfer pump supplies more oil than is consumed; a relief valve in the header feed-

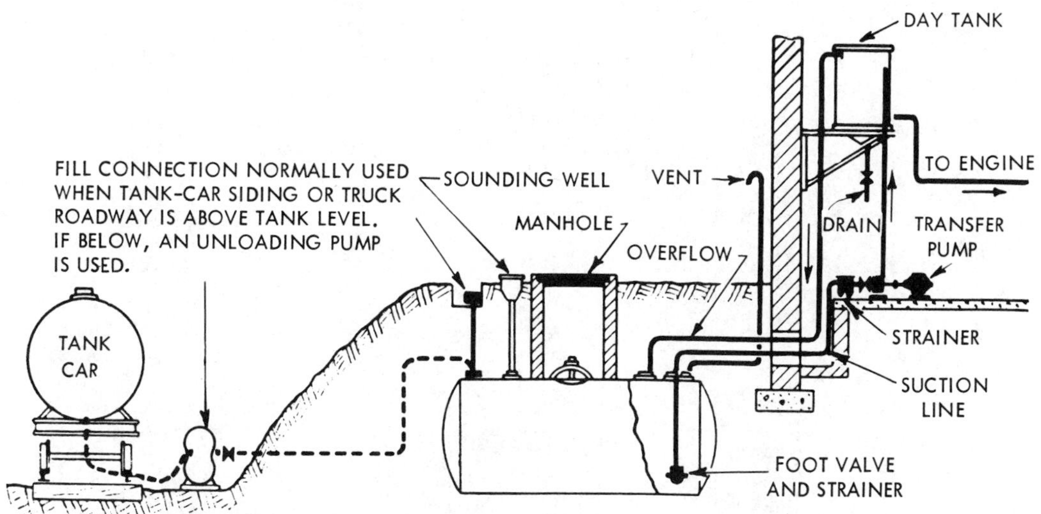

Fig. 19-13. Typical fuel storage and handling system. *(Power Magazine)*

ing the fuel-injection pumps returns the excess oil to the storage tank and maintains a constant pressure at the inlets of the injection pumps.

Storage tanks are always provided with means to determine the oil level in order that they may be refilled when necessary and sometimes, also, to measure the fuel consumption of the engines. The sounding well in Fig. 19-13 permits the use of a gage stick to find the oil level. However, more convenient readings can be obtained from float gages and pneumatic gages, which determine the depth of oil in the tank by measuring the air pressure needed to balance the oil depth. Such gages can be made remote-reading, so that the oil level may be indicated at any convenient point.

Because of the curved shape, the depth of oil in a horizontal cylindrical tank does not show directly the *quantity* of oil. For example, a one-inch change in oil depth when the tank is nearly full or nearly empty represents a much less quantity of oil than when the tank is half-full. Table 19-3 will enable you to find the capacity of a cylindrical tank when it is full, and Table 19-4 when it is partly full.

EXAMPLE 3. How many gallons will a 64″ (inside dia) tank hold if the straight part is 120″ long and two bumped heads are each 6″ long?

TABLE 19-3 CAPACITY OF FLAT-ENDED CYLINDRICAL TANKS PER INCH OF LENGTH

NOTE: A BUMPED (CONVEX) HEAD HOLDS TWO-THIRDS AS MUCH AS THE SAME LENGTH OF STRAIGHT TANK. THUS FOR A TANK WITH BUMPED HEADS, ADD TO THE LENGTH OF THE STRAIGHT PART TWO-THIRDS OF THE DEPTH OF EACH BUMPED HEAD TO GET THE LENGTH OF A FLAT-ENDED TANK OF EQUIVALENT CAPACITY

DIAMETER (IN)	GAL	DIAMETER (IN)	GAL	DIAMETER (IN)	GAL	DIAMETER (IN)	GAL
10	0.3399	30	3.060	50	8.500	70	16.66
11	0.4114	31	3.267	51	8.843	71	17.14
12	0.4896	32	3.482	52	9.193	72	17.63
13	0.5741	33	3.702	53	9.550	75	19.12
14	0.6666	34	3.930	54	9.913	78	20.69
15	0.7650	35	4.165	55	10.28	81	22.31
16	0.8700	36	4.406	56	10.66	84	23.99
17	0.9824	37	4.655	57	11.05	87	25.73
18	1.102	38	4.910	58	11.44	90	27.54
19	1.227	39	5.171	59	11.83	93	29.41
20	1.360	40	5.440	60	12.24	96	31.33
21	1.499	41	5.715	61	12.65	99	33.32
22	1.646	42	5.997	62	13.07	102	35.37
23	1.798	43	6.286	63	13.49	105	37.48
24	1.958	44	6.582	64	13.93	108	39.66
25	2.125	45	6.885	65	14.36	111	41.89
26	2.298	46	7.194	66	14.81	114	44.18
27	2.478	47	7.511	67	15.26	117	46.54
28	2.666	48	7.833	68	15.72	120	48.96
29	2.859	49	8.163	69	16.19		

SOURCE: POWER MAGAZINE

Two-thirds of 6 is 4, so the equivalent length is 120 + 4 + 4 = 128 inches. (See caption, Table 19-3)

From Table 19-3, gal per inch of length for a 64″ dia tank is 13.93 gal. The capacity of a full tank equals gal/inch of length times equivalent length. Therefore:

$$\text{Capacity of full tank} = 13.93 \times 128 = 1783 \text{ gal}$$

EXAMPLE 4. How many gallons does the same tank hold when the depth of oil is 25″? The fraction full by depth is $\frac{25}{64} = 0.39$. The volume fraction corresponding to 0.39 depth fraction, from Table 19-4, is 0.3611. Therefore:

$$\text{Content of tank} = 0.3611 \times 1783 = 643.8 \text{ gal}$$

Transfer pumps for fuel oil are almost always of the positive-displacement type, using gears, screws, or oscillating pistons. To avoid damage if the discharge line is closed, such pumps are equipped with relief valves to return excess oil from the discharge to the suction side.

Piping used in fuel-supply systems must be tight. Leaks in pressure lines are dangerous because of the fire hazard.

TABLE 19-4 CAPACITY FRACTIONS OF PARTLY FILLED, HORIZONTAL, CYLINDRICAL TANKS (FRACTION FULL BY DEPTH AND BY VOLUME)

BY DEPTH	BY VOLUME	BY DEPTH	BY VOLUME	BY DEPTH	BY VOLUME	BY DEPTH	BY VOLUME
0.01	0.0017	0.26	0.2066	0.51	0.5128	0.76	0.8155
0.02	0.0047	0.27	0.2179	0.52	0.5255	0.77	0.8263
0.03	0.0087	0.28	0.2292	0.53	0.5383	0.78	0.8369
0.04	0.0134	0.29	0.2407	0.54	0.5510	0.79	0.8474
0.05	0.0187	0.30	0.2523	0.55	0.5636	0.80	0.8576
0.06	0.0245	0.31	0.2640	0.56	0.5763	0.81	0.8677
0.07	0.0308	0.32	0.2759	0.57	0.5889	0.82	0.8776
0.08	0.0375	0.33	0.2878	0.58	0.6014	0.83	0.8873
0.09	0.0446	0.34	0.2998	0.59	0.6140	0.84	0.8967
0.10	0.0520	0.35	0.3119	0.60	0.6264	0.85	0.9059
0.11	0.0599	0.36	0.3241	0.61	0.6389	0.86	0.9149
0.12	0.0680	0.37	0.3364	0.62	0.6513	0.87	0.9236
0.13	0.0764	0.38	0.3487	0.63	0.6636	0.88	0.9320
0.14	0.0851	0.39	0.3611	0.64	0.6759	0.89	0.9401
0.15	0.0941	0.40	0.3736	0.65	0.6881	0.90	0.9480
0.16	0.1033	0.41	0.3860	0.66	0.7002	0.91	0.9554
0.17	0.1127	0.42	0.3986	0.67	0.7122	0.92	0.9625
0.18	0.1224	0.43	0.4111	0.68	0.7241	0.93	0.9692
0.19	0.1323	0.44	0.4237	0.69	0.7360	0.94	0.9755
0.20	0.1424	0.45	0.4364	0.70	0.7477	0.95	0.9813
0.21	0.1526	0.46	0.4490	0.71	0.7593	0.96	0.9866
0.22	0.1631	0.47	0.4617	0.72	0.7708	0.97	0.9913
0.23	0.1737	0.48	0.4745	0.73	0.7821	0.98	0.9952
0.24	0.1845	0.49	0.4872	0.74	0.7934	0.99	0.9983
0.25	0.1955	0.50	0.5000	0.75	0.8045	1.00	1.000

SOURCE: POWER MAGAZINE

Leaks in suction lines introduce air into the oil. This may cause transfer pumps to lose their suction. Also, air in the oil makes the engine injection pumps work irregularly and causes the spray nozzles to dribble.

Government Regulation

The National Board of Fire Underwriters has published rules covering the construction and installation of fuel-storage tanks. These rules must be fully observed in plants which are insured against fire. Local ordinances often require means for stopping engines from a point outside the engine room (in case of fire, runaway). This is usually done by installing a spring-closing, lever-operated gate valve in the fuel-supply line near the engine and running a chain from the valve to a hook outside the engine room. When the chain is released, the valve closes automatically.

Heavy-Oil Supply System

The incentive to burn heavy oil instead of light oil in diesel engines is the lower price of residual oils (such as No. 5, No. 6 or Bunker C) compared to the lighter distillate fuel oils. But heavy oils, being viscous and impure, must receive special processing if they are to be used successfully. The viscosity must be reduced (oil thinned) by heating, and most of the impurities must be removed. The processing equipment costs money and requires care to operate. Consequently, heavy oils are used mainly in large engines (as are found in ships and large tow-boats), and to a lesser extent in medium-size engines.

Fundamentally, the processing of heavy fuels (by the user) includes the following steps:

1. *Heating* the oil in the main storage tank so that it can be pumped.

2. *Purifying* the oil by means of centrifuges and filters. (A centrifuge is a high-speed rotating device which subjects the oil to centrifugal forces as much as 9000 times the force of gravity. Solid material that is heavier than the oil is thrown outward and removed.) The cleaning process usually requires further heating of the oil. Water is sometimes added to the oil and then removed in a special centrifuge; this *water-washing* removes the soluble metallic salts.

3. *Reducing the viscosity* to that required for good atomization in the engine cylinders. This requires additional heating, carefully controlled so that the fuel injectors will spray the oil properly. The best temperature is found by experiment; it is then kept constant by automatic control of the heating medium, which may be hot water, steam or electricity.

Here is an example to show the relation of oil viscosity to temperature.

To thin a particular heavy oil to the viscosity at which it can be easily pumped (2000 to 5000 Saybolt Seconds Universal) requires heating it to about 85°F. To thin it further to the viscosity which is best for centrifuging (100 to 250 S.S.U.) requires heating it to about 185°F. Finally, to reduce its viscosity to that required for efficient injection into the engine cylinder (50 to 150 S.S.U.), it must be further heated to about 210°F.

4. *Light-Oil System.* Separate tanks and piping are installed with throw-over connections so that the engine may be started and stopped on light oil. The engine must be started on light oil because only a well-warmed engine will burn heavy oil satisfactorily. Light oil is

switched into the system shortly before stopping in order to fill the piping with light oil in readiness for the next start.

Heavy-oil systems effect great savings in fuel costs but they are often quite complicated and require great care in operation. The saving in fuel cost may be partly or largely offset by increased wear of cylinder liners and piston rings, burned valves, carbon deposits and excessive smoke.

Fuel-Gas Supply System

Dual-fuel engines and spark-ignited gas engines generally obtain their gaseous fuel from a natural-gas supply pipe which is under a variable pressure that is higher than that needed by the engine. Since the pressure at the engine should be constant, one or more automatic pressure-reducing valves are used. On the other hand, if the supply-line pressure is less than the pressure needed by the engine, a compressor or booster pump is used to feed gas to the pressure-regulating valve.

Fig. 19-14 shows the elements of a typical fuel-gas supply system. Gas-vent lines and lines from relief valves must always be carried outside the building

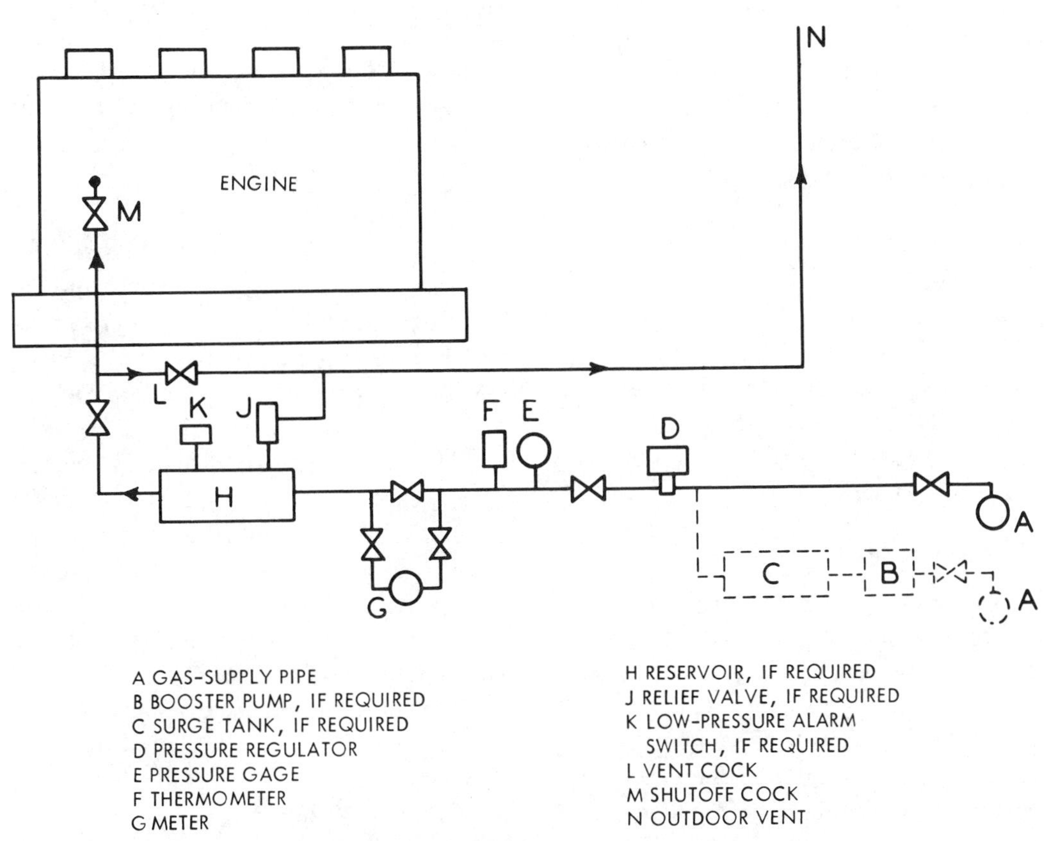

Fig. 19-14. Elements of typical fuel-gas supply system.

and never connected to drains or sewers.

In installations where the gas-supply rate is variable, such as in sewage disposal plants (Chapter 18), large gas holders are often installed to store surplus gas during periods of heavy flow.

Air-Intake System

The purpose of the air-intake system is to provide the air required for the combustion of the fuel. In the cases of two-cycle engines and supercharged engines, additional air is needed to scavenge the cylinders of the spent gases. Essentially, the air-intake system consists of piping from a source of fresh air to: (1) the air-intake manifold of a four-cycle engine, (2) the scavenging pump inlet of a two-cycle engine, or (3) the supercharger inlet of a supercharged engine. The system usually includes a cleaner to remove dust and other harmful particles and often a silencer to subdue annoying noises from air pulsations.

Piping

In order for an engine to deliver its full power and burn its fuel efficiently, it needs a full supply of air. Therefore, intake piping should be as short and as straight as possible, and bends should have a long sweep in order to reduce the resistance to air flow.

Air Cleaners or Filters

Air-cleaning devices have come into general use because they prevent the internal engine surfaces from being damaged by abrasive material which the air might carry in accidentally. They greatly extend the life of cylinder liners and piston rings by cleaning the air of most of the dust which it normally holds.

All atmospheric air is polluted to some extent with sand-like particles which increase the rate of wear. During the severe dust storms which occur occasionally in the desert section of the western United States, the air pollution may be as much as several pounds per 1000 cu ft of air. But even in clean-air localities where the air carries less than 0.2 grains (7000 grains equal one pound) of dust per 1000 cu ft, air cleaners reduce wear and ring-sticking. This is true because an engine gulps in a lot of air—from 2.2 to 5.1 cu ft/min/rated bhp, depending on the type of engine.

EXAMPLE 4. How much dust enters the cylinders of a 300-bhp, two-cycle, crankcase-scavenged engine in one year during which the engine runs 250 days of 9 hours each? The average dust content of the air (an industrial district) is 1.2 grains per 1000 cu ft. The air rate of the engine (displacement) is 3.9 cu ft/min/rated bhp.

As the amount of air taken into engine per hour = 300 bhp × 3.9 cu ft/min × 60 min/hr = 70,200 cu ft/hr, we can compute:

Dust entering engine per hour =

$$\frac{1.2 \times 70{,}200}{1000} = 84.2 \text{ grains}$$

Therefore

Dust entering engine in 1 year =

$$84.2 \times 9 \times 250 = 189{,}450 \text{ grains}$$

Since 7000 grains make one lb,

Dust entering engine in 1 year =

$$\frac{189{,}450}{7000} = 27 \text{ lb}$$

The most common types of air cleaners are: (1) dry-type filters, (2) viscous-impingement filters, (3) oil-bath filters, and (4) water sprays.

Dry-Type Filters. These clean the air by passing it through cloth or felt. When the filter element accumulates enough dirt to restrict the air flow (evidenced by an increase in the pressure drop through the filter), the surface dirt is blown off with compressed air or sucked off with a vacuum cleaner. The cloth or felt must be replaced when the pores become permanently clogged.

Viscous-Impingement Filters. These cause the air to flow a crooked path through a mass of wire or screens coated with a tacky oil. The dust particles are caught and held by the oil. To clean the filter, it is washed in fresh oil. Such filters are often made self-cleaning by mounting the filter elements on a continuously moving chain which carries the elements out of the air stream, through an oil reservoir and back into service.

Oil-Bath Filters. These filters first coat the dust particles with oil, either by blowing the air against the surface of a pool of oil or by passing the air through an oil spray. In some designs the heavier dust particles are thrown out by whirling the air through vanes; then the light particles are removed in oil-soaked metallic filter pads. The oil draining off the pads carries the dirt with it, so that the pads themselves require little cleaning.

In other designs where minimum pressure drop is essential (as for supercharged engines), the air, after passing through an oil spray, is carried into a self-cleaning filter pad to remove the dirt and then into a secondary pad to remove any remaining oil mist. The oil in the reservoir must be replaced when it becomes sludgy. Fig. 19-15 shows such a low-resistance oil-bath filter. Oil-bath filters are actually a type of viscous-impingement filter in which the air carries the oil to the filter pad and keeps it well soaked.

Water Sprays. Industrial dust in the air is sometimes washed out by passing the air through a thick curtain of water drops. This process also cools the air, not only because the water is cooler than the air, but also because some of the water evaporates and carries off heat, which is taken from the air. The heat removed by evaporating one gallon of water will cool 25,000 cu ft of air approximately 20°F.

Typical applications of cleaning air by water sprays are the power plants used to produce aluminum in the southwest. Air for combustion and also for ventilation is blown into the engine room through water sprays. In one plant, the water sprays not only remove 99 percent of the air-borne particles such as alumina dust, but also cool the air from 100°F to 78.5°F, thus making the engine room more comfortable. The combustion air, before entering the engine blowers, receives a second cleaning in dry-type filters, which also remove some of the excess moisture.

Intake Silencers

The use of an intake silencer is called for where the noise produced at the entrance to the air-intake pipe is objectionable. One way of reducing the noise is to smooth out the air pulsations by causing the air to flow through a series of passages and expansion chambers of carefully designed sizes.

Another way is to use silencers lined with sound-absorbing material, such as fire-proof hair felt separated from the air passages by perforated walls, is shown in Fig. 19-16. Sound passes through the

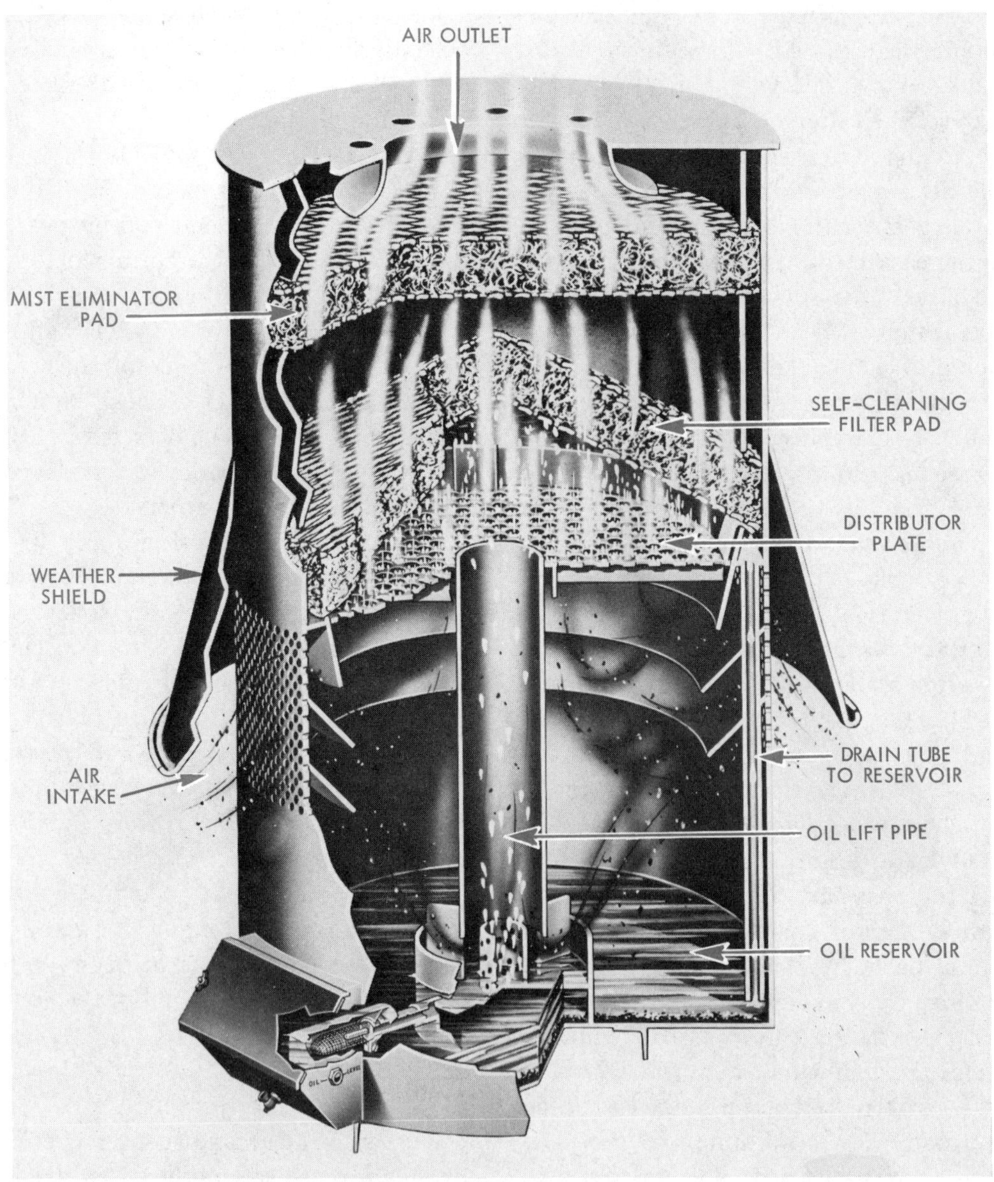

Fig. 19-15. Low-resistance, oil-bath air filter. (American Air Filter Co.)

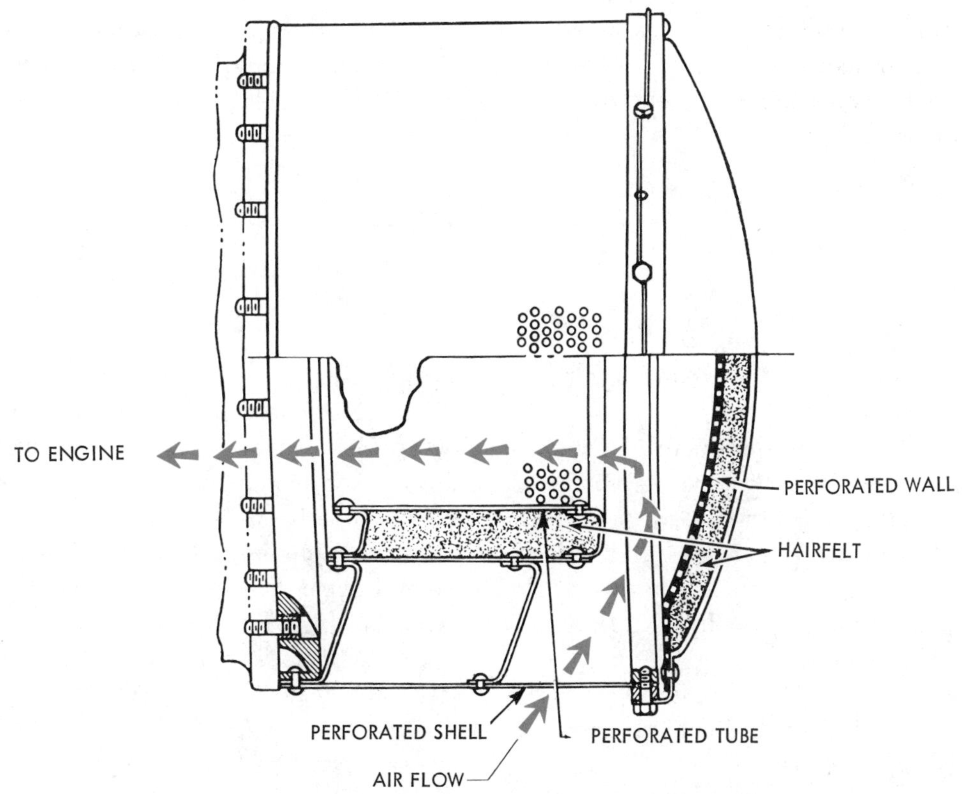

Fig. 19-16. Air-intake silencer with sound-absorbing material.

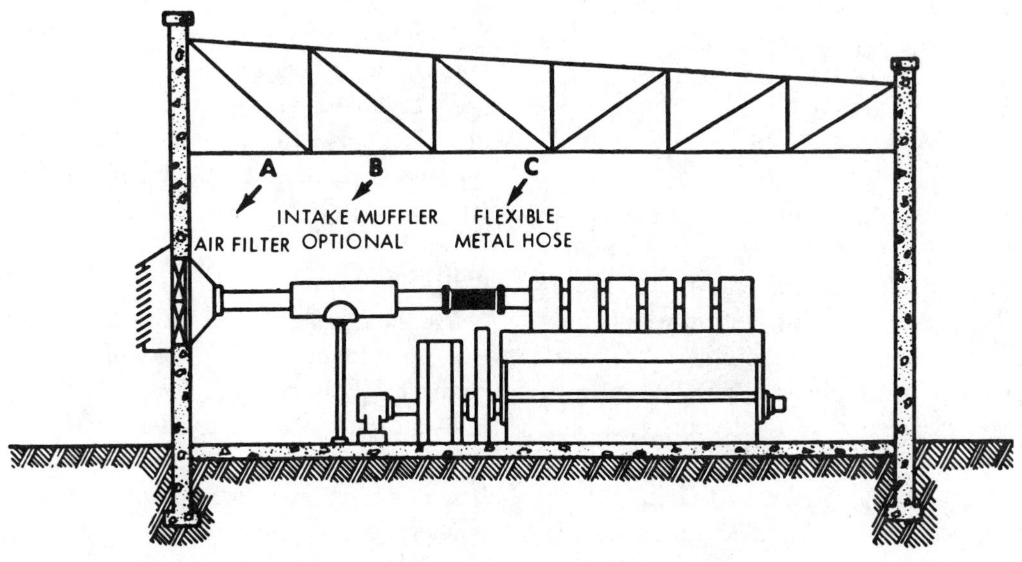

Fig. 19-17. Typical air-intake system for stationary engine. (Diesel Manufacturers' Assn.)

perforations and is partly absorbed by the felt. Intake silencers and air cleaners for small engines are often constructed as combined units.

A typical air-intake system for a medium-sized stationary engine is shown in Fig. 19-17.

Exhaust System

The purpose of the exhaust system is threefold; it is designed:

1. To carry the products of combustion to a point where the gases can be discharged without creating a nuisance.
2. To reduce the noise produced by the sharp pressure pulsations which occur when an engine cylinder releases its contents.
3. To impose the minimum back pressure on the engine. Back pressure reduces engine power.

The typical exhaust system consists of piping from the engine to the atmosphere, and a muffler or silencer. Also, for engines of moderate and large size, the system usually includes an important measuring instrument, a pyrometer, to measure the temperature of the exhaust gases at the outlet from each cylinder.

Piping

Generally, steel pipe is used to make exhaust piping. However, cast-iron pipe is sometimes used where the fuel contains a high percentage of sulfur, which may make the gases corrode steel. Brickwork or tile is unsuitable for carrying exhaust gases because it is relatively weak and cannot withstand thermal expansion, vibration, or the pressures that might occasionally result from misfiring. To reduce the back pressure, exhaust pipes should be as short and as straight as possible, and all bends should have a long sweep. Bent pipe is preferable to pipe fittings. Each engine should have its own exhaust line to the atmosphere.

Exhaust piping gets quite hot in service and expands considerably. Steel pipe expands $1/1000$ of its length for each 135°F rise in temperature. Thus, a 40-ft run of steel pipe, if heated from 80°F to 660°F, will increase its length by more than 2 inches, according to the computation:

$$\text{Expansion} = \frac{660°\text{F} - 80°\text{F}}{135°\text{F}} \times \frac{1}{1000} \times 40\ \text{ft} \times 12 = 2.016\ \text{in}$$

Therefore, where the exhaust piping is immovable, a section of flexible metallic hose or an expansion joint is generally installed close to the engine to permit the pipe to expand without straining the engine.

Mufflers

Pressure pulsations of the exhaust gases can be reduced by means of orifices, changes in direction of flow, and/or expansion chambers in mufflers. However, orifices and changes in direction tend to increase back pressure. Therefore modern mufflers use expansion chambers which are often designed according to acoustic principles so as to absorb the annoying sound waves while permitting the gas itself to flow freely. Commercial designs of mufflers have been well developed by specialized manufacturers as a result of much experimental research.

The exhaust-gas turbines used to drive turbo-superchargers are themselves fairly good mufflers, and sometimes make it unnecessary to use additional muffling devices.

Fig. 19-18. Exhaust pyrometer indicator with selector switch. (Illinois Testing Laboratories)

Exhaust Pyrometers

The more fuel burned in an engine cylinder, the hotter will be the exhaust gases. When an engine is in good mechanical condition and the load is divided equally among the cylinders, the exhaust temperatures should run about the same from cylinder to cylinder at any given load. But when the load is not balanced among the cylinders, or valves and piston rings are not tight, or nozzles are in poor condition, or some other defect is present, the temperatures will not be uniform. Therefore, an exhaust pyrometer, which measures the temperature, helps the operator keep his engine in good condition and properly adjusted.

An exhaust pyrometer consists of two elements: a *thermocouple* inserted in the exhaust passage from each cylinder, and a conveniently located *indicator* which shows the temperature. The thermocouple is simply a pair of wires of two different metals welded together. When the junction is heated, a small voltage develops proportional to the temperature. This voltage is carried by wires to the pyrometer indicator, which is a sensitive voltmeter whose scale is graduated to show temperatures directly. By means of a selector switch, one indicator can be used to show the reading of each of the thermocouples. See Fig. 19-18. On small engines, the wiring may be eliminated by using a portable indicator with exposed terminals, which the operator can place in contact temporarily with the terminals of each thermocouple.

Fig. 19-19 shows the usual range of exhaust temperatures for four-cycle and

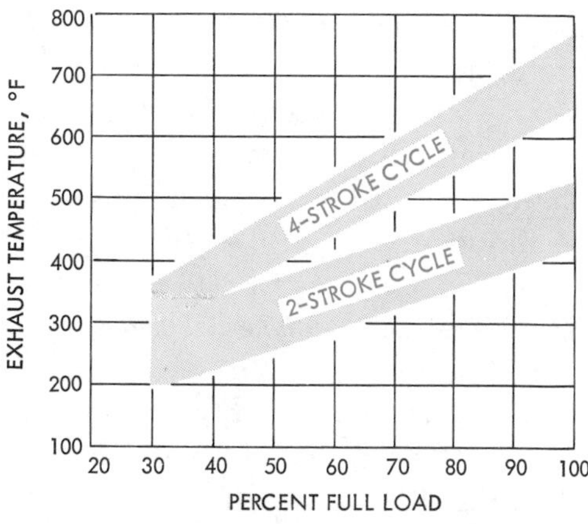

Fig. 19-19. Range of exhaust temperatures for diesels. (*Power Magazine*)

two-cycle diesel engines. Note how much hotter the engine cylinders run when the engine becomes fully loaded. For this reason, close balancing of the exhaust temperatures of all cylinders is more important at heavy loads than at light loads.

Exhaust Heat Recovery

The heat rejected in the exhaust gases is about 30 percent of the heat in the fuel (Table 19-1). Why not recover it for use where heat is necessary? This can be done, but experience has shown that ordinarily it is better to utilize first the heat in the jacket water. There are several reasons: (1) Exhaust heaters or boilers cost more than jacket-water heat exchangers. (2) Only part of the exhaust heat is recoverable because the exhaust gases should not be cooled below about 300°F; otherwise heat-transfer surfaces and piping will corrode rapidly. (3) Heat exchangers carrying exhaust gases tend to deposit soot in the exhaust passages and therefore require periodic cleaning. But where the value of the recovered heat is high enough to justify the investment and maintenance costs, satisfactory installations can be made.

Exhaust heat can be easily utilized to heat a large power plant in winter by enclosing the exhaust mufflers in chambers and blowing air over the hot surfaces of the mufflers and into the power plant.

Starting System

Diesels and gas engines are not self-starters. In order to start them, the engine crankshaft must be turned over by some outside means so as to (1) admit combustion air to the cylinders, (2) introduce the fuel, and (3) cause the mixture to fire.

In the case of diesel and dual-fuel engines, the cylinder charge, after being compressed, must be hot enough to ignite the oil fuel when it is injected at the top center. This requires that the engine be turned over with *sufficient speed*. If the engine is turned over too slowly, the unavoidable small leaks past the piston rings and also through the intake and exhaust valves (of four-cycle engines) will permit a substantial part of the air (or gas-air mixture) to escape during the compression stroke. Also, the heat loss from the compressed air to the cylinder walls will be greater at low speed because of the longer exposure.

The escape of air and the loss of heat both result in a lower temperature at the end of compression. Therefore, there is a minimum speed which the engine must attain before ignition will occur and the engine will begin firing. The starting speed depends upon the type and size of the engine, its condition, and the temperature of the air entering the engine.

Starting methods may be classified according to one of the following methods: (1) hand starting, (2) electric motor, (3) gasoline engine, (4) compressed-air motor, and the (5) compressed-air admission method.

Hand Starting

Starting done by hand, using a crank-handle as formerly used on automobiles, is practicable only for the smallest engines. It takes considerable effort to overcome the high compression pressure of a diesel engine. Often a hand valve is provided to relieve the compression at first; this enables the operator to bring the engine up to a speed at which the inertia of the flywheel can overcome the full compression pressure.

Electric-Motor Starting

Starting by means of an electric motor is accomplished in two different ways. On small engines, the common arrangement uses a starting motor similar to that used for starting automobile engines, but much more powerful, employing 24 or 32-volt storage batteries. Larger engines used to drive direct-current generators, such as on diesel-electric locomotives, may be started by using the main generator as a motor, taking current from a large 64 or 112-volt storage battery; in stationary plants where the main switchboard is always energized, the starting current may be taken from that source.

Gasoline-Engine Starting

A small gasoline engine is mounted on the diesel engine. The shaft of the gasoline engine carries a pinion which can be meshed with teeth on the diesel engine flywheel. The operator starts the gasoline engine by hand; when it has warmed up, he engages the starting gears. The diesel engine then turns over; when it fires and picks up speed, it automatically disengages the starting gears.

This starting method has the advantages of simplicity and of being able to start the diesel engine in extremely cold

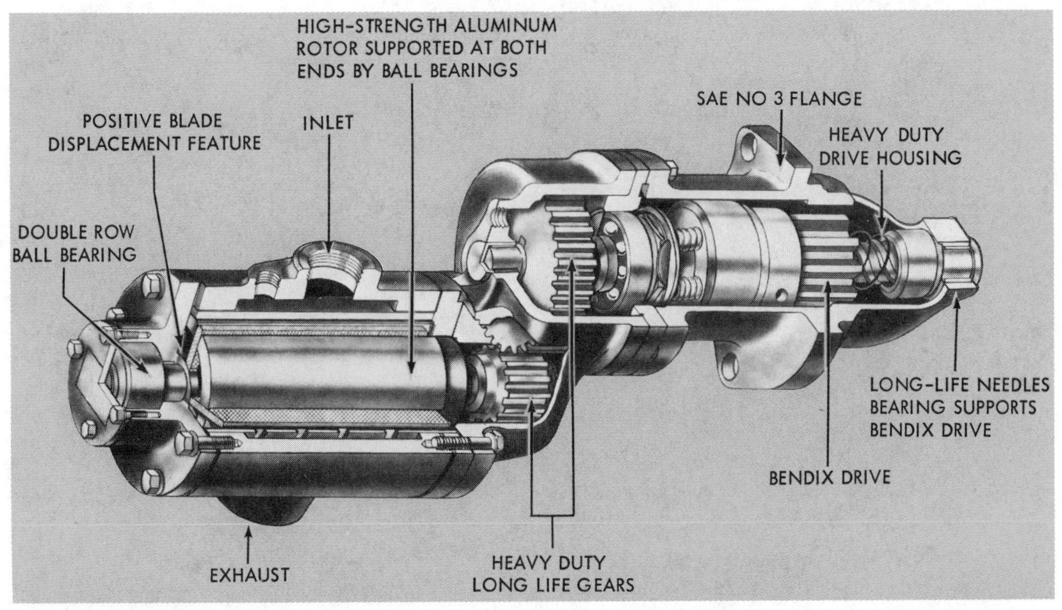

Fig. 19-20. Highway tractor air motor and transmission starter system. (Stanadyne/Hartford Division)

weather. The warm jacket water from the gasoline engine can be used to preheat the main engine, and then the gasoline engine can turn it over as long as needed.

Compressed-Air Motors

In order to eliminate the storage battery, compressed-air motors are sometimes substituted for electric motors for starting small engines. The air motor has a rotor with vanes, enclosed in a casing, and is powered by compressed air from a storage tank. Unless compressed air is available (as in a machine shop), a special compressor is needed to recharge the storage tank. For on-the-highway diesel tractors, a new type of air starter system has been developed by using the air brake compressor as an air supply to operate the system. Fig. 19-20 shows the entire air motor and starter transmission.

This system has a higher torque output than the conventional system. Fig. 19-21 shows a close-up view of the air starter shown in Fig. 19-20. This air motor will produce torque to turn an engine over at 280 rpm in 0.4 seconds, as compared with an electric starter motor which will only produce 180 rpm in 0.4 seconds.

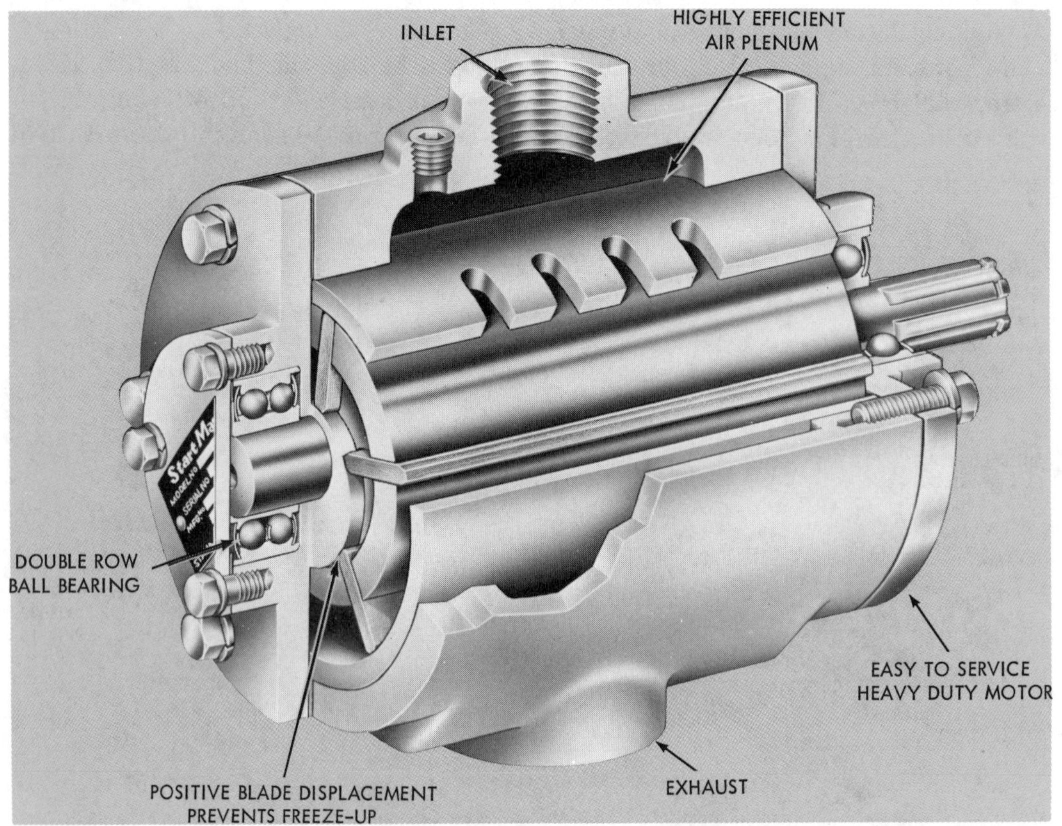

Fig. 19-21. Close-up view of an air starter motor. (Stanadyne/Hartford Division)

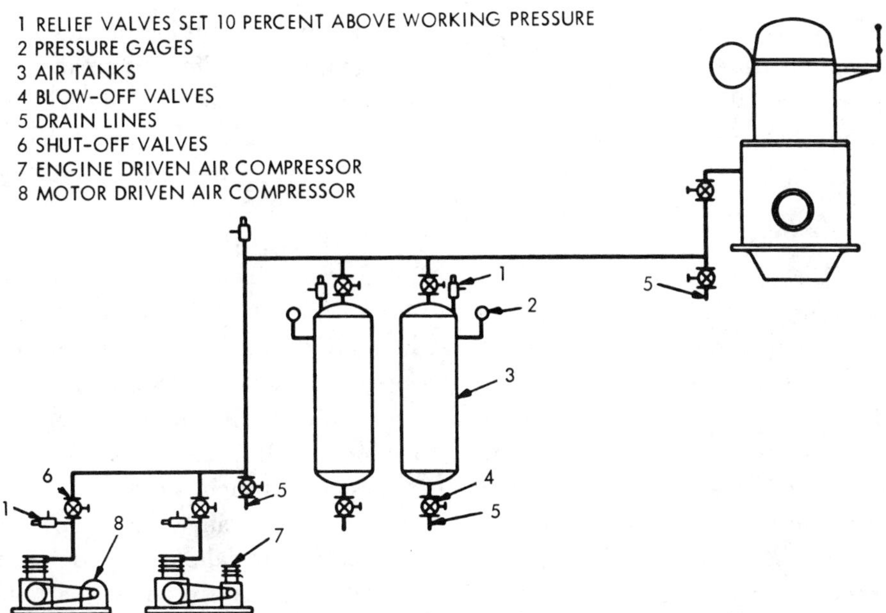

Fig. 19-22. Diagram of typical compressed-air starting system. (Diesel Manufacturers' Assn.)

Compressed-Air Admission

The starting method most commonly used on engines of moderate and large size is the compressed-air admission method. The energy needed to start the engine is stored by compressing air into one or more tanks. The compressed air is admitted through an automatic starting valve in the engine cylinder head when the piston is at top center at the start of the power stroke. The pressure of the air pushes the piston down and thus turns the engine over. When the engine is turning fast enough, the injected fuel ignites and the engine runs on its own power, whereupon the air supply is cut off.

The opening of the starting valves is timed, directly or indirectly, by cams on the camshaft. Usual starting air pressures on modern engines range from 250 to 350 psi.

A typical compressed-air starting system is shown in Fig. 19-22. Note the *two* air compressors. Ordinarily, compressor *8*, driven by an electric motor, is used to charge the storage tanks. However, if the electric power should fail, compressor *7*, driven by a gasoline or gas engine, is put into service.

Electric Ignition Systems for Gas Engines

High-compression spark-ignited gas engines need far more powerful electric ignition systems than do low-compression gas or gasoline engines. Although battery-powered systems which work like that in an auto are sometimes used, most of these engines are now fitted with magnetos, pulse-generators or electronic ignition systems.

Magneto Systems

Fig. 19-23. A magneto is a kind of electric generator, consisting of a rotating permanent magnet and stationary pole pieces wound with insulated wire. This generating winding connects with an electric breaker, or interrupter, which is operated by a cam turned by the same shaft as the rotor and is timed to open (separate) the breaker contacts at the moment the magnet is causing maximum current to flow in the generating winding.

In a *low-tension magneto system*, the resulting surge of low-tension current passes into the primary winding of a transformer (two coils of insulated wire surrounding an iron core). The primary winding, which contains comparatively few turns of wire, produces in the adjacent secondary winding of many turns a voltage about 50 times as great. The high voltage is sufficient to jump the gap of the spark plug in the engine cylinders; the resulting spark ignites the fuel mixture.

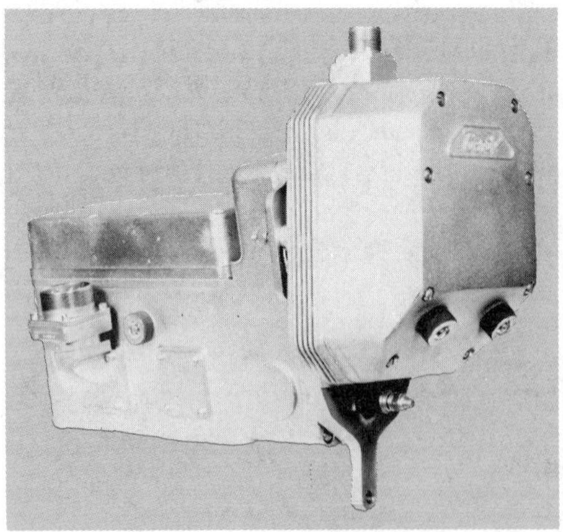

Fig. 19-23. Magneto for high-compression gas engine. (Scintilla Div., Bendix Corp.)

Fig. 19-24 is the wiring diagram of the Bendix-Scintilla low-tension magneto system for a multi-cylinder engine. It shows the fundamental principle and some additional features. Note that there is a transformer for each engine cylinder. The capacitor in the magneto section collects the energy which would otherwise be wasted in producing a heavy spark at the breaker points when they open. It interrupts the current more sharply, thus strengthening the spark in the engine cylinder.

The distributor section contains a group of contacts operated by cams which are timed to feed the low-tension current to the proper transformer. In many large engines, each cylinder is fitted with two spark plugs to ignite the mixture better. In this case, a separate transformer serves each plug. The switch shown is used to stop the engine. It grounds or short-circuits the magneto; then no current flows through the breaker contacts and therefore no high-tension current is produced.

A *high-tension magneto system* differs from the low-tension type in that the secondary winding of the transformer is placed next to the generating winding *within the magneto itself*, and there produces the high-tension current. The primary current passes through the breaker, as in the low-tension system, but the high-tension current passes through a distributor like that in your auto. From the distributor, shielded cables carry the high-tension current to each spark plug.

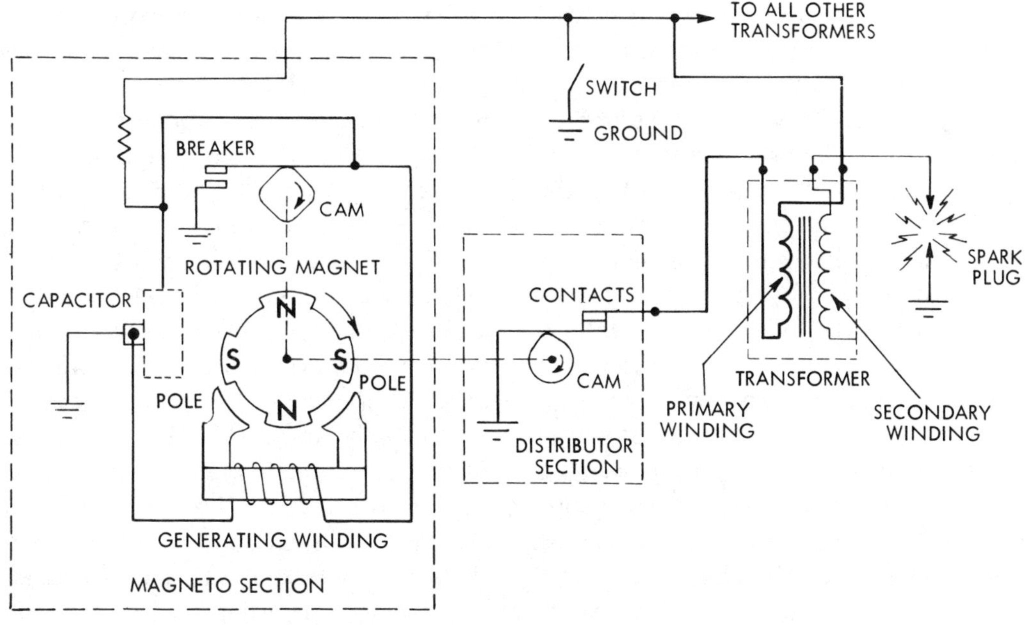

Fig. 19-24. Wiring diagram for low-tension magneto system.

The high-tension system costs less than the low-tension, but on large engines, the long shielded cables cause detrimental capacity and leakage effects.

Resistors

Ignition systems for high-compression gas engines often include a resistor in each high-tension circuit. It reduces the gradual burning away of the spark plug's metal points. The resistor is a piece of carbon with a high electrical resistance (usually about 250,000 ohms). Here's the way it works. Every high-tension circuit has electrical capacity, that is, with insulating material around the wire it stores electricity. Each time we fire a plug we must build up enough voltage in the ignition system to break down the plug gap, to start the flow of current across the gap.

While we are building up this breakdown voltage we are also filling up the electrical storage to the same voltage. In high-powered ignition systems a lot of energy is stored, far more than in your car's ignition system. There are two reasons for this: (1) the voltage is usually about 4 times as high, and the stored energy increases with the *square* of the voltage, that is, about 16 times; (2) the secondary windings of the transformer have about 2½ times as many turns of wire, with a corresponding increase in electric storage capacity. Therefore, the stored energy may be $16 \times 2\frac{1}{2} = 40$ times as great as in your car. If all this stored energy were allowed to reach the plug's points, it would burn them away in a short time. Here's where the resistor comes in—it dissipates in the carbon insulation a large part of the unwanted stored energy and thus lengthens the useful life of the spark plug points.

You may wonder why the resistor does not also reduce the voltage applied to the spark plug. It does, but only *after* the plug has fired and maximum voltage is no longer needed. Before the plug fires, the circuit is open because no current is flowing across the plug gap. The voltage is building up to the point where it will jump the gap, but the resistor has no effect on the voltage because a resistance reduces a voltage only when current is passing through it. When the rising voltage finally jumps the gap (and the resulting spark fires the charge), the gap resistance breaks down to a low value and all the stored energy is released. Now the resistor carries current and performs its task of absorbing much of the energy that would otherwise burn the plug points.

Pulse-Generator System

The electric ignition requirements of large, high-compression, lean-mixture gas engines are severe, especially when the engines run continuously at full power as in gas pipeline pumping. The pulse generator was developed to improve service reliability while providing greater sparking energy. Its power is adequate to fire two spark plugs per cylinder simultaneously in order to achieve the maximum efficiency required of these large engines. The pulse generator is highly reliable and easy to maintain because it has few moving parts and uses no breakers, distributors or cams. (A breaker is required only when starting if the cranking speed is less than about one fifth of the running speed.)

Fig. 19-25 shows the American Bosch MGC Pulse Generator. The only moving part is a powerful Alnico magnet rotor, which turns within a stationary casing or stator made of thin soft-iron rings stacked together. The stator is cut to form a number of generating teeth, one for each engine cylinder. Each tooth is encircled by a coil of wire which acts as a generating coil. When the North and South pole shoes of the rotating magnet pass a particular stator tooth, an electric current is created in the generating coil, first in one direction, then in the other, as the magnetism through the tooth changes its direction. Since the edges of the stator teeth and the pole shoes are quite sharp, the current surges are brief, hence the term *pulse*. The second of the two pulses is stronger because it is produced by reversing the magnetism; consequently, the second pulse is used to fire the spark plug.

The first pulse (which might cause preignition) is eliminated by means of a rectifier. A rectifier is a device that will pass current in one direction and will block it when it reverses its direction. The rectifier is connected across the primary winding of the ignition transformer in such a way that it permits the first, or unwanted, pulse to flow to the ground (and thus be dissipated) but blocks the second, or wanted, pulse so that it has only one place to go—through the primary winding of the transformer. Here the low-voltage pulse (about 100 volts) builds up the ignition voltage (about 25,000 volts) in the secondary winding.

There is one transformer for each cylinder head, as close as possible to the spark plug (or plugs) to reduce the length of the high-tension cable. A resistor is

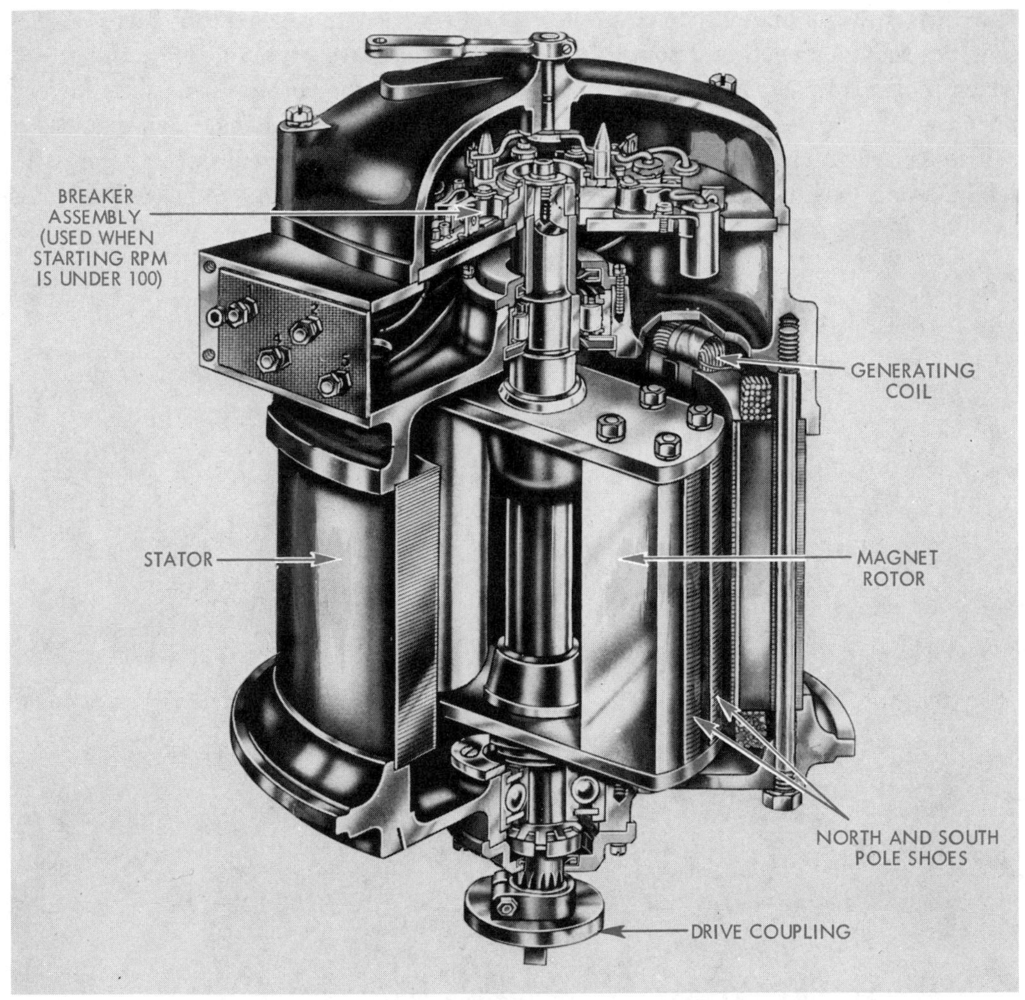

Fig. 19-25. American Bosch MGC pulse generator. (American Bosch AMBAC Industries, Inc.)

inserted in the circuit between the transformer and the spark plug to reduce the burning away of the plug's points.

Because a pulse generator has to accomplish by slow rotation what a magneto does by cam and lever action, a pulse generator has to be much larger; the usual model weighs about 100 pounds and is considerably more costly than a magneto. Consequently, pulse generators are found mostly on large engines in heavy-duty service.

Electronic Systems

The recently developed electronic ignition systems enjoy most of the advantages of pulse generator systems but being smaller and less costly they are adaptable to smaller engines. They employ transistors, which are tiny electronic switches with no moving parts. These switches control an external power source which provides the energy needed to produce the igniting spark.

The American Bosch Pulse-tronic system has only one moving part, the rotor of a small timer-generator, Fig. 19-26. Its function is to supply a tiny pulse of electricity at the correct firing time for each of the engine cylinders. Each cylinder requires a transistor control unit consisting of three transistors. The low-level pulse

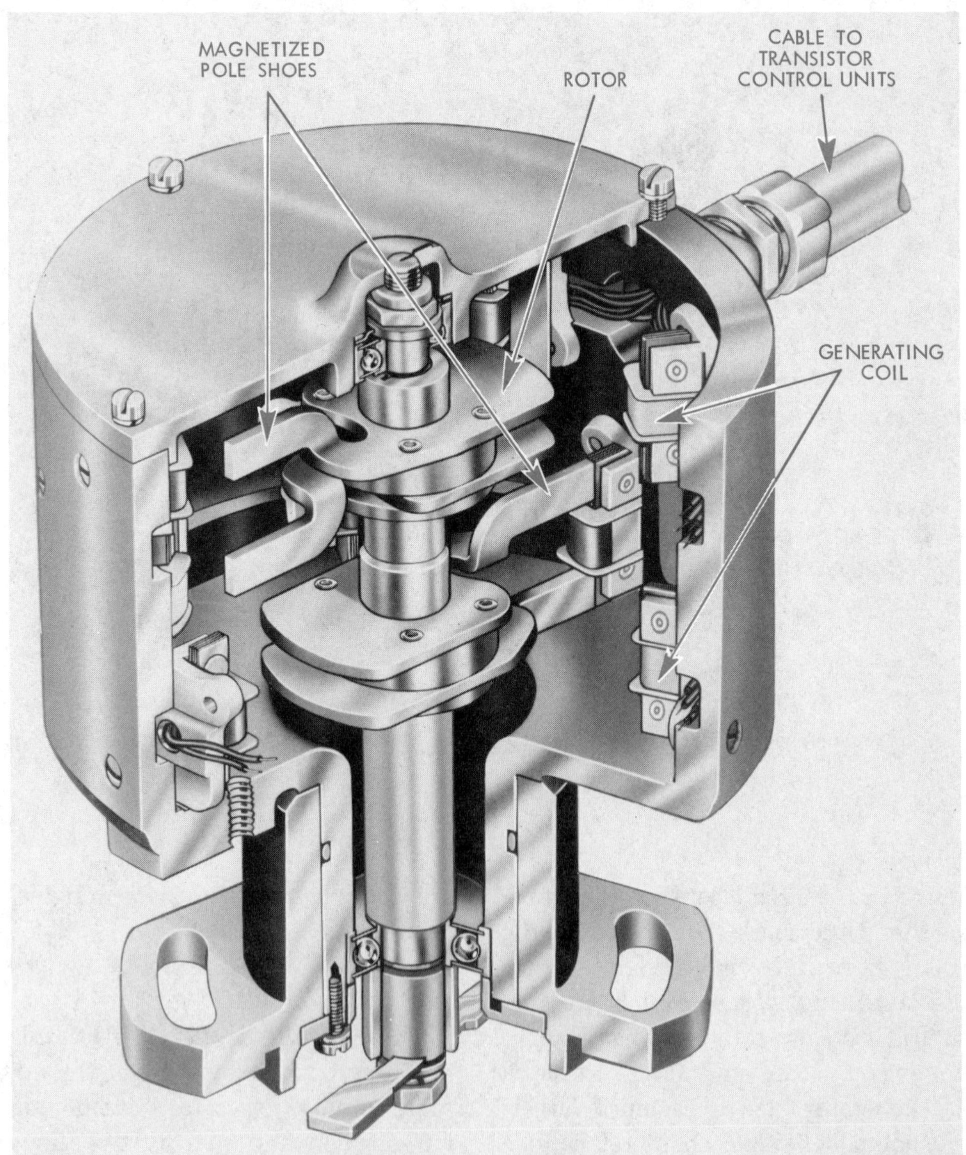

Fig. 19-26. Timer-generator of American Bosch Pulse-Tronic ignition system. (American Bosch AMBAC Industries, Inc.

from the timer-generator triggers the first transistor, which in turn triggers the other two. The latter make-and-break the circuit from the external power supply to the ignition transformer which, in turn, provides the high sparking voltage.

Let's see how a transistor works. The transistor is a small solid-state device which normally will not conduct electricity. However, when a *negative* voltage is applied, (a tiny triggering voltage will do), it becomes conductive and will pass a much larger electric current. But if the triggering voltage changes its direction and becomes *positive*, the transistor instantaneously ceases to conduct. Thus by merely changing the direction of the triggering voltage (reversing the polarity) a much larger current can be switched on and off.

This is exactly what each magnetized pole shoe of the timer-generator does; as it approaches a generating coil it produces a negative triggering voltage, causing the transistors to switch the external power to the primary coil of the sparking transformer. The moment the pole shoe passes beyond the generating coil, the triggering pulse reverses its polarity; this instantaneously switches off the primary current. The sudden interruption of primary current is similar to the opening of the breaker points in a conventional ignition system. Note that the individual generating coils for each engine cylinder take the place of the conventional distributor.

Another electronic ignition system, designed for small engines, is the Bendix-Scintilla breakerless design, Fig. 19-27. It

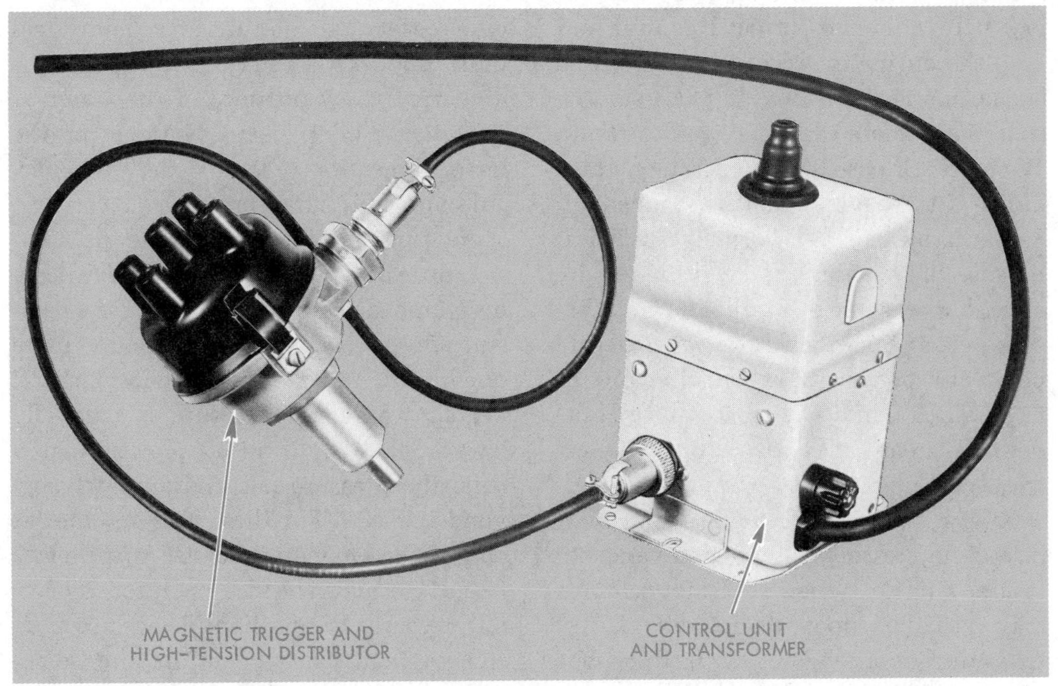

Fig. 19-27. Bendix-Scintilla breakerless ignition system. (Scintilla Div., Bendix Corp.)

differs from the American Bosch scheme described above principally in using only one transformer per engine instead of one per cylinder. The high-tension secondary current from the transformer is fed to each cylinder through a rotary distributor. As in the American Bosch scheme, an external power supply to the transformer primary is switched on and off by transistors triggered by a small magnetic device driven by the engine.

Alarm and Shutdown Systems

Safety devices for internal-combustion engines fall into two classes: (1) alarms to give warning of trouble, and (2) shutdown devices to stop the engine before damage results. Whether an alarm is sufficient, or whether the engine should be stopped automatically, depends on how closely the engine is attended and on the nature of the load it carries.

If an attendant is always near by, an alarm is generally enough. He can act promptly, either to correct the fault and keep the engine in service, or to stop the engine manually if he can't correct the condition before damage would result. It's better, if possible, to let the operator decide if the engine must be stopped.

Alarm systems are usually electrical because the warning can be given best by electric sirens, bells, or lights, and because electric annunciators conveniently point out by means of signal lamps or flags which condition is at fault. Shutdown systems may be operated electrically or by direct mechanical action.

Safety devices should receive regular inspection, testing, and maintenance to insure reliable action. A safety device may be called upon to act only a few times during an engine's entire life—but when needed, it *must* work.

Protection by Safety Devices

The operating conditions which may be protected by safety devices are some or all of the following: (1) cooling-water flow, (2) lubricating-oil flow, (3) engine speed, (4) day-tank level, (5) starting-air pressure, (6) bearing temperature, (7) exhaust temperature, and (8) unwanted gas flow.

Cooling-Water Flow. The usual indication of trouble is excessively high engine-outlet temperature. An additional indication, often used in large plants, is abnormally low pressure of the water at the engine inlet. The advantage of the latter indication is that if the water circulation fails, the pressure device responds before the temperature rises.

Lubricating-Oil Flow. If the lubricating-pump delivery falls off, the oil pressure drops to the danger point more quickly at the far end of the engine oil header where the pressure is naturally lower. Likewise, if an oil pipe springs a leak, the pressure falls more quickly beyond the leak. For these reasons, the far end of the engine oil header is the place to put an alarm device operated by low oil pressure. In addition, devices responsive to high oil temperature are sometimes used, particularly on large engines having pistons cooled by circulating oil.

Engine Speed. An overspeed stop prevents a runaway if the regular governor fails or if the load is suddenly lost on an engine not equipped with a governor. Overspeed protection is always arranged to stop the engine at once, rather than give an alarm, because an engine may speed up rapidly, and overspeed can cause severe damage.

Day-Tank Level. When fuel day tanks are filled by motor-driven pumps under automatic control, high-level and low-level alarms are sometimes used to signal any trouble with the fuel-transfer system.

Starting-Air Pressure. Starting-air tanks should always be kept at full pressure, ready for use. In plants where the precaution is justified, each tank is fitted with a pressure switch to sound an alarm if the storage pressure drops too low.

Bearing Temperature. The main bearings of most engines can easily be fitted with thermostatic elements which act when a bearing becomes hot. They either sound an alarm or stop the engine.

Exhaust Temperature. Although exhaust pyrometers are generally arranged to give only visual indications, they are occasionally fitted with electric contacts to sound an alarm or even stop the engine if a cylinder runs abnormally hot.

Unwanted Gas Flow. The flow of fuel gas should be shut off quickly whenever a gas-burning engine stops or if the supply of pilot oil to a dual-fuel engine should be interrupted. Gas shut-off valves, automatically operated, are usually supplied as standard equipment.

Electrical Alarm Systems

The major components of electrical alarm system are: (1) fault switches to measure the conditions and to open or close contacts when any condition becomes abnormal, (2) warning devices to give an alarm, (3) fault-locating devices to point out which condition is at fault, and (4) energizing devices to put the system into operation whenever the engine is running.

Shutdown Methods

Diesel engines may be shut down by: (1) stopping the fuel supply to the injection pumps, (2) stopping the action of the injection pumps themselves, (3) holding the exhaust valves open (four-cycle engines only), or (4) shutting off the air supply.

Stopping Fuel Supply. This can easily be done automatically by using a solenoid valve in the fuel-supply line close to the inlets to the injection pumps. The valve needle has a magnetic head which is placed inside a coil of wire called a *solenoid.* Electric current flowing through the coil creates a magnetic force which lifts the valve needle and permits fuel to flow. Breaking the electric circuit permits the needle to fall and close the valve, thus stopping the engine.

Stopping Injection. This common method shuts the engine down quickly by stopping the pumping action of the fuel-injection pumps, usually by shifting the fuel-control rods to their shutoff positions.

Holding Exhaust Valves Open. This is a positive way to stop a four-cycle diesel engine. Without compression in the cylinders, the fuel won't ignite, so firing stops at once. This method has the advantage of being able to stop an engine even when it is receiving excessive fuel which is out of control, as might occur, for example, if the injection-pump control jammed, or if some leak delivered oil or gas into the engine air intake.

Shutting Off Air Supply. An engine will stop if it doesn't receive oxygen to burn the fuel. Shutdown devices working on this principle close a butterfly valve in the air-intake line.

Dual-fuel engines are shut down by shutting off both the gas supply and the oil-injection pumps. Spark-ignited engines are shut down by grounding or interrupting the ignition circuit and shutting off the gas supply.

Automatic Starting and Load-Control System

A power plant in which the engines have been fitted with safety devices is already a partly automated plant; once the engines have been started they require little attention other than routine inspection and maintenance.

For full automation, additional automatic devices are needed to start the engines, apply the load, and stop them. Also, in certain cases, the power plant transmits signals to a distant supervisory point to indicate how it is operating.

How full automation is accomplished depends upon the nature of the job to be performed. For example, *emergency electric-generating* units are required to come into action quickly, by themselves, if the normal electric power supply fails. Typical applications are in hospitals, central telephone stations and microwave relay stations.

The power failure itself operates an electric relay, which in turn, causes the engine to be cranked, to fire, to come up to voltage and to take over the load from a transfer switch. Once the cycle begins, the several stages are so timed and interlocked that one stage must be completed successfully before the next stage starts. When the regular power returns, the relay throws the transfer switch back and stops the engine.

Fully-automated *multiple-unit electric-generating plants* include additional devices for starting engines one by one in accordance with the line load, and for putting a spare engine into service if any running engine gives trouble. The control devices are electrical.

Gas engine compressor units used on gas pipelines are often fully automated. The Cooper-Bessemer En-tronic control system operates pneumatically, using either air or gas. Fig. 19-28 shows a control cabinet which (1) starts and stops the engine, (2) turns the gas compressor valves on and off to load and unload the engine, and (3) provides safety shutdown and indicating. Once the signal is given (either on the spot or by remote control) to start the engine, the control system automatically puts the engine through the following starting sequence, checking that each operation is accomplished before the next one begins:

1. Activates the safety protection.

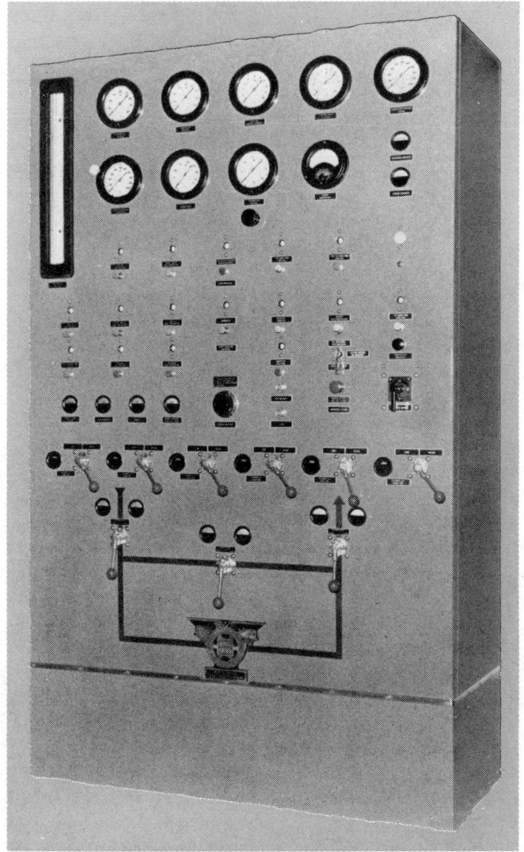

Fig. 19-28. Cooper-Bessemer En-Tronic control cabinet. (Cooper-Bessemer Corp.)

2. Checks engine conditions to be sure engine is safe to start.
3. Opens the starting air valve and rolls engine.
4. When engine begins to roll, activates purge timer.
5. After engine has rolled long enough to purge cylinders and exhaust piping of unburned gas, and after engine has reached adequate starting speed, activates the ignition.
6. Opens gas valve partially to admit starting gas to engine cylinders.
7. When engine fires, opens gas valve further to gradually increase engine speed until it comes under governor control.
8. Closes starting air valve at proper engine speed during acceleration.

Automatic control of the suction, discharge and by-pass valves of the gas compressor is integrated with the engine starting sequence so that the compressor is unloaded when the engine starts and the compressor begins to pump gas after the engine attains operating speed.

Checking On Your Knowledge

The following questions give you the opportunity to check up on yourself. If you have read the chapter carefully, you should be able to answer the questions. If you have any difficulty, read the chapter over once more so that you have the information well in mind before you go on with your reading.

DO YOU KNOW

1. Why is lubrication a difficult problem in large diesel engines?
2. How are the problems of complete lubrication of large diesel engines solved?
3. How does a mechanical forced-feed lubricator operate?
4. Explain the three ways in which lubricating oil is treated.
5. Identify 3 major jobs of the cooling system on large diesel engines.
6. Compare the heat transferred to the cooling system in a 4-cycle unsupercharged and supercharged engine.
7. Explain and diagram an open-type cooling system.
8. Identify and explain the principle of four closed-type cooling systems.
9. Explain the principle of four types of air cleaner systems used on large diesels.
10. Explain the purpose and the principle of using exhaust pyrometers in large diesel engines.
11. List five different methods of starting diesel engines.
12. Identify eight different safety devices for diesel engines.

Operation and Maintenance

Chapter

20

The purpose of this chapter is to acquaint you with *basic principles* of operating and maintaining diesel and gas engines. Engines of different types and makes differ greatly in details. Therefore, for specific instructions regarding a particular make of engine, you should refer to the engine's builder's manual. But the same underlying facts, for the most part, apply to every engine; learning these facts will give you the basic picture. This will help you better understand such instruction books when you need to use them.

Operating the engine is the normal routine process of starting it, running it, stopping it, and keeping records. Maintaining the engine is keeping it in good mechanical condition so that it will operate reliably and efficiently. This also requires keeping records.

Fundamental service problems of diesel engines often originate from using too small an engine for the required power needs. Also increased state and federal standards on emission control will require better service. Trouble-shooting and how the test instruments are used to diagnose problems are important parts of this chapter.

Operation

The key to good operation is *regularity*—establish sound procedures and stick to them. Follow a carefully worked-out routine for even such apparently simple operations as starting and stopping the engine. Determine, from the information given in this volume, from the engine builder's instructions, and from your own experience, the various items that need regular attention during normal operation. Set up a procedure that insures attention to those items.

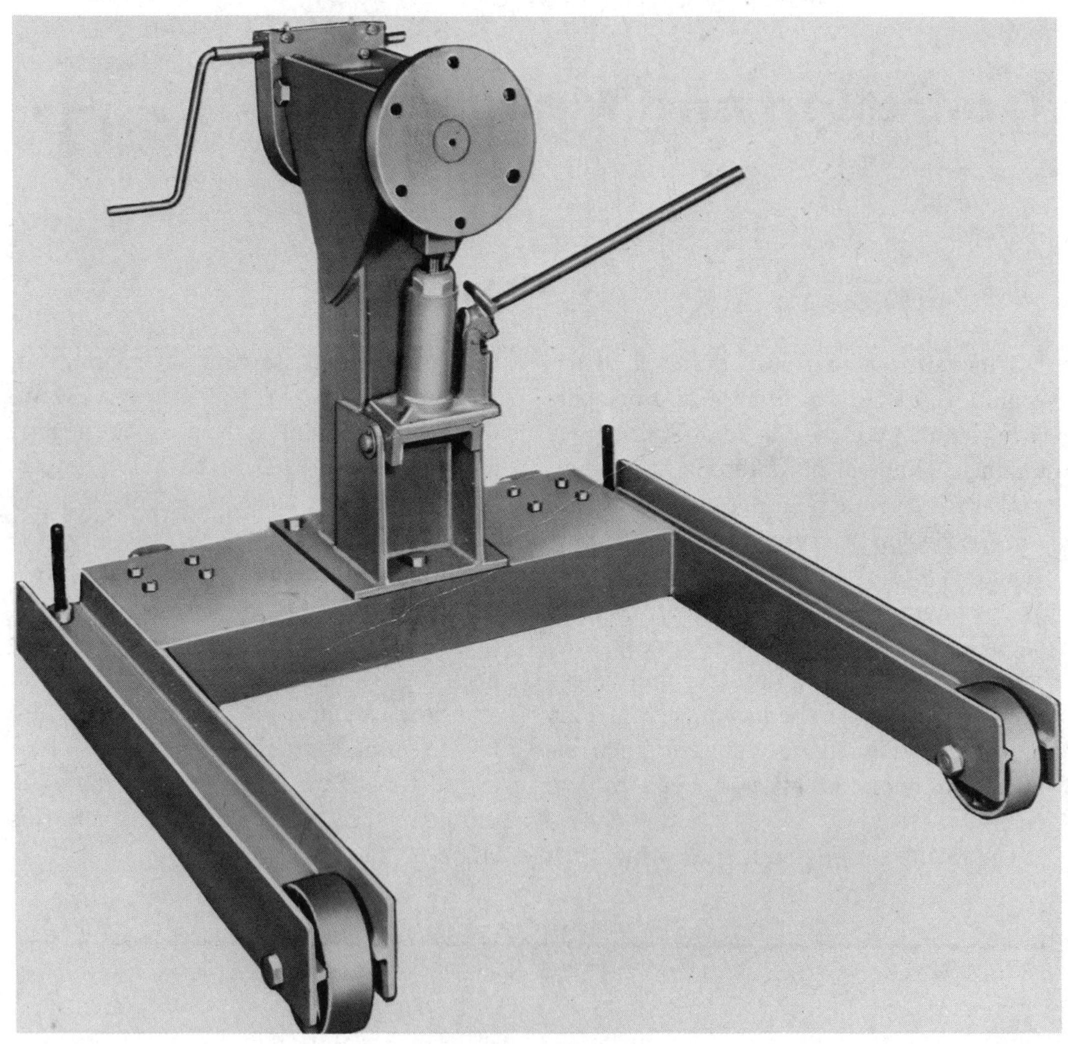

For the well equipped shop.

Don't overlook the auxiliary equipment. The engine itself, though the major part of the power plant, is not all of it. Even if the engine itself is well cared for, it can operate well only if care is also given to the auxiliary equipment which puts it into motion and which supplies it with air, fuel, cooling water, and lubricating oil.

You will find a list of operating pointers which emphasize the important factors to be considered in starting, operating, and shutting down diesel and gas engines in the section which follows.

Adequate records, whether a simple log kept in the operator's notebook or a carefully designed set of forms, play an invaluable role in plant operation. Keeping such records means that the operator must regularly check pressures, temperatures, load, and other vital factors. Merely making a record is not enough, however. The record must be used. If it is a good record in the hands of a good operator, it will prove its usefulness daily.

Operating Procedures

The following pointers on starting the engine, normal operation, and stopping the engine will appear elementary to experienced operators. However, they should prove useful as a reminder and as a guide to men new to diesel and gas engines.

STARTING THE ENGINE

Follow the manufacturer's recommended procedure, paying particular attention to the following items.

Positioning. Check for mechanical interference by turning the engine through at least one full cycle (two revolutions for a four-cycle engine).

If necessary, *spot* the engine by putting it in the crank position at which it will start. Spotting is unnecessary for an engine cranked by electric motor, rotary air motor, or gasoline engine. Spotting is also unnecessary for an engine equipped with air-starting valves if the number of cylinders is six or more in the case of four-cycle engines, or four or more in the case of two-cycle engines. Spot other air-starting engines by so positioning the crankshaft that one of the pistons is about 15 degrees past top center and its starting valve is ready to open as soon as the starting air is applied. The latter is determined by checking the *camshaft position.*

Cooling System. Make sure that the cooling system is full of water. Open all valves which may interfere with free circulation. If the water is supplied by independently driven pumps or by overhead tanks, establish the flow before starting, in order to rid the system of air.

Lubricating System. Check the oil level in the crankcase or sump tank. Operate the hand pump or turn the engine over several times to insure an oil film at the bearings. If the motor is equipped with motor-driven oil pumps, start the oil flow before the engine runs. Check hand-lubricated parts and also grease and oil cups; operate mechanical lubricators by hand a few times.

Fuel System. Prime the fuel oil lines (fill them with fuel) and vent air completely. Make sure that fuel reaches the injection nozzles to insure quick starting. Dual-fuel engines must be started only on fuel oil; therefore, make sure that the gas-inlet valve is *closed.* On spark-ignited gas engines the gas-inlet valve should be kept closed until after the starting air has blown out any unburned fuel gas in the exhaust system.

Starting System. See that the electric-starting battery or the starting-air tanks are fully charged.

Electric-Ignition System. A spark-ignited engine uses a battery, magneto, pulse generator or electronic device to produce the high-tension current which fires the spark plugs. Make sure that the system is ready to spark when the engine turns over at firing speed.

Starting. There is no mystery about starting a diesel engine. Starting requires: (1) an air supply to the combustion chamber, (2) a fuel supply to the combustion chamber, and (3) compression of the air during the cranking operation.

Set the governor or hand throttle to give about half or two-thirds of full fuel-oil feed.

If the engine is provided with a preheater or a compression release, turn these devices on and off as prescribed.

If the engine does not start promptly, stop cranking to avoid unnecessary loss of starting air or battery charge; find and eliminate the cause of failure before cranking again.

Immediately after starting, check the lubricating-oil pressure, cooling-water flow and fuel supply. Watch the entire engine to see if all parts function properly. If possible, run the engine at light load until it reaches running temperature. Charge the air-starting tanks or batteries.

For dual-fuel engines normally running on gas, shift from fuel oil to gas at no load, after engine has warmed up. Make sure that pilot fuel oil is being injected; if the pilot oil fails, shut down the engine immediately to prevent misfiring.

Spark-ignited gas engines should be rolled with starting air long enough to purge the exhaust system of unburned gas. Then open the fuel-gas valve slowly until the engine fires. When the engine fires regularly, shut off the starting air and gradually open the fuel-gas valve to full open position.

NORMAL OPERATION

Inspect Engines Frequently. Look for leaks and loose fastenings, listen for mechanical noises, and check temperatures by instruments or feel. Test emergency devices occasionally.

Watch Loading. No serious harm follows underloading, but it is uneconomical. Overloading causes combustion troubles and overheating; on dual-fuel and spark-ignited engines, it also causes detonation.

Exhaust pyrometers are regular equipment on most engines, except small high-speed units. Make full use of them. If the engine is in good mechanical condition, *uniform* temperatures at each cylinder show equal division of fuel and load. *Low* exhaust temperature may indicate lack of fuel, poor spray-nozzle action, or low compression pressure, the latter probably caused by piston-ring blow-by or leaky exhaust valves. *High* exhaust temperature may mean heavy cylinder loading, delayed combustion, or scale in the water jackets. Inspect the pyrometer thermocouples regularly; if dirty, they will read low.

Regulate Jacket-Water Temperature. Keep the temperature of jacket water to the range recommended by the manufacturer. In the absence of instructions, set the outlet temperature at 130 to 180°F, depending on cylinder size (small cylinders use higher temperatures). *Regulate the water flow* to keep the temperature rise in the jackets to 10 to 20°F. This causes a rapid flow in the water spaces, prevents local hot spots, and cuts the temperature difference between the top and bottom of the cylinder. This keeps cylinder bores straight and reduces expansion strains.

Excessively high outlet temperature promotes scale formation and may break down the oil film on the cylinder walls. It may also cause spark-ignited engines to detonate.

Low temperature increases the loss of heat to the jacket water and causes incomplete combustion. It also causes condensation on the lower cylinder walls, a common cause of liner wear, particularly in gas engines.

If *raw water* is used in the engine jackets, keep the temperature at or below 140°F, and if the water is high in scale-forming substances, treat it with zeolite or use other approved methods. Scale in jackets leads to overheating which causes *growth* (permanent enlargement) of cylinders and pistons, scoring, warping, cracking, and burning. Scale also reduces the heat-transfer in heat exchangers. In addition to the scale-forming salts found in most natural waters, impurities such as oil, mud, and animal or vegetable matter cause low heat transfer and overheating of the engine.

Keeping the Lubricating System Clean. Renew filter and purifier elements regularly. Check the condition of the oil by having a sample analyzed, or renew the oil on a regular schedule. Inspect oil coolers for scale on the water side and sludge coating on the oil side.

Keep the lube (lubricating) oil at *proper temperature* (generally not over 150°F). High temperature promotes oxidation and sludging; it also tends to increase leakage from the crankcase. Investigate any increase in crankcase temperature; it may indicate a hot bearing.

Investigate immediately any unexpected *oil-pressure change*. Increased pressure generally signifies clogging of the system. *Slow* pressure fall indicates bearing or pump wear; a *sudden* pressure drop may mean a burned-out bearing.

Excessive lube oil reaching the combustion space will cause *blue smoke* in the exhaust; check the oil pressure, or the rate of cylinder feed if a mechanical lubricator is used. If the oil pressure and rate of cylinder feed are right, look for faulty piston rings or mechanical defects. Keep a record of consumption of lubricating oil in terms of rated engine horsepower-hours per gallon of oil consumed; a drop in this figure is evidence of increased oil consumption.

Check Combustion Conditions. Do this on larger engines by taking indicator cards. If regular indication is impractical, compare the exhaust temperature and rate of fuel consumption with the normal temperature and consumption for the load at which the engine is operating. Abnormally high exhaust temperature or rate of fuel consumption means that combustion is poor and is causing waste of fuel.

Listen to the Engine. Although diesel and gas engines are inherently noisy, an observant operator is quick to notice *changes* in the normal operating sounds. Whenever an unusual noise occurs, it indicates that something is not right. Determine the cause of the noise and correct it immediately if urgent, or later if it can wait.

STOPPING THE ENGINE

Except in emergencies, follow an orderly procedure in stopping the engine. First make sure that the starting equipment is fully charged. Check all pressures and temperatures to see if they are normal. Take the load off the engine *gradually*. Let the engine idle for a few minutes. Stop a diesel engine by shutting off the *fuel injection.* Stop a dual-fuel engine by shutting off *both* the gas and the fuel injection. Stop a spark-ignited gas engine by closing the *gas-inlet valve.*

After the Engine Stops. If the pistons are cooled with oil from an independent pump, don't shut off the pump until the pistons have given off their heat. If the cooling water is supplied by an independent pump or an overhead tank, *keep the water circulating* for about 15 minutes after shutdown. If the water pump is engine-driven, run the engine at no load until the temperature falls. These precautions avoid excessive local temperatures and strains due to the heat accumulated in the metal parts.

CAUTION: Never open the doors of a totally enclosed crankcase until the engine has cooled off. Such a crankcase is likely to contain an over-rich mixture of oil vapor and air. If some engine part happens to be overheated and fresh air is admitted by opening a crankcase door, the added air may bring the mixture into the inflammable range, whereupon the overheated part may ignite it.

Performance Records

There are two kinds of performance records that will help you carry out the foregoing operating pointers: (1) operating records, and (2) accounting records.

Operating Records. These present a picture of engine conditions — cooling-water and exhaust temperatures, turbocharger pressure, lubricating-oil pressures and temperatures, electrical and other load data. It is in these records that the watchful operator spots small troubles when they can be cured inexpensively rather than later when they may mean a serious breakdown or heavy repair expense.

Accounting Records. These deal largely with: *(1)* what the plant produces in terms of kilowatt-hours, water pumped, tons of ice made, ton-miles pulled, etc., and *(2)* what the plant used in terms of fuel and lube oil, operating labor, supplies, parts, repair labor, etc. These are the records that spot inefficiencies and excessive costs. Combined with the operating records mentioned above and the maintenance records discussed earlier, they give the complete picture that the man in responsible charge must have for trouble-free low-cost performance.

Record Forms. The sample log sheet, Fig. 20-1, top and bottom, shows a typical way to keep performance records. Most of the operating data is recorded hourly; other readings are made once each shift. The form shown covers one engine only; there would be a similar sheet for each unit in the plant. The sample log would be used in an electric generating plant having several engines of several hundred kilowatts capacity each. The amount of detail required, the frequency of the readings, and the arrangement will vary from plant to plant.

Fig. 20-2 shows a sample operating log sheet for a turbocharged dual-fuel engine.

A typical accounting record sheet for an electric generating station is made up in the form of a monthly report, with space for each unit in the plant. It contains *daily entries for each unit* of: (1) kilowatt-hours generated, (2) hours of operation, (3) fuel consumption, and (4) lube-oil consumption. It also contains *daily entries for station totals* of: (1) gross output, (2) net output, (3) peak load, (4) water consumption, (5) fuel-oil consumption, (6) hours of operating labor, and (7) hours of maintenance labor. Finally, the daily items are summarized in *monthly entries* which can be compared with the same month of previous years.

Operator As Trouble Shooter

A comparatively minor engine trouble, if not recognized and remedied in its early stages, may easily develop into a major breakdown. Consequently, every diesel operator should also be a *trouble shooter*.

The successful trouble shooter must have two abilities:

1. He must be able to *recognize symptoms* of trouble when he sees, hears, smells, or feels them, or when the log readings differ markedly from normal.

2. Having decided that something is wrong with the engine, he must be able to *determine quickly what adjustments or repairs are needed.*

Follow a systematic and logical method of inspection in order to locate the cause of the trouble quickly. Apply your knowledge of *principles*. Use your head.

The following check list takes in most common symptoms, itemizes possible causes, and suggests response.

DIESEL-GENERATOR OPERATING LOG

Engine Nos.____Hp Rating:______Kw Rating:______Day ending midnight:__________Operator, 1st shift:____________Operator, 2nd shift:____________Operator, 3rd shift:____________

	1 AM	2 AM	3 AM	4 AM	5 AM	6 AM	7 AM	8 AM	9 AM	10 AM	11 AM	12 N	1 PM	2 PM	3 PM	4 PM	5 PM	6 PM	7 PM	8 PM	9 PM	10 PM	11 PM	12 M
Circulating Water System																								
Temp at engine inlet, F																								
Temp at outlet cyl 1, F																								
Temp at outlet cyl 2, F																								
Temp at outlet cyl 3, F																								
Temp at outlet cyl 4, F																								
Temp at outlet cyl 5, F																								
Temp at outlet cyl 6, F																								
Inlet and Exhaust System																								
Temp exhaust cyl 1, F																								
Temp exhaust cyl 2, F																								
Temp exhaust cyl 3, F																								
Temp exhaust cyl 4, F																								
Temp exhaust cyl 5, F																								
Temp exhaust cyl 6, F																								
Lubrication System																								
Force-feed pressure, psi																								
Crankcase temp, F																								
Temp to centrifuge, F																								
Temp from cooler, F																								

Load data																								
Kwhr generated, gross																								
Volts, phase A																								
Volts, phase B																								
Volts, phase C																								
Amperes, phase A																								
Amperes, phase B																								
Amperes, phase C																								
Frequency, cycles																								
Power factor, %																								

				Electric	Fuel	Lube Oil				Electric	Fuel	Lube Oil				Electric	Fuel	Lube Oil
	Unit on line at: M	Meter					Unit on line at: M	Meter					Unit on line at: M	Meter				
	Unit off line at: M						Unit off line at: M						Unit off line at: M					
	Total						Total						Total					
	Peak load, 15 min:		Hr operation:				Peak load, 15 min:		Hr operation:				Peak load, 15 min:		Hr operation:			
	Load Factor, %:		Gross kwhr per gal:				Load Factor, %:		Gross kwhr per gal:				Load Factor, %:		Gross kwhr per gal:			
	Outside air temp, F:		Barometer, in. Hg:				Outside air temp, F:		Barometer, in. Hg:				Outside air temp, F:		Barometer, in. Hg:			
	Air temp at engine inlet, F:		Press. drop at filter, in H_2O:				Air temp at engine inlet, F:		Press. drop at filter, in H_2O:				Air temp at engine inlet, F:		Press. drop at filter, in H_2O:			
	Water meter reading:		Water used:				Water meter reading:		Water used:				Water meter reading:		Water used:			
	Supplies used:						Supplies used:						Supplies used:					
	Repair parts used:						Repair parts used:						Repair parts used:					
	Men on duty:						Men on duty:						Men on duty:					
	Remarks:						Remarks:						Remarks:					

Fig. 20-1. Typical operating log for diesel engine.

ENGINE NO ______ ENGINE SERIAL NO. ______ MAKE ______ RATED HP ______ RPM ______
GENERATOR ______ GENERATOR SERIAL NO. ______ MAKE ______ RATED KW ______ VOLTS ______ AMPS ______
LUBRICATING OIL USED:
FUEL GAS USED: SHIFT 1 ______ SHIFT 2 ______ SHIFT 3 ______ TOTAL ______
CRANKCASE : SHIFT 1 ______ SHIFT 2 ______ SHIFT 3 ______ TOTAL ______ FUEL OIL USED: SHIFT 1 ______ SHIFT 2 ______ SHIFT 3 ______ TOTAL ______
CYLINDER LUBRICATION : SHIFT 1 ______ SHIFT 2 ______ SHIFT 3 ______ TOTAL ______ KW GENERATED: SHIFT 1 ______ SHIFT 2 ______ SHIFT 3 ______ TOTAL ______
USE SEPARATE SHEET FOR EACH UNIT EVERY 24 HOURS.

TIME	ON	OFF	RPM	EXHAUST TEMPERATURES										WATER PRESSURE	WATER TEMP			FUEL OIL PRESS	FUEL GAS PRESS	LUB OIL PRESSURE		TEMP		GENERATOR							EXCITER			REMARKS
				CYLINDERS						IN TURBO-CHARGER			OUT TURBO-CHG		ENG IN	ENG OUT	TURBO			ENGINE	TURBO CHARGER	LUB OIL	TURBO OIL	KW	VOLTS	AMPS PHASE			CYCLES	P F	VOLTS	AMPS		
				1	2	3	4	5	6	1	2	3	1													1	2	3						
AM 8																																		
9																																		
10																																		
11																																		
12																																		
PM 1																																		
2																																		
3																																		
4																																		
5																																		
6																																		
7																																		
8																																		TOTALS TO DATE:
9																																		LUBRICATING OIL USED:
10																																		FUEL OIL USED:
11																																		FUEL GAS USED:
12																																		KW GENERATED:
AM 1																																		OPERATORS:
2																																		SHIFT 1
3																																		SHIFT 2
4																																		SHIFT 3
5																																		APPROVED:
6																																		
7																																		Chief Engineer

Fig. 20-2. Operating log for turbocharged dual-fuel engine.

ENGINE FAILS TO START

Not enough fuel; air in fuel line	Are tanks full and valves open? Make sure pumps and piping are primed and vented. Is transfer pump working? Are filters clean?
Water or dirt in fuel	Drain fuel system and tanks; clean tanks; prime and vent pumps and lines properly.
Starting valves out of order	Make sure valve admits air just as piston passes top center. Does valve open fully? Check for starting-valve leakage, or valve stuck open.
Low compression	Inlet and exhaust valves may not seat properly. Look for stuck piston rings. Check position of compression release. Cylinder head or valve-cage gaskets may leak.
Cranking speed too low	Check starting-air pressure or battery charge. Bearings may be too tight or cylinder walls need lubrication. Check for low compression, as above. Warming jacket water, intake air, lube oil, or engine itself may help.
Fuel injection improperly timed	Make sure fuel injection occurs at or just before top center.

ENGINE FAILS TO START (SPARK-IGNITED GAS ENGINES)

Insufficient gas	Is gas line clogged or gas pressure regulator stuck?
Ignition system out of order	Is there a short circuit? Are magneto breaker contacts burned or out of time? Are spark plugs wet or do gaps need adjusting?

FAILS TO COME UP TO SPEED

Not enough fuel	Heavy fuel may require heating for free flow. Check governor or throttle controls. There may be air or water in fuel.

Fuel nozzles dirty or clogged	Inspect nozzles; clean them or replace with spares if necessary.
Injection-pump valves leak	Check valve condition; regrind or replace valve and seat assemblies if necessary.
Low compression	Check compression as described above.
Engine overloaded	If electric load, see that switches are open. If mechanical load, open clutch, unload compressor, etc.
Too much friction	Bearings may be too tight. Lubrication may be inadequate. Check for overheating, metal discoloration, etc.

CYLINDERS MISS

One or more cylinder miss	Locate faulty cylinder by opening each fuel line in turn or by cutting out individual injection pumps. If cylinder is firing, speed drops when fuel is shut off; if speed stays the same, that cylinder is at fault. Missing cylinder can also be detected by opening exhaust indicator connection for each cylinder.
Valves at fault	If valves are sticking, put mixture of kerosene and thin lube oil (or commercial valve-freeing fluid) on stems until basic source of trouble can be remedied. Look for broken valve and replace if found. Check for leaky valves and regrind.
Lack of fuel	If fuel nozzle is plugged, remove and clean. Check fuel-pump valves for sticking. Look for air in fuel or pump; look for water or dirt in fuel. Check for plugged filters or strainers. In gas-burning engines, check gas supply pressure.
Missing stops after engine warms up	If one or more cylinders fail to fire until engine has run for some time, cause may be low compression or presence of water. Look for leaky or sticking valve or stuck rings. Check for head or jacket leaks; small amount of water in cylinder may prevent ignition until heat evaporates the water.
Erratic missing, all cylinders	Check for conditions affecting all cylinders—water or dirt in fuel, plugged air filters, fuel of poor burning quality.

Two-cycle engines	Check scavenging-air pressure, inspect scavenging-air valves, also air-suction valves to crankcase in crankcase-scavenged engines. Look for carbon deposits on exhaust ports which might cause back pressure.
Dual-fuel engines	Check pilot-oil injection. If running at partial load, see if intake air is throttled.
Spark-ignited high-compression gas engines	Check electric-ignition system. Make sure mixture ratio is properly adjusted. Check spark timing.

FAILS TO CARRY LOAD

Too much friction	Possible causes include: inadequate lubrication, bearings too tight, piston seizure. Stop engine, rotate it with bar or turning gear, and try to find which is cause.
Low compression	See pointers under this heading above.
Engine overloaded	Defective or inaccurate electric meters may conceal actual overload. If overload actually exists, drop load to normal.
Lack of fuel	See pointers under this heading above.
Lack of air	Check pressure drop in air filters and clean if required. Look for any other restrictions in air-intake system. Was engine properly derated for altitude?
Excessive back pressure	Inspect exhaust piping and silencers for restrictions caused by carbon deposits. In two-cycle engines, check for pulsations that might cause backflow of exhaust gas.

ENGINE STOPS

Lack of fuel	See pointers under this heading above.
Engine overloaded	See pointers under this heading above.
Lubrication failure	Look for break in piping or pump failure. Make sure system is filled to proper level.

Mechanical interference	Crank engine over to see if it is free. **Don't** remove crankcase doors until crankcase has cooled (to avoid danger of explosion). Check bearings and pistons. If piston seizure stopped engine, inspect piston and liner; mild scoring can be corrected with an oil stone; severe scoring means reboring or renewal of parts.

ENGINE KNOCKS

Injection nozzles at fault	If injector valves stick, clean them or install new ones. Check for broken valve springs.
Fuel timing wrong	If rate of combustion rise is too rapid, engine will knock. Fuel knocks are sharp pings. Indicate engine; check maximum combustion pressure and note rate of combustion-pressure rise. Adjust timing of fuel-injection pump, or fuel-timing valve on common-rail engines.
Inlet or exhaust valve sticking	Lubricate stem with mixture of kerosene or thin lube oil. If condition is not severe, turning stem by hand may free valve.
Improper fuel; dirt or water in fuel	Fuel with poor ignition qualities may cause erratic knocking in one or more cylinders, or more or less steady knocking in all. Erratic knocking may indicate water or dirt in fuel or presence of air in fuel lines.
Gas-burning engines	Detonation may be caused by overload. Compression may be too high for fuel used, especially if it contains considerable hydrogen. Spark may be timed too early. Intake air or jacket water may be too hot.

MECHANICAL NOISES

Pounding	Check for loose crankpin bearing by lifting connecting rod with jack. Check for excessive clearance in wristpin bearing by putting crank on bottom center and prying up on piston. Piston may tip in cylinder; check clearance and, if excessive, fit new liner or piston.

Knocking — If head gasket is of wrong thickness or bearings are worn, piston may strike inlet and exhaust valves. Excessive valve-tappet clearance may cause knocking. Check entire valve mechanism for looseness or play.

ENGINE OVERHEATS

Engine overloaded — See pointers under this heading above. If one cylinder overheats, it may be receiving more than its share of fuel.

Poor cooling — Check cooling-water temperature and rate of flow. Look for scale, mud, or other obstructions in jackets of cylinders and water spaces of heads. Inspect heat exchangers, cooling towers, radiators.

Late combustion — If fuel is injected too late or burns too slowly, water and exhaust temperatures will rise; therefore check timing. Indicate engine and study cards for late ignition and after-burning.

Lubrication inadequate — Lube-oil pressure may be too low; check relief valve for sticking or leakage. Inspect and clean strainers and filters. Loose or burned-out bearing may cause loss of pressure. Check piping for leaks. Lubricating oil may not be suited to engine.

Hot bearings — If lubrication is adequate, check condition of bearings, looking for tightness, poor fit, or misalignment causing binding.

SMOKY EXHAUST

Engine overloaded — See pointers under this heading above.

Poor combustion — Check grade of fuel used. Inspect action of entire fuel-injection system. See that each cylinder receives equal share of fuel. Check for clogged air-intake filters. Indicate engine and study cards for combustion irregularities.

Excessive lubrication — Blue smoke in exhaust often indicates excessive lubrication. Check oil pressure and rate of cylinder feed. If oil feed is right, check for faulty piston rings or other defects.

Fundamental Problems

Although strict operating principles and procedures for care of diesel engines are absolutely necessary for all sizes and types, the new and severe demands placed on this form of motive power have caused an increased amount of technical maintenance problems.

Demands for increased horsepower without increasing size or weight is of great concern to the users and manufacturers of (on the highway) diesel truck tractors, farm tractors, and harvesters as well as many other kinds of mobile equipment with high torque and horsepower requirements.

Application Problems

The value of supercharging is still a debatable question with some manufacturers. However, with increased research and greater experience, many builders have found with the proper additions of accessories, supercharging can be an approach to increasing power without increasing weight, size, or maintenance problems.

John Deere is one example of a company who have designed a basic tractor package to carry 117, 133, 141, and 175 horsepower without changing to any extent the physical engine size (Figs. 20-3, 20-4 and 20-5).

They begin with a basic variable-speed six-cylinder diesel engine with a 404 cu in displacement. As a naturally-aspirated engine, it produces 95 horsepower as measured at the power take-off (PTO). The parts in this engine are illustrated in Fig. 20-3.

Beginning with an engine block which is designed to take stresses and strains more than double its rating, the supporting components—lubrication, oil cooling, fuel filtering and delivery, engine cooling, air cleaning, and others, are sized to match the requirements of the engine without adding excessive bulk.

Fig. 20-4 shows the same 404 cu in size block but it is turbocharged. Turbocharging pretty well rules out *cubic inches* as a modern-day measure of engine capacity, except with naturally-aspirated engines. This engine develops 116 PTO horsepower. However, a number of design steps are taken to reduce the potential maintenance problems so they are equivalent to the naturally-aspirated engine described in Fig. 20-3.

Although the physical size of the engine remains the same, the head bolts, gaskets, lubrication and cooling systems, air cleaning and fuel-delivery systems are matched to fit the demands of higher horsepower.

In fact the only items remaining from the engine in Fig. 20-3 which remain *exactly the same* are the manifold, fuel filter, and valve springs.

Fig. 20-5 shows the same 404 cu in size engine but it is turbocharged and intercooled. This engine develops 135 horsepower at the power take-off because the air which has been heated by turbocharger compression is cooled by the in-

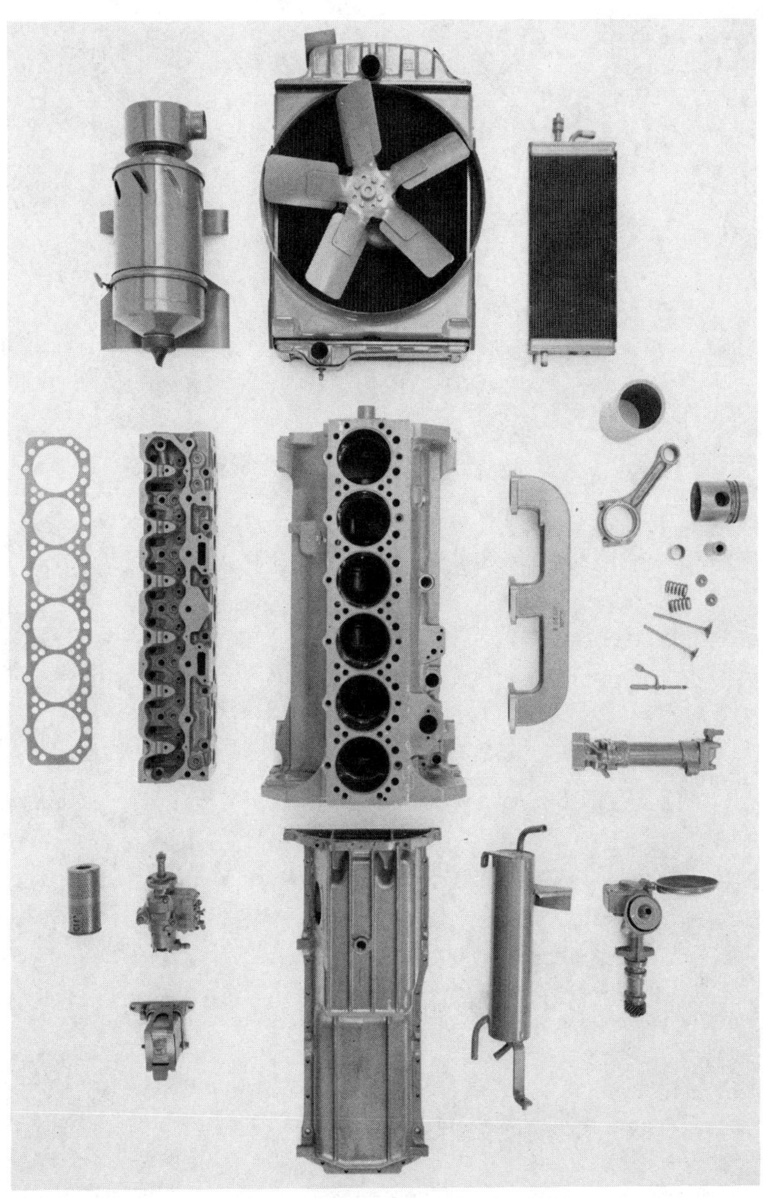

Fig. 20-3. Basic engine package: parts of a 404 cu in naturally-aspirated diesel tractor engine with basic components for producing 95 PTO hp at the power take-off. (Deere & Company)

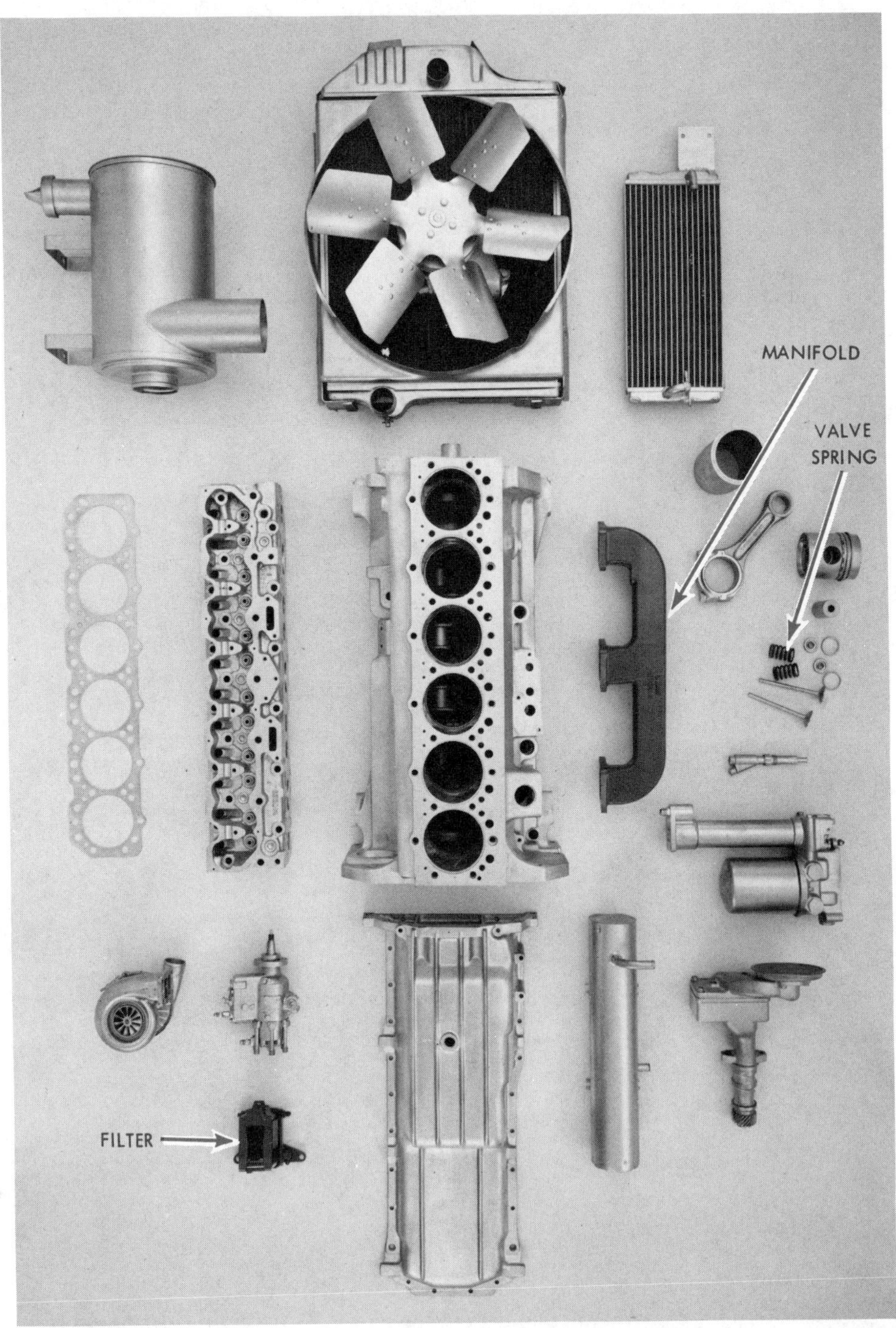

Fig. 20-4. Basic engine package: turbocharged, with component variations to produce 116 PTO hp. Parts remaining from first set are indicated by callouts, namely: film, manifold valve spring. (Deere & Company)

tercooler. By cooling, the air is further reduced in volume so that more of it can enter the combustion chamber. Thus, more fuel can be burned more completely, providing both additional power and economy. In addition, this cooling volume of air cools valves, pistons, cylinder walls, and turbocharger blades, many times per minute, extending the life of these parts.

There are only two parts remaining in Fig. 20-5 which were used in Fig. 20-3, and approximately half the parts are retained from Fig. 20-4.

The point to be made from the previous illustrations and discussion is that it is possible to maintain long maintenance-free engine life, while doubling the horsepower for the same engine size, if the component parts are matched to meet the increased demands.

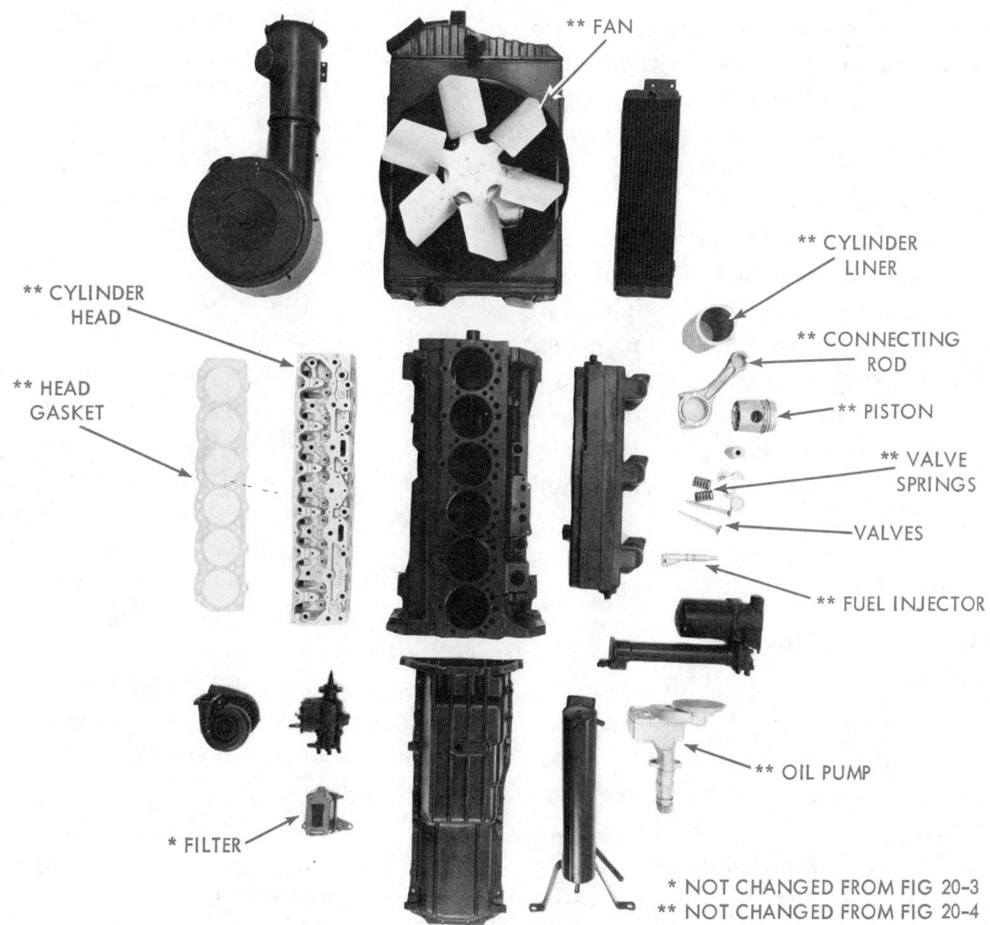

Fig. 20-5. Basic engine package: turbocharged and intercooled, with component variations to produce 135 PTO hp. Parts not changed from Fig. 20-3 are indicated by *; and those not changed from Fig. 20-4 are indicated by **. (Deere & Company)

Environmental Problems

Due to Federal and State legislation, perhaps the greatest problems facing the manufacturers and maintenance organizations of diesel engines are the problems concerned with ecology.

Although diesel-vehicle tests on grams-per-mile have not been set, we do know diesel engines emit hydrocarbons, oxides of nitrogen, carbon monoxide, carbon dioxide, and smoke, which is made up of carbon particles, and possibly some hydrocarbons.

In comparison to the late model gasoline-powered automobile, diesel engines emit only 50 to 65 percent of the hydrocarbons, 1.6 to 3.4 percent of the carbon monoxide, and although not completely researched, somewhere between 42 to 75 percent of the oxides of nitrogen.

So far, the Federal government has been more concerned with diesel smoke standards than the other emission problem areas related with this kind of engine. However, the time is probably not far off when nitrous oxides, carbon monoxide and hydrocarbon emissions from diesel engines will have to be below minimum standards.

Diesel Smoke Emissions

Since the 1970 Federal diesel smoke standards went into effect, there has been a serious problem for manufacturers, users, and agencies concerned with highway vehicles to reduce visible exhaust smoke. It should be remembered that the smoke characteristics of naturally-aspirated and of turbocharged engines vary considerable under different loads and speeds. For this reason, emission testing, especially smoke emission, is evaluated as a maintenance problem, and in the Research Division, as a design problem. Similar testing equipment is used in both divisions.

Instruments for measuring diesel smoke may be classified as:

1. *Opacimeter*, full-flow type, measured opacity (opacity means opaqueness or quality or state of a substance which renders it impervious to the rays of light—density of exhaust smoke.)

2. *Filtering systems*— sample soot particles are collected on a filter paper medium as exhaust is forced through.

The filter may be evaluated by comparison with standard samples which range from white to gray to black.

The Clayton Manufacturing Company has developed a diesel smoke opacity test-cycle *preliminary procedure* for diesel-powered on-the-highway tractors. It utilizes the opacity smoke meter and a chassis dynamometer.

Since the Federal standards are concerned only with new engines coming off the production lines, they hope that the chassis dynamometer test-cycle will encourage states to establish understandable regulations which include engines rebuilt, repaired, or adjusted in the field. It seems proper to include the engine smoke level after they are placed in service, not just the new engine off the assembly line. Some states are moving in this direction, but there is so far no uniformity among state regulations.

The procedure, as set forth by Clayton, of the chassis dynamometer opacity test-cycle is as follows:

A graphical illustration of the Clayton Under-load Diesel Exhaust Opacity test cycle is shown in Fig. 20-6.

CLAYTON CHASSIS DYNAMOMETER OPACITY TEST*

A. Warm-Up Time—5 Minutes

1. With truck properly positioned on chassis dynamometer:

(a) operate in direct transmission gear at approximately 60 percent of engine horsepower at 80 to 85 percent at rated engine speed.

Example: Engines rated at 250 hp at 2100 rpm—dynamometer load about 150 rpm @ 1700 to 1800 engine rpm

2. After 2 or 3 minutes of warm-up and after engine temperature is at normal operating range and shutters are open, load dynamometer until engine is developing its full power at 95 to 100 percent of rated governed speed.

Example: Engines rated at 2100 rpm should be loaded to full power @ 2000 to 2100 rpm

3. Caution: LEAVE LOAD ON DYNAMOMETER.

B. Idle—Time 2 to 3 Minutes

1. Take transmission out of gear and allow engine to operate in neutral at normal idle rpm

2. Calibrate opacity meter and install sensor unit on exhaust stack

3. Measure and record exhaust smoke opacity with engine operating at idle

C. Operation Smoke Opacity Check

1. Clutch engagement simulation—2 to 3 seconds. Raise engine speeds approximately 200 rpm above normal idle; hold for 2 or 3 seconds while measuring and recording smoke opacity

NOTE: The gear selection for this 5-second acceleration to 85 percent of rated engine speed will depend on the type of transmission and the number of gear splits. Too low a gear may make it difficult to maintain a suitable acceleration; the acceleration rate may be too fast. Too high a gear may result in too slow an acceleration rate.

Example: Based on some of the vehicles tested to date, it was found—for a vehicle equipped with a six-speed automatic transmission, the 4th gear provided easy operation during the 5-second acceleration to 85 percent of engine speed—for a vehicle equipped with a manual eight-speed box (8th direct), 6th gear provided a smoother acceleration in the prescribed time.

2. Number One acceleration mode—time 5 seconds
With transmission in selected gear, advance throttle setting to provide acceleration to 85 percent of rated engine rpm in 5 seconds. During this acceleration, measure and record smoke opacity.

Caution: DO NOT CHANGE DYNAMOMETER SETTING.

3. Decrease engine speed and change gears to direct

NOTE: Because of the low inertia of the vehicle on the dynamometer compared to that of a fully loaded vehicle on the highway, it may be necessary to decrease engine speed close to idle to change gears on a manual shift box. Thus, after changing to direct drive, it is advisable to bring engine speed to about 60 percent of rated governed speed before proceeding with the next acceleration step of the cycle.

4. Number Two acceleration mode—time 10 seconds
With transmission in direct drive gear and operating against the present dynamometer load, advance throttle to provide acceleration from 60 percent of rated governed speed to 95 or 100 percent of rated governed speed in 10 seconds. Measure and record smoke opacity during acceleration cycle.

5. Lug-down mode—time 35 seconds
With transmission still in direct gear, load dynamometer to lug engine down to 100 rpm below rated governed speed.
Measure and record smoke opacity.
Repeat lug-down in 100 rpm increments. Measure and record smoke opacity for each 100 rpm lug-down to the specified speed for peak torque of the engine, or 60 percent of rated governed engine speed (whichever is higher)

NOTE: With automatic transmissions, it may not be possible to lug engine down to specified engine peak-torque speed or 60 percent of rated engine speed without down-shifting. For this type of vehicle, lug engine down to point just before transmission down-shift. This may allow lug-down to only 75 percent of rated engine speed.

NOTE: To determine lug-down points and rate the lug down, subtract specified peak torque rpm or 60 percent of rpm from rated governed speed. Determine number of 100 rpm increments and divide into the 35 seconds allowed.

Example: Engine rated governed speed ....2100 rpm
Specified peak torque speed ....1200 rpm
Difference .... 900 rpm

This provides nine 100 rpm increments

35 ÷ 9 = approximately 4 seconds per 100 rpm to lug down.

Measure and record smoke opacity

NOTE: Early tests have indicated greater ease by lugging engine down in 200 rpm increments. This allows 8 seconds per increment to load dynamometer and to measure and record smoke opacity. The difference in results to date appears insignificant.

MAXIMUM SMOKE OPACITY LIMITS: (per 1970 Federal procedures)

Number one acceleration mode	40 percent
Number two acceleration mode	40 percent
Lug-down time	20 percent

TOTAL TIME:

Warm up	(5 minutes)—300 seconds
Idle	(2 minutes)—120 seconds
Idle plus 200 rpm	2 seconds
Number one acceleration mode	5 seconds
Deceleration from number one acceleration mode	5 seconds
Number two acceleration mode	10 seconds
Lug-down mode	35 seconds
	477 seconds (about 8 minutes)

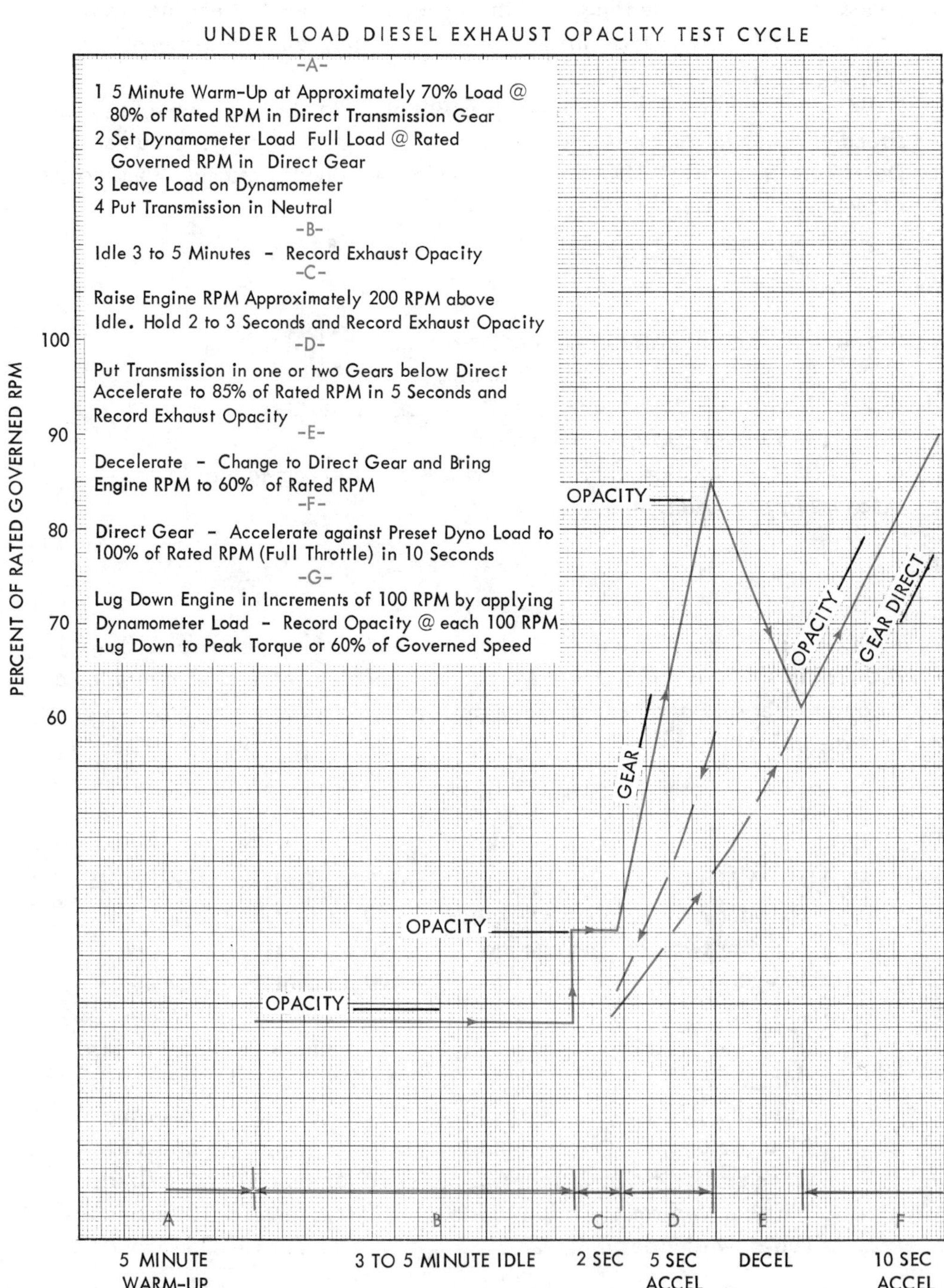

Fig. 20-6. Clayton under-load diesel exhaust opacity test. (Clayton Manufacturing Co.)

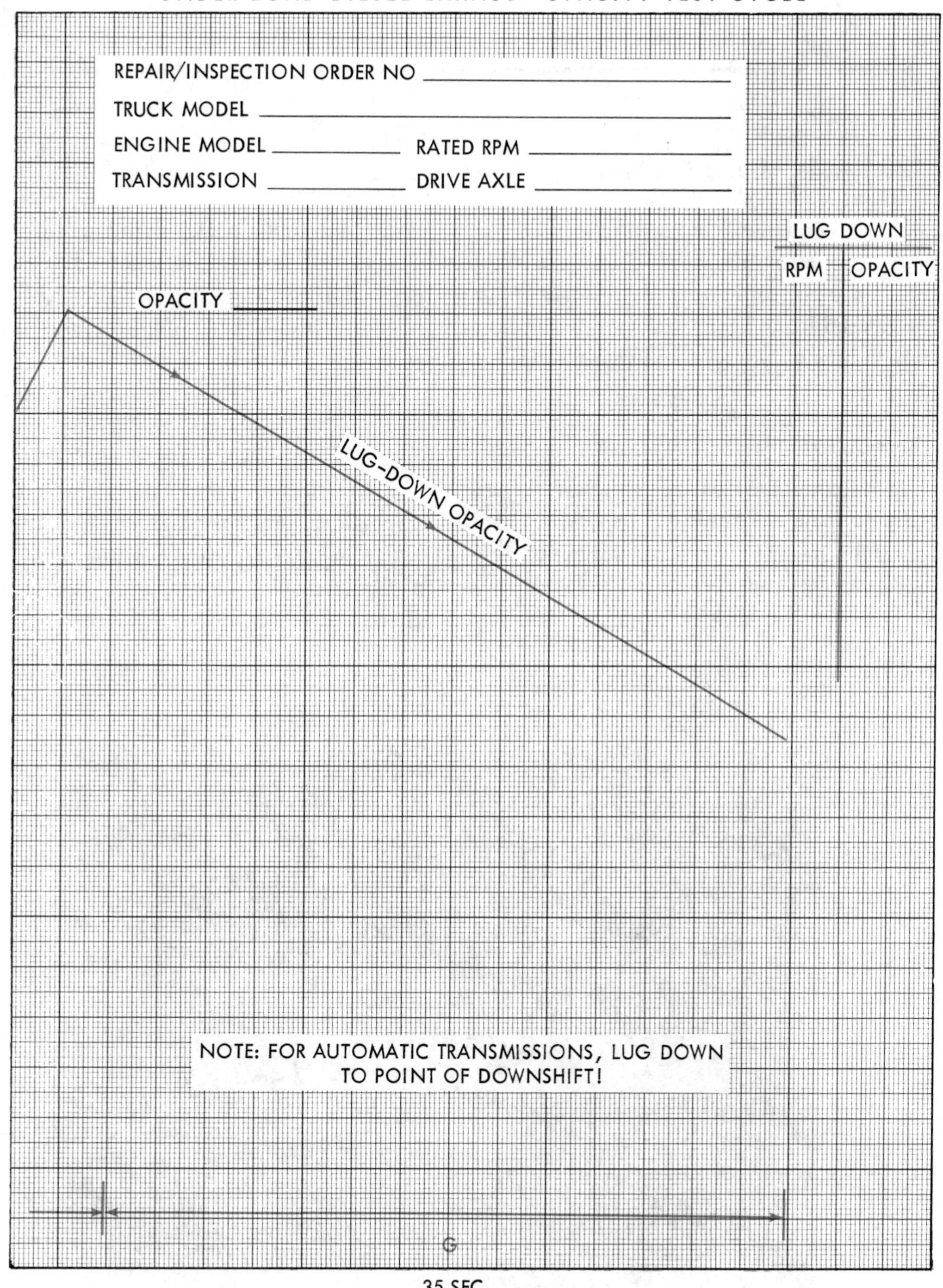

Fig. 20-6. Cont'd.

Diagnosis of the Problem

Two conditions must be fulfilled before problems in diesel engines can be diagnosed. Proper test equipment must be available, and the trouble-shooter must know how to analyze the test results.

Testing Equipment

Diesel engine testing equipment parallels to some extent the test equipment used in research and development divisions.

Pressure sensing equipment for diagnosis will include the cylinder compression tester and other assorted pressure gages for testing exhaust back-pressure, intake manifold pressure, fuel pump pressure, and others.

Quite often this kind of field testing equipment may be found as part of an engine diagnosis kit shown in Fig. 20-7.

The cylinder compression tester is an important instrument of this group because it can give the operator of a comparison of cylinder pressures at cranking speed.

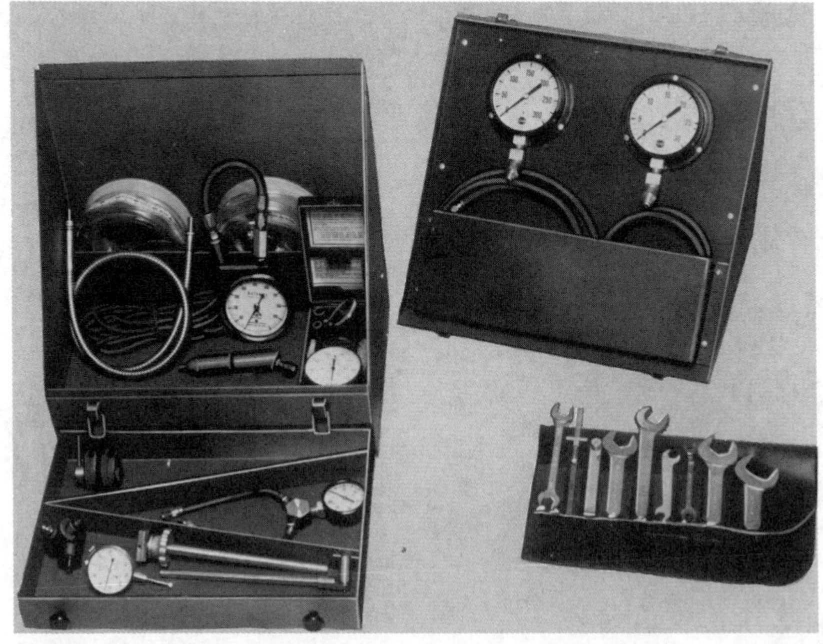

Fig. 20-7. Diesel engine diagnosis kit for field testing. (Bacharach Instrument Co.)

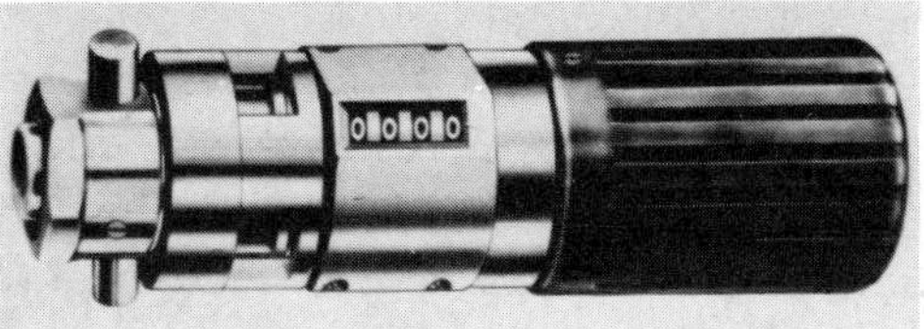

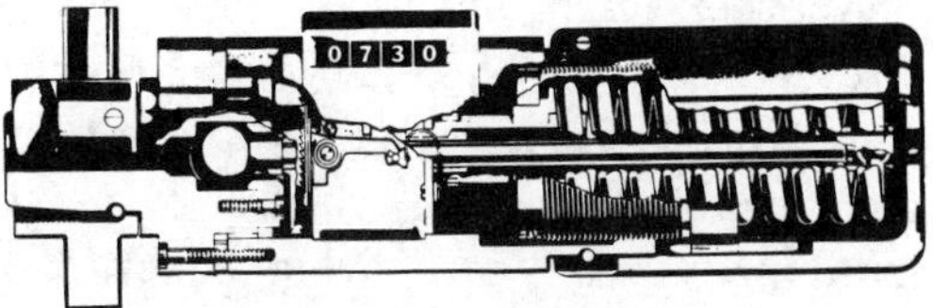

Fig. 20-8. Premax direct reading engine pressure indicator, range 125 to 1700 psi. (Bacharach Instrument Co.)

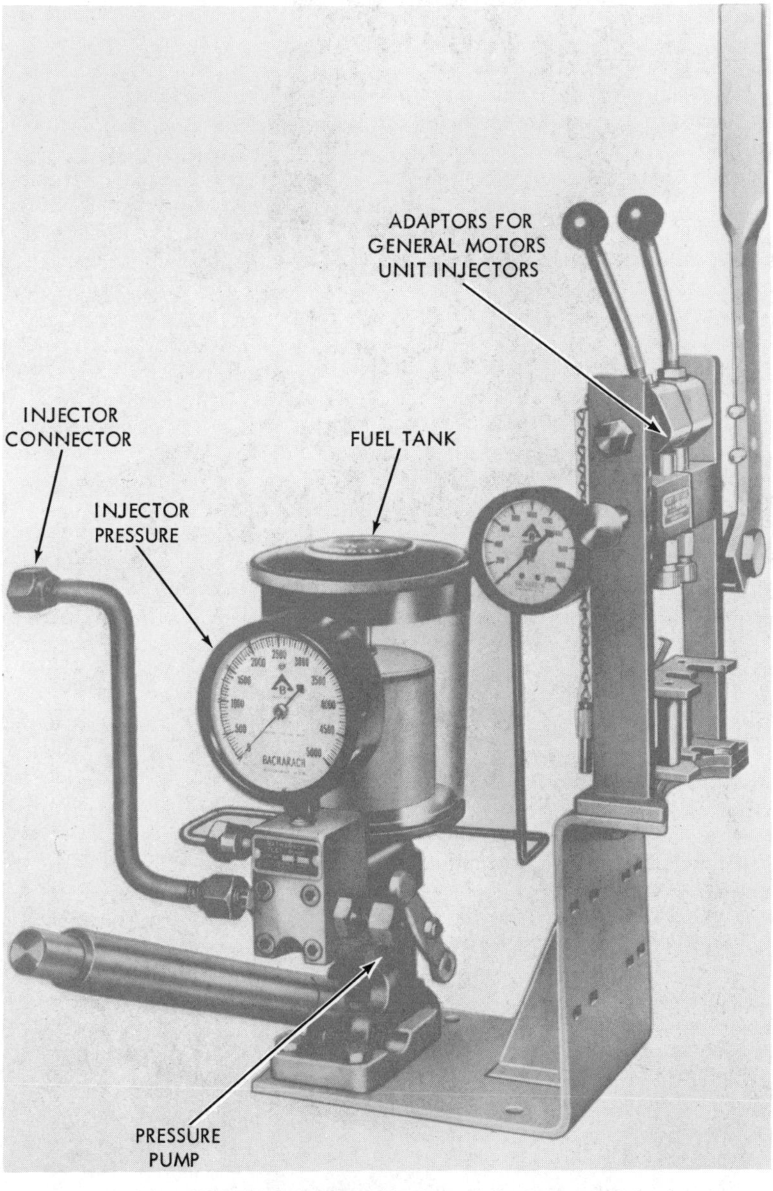

Fig. 20-9. Injector tester. (Bacharach Instrument Co.)

Another important diagnosis test instrument is the *Premax Engine Pressure Indicator*. This instrument will give a direct reading for each cylinder of maximum pressures under firing conditions. See Fig. 20-8.

The Chronomatic Engine Pressure Indicator described in Chapter 17 under *Pressure Measurement*, can be an invaluable diagnosis instrument for studying the

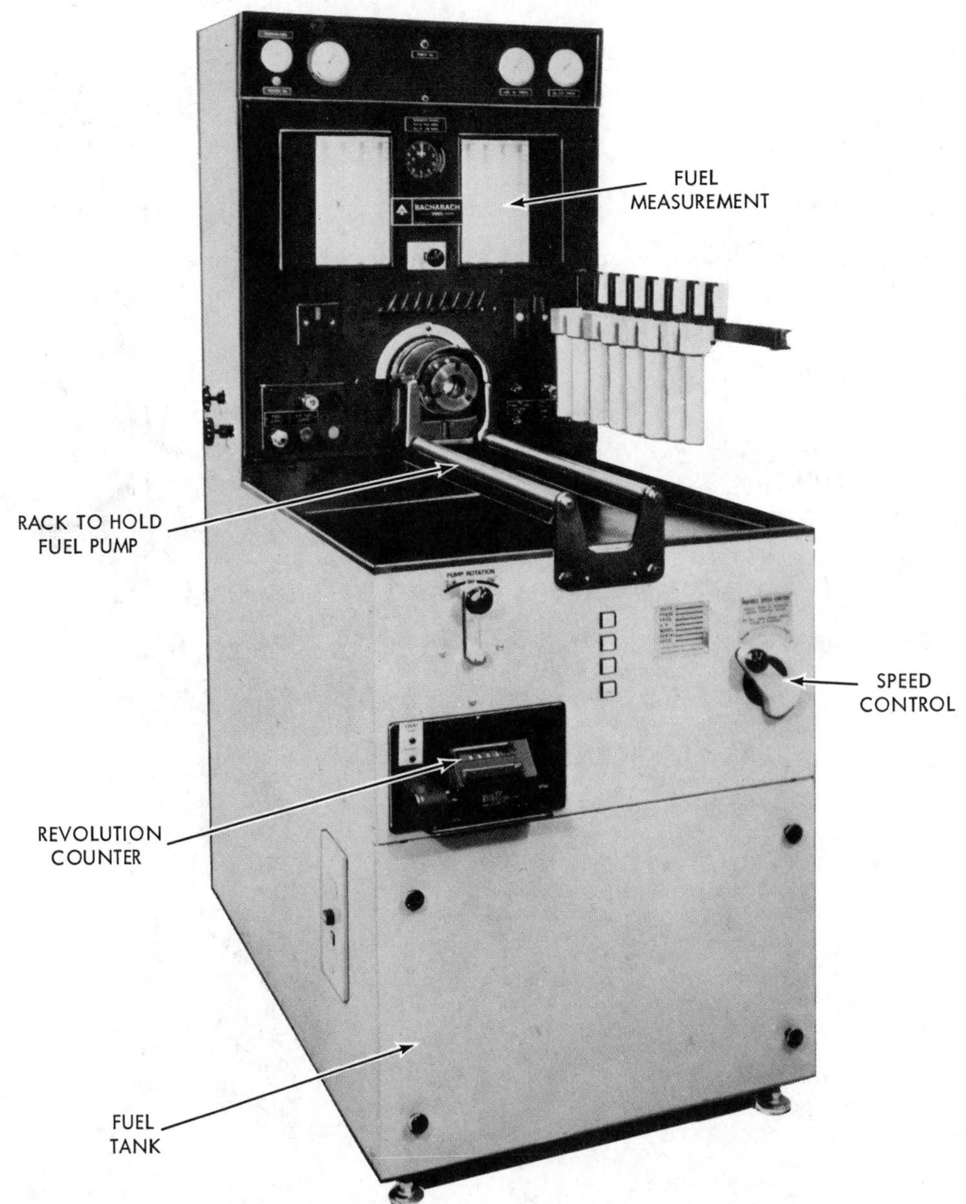

Fig. 20-10. Model U-7500A fuel injection pump test stand. (Bacharach Instrument Co.)

firing cycle on an engine.

The injector tester in Fig. 20-9 will test the pop-off pressure of a fuel injector and also the quality of the pattern of spray.

The fuel injection pump test stand illustrated in Fig. 20-10 can be adapted to any fuel pump with the proper accessories. It will measure the amount of fuel pumped from each injection discharge of a complete pump cycle for a predetermined amount of time. Without this equipment it is impossible to diagnose pump troubles.

Some of the common conditions which cause injection pump problems are shown in Fig. 20-11. These plungers are from General Motors unit injectors. In each case they would require replacement of the unit injector.

The chassis dynamometer is a diagnosis instrument in itself. In order to determine problems with a diesel engine, it must be run under various load conditions and at various speeds to see how it performs.

Since diesels are rated as *so much horsepower at a given speed,* a comparison between the rated specifications and the actual specifications can be obtained on a chassis dynamometer.

Also, by the use of certain test equipment utilizing the chassis dynamometer, various systems of the engine can be

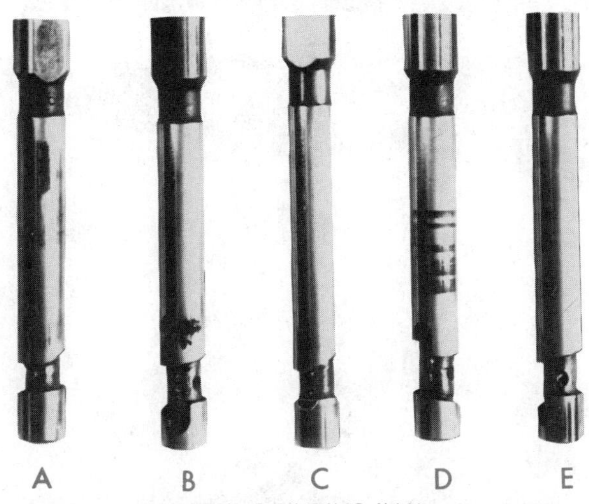

A TIGHT RACK CAUSING BINDING IN UP-AND-DOWN MOVEMENT

B DIRT IN FUEL. THEIS SHOWS ADVANCED STAGES OF ABRASIVE MATTER IN FUEL

C CHIPPED AT LOWER HELIX

D HIGH PRESSURE SCORING CAUSED BY A PLUGGED TIP OR WRONG SIZE TIP BEING INSTALLED

E THE CONDITION SHOWN CAN BE CAUSED BY EITHER LACK OF FUEL AT HIGH SPEEDS OR WATER IN FUEL

Fig. 20-11. Unit injector plunger problems. (General Motors Corporation)

tested and eliminated as a problem. Fig. 20-12 shows two tandem axle diesel trucks being tested in the top part. The lower part of the illustration shows a closer view of the dynamometer without the trucks. Another kind of dynamometer testing facility with a Mack truck in position is illustrated in Fig. 20-13.

Fig. 20-12. (top) Two tandem-axle trucks being tested on an 800 hp chassis dynamometer. (bottom) A closer view of the chassis dynamometer without the truck. (Clayton Manufacturing Co.)

Fig. 20-13. Dynamometer testing facility.

Checking On Your Knowledge

The following questions give you the opportunity to check up on yourself. If you have read the chapter carefully, you should be able to answer the questions. If you have any difficulty, read the chapter over once more so that you have the information well in mind before you go on with your reading.

DO YOU KNOW

1. What are some reasons why a diesel engine engine is potentially more difficult to start than a gasoline engine?
2. What is meant by *spotting* a large diesel engine for starting?
3. List some common checks to make before starting a diesel engine.
4. List the common checks to make on a diesel engine when it is operating.
5. What are the demands placed upon the smaller diesel engines such as those used in trucks, farm tractors and mobile power units?
6. What is being done by manufacturers of small diesels to increase horsepower per pound of weight?
7. What are the emission problems created by the diesel engine?
8. Why is diesel emission test conducted under various loads and speeds?
9. How can the chronomatic pressure indicator be used as a diagnosis instrument?
10. What are the necessary requirements of a successful diesel engine trouble-shooter?

Glossary

A

absolute pressure: Pressure in psi above zero pressure (or complete vacuum) that amounts to the weight of air in pounds per square inch at sea level under normal barometric conditions. That is 14.7 psi.

absolute temperature: In theoretical calculations related to gases, the *absolute* or Rankine scale is used. On the Fahrenheit scale, the absolute temperature is −460°F.

AC: Alternating current.

acceleration: Change of velocity or rate of such change with respect to speed, direction, or both.

adiabatic processes: A change of state, or process on a gas in which the gas neither receives nor rejects heat while it expands or is compressed.

advance: To cause to happen earlier, as in advance of timing of an internal combustion engine. Usually in degrees of crank-angle before top dead center of the piston.

air-cell: A space provided in the piston or cylinder to trap air during the compression stroke. It later flows out into the combustion chamber.

air cleaner: A mechanism for filtering and removing dust, moisture, and other foreign matter from the air before it reaches the engine.

ambient: The air around us.

antiknock: Prevention of premature combustion which causes knock.

API gravity: A scale adopted by the American Petroleum Institute to identify the specific gravity of various oils and fuels. Most diesel fuels fall within a range of 20 to 40 degrees API.

aspirated: To draw into or from a cylinder by suction. The normally aspirated engine operates only on its ability to create a vacuum in the cylinder to bring in air, as compared to the turbocharged engine which pumps in air.

atmospheric pressure: The weight of air measured at sea level; about 14.7 lbs. psi.

atomize: To break up in very fine particles. However, it is still a liquid suspended in a gas.

B

backfire: The ignition of a fuel-air mixture in the intake manifold.

back pressure: A resistance to free flow of gases, as from a restriction in the exhaust system.

ballhead centrifugal: A pair of governor flyweights which are rotated by the engine.

bedplate: The lower part of an engine which rests on the foundation.

blow-by: The leakage of cylinder pressure past the piston into the crankcase.

blower: Refers to an air pump to supercharge engines.

bore: The width or diameter of an engine cylinder.

brake horsepower (bhp): The useful output of an engine as measured at the flywheel.

brake mean effective pressure (bmep): The average pressure acting throughout the entire power stroke which is necessary to produce brake horsepower of an engine.

brake thermo efficiency: The ratio of the useful power output of an engine to the fuel energy per unit time.

British thermal unit (Btu): The amount of heat required to raise the temperature of one pound of water one degree Fahrenheit.

C

calibrate: To check and adjust the graduations of any measuring device.

calorie: Metric measurement of amount of heat required to raise one gram of water from zero degree to one degree Centigrade. 251,996 calories = 1 Btu.

calorimeter: An instrument which measures the amount of heat given off by a substance when it is burned.

cam: An eccentric which changes rotary motion to linear motion.

camshaft: A shaft containing lobes or cams which operate valves of engines.

carbon dioxide: A gas obtained from the complete burning of a fuel—(CO_2) is composed of 1 atom of carbon and 2 atoms of oxygen.

carbon monoxide: A gas formed with incomplete burning of fuel due to lack of sufficient oxygen.

center of gravity: That point at which the weight of the body may be represented as a single force.

centrifugal force: A force which causes a body to move away from its center of rotation.

cetane ($C_{16}H_{34}$): A straight chain hydrocarbon used as a standard in testing the ignition quality of diesel fuel.

cetane number: A percentage number indicating the ignition quality of fuel.

clearance volume: With the piston at top dead center, it is the remaining space which holds the gases at full compression.

clutch: A mechanism for connecting and disconnecting an engine from the power train.

combustion chamber: The volume of space above the piston head when the piston is at top dead-center.

compression pressure: The pressure of the air charge at the end of the compression stroke.

compression ratio: The ratio of volume of the charge at the beginning of the compression stroke to that at the end of the compression stroke.

connecting rod: The connection between the piston and the crankshaft.

constant pressure combustion: An ideal combustion process in a diesel engine which holds cylinder pressures approximately the same from top-dead-center through a portion of the expansion stroke.

constant volume combustion: An ideal combustion process in carbureted automotive engines. The burning extends from 10° to 20° before TDC and ends 18° to 28° past TDC and promotes burning at nearly constant volume.

crankcase: The housing within which the crankshaft is supported and rotates.

crankshaft: The main shaft of an engine which, turned by the connecting rods, changes the reciprocating motion of the pistons to rotary motion in the power train.

cycle: A series of events that repeat themselves in a regular sequence. Four-stroke cycle: intake, compression, power, and exhaust.

cylinder: A space in which the piston moves.

cylinder block: The solid casting which includes the cylinder and water jackets (cooling fins in the case of an air-cooled engine). That part of the engine which includes the cylinder spaces.

cylinder head: A detachable portion of an engine which covers the upper ends of the cylinders and includes the combustion chamber. In the case of overhead valve engines, it also includes the valves.

cylinder liner: A cylindrical lining, either *wet* or *dry,* which is inserted in the cylinder block in which the piston slides.

D

DC: Direct current

dead center: Center of upper or lower position of the piston within the cylinder, as top or bottom dead center.

detonation (knocking): A spontaneous combustion of the remaining portion of the charge of fuel and air causing knocking due to rapid high pressure rise on a gasoline engine.

diesel: An engine designed to convert the chemical energy of heavier fuel oils into mechanical power. The fuel is ignited by the heat of the air which is compressed by the piston within the cylinder head.

displacement: The volume displaced by the movement of a piston as it moves from one end of its stroke to the other.

distillation: Separation of more volatile parts of a liquid from those less volatile by vaporization (boiling) and then condensing (cooling).

dribbling: Slow loss of fuel from injection nozzle tip after cut-off.

droop: (of flyweights): Sometimes referred to as steady state speed regulation. It is the control of speed variation from no load to full load.

E

eccentric: The degree of being off a central axis. The distance between the center of an eccentric and its axis is its throw.

efficiency: Ratio of output over input.

electromotive force (emf): Voltage from battery, generator, or thermocouple.

emissions: Gases produced from exhaust, crankcase, and fuel tanks.

energy: Capacity for doing work.

engine de-rating: Reducing the standard horsepower and speed ratings on an engine because of the kind of service it performs. For example, an intermittent rating will be higher than a continuous rating on an engine.

engine displacement: The sum of the displacements of the individual cylinders which compose the engine.

exhaust emissions: The products of combustion that are discharged from the engine exhaust.

exhaust gas analyzer: An instrument which measures the quantities of the gases which form the exhaust and thereby determine the engine's efficiency.

exhaust pyrometer: An instrument used to measure the temperature of exhaust gases.

F

filter: A device to remove dirt from oil, air, or water.

flash point: Temperature (in °F) to which oil must be heated before the oil vapor will ignite when a flame is passed over it.

foot-pound: A unit of work. Raising one pound a distance of one foot.

force: Defined as a push or pull acting on an object.

friction: Resistance to relative motion between two contacting bodies.

friction horsepower: Power consumed within an engine from friction of its moving parts.

fuel: A substance consumed to produce energy. In the case of a diesel engine, it is fuel oil obtained from petroleum.

fuel injector: A device which sprays fuel into the combustion chamber.

fuel pump: A pressure developing unit which supplies fuel to the injector.

G

gas: A substance that changes volume and shape according to temperature and pressure applied to it. Often a combustible substance.

governor: A device designed to control the speed of an engine within specified limits.

H

heat: A form of kinetic energy associated with the motion of molecules and capable of being transmitted by conduction, convection, or radiation.

heat exchanger: A device which removes heat from a heat producing system, or which utilizes waste heat for useful purposes.

heat loss: Heat from burning fuel that is lost in the cylinder without doing useful work.

heat value: Heat content of the fuel in Btu's per pound or gallon.

horsepower: A unit of power, the rate of doing work: 1 hp = 33,000 ft-lbs per minute.

hunting: Erratic variation of the speed of the governor when it overcompensates for speed changes.

hydrocarbon: A compound composed entirely of hydrogen and carbon. Petroleum is a mixture of hydrocarbons.

I

ignition: The combustion of the fuel mixture in the combustion chamber.

ignition delay: The time from initial injection of fuel and air to actual ignition.

indicated horsepower: Power developed in a cylinder without friction or auxiliary units. Can be computed from engine dimensions, speed, and indicated mean effective pressure (imep).

indicator: Instrument to investigate firing pressures in a cylinder.

inertia: Tendency of a body to remain at rest or to maintain its existing motion.

injector: A device which injects fuel oil into the combustion chamber of an engine against the pressure of air within the chamber.

inlet port: The opening through which the air enters the cylinder of a two-cycle engine.

inlet valve: The valve which permits air to enter the cylinder of an engine.

intercooler: Intake manifold cooled by circulated water.

internal combustion engine: An engine which is designed to derive its power from fuel burned within the engine.

isochronous governor: Maintains constant engine speed regardless of changes in the load being carried.

J

journal: That part of a shaft or axle in contact with the bearing.

K

kilometer: A metric measurement of distance. One kilometer equals 0.6214 mile.

kinetic energy: Energy of a moving body due to its mass and velocity.

L

lapping: A process of fitting one surface to another by rubbing them together with an abrasive dust between the two surfaces.

linkage: Any series of rods, pedals, and levers used to transmit motion from one unit to another.

lubricant: A material which reduces friction between surfaces.

M

manifold: Pipes connecting a series of openings or outlets to a common opening. An exhaust opening.

manometer: An instrument used to measure pressure; a U-tube half filled with mercury or water.

mean effective pressure (mep): The mep of a cycle or stroke of a heat engine is the average net pressure in pounds per unit area that operates on the piston throughout its stroke.

mechanical efficiency: The ratio of brake horsepower output of an engine to the indicated horsepower for the cylinders.

meter: Metric unit of length equal to 39.37 inches.

molecular weight: The atomic weight of an element times the number of atoms in a molecule of the element.

N

natural gas: A colorless and odorless gas produced along with crude oil or obtained directly from gas wells.

nickel steel: A small percent of nickel alloyed with steel to form a heat and corrosion resistant alloy.

nitrogen: An inert gas that makes up about three-fourths of the atmosphere by volume.

nitrous oxide: A harmful gas (emissions) formed by high temperature combustion.

normally aspirated: An engine without a supercharger. At the start of compression it operates on a cylinder air charge at a pressure very near to or slightly below atmospheric.

O

octane number: Burning quality of a fuel when compared with a standard reference fuel by percent of iso-octane.

opacity: Opaqueness or quality or state of a body which renders it impervious to the rays of light, such as the density of exhaust smoke.

oscilloscope: A grid screen which shows the pattern or wave form of spark ignition action.

overhead valve: An engine designed with the valves housed within the cylinder head. Also known as a valve-in-head engine.

P

petroleum: A natural hydrocarbon, yellow to black in color, a thick flammable mixture found within the earth and processed for its various fractions including natural gas, gasoline, diesel fuel, kerosene, lubricating oil, etc.

piston: A cylindrical part closed at one end which maintains a close, sliding fit in the engine cylinder. It is connected to the connecting rod by a piston pin. The force of the expanding gases against the closed end of the piston forces the piston down in the cylinder, causing the connecting rod to move, which rotates the crankshaft.

piston displacement: The volume of air moved or displaced by movement of the piston as it goes from bottom dead center to top dead center.

piston head: The closed portion of the piston above the piston rings.

piston ring: A split expansion ring placed in the grooves of the piston; prevents passage of air from the combustion chamber to the crankcase.

piston skirt: That portion of the piston below the piston pin which is designed to absorb the side movement of the piston.

plunger: Long rod or piston of a single acting injection pump.

port: The opening in a cylinder head or cylinder through which air or exhaust gases pass.

pour point: The pour point of an oil is that temperature at which oil will just begin to flow under a prescribed condition.

power: The rate at which work is done.

pre-combustion chamber: A chamber in the cylinder head of some diesels where some fuel is injected, ignited, and partly burned before being forced out into the main chamber.

pre-ignition: A common fault with a spark-ignition engine where the mixture begins to burn prior to ignition, usually caused by an overheated part in the combustion chamber.

pressure: Force per unit area, usually expressed in pounds per square inch (psi).

prony brake: A machine for measuring the brake horsepower of an engine.

push rod: A connecting link in an operating mechanism, sometimes a cam operated rod that operates the rocker arm which opens an engine valve.

R

radial engine: An engine which has its cylinders arranged in a circle around the crankcase.

rocker arm: An arm used to change upward motion of the cam-operated push rod to downward motion to open an engine valve.

S

safety valve: A valve designed to open and relieve the pressure within a container when the pressure exceeds a set value.

Saybolt viscosimeter: An instrument used to measure flowability or viscosity of oil.

scavenging: Removing the exhaust gases from the cylinder, usually by a flow of air.

scavenging efficiency: In a two-cycle engine, the ratio of a new air charge, trapped in the cylinder, to the total volume of air and exhaust gases in the cylinder at port closing position.

seat: A machined surface upon which another engine part rests, such as the surface upon which a valve face rests.

seize: Grab of tightly fitted surfaces, inability to move due to heat expansion of metals.

semi-diesel: A name given to oil engines, usually lower in compression than a diesel, which use a hot surface to ignite the fuel mixture.

sensitivity: The percent of speed change required to produce a corrective movement of the governor or other speed control mechanism of an engine.

sewage gas: A gas produced from sewage sludge as it decomposes at a sewage disposal plant.

silencer: A kind of muffler.

solenoid: An electro-mechanical device designed to move a mechanical linkage when a current is passed through the coil.

specific fuel consumption: Fuel consumption per hour divided by brake horsepower expressed in pounds per horsepower.

specific gravity: Weight of a solid or liquid at 60°F compared to an equal volume of water at that temperature. Water has a specific gravity of 1.0. For gas, it is equal volume of air at the same temperature and pressure.

specific heat: Amount of heat required to raise the temperature of 1 pound of a substance 1 degree Fahrenheit.

speed changer: A device for adjusting the speed governing system to change the engine speed.

speed drift: A very gradual deviation of the governed speed from the desired speed.

speed droop: Governor control of engine speed variation from no load to full load.

supercharging: Pressurizing the inlet air above atmospheric pressure.

T

tachometer: An instrument which measures the rotational speed of an engine in revolutions per minute.

thermocouple: Two dissimilar wires, joined at one end, and used to measure temperature differences.

thermodynamics: A study of heat as it affects materials; the dynamics of molecules.

thermostat: A control valve used in a cooling system to control the flow of coolant when activated by a temperature signal.

torque: A twisting effort which produces rotation. Torque is measured in ft-lbs or in-lbs.

torsional vibration: Oscillating twisting vibration caused by inertial loading on each end of a shaft.

turbine: A wheel upon which a series of angled vanes are fixed so that a moving fluid will impart rotational movement to the wheel.

turbocharger: Introduction of fluid or gas in a turbulent manner. Used with reference to movement of air in the cylinder and combustion chamber.

turbulence: Rapid mixing and swirling of fuel, air, and gases within a combustion chamber. Improves engine performance and efficiency.

two-stroke cycle engine: An engine requiring only one complete revolution of the crankshaft to complete a cycle of combustion events.

U

U. S. gallon: Common measure of volume which equals 231 cubic inches. One gallon of water weighs 8.33 pounds, or 3.7775 kilos. One gallon of diesel fuel weighs from 6.5 to 7.5 pounds, or 2.9484 to 3.402 kilos.

V

vacuum: Pressure below atmospheric pressure.

valve: A device designed to open and close the entrance to an open line or tube.

velocity: Rate of motion at any instant. Measured in feet per minute (fpm) or revolutions per minute (rpm).

viscosity: Flowability or internal resistance to flow.

viscosity index: A number given to a lubricating oil pertaining to its viscosity with respect to temperature.

W

wet sleeve: A cylinder designed so the coolant contacts a major portion of the sleeve.

work: Mechanical work is done when some object is moved against a resisting force. It is the product of force, which acts to produce displacement of body and the distance through which body is displaced in the direction of force.

wrist pin: Piston pin which connects the connecting rod with the piston.

Index

A

absolute pressure, 153, 476
absorption unit, 341, **342**
acceleration, 141
acidity, 135
accounting records, 451
adhesion, 126
adjustable speed droop, 334
(see speed droop)
advantages of diesel engine,
41-42, 477
A-frame engine block, 65, **92,**
93, 363
air cell system, 291, **291,** 476
air compressors, 13
air-fuel ratio, 160-161
air intake system, 217-219, 380,
424-428, 478
cleaners and filters, 217
nozzle, 380
air pump, 35 (also see air intake)
air pollution standards, 366,
464, 467
California, 366
exhaust emissions, 478
Federal, 464, 467
(see opacity tests)
alarm systems, 440-441
altitude ratings, 199
annealing, 69
APE pump, 244-245, **245,** 247,
248, 249
delivery valve, 247
installation, 250
metering principles, 247, **248**
operating principles, 244
plunger types, 247, **249**
timing, 247, 252
API gravity, 167, 170-171, 476
area, 139-140
ash content, 170
atmospheric pressure, 4-5, 476
automotive diesel governor, 335
automotive service, 9
automotive starting, 442-443
emergency electric generating
units, 442
gas engine compressor units,
442
automatic valve-lash adjustors,
214
auxiliary systems, 402, 446

B

back-hoe, 14, **15**
back-pressure, 457
balancer shafts, 121, **121**
ball-arms, **299, 300**
ball-head, 298, 476 (see
centrifugal ball-head)
base (see bed plate), 49
basic units of measurement,
138-139
barometer pressure, 4, 153,
(see pressure) **199**
bearings, 27, 50-51, 122-123
crank pin, 50
main, 51, 56
precision, 123
wrist pin, 50
bed-plate, 49, 91, **92**
blowers, 35, **36,** 48, 227-231, 276
impellers, 231
positive displacement, 227,
229
rotor, 231
rotary, **36,** 38
scavenging, 227
(see supercharger), 229
V-8, **230**
Bmep (brake mean effective
pressure), 182, 360-361
relation to torque, 183, 361
bottom-dead-center (BDC),
351-352
brake horsepower, 178-180, 183,
476
prony brake, **179**
brake thermal efficiency,
185-187
bhp-hr, 186
Btu (British thermal unit), 147,
476
bull-dozer, 13
buses, 43

C

California air pollution
standards, 366
cam, 56, 477
follower, **214**
overhead, **53**
camshaft, 48, 52, 56, **57,**
212-213, **213,** 477
drive, 212, **213**
carbon compounds, 158-159
dioxide, 158
monoxide, 159
carbon dioxide, 158, 477
carbon monoxide, 159, 477
carbon residue, 134, 167, 169
centerframe, 91
centrifugal ball-head, 298, **298,**
299, 300, 305, 476
cetane number, 167, 172-173,
477
chassis dynamometer, 341, **343,**
473-474, (see dynamometer)
hydraulic type, **343**
chemical terms, 156-157
atomic weights, 157
reactions, 158-160
chronomatic engine-pressure,
indicator, 349, **349, 351,** 470
cleaner, 217-218, **235** (see filters)
Constantin, 348
Chromel, 348
cleaners, 217
clearance volume, 188
cohesion, 126, 128
combustion, 156, 159-160, 193,
283-286, 450
requirements, 285
combustion chambers, 286-289,
477
air cell system, 291, 476
Lanova energy cell, 292, **293**
open type, 286, **287,** 288
pre-combustion type, 51, 59,
60, 61, 289, **290**
Ramsey energy cell, 293, **293**
turbulence type, 74, 289, **289**
with swirl cup, **287**

Index

combustion space, 88
composition limits, 370
compressed air motor, 432
compression, 25, 29, 34, 189, 350
diagram, **350**
ignition, 44, 285
ratio, 34, **34,** 189, **189,** 477
compression diagram, **350**
compression ignition, 44, 285
engine, 285
compression ratio, 34, **34,** 189, **189,** 477
compression rings, 58, **110**
compression stroke, 30, 34-35, **37**
compressor, 24
conformable rings, 113
connecting rod, 29, 44, **57, 59,** 114, **114, 477**
hinged-strap design, 115, **116**
marine big end, **115**
Conrod ratio, **352**
conservation of energy, 147
construction machinery, 43
control system, 339, **443**
conversion processing of oil, 166-167
catalytic cracking, 166
fluid catalytic process, 166
Houdry process, 166
polymerization, 167
thermal cracking, 166
cooling systems, 409-418, 447
by-pass systems, 416-417
climactic conditions, 411, 418
closed circuit, 412-414, **413**
open, 411-412
steam systems, 414-416, **415**
water purity, 411, 418
cotton picker, 12
counter weighting, 121
crankcase, 51, **91,** 477
crankcase pressure, 356-357
crankcase scavenging, 38, **38, 222**
crank mechanism, 28, **28**
crankpin, 50, 55
bearing, 50
box, **116**
hinged strap design, 83
crankshaft, 30-31, 47, **51,** 55, **56,** 120-121, 477
counterweighting, 121
revolution, 35
scavenging, **38**
crankshaft scavenging, 38, **38**
cycle, 29-30, 477
four-cycle (see four stroke cycle)
two-cycle
cylinder, 4, 25, 27, 36, 47, 93, 477
arrangement, 68
block, **90,** 92
dry type, **94**
head, 49, **50,** 52, 54, 96-97, **97,** 210
liner, 93
sleeve, 50
wet type, 76, 78, **94, 95**
cyclinder block, **90, 92,** 477
cylinder head, 48, **50, 52,** 54, 86-97, **97, 210,** 477
cylinder liner, **55**

D

dead band, 303
delivery valve, **215,** 259
function, 259
DEMA ratings, 198-200, **346** (see standard ratings)
derived units of measurement, 139
detergent oil, 131
detonation, 370, 376-377, 389-390, 477
factors, 390
diesel characteristics, 1, 4, 477
advantagcs, 40-42
disadvantages, 42
(see classification), 65
diesel engine design, 65, 363-368
Diesel fuel properties, 128, 135, 167-170
API gravity, 167, 170
ash content, 167, 170
carbon residue, 134, 167, 169
flash point, 135, 167, 169
heating value, 167-168, 174
ignition quality, 167, 171
pour point, 135, 167-169
specific gravity, 167, 170
sulfur content, 167, 170
viscosity, 128, 134, 167, 169
volatility, 167, 169
water and sediment, 135, 167, 170
diesel-electric, 43
diesel engine, 27, 43-44, 47, 65-70
basic parts, **27**
design, 65
operating cycle, 65
use, 43-44, 70
diesel fuel, 41, 167
properties, 167
diesel exhaust, 41
Diesel (Rudolph Diesel)
1892 German patent on diesel engine, 44, 338
disadvantages of diesel engine, 42
distribution type fuel injection pump, 255-261, **255**
double acting engines, 67, 105
drill, 13
dragline, 13
dry filter, 219, 425
dual fuel engines, 173, 203, 370-371, 379-388, **383,** 457
features, 379
duplex oil-injection pump, 386
dynamometer test cell, 339-341, **339,** 465-467, **475**
chassis type, 341, **343,** 465, **473, 474**
engine only, 341
hydraulic type, 341
PTO type, 341, **342**
(see opacity test), 364-367

E

efficiency, 183, 185-189, 478
indicated thermal, 185
mechanical, 183-184
thermal, 188-189, 338
volumetric, 186-188
electric generator, 13-14, **13-14,** 19, **21**
electric ignition, 390, **391,** 433-440
electronic system, 437-440, **439**
magneto system, 434-435
pulse generator system, 436-437, **438**
resistors, 435
electric load-sensing governors, 296, 298, 335
electric speed-governors, 298, 331-332
electronic system, 437-440
emergency electrical units, 22
energy, 146-148, 478
conservation, 147
internal, 147
kinetic, 146-148
potential, 146
transfer, 147
units, 147
energy cells, 291
engine dynamometer, 341, 465-467, 475
(see dynamometer)
engine rating, 193, 478
bmep, 193
capacity, 193
engine torque, 180-181, **181,** 341
(see torque)
engine structure, 89
exhaust, 172, 207-208
back-pressure, 355, **356**
pyrometer, 429
stroke, 31, 37
valves, 209
exhaust back-pressure, 355-**356**
exhaust emissions, 478

exhaust heat recovery, 430
exhaust pyrometers, 429, 448, 478
exhaust stroke, 31, **37**
exhaust systems, 209, 428-430
  mufflers, 428-429
  piping, 428
exhaust valve, 31
explosive limits, 375-378
explosive mixtures, 375
external combustion, 25

F

farm power equipment, 10
  cotton picker, 12
  harvester, **11**
  planter, **11**
  tractor, **10**
fatigue failure, 117
Federal regulations, 423, 464, 467
  air pollution standards, 464, 467
  fuel storage tanks, 422
filters, 217, 219, **235,** 478
  dry type, 218, 219, 425
  oil bath, **219, 220,** 425, 426
  water sprays, 425
  wet type, 219, 425
firing pressure, 89
fixed throttle, 296
flammability, 371, 375, 378
  limits, 375
  (see lean rich mixtures)
flash point, 135, 167, 169, 478
flat cylinder arrangement, 68
floating piston, 119
flywheel, 31-32, 48, 52, 123, 297
foot-pound, 147-478
force, 139, 478
four stroke cycle engine, 5, **5,** 25, 29-34, **32-33, 47,** 65, 71, 84, **85,** 90, **208,** 216-217, **381, 383, 394,** 395
  balancer shafts, **121**
  compression, 30
  exhaust, 31
  intake, 30
  power, 31
frame, 50, **54, 56,** 89, 363
  A-frame, 65, **92,** 93, 363
  automotive type, 89
  two piece, 91
  welded, 93
friction, 125-126, **126,** 478
  rolling, **125**
  sliding, 125, **126**
friction horsepower, 176, 183, 478
fuel consumption, 185, 202
fuel control valve, 305
fuel-gas supply system, 423, **423**
fuel injection nozzle, 31, 48, 281
fuel injection pump, 3, 48, 58, **60,** 243-245, **275**
  APE, 244-245
  constant stroke, 244
  injector, 29, **60**
  jerk pump type, 72, 80, 243-244, **245,** 269, 298
  line, 60
  multiple plunger, 72, 80, 243-244, **245,** 269, 298
fuel injection system, 242-243
  atomization, 243
  distribution, 243
  mechanical, 243
  metering, 243-244, 248, 267
  multiple plunger, 243-244
  pump timing, 244
  rate control, 243
  solid, 243, 286
fuel injection valve, 59, **61**
fuel injector, 29, **60,** 269-272, 478
  pencil type, 76
  unit type, 269-273, **273,** 274-278
fuel mixtures, 371, 378, 375, 390
  lean, 371, 378
  rich, 375, 390
fuel supply system, 419, **419,** 448
  cleaning, 419, 424-425
  day tanks, 419
  fuel-gas, 423, **423**
  heating, 419
  heavy oil, 422
  pumps—low pressure, 253
  storage, 419
fuel supply pump, 253-254, **254**
  low pressure, 253
fuel supply system, 419
fuel transfer pump, 257-261, 266
  (see transfer pump)

G

gage pressure, 153
  (see barometric pressure)
gas burning engines, 374-378
gas compressor, **23**
gas-diesel engines, 370-374, 373
gaseous fuels, 173
  by-product gases, 173, 174
  manufactured gases, 173
  natural gas, 173, 174
gases, 152, 478
  (see gas law)
gas laws, 154-155
gas-metering valve, 387-388, **387**
gasoline engine, **3,** 39, 161, 464
gear pump, 403
governor, 20, 48, 58, 60
  dead band, 303-304
  defined, 296
  modifications, 321-327, 387
  promptness, 303-304
  stability, 303
  system, 60
  use, 333
  work capacity, 304
governors, 295-337, 478
  constant speed, 295, 297, 321
  centrifugal, 261
  electric load sensing, 296, 298
  electric speed, 298, 331, 332
  hydraulic, 295, 299, 309-316
  isochronous, 303, 311, 316-320, **317, 318, 319, 320**
  load control, 297, 330-331, **331**
  load limiting, 295, 297, 330
  mechanical, 295, 299, 305-308
  overspeed, 296-297, 329
  pressure regulating, 296, 298, 329
  speed, 296, 298, 329
  speed limiting, 295, 297, 302
  throttle, 48, 298
  torque control, 298, 329
  torque converter type, 337
  two-speed, 327
  variable speed, 295, 297, 327
governor system, **60,** 295
grader, 13
gravity, 136

H

harvester, 11
heat, 149-152, 478
  conduction, 151
  convection, 151-152
  exchanger, 339, 347
  flow, 150-151
  high/low values, 162, 167-168, 174
  radiation, 151
  rejection, 346
  specific, 150
  transfer, 151-152
  treatment, 69
heat balance, 409-411, **409**
heat consumption, 203-204
heat exchanger, 339, 347, 478
heat flow, 150-151
heat quantities, 161-162
heat rejection, 346, 347
heat transfer, 151
heat treatment, 69
heat values, 162, 167-168, 174, 478
  (heating values)

Index

high-compression gas engines, 24, 370-371, 399-401, **400**
high-compression spark ignited gas engines, 370, **392, 393, 396, 397**
high-speed diesel, **2,** 4
Hornsby-Ackroyd, production engine in England and U. S. (1893), 44, 45
  (see Stuart)
horsepower, 176-184, 478
  brake, 178-180, 183, 476
  friction, 183
  indicated, 176-177, 183-184
  ratings (see standard ratings)
horsepower-hour (hp-hr), 147
hydraulic (relay) governor, 295, 299, 309-316
  elementary type, 310, **310**
  servo-motor, 295, 310
  with permanent speed droop, 311
hunting, 295, 303, 311, 478
  in elementary hydraulic models, 311

I

ignition limits, 375
ignition quality, 167, 171
  cetane number, 167, 172-173, 477
  delay, 171
ignition temperature, 370, 375
impeller, 231
indicated horsepower, 176-177, 183-184, 349, 479
  indicator cards, **177**
indicated thermal efficiency, 185
inertia (inertial forces), 4, 89, 194-198, **196,** 478
  piston speed, 197
injection nozzle, 96, 278-281, 458
  multiple hole type, 279-280
  pintle type, 279-280
  pencil type, 281, **281**
  single hole type, 279
injection pump, 263, **263,** 472
  test stand, 473
injector, 29, **60,** 270, 479
  plunger, **473**
  rack and gear, **270**
  tester, **471**
inlet manifolds, 221
inlet ports, 35, 479
inlet valve, 30, 479
in-line engine, **50, 67, 225**
  scavenging, **225**
instrumentation, 339, 347
  absorption unit, 341, **342**
  chronomatic drum indicator, 349-350
  dynameter, 340, **343,** 465-467
  galvanometer, 353
  manometer, 355-357, 360
  millivolt pyrometer, 357
  oscilloscope, 354
  planimeter, 351
  pressure indicator, 352-355, 470
  test cells, 339
  thermocouple, 348, **348**
  water flow meter, 357
intake, 200, 207-208
  air coolers, 201, 234-236
  manifold, 234-236
  silencer, 425-426
  stroke, 30, 35, **37**
  system, 207, 217
  valves, 209
intake manifold intercooler, 234-236, **236,** 462, 479
intake silencers, 425-426
intake stroke, 30, 35, **37**
intake valves, 209
intercooler, 235, **235,** 462, 479
  efficiency, 357-360
internal combustion, 25, 39
internal combustion engines, 25, 39, 370, 402, 479
  dual-fuel, 173, 203, 370-371, 378-388, **383**
  gas-diesel, 370-374
  high-compression gas engines, 24, 370, 399-401, **400**
  high-compression spark-ignited, 370-371, **392, 393, 396, 397**
internal energy, 147
isochronous hydraulic governor, 295, 303, 316-321, **317, 318, 319, 320,** 479
  pressure compensation, **326**
  with permanent speed droop, 321-322, **323, 324, 325**

J

jet air starting, 240
jerk-pump (multiple plunger), 72, 80, 243-244, 269, 298

K

Kilowatt-hours (kw-hr), 47
kinetic energy, 146-148, 479
knocking, 172, 459
  (fuel knock)

L

Lanova energy cell, 292, **292**
lean mixture, 371, 378
length, 139
liner, 54, **54-55, 94,** 95
  dry, **94,** 95
  materials, 96
  sealing, 95
  wet, **94,** 95
load control governor, 330-332, **331,** 442
load decrease, 306-307, **307,** 314
load increase, 306, **306,** 313
load-limiting governor, 295, 297, 330
load sharing, 334
locomotive, 15, **16,** 43, **81,** 82
logging equipment, 43
lost heat, 190
low-compression gas engines, 377
lubricant, 128-129, 479
  contamination, 129
  deterioration, 129
lubricating oil treatment, 406-407
  batch, 406
  centrifuging, 407
  continuous, 406
  continuous by-pass, 406
  filtering, 407
  settling, 407
lubrication, 125, 131, 195
  friction, 125-126, **126,** 478
  lubricants, 128
  principles, 125, 131
  (see oils)
  systems, **132,** 133, 402, 447, 449
  treatment, 406
lubrication systems, 133, 403-406, 409, 447
  cooling, 409-418
  mechanical forced feed, 403-405
  oil coolers, 406
lubricators, 404-405

M

main bearings, 51, 56
magnets system, 434-435
manometer, 355-357, 360
  mercury filled, 357, 360
  (see U-tube manometer)
  water filled, 356
marine service, 17, **17, 19,** 43
  barges, 17
  dredge, **17**
  ferry boat, 43
  icebreakers, 17
  motor ship, 17, 43
  naval vessels, 43
  towboat, 18, **18**
  tugs, 17, 34

mass and weight, 144
mean effective pressure (MEP), 351-352, 479
mechanical drive, 23, 239-240
  cotton gin, 23
mechanical efficiency, 183-184
  (see efficiency)
mechanical governors, 295, **299,** 305-307
  advantages, 307
  disadvantages, 307
  Pickering, 307, **308**
metering principles, 243-244, 247, **248,** 267
metric system, 124, 137, 139, 479
  meter-gram-second, 139
mobile service, 13
  air compressors, 13
  back hoe, 14, **15**
  bull-dozer, 13
  drills, 13
  drag line, 13
  grader, 13
  portable electric generators, 13, **13,** 14, **14**
  power hoists, 13
  power shovels, 13, **14,** 43
  scrapers, 13
  tractors, 10, 13, 43
moving parts, 99
  balancer shafts and vibration damper, 121, **121**
  bearings, 122
  flywheels, 52, 123, 297
  pistons, 47, **57,** 99-106
  rings, 106-116
  wristpins, 50, 117-119
mufflers, 429
  (see silencers)
multiple plunger injection system, 243
multiple plunger pump, 72, 80, 243-244, **245,** 269, 298
  metering, 244
  plungers, 244
  pressures, 244
  timing, 244
municipal power plant, 20

N

neutralization number, 135
normally aspirated engines, 344, **344,** 385, 460, **461,** 479
  (naturally aspirated)

O

octane number, 479
oil, 128-136, 405-407
  characteristics, 133-136
  coolers, 406
  detergent, 131
  occurrence, 165
  oxidation, 129
  properties, 133-136
  refining, 165, 406
  (see lubricant), 129
  storage, 403, 405
  sump, 404-405
oil and gas fuels, 157, 164-165
  storage, 405-406
oil bath filters, 219, 425, **426**
oil control rings, 111
oil coolers, 406
oil film wedge, 127, **127**
oil injection system, 386, **386**
oil properties, 134-136
  acidity and neutralization number, 135
  carbon residue, 134
  color, 136
  flash point, 135
  gravity, 136
  pour point, 135
  precipitation number, 135
  (see diesel fuel properties)
  viscosity, 134
  water and sediment, 135
oil pumps, **133,** 386
oil refining, 165-167
  conversion, 166
  distillation, 166
  types, 165
opacity test, 464-467, **468,** 479
  (see testing equipment)
open combustion chambers, 286-287, 288
operating cycle, 65
operating records, 450-451, **453, 454**
  procedure, 447, 451
performance records, 450-451, **453, 454**
  accounting, 451
  record forms, 451
opposed piston engine, **66, 83, 224**
  gas engine, 390
oscilloscope, 354, 479
overhead cam, 53
overspeed governor, 296-297, 329
  trip, 329
oxidation, 129
  inhibitors, 131

P

pan, 49
pencil type fuel injector, 76-77
perfect mixture, 371
permanent speed droop, 295, 302, 311, **315**
  advantages, 314
  disadvantages, 315
Pickering mechanical governor, 307-309, **308**
piezoelectric pressure indicator, 352-355, **353, 354, 355**
pilot oil, 371, 387
  control, 387
piston, 47, **57,** 99-106, 197, 480
  action, 67
  assembly, 58
  cam-ground, 102, **102**
  composite, 103
  cooling, **103, 104, 105**
  crosshead, 68, 103, **104**
  crown, **100,** 103
  double-acting, 67
  materials, 101
  opposed-piston, 67
  rings, (see rings)
  single-acting, 67
  skirt, 100
  speed, 197
  trunk, 103
  uncooled, 99
piston speed, 197-198
plunger, 250, 301, 480
  rotation, 301
plunger type pump, 247, **249, 264**
port, 35, 48, 480
port exhaust, 38
port scavenging, 223, **223**
portable electric generator, 13-14, **13, 14**
positive displacement blower, 227, **229**
potential energy, 146
power rating, 193, 201-202, **201, 203,** 345, 379
  (see standard ratings)
pour point, 135, 167, 169, 480
power, 4, 145-146, 176-181
  brake, 117-180
  friction, 180
  indicated, 176-177
  rating, 201, **201**
  torque, 180-182, 341, 344, **344, 345,** 360-362
power hoist, 13
power shovel, 13, **14,** 43
power stroke, 31, **37**
precipitation number, 135
precombustion chamber, 51, 59, **60-61,** 289, **290**
pre-ignition, 370, 376
pressure, 142, 152-153, 352-353
  absolute, 153, 476

gage, 153
indicators, 352-353, **353**
of gases, 152-153
pressure compensation, 325-326, **326**
pressure indicators, 352-353, **470**
piezoelectric, 352, 353
pressure lubrication system, **130, 131**
pressure-regulating governors, 329-330
pressure-time diagram, 349, **354, 367**
relation to MEP, 351
prony brake, 179
pulse generator system, 436-437
pump, 3, 48, 58, **60**, 226, 243-248, **249, 264**
APE, 244, **245**
fuel injection, 3, 48, 58, **60**, 243-245
gear, 402-403
plunger type, 247, **249**, 264
pressures, 244
reciprocating, 227-228, 405
scavenging, 226-227
transfer, 256-261
pumping principles, 246, 266
push rod, 214
pyrometer, 429-430, **429**, 448

R

rack and pinion, **301**, 302
radial cylinder arrangement, 69
radial engine, 69, 393, **400, 480**
railroad service, 15, **16**, 43
diesel-electric, 15
Ramsey energy cell, 293, **293**
ratings (see power rating), 201, 345
reciprocating pump, **227-228**, 405
resistors, 435
rich mixtures, 375, 390
rings, 106-113
compression type, 106, **107**, 108, **108**
conformable, **112**, 113
for two-cycle engines, 113, **113**
oil-control type, 110-112, **111, 112**
Keystone, 108
seal type, 109-110, **110**
rocker arm, 214, 480
rod bolts, 116
rotary blower, **36**, 38
rotor, 231
rotary motion, 140
rotor, 231, **231, 232**

S

safety devices, 440-442,480
alarm systems, 440-441
shut-down systems, 440-442
valve, 480
safety precautions, 387-388, 440-442
Federal regulations, fuel storage, 422
scavenging system, 88, 221-225, **222**, 480
blowers, 221, 227-231
crankcase, 38, **221-222**
in-line engine, 50, **67**, 225
opposed piston engine, **66, 83, 224**
port, 222, **223**
power piston, 221
pump, 221, 226-227
uniflo, 88
valve type, 223, 225
V-type engine, 226
scraper, 13
seal rings, 58, 109, **110**
(see compression rings)
self-ignition, 371, 375-377, 379
defined, 379
servo-motor, 295, 310
sewage gas, 400-401, 480
shaft, 27
(see crankshaft)
shut-down systems, 440-441
short run engines, 22
stand-by (emergency) units, 22
shut-down methods, 441-442
hold exhaust valves open, 441
stop fuel supply, 441
stop injection, 441
silencer, 425, **427**, 480
(see mufflers)
single acting engine, 67
six cylinder engine, **49, 77, 365**, 460
crankshaft, 120
smoke emission standards, 464-467, **468**
Federal, 464
solid fuel injection, 243, 286-288
spark-ignited engines, 173, **391, 392**, 457
spark ignition, 44, 173
specific gravity, 167, 170, 481
speed adjuster, 300-301, **300**, 333, 480
(load adjuster)
speed droop, 295, 302, 311, 480
lever, 312, **312, 313**
permanent, 295, 303, 311, 315
temporary (compensation), 295, 383
speed governor, 296, 298
speed regulation, 295, 302
spray nozzle, 28
stand-by units, 22
standard ratings (DEMA), 198-200, **346**
continuous, 200, 202, **346**
de-rating, 200
elevated temperature, 198, 200
gross, 345, **346**
high altitude, 198, **199**
intermittent, 202, 345, **346**
maximum, 200, 202
minimum, 200, 202
starting systems, 430-439, 448
automatic, 442-443
air motor, **431**
breakerless, 439
compressed air admission, 433
compressed air motor, 431
electric motor, 431, 433
electronic, 437, **438**
gasoline engine, 431
hand, 431
magneto, 434-435, **434**
pulse generator, 436, **437**
resistors, 435
stationary engines, 91
cast frames, 91
stationary parts, 88-92
stationary service, 19, 44, 91-92
electric generators, 13-14, **13, 14**, 19, **21**
hospital stand-by, **21**
municipal power plant, **20**
steam engines, 3, 25
storage, 403, 405
stroke, 30-31
compression
exhaust
intake
power
Stuart (Herbert Ackroyd Stuart), 1888 patent on hot wall diesel, 44-45
sulfur, 159, 167, 170
content, 170
sump, 404-405
wet and dry, 405
supercharging systems, 236-240, **238, 239**, 460, 481
supercharger, 5, **6**, 8, 35, 200, 207, 228-233, 345
centrifugal type, 6, 230
positive displacement rotary blowers, 229
Roots type, 5, **229**
system, 236-240, **238-239**
three-lobed, **330**

two-lobed, 231
supercompression, 377
swirl cup, **287**

T

taxi, 43
temperature, 144, 410-411
test cells, 339-341, **339,** 362, 465-467, **475**
testing equipment, 470-474
thermal efficiency, 188-189, 338
thermocouple, 348, 481
Chromel, 348
circuit, 348
Constantin, 348
three cylinder engine, 75
throttle (see governor), 48, 298
through bolts, **92**
time, 139
timing, 62-63, **252**
timing gears, **62,** 63
top-dead-center (TDC), 36, 350, 352
torque, 180-182, **181,** 341, 344, **344,** 345, 360-362, 481
relation to Bmep, 182-183
(see engine torque)
torque and horsepower curves, **344, 345**
torque-converter governor, 329
tractor, 10, 13, 43
transfer energy, 147
transfer pump, 257-261
charging cycle, 258
discharge cycle, 259
fuel cycle, 257-258
troubleshooting, 451-459
problem-solution, 455-459
procedures, 455-459
trucks, 43
trunk piston, 67, 86
turboblower, 235
turbocharging systems, 238-240, **238-239,** 355, 460
constant-pressure, 237
pulse (Buchi), 237
turbocharger, 7, **7, 17, 49, 63, 109, 130,** 207, **233, 235,** 237-240, 355, 361, **363, 364, 365,** 397, 481
systems, 237-240, **238, 239,** 355
with wheels, 234
turbocooler, 393, 395, 397
system, **398**
turbulence, 481
turbulence chamber, 74, 289
two piece frame, 91
two-speed mechanical governor, 327, **328**
two stroke cycle engine, 6, **9,** 25, 29, 35-38, **36, 37,** 48, 66, 78-79, 85, **86,** 380, 457, 481
rings, 113
turbocharging system, 238-240, **238-239**

U

uniflow scavenging, 225
unit injector, 269-273, **273,** 274-278
for V-6 and V-8 engine, 278
pressure-time principle, 274, **276-277**
units of measurement, 138-148
acceleration, 141
area, 139-140
basic, 138-139
derived, 139-148
energy, 146-148, 478
mass and weight, 144
power, 4, 145-146, 176-181
pressure, 142
rotary motion, 140
temperature, 144
velocity, 140, 481
volume, 140
work, 145, 481
uses of diesel engines, 43
U-tube manometer, 354

V

vacuum, 5, 481
partial, 5
(see pressure gage)
valve, 28, 48, 54, 209, 481
construction, 209
gear, 209, 212
guide, **53**
lash adjustors, 215
requirements, 209
springs, **52-53**
stems, **52**
timing, 216
type, 226
valve gear, 208, 212
lash adjustors, 214, **215**
springs, 215
valve-in-head exhaust, 35
valve scavenging, 223, **223**
valve type, 226
variable speed governor, 327
V-cylinder arrangement, **51,** 68, **68, 76,** 82, **86, 93**
twelve and sixteen, 82
velocity, 140, 481
vibration damper, 121
fluid type, **122**
viscosity, 128, 134, 167, 169, 481
volatility, 167, 169
volume, 140
gas, 154
volumetric efficiency, 186-188
V-type engine, **226**

W

water and sediment, 135, 167, 170
purity, 411, 418
water jacket, 89, 97, **97,** 449
coolant passages, 97
welded frames, 93
wet filter, 219, 425
work, 145, 481
wristpin, 50, 87, 117-119, **117,** 481
bearing, 50
full floating design, 117, **118**